T0140312

Advances in Intelligent Systems and Computing

Volume 948

The series "Advances in Intelligent Systems and Computing" contains publications on theory, applications, and design methods of Intelligent Systems and Intelligent Computing. Virtually all disciplines such as engineering, natural sciences, computer and information science, ICT, economics, business, e-commerce, environment, healthcare, life science are covered. The list of topics spans all the areas of modern intelligent systems and computing such as: computational intelligence, soft computing including neural networks, fuzzy systems, evolutionary computing and the fusion of these paradigms, social intelligence, ambient intelligence, computational neuroscience, artificial life, virtual worlds and society, cognitive science and systems, Perception and Vision, DNA and immune based systems, self-organizing and adaptive systems, e-Learning and teaching, human-centered and human-centric computing, recommender systems, intelligent control, robotics and mechatronics including human-machine teaming, knowledge-based paradigms, learning paradigms, machine ethics, intelligent data analysis, knowledge management, intelligent agents, intelligent decision making and support, intelligent network security, trust management, interactive entertainment, Web intelligence and multimedia.

The publications within "Advances in Intelligent Systems and Computing" are primarily proceedings of important conferences, symposia and congresses. They cover significant recent developments in the field, both of a foundational and applicable character. An important characteristic feature of the series is the short publication time and world-wide distribution. This permits a rapid and broad dissemination of research results.

**** Indexing: The books of this series are submitted to ISI Proceedings, EI-Compendex, DBLP, SCOPUS, Google Scholar and Springerlink ****

More information about this series at http://www.springer.com/series/11156

Alexei V. Samsonovich
Editor

Biologically Inspired Cognitive Architectures 2019

Proceedings of the Tenth Annual Meeting of the BICA Society

Editor
Alexei V. Samsonovich
Moscow Engineering Physics Institute
(MEPhI), Department of Cybernetics
National Research Nuclear University
(NRNU)
Moscow, Russia

ISSN 2194-5357 ISSN 2194-5365 (electronic)
Advances in Intelligent Systems and Computing
ISBN 978-3-030-25718-7 ISBN 978-3-030-25719-4 (eBook)
https://doi.org/10.1007/978-3-030-25719-4

This Springer imprint is published by the registered company Springer Nature Switzerland AG
The registered company address is: Gewerbestrasse 11, 6330 Cham, Switzerland

Preface

This volume documents the proceedings of the 2019 Annual International Conference on Biologically Inspired Cognitive Architectures (BICA), also known as the Tenth Annual Meeting of the BICA Society. BICA 2019 was held at the Microsoft Main Campus in Seattle, Washington, USA, on August 15–18. In addition to BICA Society, the conference was sponsored by AGI Laboratory, Microsoft, and Boston Consulting Group. By the time when this volume was sent to production, we had 141 submissions, of which 79 were included here after a careful peer review.

BICA are computational frameworks for building intelligent agents that are inspired from biological intelligence. Since 2010, the annual BICA conference attracts researchers from the edge of scientific frontiers around the world. It contrasts major conferences in artificial intelligence (AI) and cognitive science and neuroscience by offering informal brainstorming atmosphere and freedom in conjunction with greater publication venues to ambitious ideas.

Researchers understand "biological inspirations" broadly, borrowing them from psychology, neuroscience, linguistics, narratology, design, and creativity studies in order to advance cognitive robotics and machine learning, among other hot topics in AI. The "filter" is the question of whether your contribution may help us make machines our friends or understand how the mind works. With our steady growth over 10 years and unmatched sociocultural programs, we anticipated the new unseen success of BICA in Seattle in 2019.

All BICA conferences and schools, including BICA 2019, are organized and sponsored by BICA Society. In this year, the society celebrates its 10-year anniversary and therefore deserves special attention here. BICA Society (Biologically Inspired Cognitive Architectures Society) is a scientific non-profit 501 (c)(3) organization, whose mission is to promote and facilitate the many scientific efforts around the world in creating a real-life computational equivalent of the human mind. We call this task the BICA Challenge. BICA Society brings together researchers from disjointed fields and communities who devote their efforts to solving the challenge, despite that they may "speak different languages." This is achieved by promoting and facilitating the transdisciplinary study of cognitive architectures, and in the long-term perspective, creating one unifying widespread framework for the human-level cognitive architectures and their implementations.

With this scientific agenda in mind, over 10 years of its existence, BICA Society organized or co-organized 10 international conferences (the BICA conferences), 2 schools, and 2 symposia. These events were hosted in various countries around the world, including the USA, Italy, France, Ukraine, Russia, and the Czech Republic. In organizing these events, BICA Society collaborated with several universities around the world.

BICA Society created and is maintaining an online database on cognitive architectures (http://bicasociety.org/mapped/), which includes as a component one of the world's largest databases of freely accessible video records of scientific presentations about cognitive architectures and related topics (over 500 titles: http://vimeo.com/bicasociety).

In addition, BICA Society initiated the creation of several mainstream periodic publication venues, including the Elsevier journals "Biologically Inspired Cognitive Architectures" and "Cognitive Systems Research (CSR)—Special Issue Series on BICA*AI". Plus, BICA Society regularly publishes its proceedings in other editions dedicated to BICA, including BICA volumes in the Elsevier journal "Procedia Computer Science" and, of course, the Springer book series "Advances in Intelligent Systems and Computing."

It should be mentioned that in this year, the BICA conference and the associated with its CSR journal track are changing the name from BICA to BICA*AI: Brain-Inspired Cognitive Architectures for Artificial Intelligence. The following topics at the front edge of science and technology, as before, are in the focus of this and future BICA*AI conferences, books, and journals:

- Artificial social–emotional intelligence;
- Active humanlike learning and cognitive growth;
- Narrative intelligence and context understanding;
- Artificial creativity and creation of art by AI;
- Goal reasoning and true autonomy in artifacts;
- Embodied intelligence and cognitive robotics;
- Synthetic characters, HCI, and VR/MR paradigms for AI;
- Language capabilities and social competence;
- Robust and scalable machine learning mechanisms;
- The role of emotions in artificial intelligence and their BICA models;
- Tests and metrics for BICA in the context of the BICA Challenge;
- Interaction between natural and artificial cognitive systems;
- Theory-of-Mind, episodic, and autobiographical memory in vivo and in vitro;
- Introspection, metacognitive reasoning, and self-awareness in BICA;
- AI and ethics, digital economics, and cybersecurity;
- Unifying frameworks, standards, and constraints for cognitive architectures.

Works of many, yet not all, distinguished speakers of BICA 2019 are included in this volume. Among the participants of BICA 2019 are top-level scientists like Peter Boltuc, Antonio Chella, Steve DiPaola, Ricardo Gudwin, Magnus Johnsson, Deepak Khosla, Robert Laddaga, John Laird, Antonio Lieto, Paul Robertson, Junichi Takeno, Rosario Sorbello, Paul Verschure—to name only a few of the famous names at BICA 2019. Other names may not be well known; yet it has been always our tradition to treat everyone equally. As a part of our treatment, during the last 4 years we were issuing notable scientific awards to many BICA participants every year.

It was my great pleasure to work as the BICA 2019 General Chair together with the Co-Chair, David J. Kelley, and with all the participants and committee members. Particularly, I would like to thank all members of the organizing committee and program committee for their precious help in reviewing submissions and in helping me to compose the exciting scientific program. Last but not least, my special thanks go to the publisher Leontina Di Cecco from Springer, who made this publication possible.

June 2019

Alexei V. Samsonovich
General Chair of BICA
Conference Series

Organization

Program Committee

Kenji Araki	Hokkaido University, Japan
Joscha Bach	MIT Media Lab, USA
Paul Baxter	Plymouth State University, USA
Paul Benjamin	Pace University, New York, USA
Galina A. Beskhlebnova	SRI for System Analysis RAS, Russia
Tarek R. Besold	Alpha Health AI Lab, Telefonica Innovation Alpha, Spain
Jordi Bieger	Reykjavik University, Iceland
Perrin Bignoli	Yahoo Labs, USA
Douglas Blank	Bryn Mawr College, USA
Peter Boltuc	University of Illinois Springfield, USA
Jonathan Bona	University of Arkansas for Medical Sciences, USA
Mikhail Burtsev	Moscow Institute of Physics and Technology, Russia
Erik Cambria	Nanyang Technological University, Singapore
Suhas Chelian	Fujitsu Laboratories of America, Inc., USA
Antonio Chella	Dipartimento di Ingegneria Informatica, Università di Palermo, Italy
Olga Chernavskaya	P.N. Lebedev Physical Institute, Russia
Thomas Collins	University of Southern California (Information Sciences Institute), USA
Christopher Dancy	Penn State University, USA
Haris Dindo	Computer Science Engineering, University of Palermo, Italy
Sergey A. Dolenko	D.V. Skobeltsyn Institute of Nuclear Physics, M.V. Lomonosov Moscow State University, Russia

Alexandr Eidlin	National Research Nuclear University MEPhI, Russia
Jim Eilbert	AP Technology, USA
Thomas Eskridge	Florida Institute of Technology, USA
Usef Faghihi	University of Quebec in Trois Rivieres, Canada
Elena Fedorovskaya	Rochester Institute of Technology, USA
Stan Franklin	Computer Science Department & Institute for Intelligent Systems, University of Memphis, USA
Marcello Frixione	University of Genova, Italy
Salvatore Gaglio	University of Palermo, Italy
Olivier Georgeon	Claude Bernard Lyon 1 University, France
John Gero	University of North Carolina at Charlotte, USA
Jaime Gomez	Technical University of Madrid, Spain
Eva Hudlicka	Psychometrix Associates, USA
Dusan Husek	Institute of Computer Science, Academy of Sciences of the Czech Republic
Christian Huyck	Middlesex University, UK
Ignazio Infantino	Consiglio Nazionale delle Ricerche, Italy
Eduardo Izquierdo	Indiana University, USA
Alex James	Kunming University of Science and Technology, China
Li Jinhai	Kunming University of Science and Technology, China
Magnus Johnsson	Lund University, Sweden
Darsana Josyula	Bowie State University, USA
Kamilla Jóhannsdóttir	Reykjavik University, Iceland
Omid Kavehei	University of Sydney, Australia
David Kelley	Artificial General Intelligence Inc., USA
Troy Kelley	U.S. Army Research Laboratory, USA
William Kennedy	George Mason University, USA
Deepak Khosla	HRL Laboratories, LLC, USA
Swathi Kiran	Boston University, USA
Muneo Kitajima	Nagaoka University of Technology, Japan
Valentin Klimov	National Research Nuclear University MEPhI, Russia
Unmesh Kurup	LG Electronics, USA
Giuseppe La Tona	University of Palermo, Italy
Luis Lamb	Federal University of Rio Grande do Sul, Brazil
Leonardo Lana de Carvalho	Federal University of Jequitinhonha and Mucuri Valleys, Brazil
Othalia Larue	University of Quebec, Canada
Christian Lebiere	Carnegie Mellon University, USA
Jürgen Leitner	Australian Centre for Robotic Vision, Australia
Simon Levy	Washington and Lee University, USA

Antonio Lieto	University of Turin, Italy
James Marshall	Sarah Lawrence College, USA
Olga Mishulina	National Research Nuclear University MEPhI, Russia
Steve Morphet	Enabling Tech Foundation, USA
Amitabha Mukerjee	Indian Institute of Technology Kanpur, India
Daniele Nardi	Sapienza University of Rome, Italy
Sergei Nirenburg	Rensselaer Polytechnic Institute, USA
David Noelle	University of California Merced, USA
Andrea Omicini	Alma Mater Studiorum–Università di Bologna, Italy
Marek Otahal	CIIRC, Czech Republic
Alexandr I. Panov	Moscow Institute of Physics and Technology, Russia
David Peebles	University of Huddersfield, UK
Giovanni Pilato	ICAR-CNR, Italy
Roberto Pirrone	Università degli Studi di Palermo
Michal Ptaszynski	Kitami Institute of Technology, Japan
Uma Ramamurthy	Baylor College of Medicine, USA
Thomas Recchia	U.S. Army ARDEC, USA
Vladimir Redko	Scientific Research Institute for System Analysis RAS, Russia
James Reggia	University of Maryland, USA
Frank Ritter	The Pennsylvania State University, USA
Paul Robertson	DOLL Inc., USA
Brandon Rohrer	Sandia National Laboratories, USA
Paul S. Rosenbloom	University of Southern California, USA
Christopher Rouff	Near Infinity Corporation, USA
Rafal Rzepka	Hokkaido University, Japan
Ilias Sakellariou	Department of Applied Informatics, University of Macedonia, Greece
Alexei V. Samsonovich	National Research Nuclear University MEPhI, Russia
Fredrik Sandin	Lulea University of Technology, Sweden
Ricardo Sanz	Universidad Politecnica de Madrid, Spain
Michael Schader	Yellow House Associates, USA
Howard Schneider	Sheppard Clinic North, Canada
Michael Schoelles	Rensselaer Polytechnic Institute, USA
Valeria Seidita	University of Palermo, Italy
Ignacio Serrano	Instituto de Automatica Industrial - CSIC, Spain
Javier Snaider	FedEx Institute of Technology, The University of Memphis, USA
Donald Sofge	U.S. Naval Research Laboratory, USA
Meehae Song	Simon Fraser University, Canada
Rosario Sorbello	University of Palermo, Italy

Contents

Narratological Formulation of Story-Form Memory Construction: Applying Genette's Narrative Discourse Theory

Taisuke Akimoto(✉)

Kyushu Institute of Technology, Fukuoka, Japan
akimoto@ai.kyutech.ac.jp

Abstract. An episodic memory is generally defined as a memory that enables the recollection or remembrance of past events or experiences. It is not assumed as a copy of the past events themselves and is instead assumed as corresponding to mentally encoded or constructed information. However, the mechanism that constructs episodic memory is not systematically formulated in previous studies on cognitive architectures and systems. In this study, the term "story" is used, rather than episodic memory, to refer to a mental representation of temporally and linguistically organized events and entities. The main difference between a story and an episodic memory is that a story involves the *form* of the mental representation, while the definition of an episodic memory is generally based on the *function* (i.e., it enables the remembering of past events). A story is considered as a uniform mental representation involving episodic memory, current situation, prospective memory, planned or imagined future, and fictional story. In order to ensure a systematic formulation of the story-form memory construction, this study introduces Genette's narrative discourse theory, a representative work in narratology. Genette's theory is characterized by its systematicity. He provided a hierarchical classification of terms to describe a narrative structure, with particular focus on *how* a narrative is structured on a text as opposed to *what* is told. In this study, Genette's narratological terms are analogically translated into methods of a story-form memory construction in a cognitive system.

Keywords: Narrative intelligence · Story · Episodic memory · Narratology · Genette · Narrative discourse theory · Story-centered cognitive system

1 Introduction

In psychological studies on memory, an episodic memory is generally defined as a memory that enables the recollection or remembrance of past events or experiences [1, 2]. It is not assumed as a copy of the past events themselves and is instead assumed as corresponding to mentally encoded or constructed information including reinterpretation, sensemaking, abstraction, generalization, errors, and forgetting. However, the mechanism that constructs episodic memory is not systematically formulated in previous studies on cognitive architectures and systems. For example, in the Soar cognitive architecture, an episodic memory is considered as sequential snapshots of the agent's working memory [3]. Faltersack et al. [4] presented an episodic memory

A. V. Samsonovich (Ed.): BICA 2019, AISC 948, pp. 1–10, 2020.
https://doi.org/10.1007/978-3-030-25719-4_1

structure that involves a hierarchy from lower primitive elements to higher compound elements. In the ICARUS cognitive architecture, an episodic memory is encoded via generalization based on similarity with previous episodes [5]. The aforementioned episodic memories involve only few linguistic aspects. On the other hand, Bölöni [6] presented an episodic memory model including conceptual information. Recently, several studies focused on narrative structure in an episodic memory. Anderson [7] discussed the mental processes of constructing an episodic memory including temporal segmentation of event and creating a relationship between events. León [8] formalized the relational structure of a narrative memory that connects narrative objects in terms of kernels and satellites. Although the aforementioned studies captured a partial aspect of episodic memory, it is necessary to explore a unified computational theory of episodic memory construction.

In this study, the term "story" is used, rather than episodic memory, to refer to a mental representation of temporally and linguistically organized events and entities. The main difference between a story and an episodic memory is that a story involves the *form* of the mental representation, while the definition of an episodic memory is generally based on the *function* (i.e., it enables the remembering of past events). As described in Akimoto [9], a story is considered as a uniform mental representation involving episodic memory, autobiographical memory, current situation, prospective memory, planned or imagined future, and fictional or hypothetical story. They include long-term, short-term, and working memories. Hence, the generative cognition of story constitutes a common basis for a cognitive system or an integrative autonomous artificial intelligence. The importance of story cognition in artificial intelligence was explored by Schank [10, 11]. Winston's strong story hypothesis also argues for the generality of story in an intelligence [12].

In order to ensure a systematic formulation of the story-form memory construction, this study introduces Genette's narrative discourse theory [13]. It is a representative work in narratology (which is a discipline that examines fundamental structures, principles, and properties of narratives). Narratological theories are applied in artificial intelligence studies on narrative generation [14–16], interactive narrative [17], narrative analysis [18], and cognitive architecture [7–9, 19, 20].

Genette's narrative discourse theory is characterized by its systematicity. He provided a hierarchical classification of terms to describe a narrative structure, with particular focus on *how* a narrative is structured on a text as opposed to *what* is told. The aim of this study involves formulating the methods of story-form memory construction via analogically applying Genette's systematic theory.

The rest of the paper is organized as follows: Sect. 2 provides an overview of Genette's narrative discourse theory including the analytical framework and structural categorization. In Sect. 3, Genette's analytical framework is translated into the structure of a cognitive system. In Sect. 4, Genette's categories are reinterpreted as methods of story-form memory construction. Finally, Sect. 5 concludes the study.

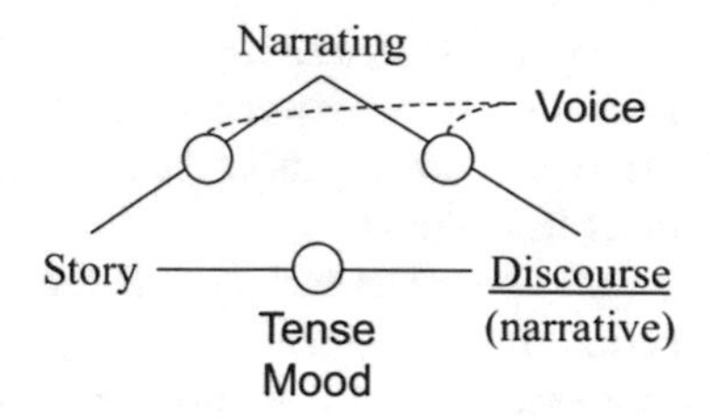

Fig. 1. Analytical framework of the narrative discourse theory.

2 Genette's Narrative Discourse Theory

This section provides a brief overview of Genette's narrative discourse theory [13]. The theory was developed based on analyzing a novel *À la recherche du temps perdu* (*In Search of Lost Time*) written by Marcel Proust. The basic analytical framework is explained in Sect. 2.1. Subsequently, categorization of narrative discourse structure is described in Sect. 2.2.

2.1 Analytical Framework

In order to deal with the structural analysis of a narrative, Genette distinguished a narrative into three aspects, namely *discourse* (*narrative*), *story*, and *narrating*. A discourse refers to the text of a novel or narrative itself. A story refers to the content information recounted in the discourse. Thus, it contains chronologically organized events. Narrating refers to the action of producing a narrative by a *narrator(s)* and *narratee(s)*. In this context, a narrator and narratee correspond to persons inscribed in the narrative text and not the author and reader of a narrative. In the aforementioned three aspects, only a discourse exhibits a materiality while a story and narrating are interpreted from the discourse. Hence, the object of the analysis corresponds to a discourse.

The categorization of structural properties of a narrative discourse is constructed based on the relationship among the discourse, story, and narrating. Specifically, it consists of the following three broad categories (see also Fig. 1): tense refers to the relationships between the temporal aspects of a discourse and story; mood refers to the modalities of expressing the story in the discourse; and voice refers to the situation of narrating, in the relationships with the discourse and story.

2.2 Categorization of Narrative Discourse Structure

Genette established hierarchically organized terms to describe a narrative structure. The terms are arranged as subcategories of the aforementioned three broad categories. The hierarchical categorization of structural terms and their brief explanations are provided below.

- Tense: Relationships between the temporal aspects of a discourse and story.
  - Order (anachronies): Relationships between the chronological order of events in a story and the order in which the events are recounted in a discourse. There are two major subcategories as follows:
    - Analepsis: Going back to past events from a present temporal position via a flashback, recollection, or other methods.
    - Prolepsis: Going forward to future events from a present temporal position via a flashforward, prediction, or other methods.
  - Duration (anisochronies): Relationships between the length of a discourse (for e.g., words, lines, and pages) and amount of time of the recounted events (for e.g., seconds, minutes, hours, days, months, and years).
    - Summary: A sequence of events is briefly recounted via relatively short text.
    - Scene: A sequence of events is recounted in detail via relatively long text.
    - Pause: The story's temporal progress is stopped in a discourse via the narrator's commentaries or descriptions of objects, things, or situations.
    - Ellipsis: A certain time period in a story is explicitly or implicitly omitted in the discourse (for e.g., "some years passed").
  - Frequency: Relationships of frequency between the occurrences of an event(s) in a story and recounted number in a discourse.
    - Singulative narrative: An event is recounted once, i.e., a normal narrative.
    - Repeating narrative: The same event is recounted two or more times.
    - Iterative narrative: Similar events are recounted once (e.g., "every day of the week I went to bed early").
- Mood: Modalities of expressing a story in the discourse.
  - Distance: A mode of regulating narrated information in terms of the degree of the narrator's mediation. The notion can be understood as a contrast between *mimesis* or *showing* (smaller distance) and *diegesis* or *telling* (larger distance).
  - Focalization: A mode of regulating narrated information based on a choice or not of a restrictive perspective. The three basic types of focalization are listed below.
    - Zero focalization: A nonfocalized discourse without information regulation based on a specific perspective.
    - Internal focalization: A discourse is composed from a character's perspective including the character's mental actions.
    - External focalization: A discourse presents only the external behavior of characters (i.e., characters' mental actions are not recounted).
- Voice: Situations of narrating in the relationship with the discourse and story.
  - Time of the narrating: Relationships between the temporal positions of narrating and the narrated story.
    - Subsequent: A past-tense narrative.
    - Prior: A predictive narrative (generally presented in the future tense).
    - Simultaneous: A present-tense narrative.
    - Interpolated: Temporal position of the narrating is interpolated between the moments of action.

- Narrative level: The notion can be understood as the nesting structure of narrating and is explained as "a narrative narrated within a narrative." A first-level narrative is produced in its external level (extradiegetic position).
- Person: Relationships between a narrator and characters in the story.
  - First-person: A narrator appears in the story as a (central) character.
  - Third-person: A narrator does not appear in the story.

3 Translating the Analytical Framework into a Cognitive System

This section translates the relationship among discourse, story, and narrating into a cognitive system. Figure 2 illustrates an overview of the framework. First, the three aspects of a narrative are replaced into representational and procedural elements of a cognitive system: the story corresponds to a mental representation containing information of structured events; discourse corresponds to an expressive structure of a narrative; and narrating corresponds to the action of producing stories and discourses. With respect to the aspect of narrating, the narrator essentially corresponds to the agent itself. The narratee is unspecified although it corresponds to an environment or an objective that directs the production of a story and discourse.

In the relationship with physical and social environments, a story that forms a temporal structure of a current situation constitutes a foundation of a higher-level action–perception system [9]. Conversely, narrative-communication with another person or agent is always mediated via a discourse.

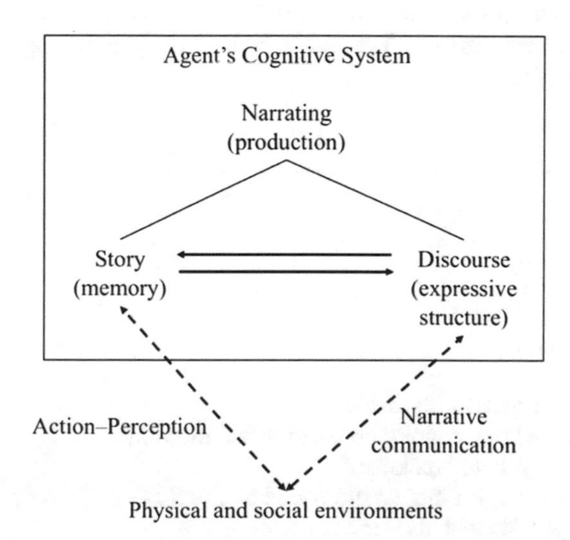

Fig. 2. Framework of story construction in a cognitive system.

Genette's theory does not clarify whether a story refers to events and things themselves (in an external world) or a mental representation of events and things. In this study, a story is clearly positioned as a mental representation. Hence, it can be considered that a discourse as well as a story is linguistically constructed. The aforementioned perspective blurs the boundary between a discourse and story. However, the matter appears to constitute the essential nature of the mutual relationship between the memory (story) and expression (discourse). It captures a cyclic relationship exists between a story and discourse, i.e., a story is reconstructed via structuring a discourse based on that story.

4 Translating the Narratological Categories into Story Construction Methods

In this section, the categories of narrative structure are translated into methods of story construction in a cognitive system. In dealing with the issue, we focus on processes wherein a relatively raw (simple) story is reconstructed by narrative methods. In order to provide a formal description of the methods, we define a minimal representation formalism of a memory containing stories.

A memory M denotes a tuple $< S, T >$, where S denotes a set of stories $\{s_1,\ldots, s_\mathrm{n}\}$ and T denotes a set of times $\{t_1,\ldots, t_m\}$. Time t_i contains one or more stories: [*time* $\{s_i\}$]. A story s_i is represented by a tuple $< E, R >$, where E denotes a sequence of events $(e_1,\ldots, e_l)$ and R denotes a set of relationships $\{r_1,\ldots, r_k\}$. An event e_i represents a character's action or a stative information including description and commentary. An event can contain a sequence of two or more sub-events: $[e_i\ (e_1,\ldots, e_j)]$. A relationship r_i relates two events: [*relationship* (e_p, e_q)]. Figure 3 shows an example of memory containing three simple stories. The example is subsequently used to explain story construction methods.

$M = \{S\{s_1, s_2, s_3\}, T\{t_1, t_2\}\}$
$s_1 = \{E(e_{1,1}, e_{1,2}, e_{1,3}), R\{\}\}$
$s_2 = \{E(e_{2,1}, e_{2,2}, e_{2,3}), R\{\}\}$
$s_3 = \{E(e_{3,1}, e_{3,2}, e_{3,3}), R\{\}\}$
$t_1 =$ [yesterday $\{s_1, s_2\}$]
$t_2 =$ [today $\{s_3\}$]

Contents of stories (in English):
s_1: "(Yesterday) $e_{1,1}$: I woke up early in the morning. $e_{1,2}$: I took a walk at a beach. $e_{1,3}$: I ate breakfast."
s_2: "(Yesterday) $e_{2,1}$: I met an old friend at a park. $e_{2,2}$: We went to a coffee shop. $e_{2,3}$: We chatted over our coffee."
s_3: "(Today) $e_{3,1}$: I woke up early in the morning. $e_{3,2}$: I took a walk in a park. $e_{3,3}$: I ate breakfast."

Fig. 3. Example memory.

In this study, the order is not applied to the construction method for a story itself; instead, it exhibits a close relationship with a memory organization. Therefore, we start with duration, and the relevance of temporal ordering to a memory is discussed in the last subsection.

4.1 Duration

The four subcategories of duration (i.e., summary, scene, pause, and ellipsis) are translated as follows.

Summary and scene correspond to a change in the level of detail of a sequence of events. In summary, two or more events are abstracted into a comprehensive event. For example, s_2 changes to $(e_{2,1}, [e_{2,x}\ (e_{2,2}, e_{2,3})])$, where $e_{2,x}$ (e.g., "We had coffee at a coffee shop.") is newly generated from $e_{2,2}$–$e_{2,3}$. If the sub-events ($e_{2,2}$–$e_{2,3}$) are forgotten, then s_2 corresponds to $(e_{2,1}, e_{2,x})$.

In scene, an event is detailed into two or more sub-events. For example, s_2 changes to $(e_{2,1}, [e_{2,2}\ (e_{2,2a}, e_{2,2b}, e_{2,2c})], e_{2,3})$, where $e_{2,2a}$–$e_{2,2c}$ (e.g. "We walked to a coffee shop. We entered the coffee shop. We ordered two cups of coffee and two shortcakes.") correspond to newly generated events from $e_{2,2}$.

Pause linguistically composes of a situational or scenery information or commentary that is associated with an event. For example, s_1 changes to $(e_{1,1}, e_{1,2}, e_{1,2d}, e_{1,3})$ where $e_{1,2d}$ refers to a scenery description of the beach (e.g., "The sea was sparkling in the morning sun.").

Ellipsis can be interpreted as a choice of information that is discarded from a story. For example, if $e_{2,2}$–$e_{2,3}$ in s_2 are discarded (consciously or unconsciously), then the story shortens to $(e_{2,1})$, "Yesterday, I met an old friend at a park."

4.2 Frequency

Singulative narrative is a normal form of story construction. The act of repeating can be considered as a memory rehearsal although it does not produce a copy of an event into a story. Thus, only iterative narrating can be translated into a method of memory construction from the subcategories of frequency. In this method, two or more similar stories or events are compressed into a single story. For example, s_1 and s_3 can be transformed into a compressed story s_4, "(Every morning recently) $e_{4,1}$: I wake up early. $e_{4,2}$: I take a walk. $e_{4,3}$: I eat breakfast." The story is contained in a new time t_3: [every-morning-recently $\{s_4\}$]. The original stories can be forgotten, and in this case $M = \{S\{s_2, s_4\}, T\{t_1, t_3\}\}$.

4.3 Distance

Example stories in Fig. 3 contains only a sequence of simple events without relational information. Given that the stories involve only a few mediations of the narrator and that they appear to constitute *showing* as opposed to *telling*, it is assumed that the stories are created from a relatively small distance. A basic method to increase the distance involves adding a *relationship* between story elements because a relationship is considered as an immaterial object created by a narrator's mentality. Conversely, decreases in the relational information from a story decrease the distance.

We provide several examples of the idea. If a causality between events is added into s_1, then it changes to $\{E(e_{1,1}, e_{1,2}, e_{1,3}), R\{r_1[\text{reason}\ (e_{1,1}, e_{1,2})]\}\}$, which implies "$e_{1,1}$: *Because* I woke up early in the morning, $e_{1,2}$: I took a walk at a beach. $e_{1,3}$: I ate breakfast.". A creation of a new event to provide a reason for an action increases the distance to s_1, e.g., $\{E(e_{1,1}, e_{1,2r}, e_{1,2}, e_{1,3}), R\{r_1[\text{reason}\ (e_{1,2r}, e_{1,2})]\}\}$ which implies for example, "… $e_{1,2r}$: *Because the morning sun was calling me*, $e_{1,2}$: I took a walk at a beach….".

Another method to increase the distance is to add a commentary about a story element (e.g., "… $e_{1,2}$: I took a walk at a beach. $e_{1,2c}$: I believe that walking is good for mental health….").

4.4 Focalization

In a story construction process, focalization works as a perspective-based regulation of information to be contained in the story. In an agent's cognitive system, a natural focalization considers the internal focalization of the agent itself. Hence, an important issue in focalization involves taking a different perspective from the agent's internal perspective. It generates a different version of a story. For example, a different version of s_2 (i.e., $s_{2'}$) is produced via simulating the internal focalization of the old friend, e.g., "$e_{2',1}$: An old friend met me at a park. $e_{2',2}$: He felt nostalgic about me. …". A perspective is distinguished from the narrator, and thus the first-person character does not change in this version.

4.5 Voice

In Genette's theory, voice considers the aspect of narrating and not the structure of a discourse and the story itself. As shown in Fig. 4, subcategories of voice (i.e., narrative level, time of the narrating, and person) can be translated into a set of elements to form a meta-story structure via interpreting the notion as background conditions in the cognitive action of constructing a story.

Narrative level corresponds to an essential aspect of the meta-story structure to create a division between a story and cognitive action to produce the story. It is represented via a nesting structure and a first-level story is produced in its external (extradiegetic) level by the narrator.

Time of the narrating corresponds to the temporal relationship between a story's temporal position and present time of producing or remembering the story. The information is essentially reflected in the linguistic tense (i.e., past, future, or present) when a story is expressed as a narrative.

Person corresponds to whether the *narrator* appears in a story as a *character*. In an agent's cognitive system, the basic narrator corresponds to the agent itself. However, another person or agent (as a character in a story) can become the narrator of a second-level story. For example, if $s_{2'}$ (the friend's internal focalization presented in Sect. 4.4) is nested as the friend's story, then it changes to "I met K at a park. I felt nostalgic about K. …" (K refers to the original "me").

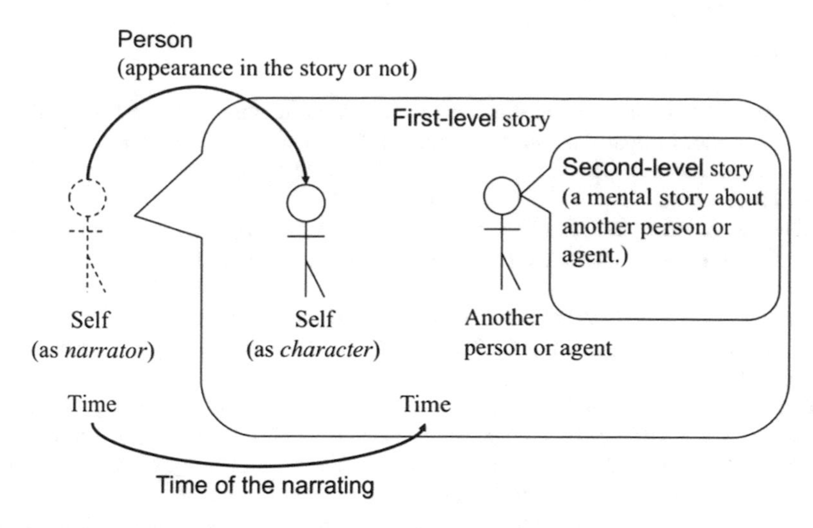

Fig. 4. Meta-story structure including narrative level, time of the narrating, and person.

4.6 Order

Events in a story are chronologically organized, and thus the temporal order of events can be essentially manipulated only in the discourse dimension. Hence, we do not apply temporal ordering to a story construction method. However, the mechanism of forming a connection between events or stories over the chronological order and a temporal gap is relevant to the organization of stories in a memory. Therefore, we provide a short consideration on a structure underlying temporal ordering in a discourse. We use only one example that associates s_2 with an event in s_3 via a flashback: ($e_{3,1}$, $e_{3,2}$, ($e_{2,1}$, $e_{2,2}$, $e_{2,3}$), $e_{3,3}$). The structure is expressed as "Today, $e_{3,1}$: I woke up early in the morning. $e_{3,2}$: I took a walk in a park. (Yesterday, $e_{2,1}$: I met an old friend at the park. $e_{2,2}$: We went to a coffee shop. $e_{2,3}$: We chatted over our coffee.) $e_{3,3}$: I ate breakfast.". The structure can be explained as an association mediated by the "park."

5 Concluding Remarks

In search of a unified computational theory of story-form memory construction, this study analogically translated Genette's narrative discourse theory into story construction methods and meta-story structures in a cognitive system. These narratological methods and structures are relevant to various cognitive issues such as temporal segmentation, generalization, theory of mind, metacognition, and sociocultural aspect of cognition. A story involves an integrative information structure, and a story-centered approach to a cognitive system exhibits an advantage to capture the integrative working of a mind. The presented theory acts as a foundation to develop a story-centered cognitive system or architecture.

Acknowledgments. This work was supported by JSPS KAKENHI Grant Number JP18K18344.

References

1. Tulving E (1983) Elements of episodic memory. Oxford University Press, Oxford
2. Tulving E (2002) Episodic memory: from mind to brain. Ann Rev Psychol 53:1–25
3. Nuxoll AM, Laird JE (2012) Enhancing intelligent agents with episodic memory. Cogn Syst Res 17–18:34–48
4. Faltersack Z, Burns B, Nuxoll A, Crenshaw TL (2011) Ziggurat: steps toward a general episodic memory. In: Advances in cognitive systems, AAAI 2011 fall symposium, pp 106–111
5. Ménager DH, Choi D (2016) A robust implementation of episodic memory for a cognitive architecture. In: Proceedings of the 38th annual conference of the cognitive science society, pp 620–625
6. Bölöni L (2011) An investigation into the utility of episodic memory for cognitive architectures. In: Advances in cognitive systems, AAAI 2011 fall symposium, pp 42–49
7. Anderson TS (2015) From episodic memory to narrative in a cognitive architecture. In: Proceedings of 6th workshop on computational models of narrative, pp 2–11
8. León C (2016) An architecture of narrative memory. Biologically Inspired Cogn Architectures 16:19–33
9. Akimoto T (2018) Stories as mental representations of an agent's subjective world: a structural overview. Biologically Inspired Cogn Architectures 25:107–112
10. Schank RC (1982) Dynamic memory: a theory of reminding and learning in computers and people. Cambridge University Press, New York
11. Schank RC (1990) Tell me a story: a new look at real and artificial memory. Charles Scribner's Sons, New York
12. Winston PH (2012) The right way. Adv Cogn Syst 1:23–36
13. Genette G (1980) Narrative discourse: an essay in method (Lewin JE, Trans.). Cornell University Press, Ithaca. (Original work published 1972)
14. Gervás P, Lönneker-Rodman B, Meister JC, Peinado F (2006) Narrative models: narratology meets artificial intelligence. In: Proceedings of satellite workshop: toward computational models of literary analysis, 5th international conference on language resources and evaluation, pp 44–51
15. Lönneker B (2005) Narratological knowledge for natural language generation. In: proceedings of the 10th European workshop on natural language generation, pp 91–100
16. Ogata T (2016) Computational and cognitive approaches to narratology from the perspective of narrative generation. In: Ogata T, Akimoto T (eds) Computational and cognitive approaches to narratology. IGI Global, Hershey, pp 1–74
17. Montfort N (2007) Generating narrative variation in interactive fiction. Doctoral Dissertation, University of Pennsylvania
18. Mani I (2013) Computational modeling of narrative. Morgan & Claypool, San Rafael
19. Szilas N (2015) Towards narrative-based knowledge representation in cognitive systems. In: Proceedings of 6th Workshop on Computational Models of Narrative, pp 133–141
20. Samsonovich AV, Aha DW (2015) Character-oriented narrative goal reasoning in autonomous actors. In: Goal Reasoning: Papers from the ACS Workshop, Technical Report GT-IRIM-CR-2015-001, pp 166–181

Story-Centric View on the Mind

Taisuke Akimoto(✉)

Kyushu Institute of Technology, Fukuoka, Japan
akimoto@ai.kyutech.ac.jp

Abstract. The basic assumption of the present study is that the essence of the human mind is to generate stories by interacting with environments, or to interact with environments by generating stories. In this context, a story refers to a mental representation of an individual's subjective world including the past, present, future, and fiction. This assumption leads us to a consistent and plausible understanding of the human mind and realization of a human-like artificial intelligence. In this paper, I present an exploratory computer-oriented theory on the mind by ways of a story-centric view. The theory comprises two concepts. First, the mind is described as an interactive story generation system between the narrator-self and a story that has the power of self-organization. Second, the interrelationship among stories is put in focus, and the generative process of stories is described in terms of the mutual actions between them. Furthermore, these two concepts are adapted to characterization of the conscious and unconscious mind.

Keywords: Story-centered cognitive system · Story generation · Memory · Conscious and unconscious mind

1 Introduction

A major problem in studies on cognitive systems or cognitive architectures [1–3] is to find computational accounts of the integrative and complex workings of the mind. Hence, it is important to develop common theories, principles, models, and frameworks for diverse cognitive functions or phenomena.

From this perspective, I consider the essence of the human mind to be the generation of stories by interaction with environments, or the interaction with environments by generating stories. This assumption will lead to a consistent and plausible understanding of the human mind and realization of a human-like artificial intelligence.

In this paper, I present an exploratory computer-oriented theory of the mind by ways of a story-centric view. The presented theory is constructed from two perspectives. First, the mind is described as the interaction between a story and the narrator who observes and manipulates the story. Second, the mind is described in terms of the mutual actions between stories.

A. V. Samsonovich (Ed.): BICA 2019, AISC 948, pp. 11–14, 2020.
https://doi.org/10.1007/978-3-030-25719-4_2

2 Mind as Interactive Story Generation

The mind can be interpreted as an interactive story generation system. This concept is derived from the following three assumptions:

- A story is a mental representation of an individual's subjective world [4]. A story is a chronological organization of events including relevant entities. It is a universal form of mental representation involving the past, present, future, and fiction.
- The "self" is divided into two concepts: a *character* in a story and a *narrator* who observes and produces the story from the outside [5].
- A story has the power of self-organization [6]. This claim is derived from the structural complexity of a story. A story involves complicated whole–part and part–part interdependencies. The generation of such a complex structure is difficult to explain on the basis of a centrally controlled system. Instead, this complexity can be understood as a typical complex and distributional system.

An agent interacts with the external environment by organizing the world in the form of a story. This generative process is driven by two powers. On the one hand, the narrator-self can be positioned as the observer and producer of a story. On the other hand, a story also generates its own structure by the power of self-organization. This dual nature of story generation can be interpreted through the analogy with an interactive narrative system [7] (where the user experiences a dynamic scenario by interaction with a computer-based narrative environment). In the mind, the generative interaction occurs between the narrator-self and a story that has the power of self-organization. Figure 1 illustrates the above-described concept. The story in this figure assumes the representation of the situation faced by the agent, including past events and future predictions, plans, and goals.

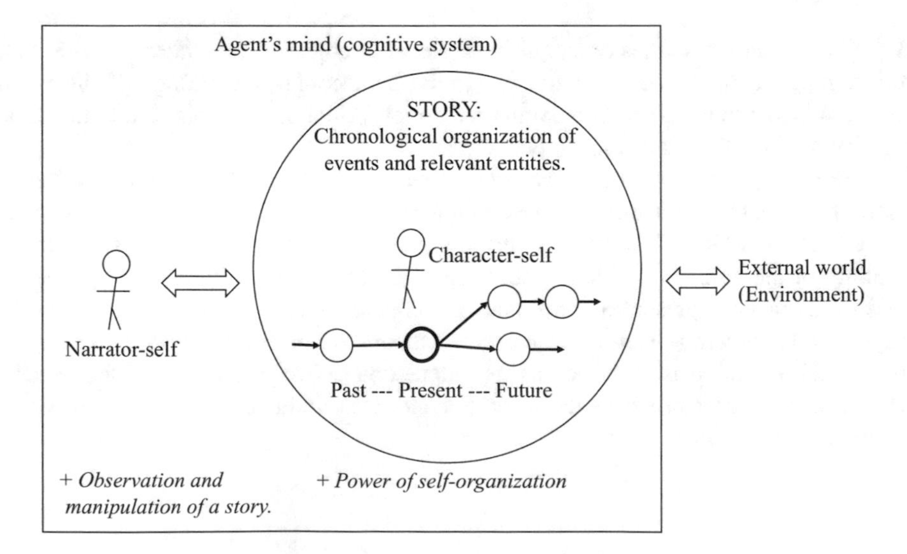

Fig. 1. Mind as interactive story generation.

3 Mind as Mutual Actions Between Stories

I make two assumptions with regard to the background knowledge underlying the generation of a story:

- A human's memory contains numerous and various stories including the past, future, and fiction [4]. These stories are accumulated throughout life.
- These stories provide the knowledge or material for composing a new story [8]. This claim is rooted in the concepts of analogy, case-based reasoning, conceptual blending [9], etc.

From this perspective, the interrelationship among stories is a crucial aspect of the mind. According to the story-centric view, Fig. 2 illustrates a conceptual diagram of the mental structure focusing on the interrelationship among stories. The figure assumes the situation in which the mind generates a future story (as a plan, prediction, or dream). In this situation, the memory of past experiences provides a basis for imagining a realistic future. A fictional story may also act on the future, for example, a child can have a fantastic dream brought about because of the influence of animated movies.

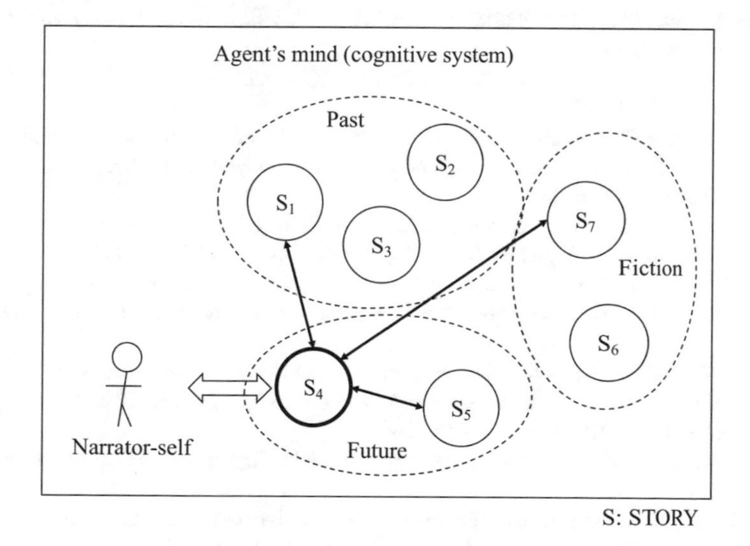

Fig. 2. Mind as mutual actions between stories.

The presented diagram can be generalized in terms of mutual actions between stories. This concept captures the complex work of the mind, as exemplified below:

- Composing a future story based on past experiences and fictional stories.
- Reconstructing a past story based on culturally shared stories obtained from literary works, ethical or religious narratives, etc.
- Interpreting and appreciating a fiction (literary work) based on real experiences and its relationship with other fictions.

4 Conscious and Unconscious Aspects of the Mind

Organizing the world into the form of a story is similar to the work performed by consciousness. However, one can assume that a large part of the story generation process is an unconscious automatized process. Indeed, the human being is unaware of how past experiences are encoded and stored into their memory, and how they are retrieved or recollected. The detailed mental process of creating or reading a fiction is also almost impossible to explain. However, the human being generally possesses intuition with regard to controlling of the self. In the theory presented in this paper, the relationship between the conscious and unconscious aspects of the mind is explained as the interaction between the narrator-self and a story's self-organization, including its relationships with other stories.

Acknowledgments. This work was supported by JSPS KAKENHI Grant Number JP18K18344.

References

1. Langley P, Laird JE, Rogers S (2009) Cognitive architectures: research issues and challenges. Cogn Syst Res 10:141–160
2. Samsonovich AV (2012) On a roadmap for the BICA challenge. Biologically Inspired Cogn Architectures 1:100–107
3. Laird JE, Lebiere C, Rosenbloom PS (2017) A standard model of the mind: toward a common computational framework across artificial intelligence, cognitive science, neuroscience, and robotics. AI Mag 38(4):13–26
4. Akimoto T (2018) Stories as mental representations of an agent's subjective world: a structural overview. Biologically Inspired Cogn Architectures 25:107–112
5. Akimoto T (2019) Narratological formulation of story-form memory construction: applying Genette's narrative discourse theory. In: Proceedings of the tenth annual meeting of the BICA society
6. Akimoto T (2018) Emergentist view on generative narrative cognition: considering principles of the self-organization of mental stories. Adv Hum Comput Interact Spec Issue Lang Sense Commun Comput 2018, Article ID 6780564
7. Riedl MO, Bulitko V (2013) Interactive narrative: an intelligent systems approach. AI Mag 34 (1):67–77
8. Akimoto T (2019) Key issues for generative narrative cognition in a cognitive system: association and blending of stories. In: Story-enabled intelligence, AAAI 2019 spring symposium
9. Fauconnier G, Turner M (2002) The way we think: conceptual blending and the mind's hidden complexities. Basic Books, New York

Biologically Inspired Physical Model of the Vocal Tract for the Tasks of Recognition of the Current Psycho-Emotional State of a Person

Alexander M. Alyushin[1,2(✉)]

[1] National Research Nuclear University MEPhI (Moscow Engineering Physics Institute), Kashirskoe shosse 31, Moscow 115409, Russia
Alyshin@list.ru

[2] Plekhanov Russian University of Economics, 36 Stremyanny per., Moscow 117997, Russia

Abstract. The urgency of the development and use of a physical biologically inspired model of the human vocal tract (VT) of a person for solving a complex of tasks on recognizing the level of stress based on the processing of speech information is shown. As an advantage of such a biologically inspired model, compared with other types of existing VT models, the possibility of studying the complex effects of nervous tension on speech parameters was highlighted. The volatility of the fundamental frequency (FF) is highlighted as such an informative parameter. The structure of the VT physical model is considered, as well as an example of its practical application. It is shown that improving the reliability of the so-called speech signature technology (SS), which is used to protect important paper documents from counterfeit and falsification, is one of the promising areas of application of the developed model.

Keywords: Biologically inspired model · Stress recognition · Speech signature

1 Introduction

Information technologies based on speech information processing are widely used nowadays in almost all spheres of human activity. These technologies are of particular importance in the field of information security. Thus, the SS technology actively developed recently is an additional level of protection of important documents from their fakes and falsification [1].

Such a signature allows you to increase the level of security, first of all, documents presented in paper form. The SS technology assumes the inclusion of significant fragments of the speech of its author, represented graphically using sonograms, in the document. Reproduction and analysis of speech fragments by the recipient of the document makes it possible to identify his personality and the authenticity of the main provisions of the document. This technology of protection of paper documents does not limit the ability to copy, replicate and scan [1].

A. V. Samsonovich (Ed.): BICA 2019, AISC 948, pp. 15–21, 2020.
https://doi.org/10.1007/978-3-030-25719-4_3

SS technology also makes it possible to ensure that the author of the document, when compiled, was in a normal sane state and did not sign it under duress. To do this, in the SS, it is necessary to preserve and transmit information about the voice features that most characterize a person's current psycho-emotional state. In this regard, it should be noted that in terms of diagnosing the psycho-emotional state, voice information takes the second place in informational content after video information about a person's face.

The conducted research identifies a number of informative features of speech in its psycho-semantic, paralinguistic, psycholinguistic, pragmatic and syntactic analysis [2]. Taking into account, as a rule, the time-limited duration of the SS, the paralinguistic analysis of speech is most appropriate. Such an analysis usually includes an assessment of the rate of speech, its loudness, melody, the dynamics of changes in timbre, jitter of FF. It is possible to determine the gender characteristics of the author of the document. Jitter (variability) of FF is widely used in practice, including for purely medical purposes, to identify a number of serious diseases [3].

From the point of view of human physiology, the so-called tremor of the voice, which is manifested, above all, in the FF tremor, is caused by involuntary muscle tension during stress. First of all, these are the muscles of the vocal cords, the muscles of the neck and the face. The influence of the state of these muscles on the level of tremor of the voice is quite difficult and ambiguous. The individual features of body shaping are one of the factors that aggravate the study of the mechanisms of tremor.

In practice, various VT models are used, allowing simulation to be carried out with some degree of approximation of the mechanisms by which voice tremor occurs and its conditional state of stress. The main problem with the use of such models is the need for verification, which implies a rather large amount of experimental research involving the necessary contingent of people.

The development of a model of human VT, which allows for a complex of studies aimed at identifying patterns of voice trembling at various levels of stress, is an urgent task at present. The application of such a model in practice will reduce the time and material costs in the development of algorithms and tools for reliable determination of the level of stress. This model is of particular importance in the implementation of SS technology with the support of the function of assessing the psycho-emotional state of the author of the document.

The aim of the work is to develop a biologically determined physical model of human VT, focused on the study of the mechanisms of tremor of the gloss.

2 Biological Inspiration and the Essence of the Proposed Approach

The approach developed is based on the following biologically inspired principles:

- the hearing aid and the human brain allow you to simultaneously perceive and analyze the spectral composition of acoustic signals in real time;
- the person's nervous tension is manifested both in the tension and trembling of the vocal cords muscles, and of the muscles of the neck and face, which has a complex

effect on FF and, ultimately, manifests itself in characteristic modulations of the entire spectrum of overtones;
- the constitution, age, gender and health conditions (inflammation, swelling of the throat) are the main factors influencing the degree and nature of stress.

In Fig. 1a shows the model of the main recognition part of the human hearing aid. The so-called cochlea of the inner ear 1, having a throat 2, is an acoustic wave guide in which neurons 3 are located with resonating hairs 4.

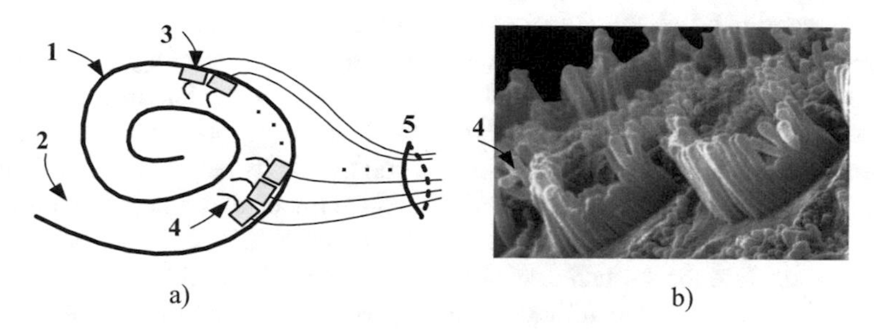

Fig. 1. Model of the human hearing aid.

The parameters of the cochlea waveguide make it possible to perceive oscillations in a wide frequency range of 25 Hz–20 kHz. The auditory nerve 5 transmits information to the appropriate brain center for detailed processing. The huge number of hairs 4 in the cochlea (Fig. 1b [4]) allows simultaneous parallel perception of the parameters of all harmonics of the speech signal. This circumstance is important and is taken into account when developing the human VT model.

From practice, it is known, for example, that at a high level of stress, the voice of a person becomes "metallic". At the same time, in its spectral composition changes, including the energy balance between the individual harmonics, which is clearly perceived by the audience.

The complex nature of the effects of nervous tension must be considered when developing VT models. From this point of view, the main requirements for such models include:

- the ability to simulate the effects of stress and muscle tremor of the vocal cords, neck and face under stress;
- the presence of parameters regulating the degree of such a stress effect;
- the possibility of changing the parameters of the VT model responsible for the individual constitutional characteristics of the person's body;
- focus on the synthesis of vowel sounds, the tremor of which, to the greatest extent, conveys a stressful state;
- ease of implementation, clarity, the possibility of modifying and developing the VT model to describe other effects.
- The complex effect of nervous tension is to some extent taken into account when developing VT models. Among these models should be highlighted:

- mathematical models based on differential equations describing the propagation of acoustic oscillations [5, 6];
- models describing the processes of mechanical oscillations [7];
- physical models [8].

The most complete listed requirements satisfy the physical model. Such models implement the mechanisms of propagation, reflection and resonance of actually existing acoustic oscillations, which makes it possible to model effects that are difficult to give a precise mathematical description.

Studies conducted at the Department of Electrical and Electronic Design at the Japanese University Sophia [8] confirmed the effectiveness of using a physical model that allows synthesizing the main vowel sounds of the Japanese language. The model is an acoustic channel completely identical to the human VT geometry, containing resonant laryngeal cavity, pharynx, mouth, nose, and the cavity between the teeth and lips. The model is tunable. It may contain a different number of fragments with holes of different diameters, which permits to imitate a person's VT 10–20 cm long. For example, to simulate the formation of the Japanese vowel sound "o", you must use 14 fragments with holes of the following diameters: 14 mm, 22 mm, 26 mm, 34 mm, 34 mm, 38 mm, 34 mm, 28 mm, 22 mm, 16 mm, 14 mm, 16 mm, 20 mm, 25 mm.

The disadvantages of the physical model include the limited functional possibilities for studying the effects of the modulation of vowel sounds caused by the tremor of the muscles involved in the speech-forming process under stress. The main factor limiting the functionality is the rigid construction of the model fragments.

The developed biologically inspired physical model of VT is adaptive. Its fragments are made of soft, easily deformable material (silicone). Moreover, each fragment has a built-in mechanism for its deformation (compression) with computer control. This makes it possible to simulate the static and dynamic effects of the influence of muscle activity on the main components of VT. The developed VT physical model allows one to satisfy all the requirements listed above in order to take into account the complex nature of the effects of nervous tension on the gloss tremor.

3 Practical Use of the Developed Biologically Inspired Physical VT Model

In Fig. 2 shows an example of the practical use of the developed model as part of an experimental setup for the study of acoustic manifestations of a person's stress state.

In the example presented, the model is used to synthesize the vowel sound "a". For this, 19 fragments with characteristic window sizes were used. Filling the fragments with plastic material makes it possible to modulate the size of windows under the influence of forces created by actuators F. Taking into account the low-frequency range of action of forces imitating muscular effects on the VT under stressful conditions of varying degrees, electromagnetic performing devices were used.

Controlling actuators using an interface (I) connected to a personal computer (PC) allows you to simulate the tremor of various muscles involved in the process of speech formation.

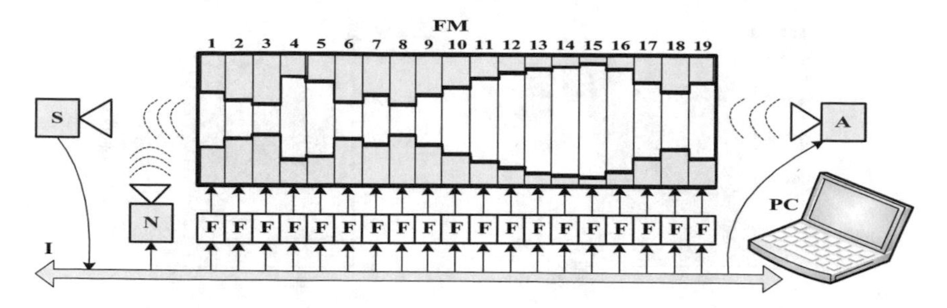

Fig. 2. An example of the use of the developed biologically inspired model.

In order to study the effect of external acoustic noise on the recognition of a tremor in a person's voice, an additional acoustic emitter (N) was used in the experimental setup. This emitter allows you to mix in the signal generated by the FM physical model when it is activated by periodic signals from the source A, the signal interference of a given spectrum and power level. The resulting acoustic signal is recorded by receiver S.

Practical use of the developed model allows solving the following tasks:

- study of the influence of the peculiarities of the constitutional structure of the human body on the information content of various overtones under stress of various degrees;
- the study of the influence of the level and spectral composition of external noise on the accuracy of the determination of FF and its instability;
- study of the degree of influence of the tremor of various muscles that affect the work of VT,
- study of the informativeness of various vowel sounds from the point of view of the reliability of recognition of stress at various levels;
- justification of the optimal set of overtones of vowels, the most informative in terms of recognizing the current level of stress.

4 Experimental Approbation of the Approach

The developed physical model was used to determine the most informative set of overtones that provide reliable recognition of the stress level in the presence of external noise of varying intensity for two vowel sounds of the Russian language.

In Fig. 3 shows examples of sonograms synthesized using the developed model in the presence of a noise component (gray background between brighter lines corresponding to overtones).

It is clearly seen that for the first vowel sound (Fig. 3a), the overtones F14, F19-F21 should be the most informative. Although their intensity loses to the lower-frequency components of F0-F2, the manifestation of tremor is much more pronounced. Similarly, for the second vowel sound (Fig. 3b), the most informative overtones are F14 and F15. The presented examples correspond to the modeling of the complex effect of muscle tremor under stress.

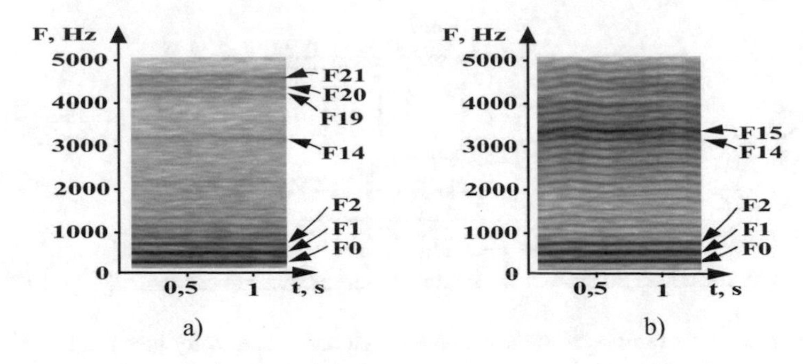

Fig. 3. Examples of sonograms of vowel sounds, obtained using the developed model.

5 Conclusion

The developed VT physical model is biologically inspired and is intended for use in research and training purposes.

The model makes it possible to investigate the effects of a complex effect of a stressful state of a person on VT, which makes it possible to identify the most informative group of speech parameters that reliably identify the state and level of stress.

The study was carried out in the framework of project No. 25.2911.2017/4.6 "Testing technology to reduce the risk of occurrence and mitigate the consequences of man-made disasters by minimizing the human factor's impact on the reliability and trouble-free operation of nuclear power plants and other hazardous facilities" in accordance with the competitive part of the MEPhI project for 2017–2019.

References

1. Alyushin AM (2018) Assessment of a person's psychoemotional state by his signature on the document as the basis for his disavowal. Voprosy Psikhologii 2:133–140
2. Alyushin VM (2018) Visualization of the quality (state) of the psychological climate in the team based on the speech interaction acoustic information processing. Sci. Vis. 10(3):72–84
3. Rusz J, Cmejla R, Ruzickova H, Ruzicka E (2011) Quantitative acoustic measurements for characterization of speech and voice disorders in early untreated Parkinson's disease. J Acoust Soc Am 129(1):350–367
4. LNCS Homepage. https://www.latimes.com/science/sciencenow/la-sci-sn-gene-editing-deafness-20171220-story.html. Accessed 27 Mar 2019
5. Donaldson T, Wang D, Smith J, Wolfe J (2003) Vocal tract resonances: preliminary study of sex differences for young Australians. Acoust. Aust. 31:95–98
6. Ramsay G (2001) Quasi-one-dimension model of aerodynamic and acoustic flow in the time-varying vocal tract: source and excitation mechanisms. In: Proceedings of the 7th European conference on speech communication and technology, 2nd interspeech event, Aalborg, Denmark, September 3–7, vol 1, pp 43 − 46

7. Lucero, J.: Simulations of temporal patterns of oral airflow in men and women using a two-mass model of the vocal folds under dynamic control. Acoustical Society of America, 1362 – 1372 (2005)
8. Arai T, Usuki N, Murahara Y (2001) Prototype of a vocal-tract model for vowel production designed for education in speech science. In: Proceedings of the 7th european conference on speech communication and technology, 2nd Interspeech Event, Aalborg, Denmark, September 3–7, vol 4, pp 2791 − 2793

Ontology-Based Modelling for Learning on Bloom's Taxonomy Comprehension Level

Anton Anikin(✉) and Oleg Sychev

Volgograd State Technical University, Volgograd, Russia
anton@anikin.name, poas@vstu.ru

Abstract. The taxonomy of educational objectives allows specifying the level of assessments and making the learning process gradual progress over subsequent levels. Out of six levels for the cognitive domain, the most problematic level for e-learning task is the comprehension level because it includes such tasks as interpreting, summarizing and inferring which is typically manually graded. The lack of comprehension-level assessments in e-learning leads to students attempting to solve more difficult tasks without developing comprehension skills, which complicates the learning process.

The article discusses using subject domain representation in the form of an ontology model which can be used to visualize the concepts and their relations for students, serve as a basis of a question-answering system for assessing student's comprehension of the subject domain, generating a natural-language explanation of the student's mistakes and creating adaptive quizzes. The requirements for the ontology to capture the comprehension level of the subject domain are formulated. The article discusses various application for such ontological models in the learning process.

Keywords: Ontology · E-learning · Bloom's taxonomy

1 Introduction

Current trends in learning favor using of e-learning to improve the learning process, provide the adaptability (search of educational resources of different levels of complexity, covering various topics of the course domain, with different didactic roles) to form the learner competencies. In order to achieve this goal, studies are been conducted on modeling the subject area of learning courses, educational resources and learners knowledge, and competencies [3,10]. An important task in

This paper presents the results of research carried out under the RFBR grant 18-07-00032 "Intelligent support of decision making of knowledge management for learning and scientific research based on the collaborative creation and reuse of the domain information space and ontology knowledge representation model".

A. V. Samsonovich (Ed.): BICA 2019, AISC 948, pp. 22–27, 2020.
https://doi.org/10.1007/978-3-030-25719-4_4

this approach is to assess the current knowledge and competencies of the learner in the course domain.

To solve this problem, it is beneficial to implement systems for questions generation and student's answers assessment. This will significantly improve the quality and effectiveness of the testing systems and the adaptability of the learning systems, reduce the participation of the teacher both in the preparation of test tasks and in learners answers assessment.

2 Bloom's Taxonomy of Educational Objectives and Educational Quizzes

The taxonomy of educational objectives developed by a committee of educators chaired by Benjamin Bloom [5] is widely used in education and educational research to categorize goals of educational assessments [9]. It allows educators to systematize their exercises to gradually develop the learners' knowledge and skills.

Educational quizzes with automatic feedback to the learners free educators from routine checks of the learners' exercises and providing individual feedback on the errors they made. It allows learners to get immediate feedback and do more exercises because they are no longer limited by the educator's time. This adaptive learning approach influenced the quick growth of online learning in the last decades [12].

However, not all categories of Bloom's taxonomy are easy to assess automatically, even in highly formalized subject domains like mathematics and programming. The knowledge level is easy to test automatically because its goal is to just recollect definitions and facts. The application and synthesis levels are easy to test for subject domains where you can give learners a task they can perform and write the results: writing or modifying a program or its part, devise a formula and so on. The analysis and evaluation levels can be testing by asking learners to perform analysis that yields definite numeric or string results and comparing them.

Of all the levels in Bloom's taxonomy, comprehension (or understanding) level tasks is the hardest to automate because typically involve stating the knowledge in a learner's own words, translation, extrapolation, and interpretation. These are tasks are hard to check automatically, which often leaves comprehension less used level in online learning courses. That is unfortunate because its the second level in the cognitive domain, and the lack of exercise for it often pushes learners to perform higher-level tasks without proper comprehension of the subject domain, which leads to frequent errors and frustration.

Automating the evaluation of comprehension-level assignments requires building a formal model of comprehension of the subject domain. One of the most relevant model for this is ontology [4]. If the comprehension will be captured in an ontology, a question generator to assess learner's comprehension can be created on the base of it.

3 Ontology-Based Questions Generators

According to [7], the ontology-based question generation approaches are divided into the three main strategies:

1. the class-based strategy, based on the relations between the classes and entities of the ontology;
2. the terminology-based strategy, based on the relations between the classes and sub-classes (taxonomic relations);
3. the property-based strategy, based on the properties in the ontology of different kind (objects, datatype and annotations).

In [2], an ontology-based approach to construct multiple-choice questions (MCQ) with controlled difficulty is described. The work states that all of Bloom's levels are covered. On the other hand, the examples of the questions provided are based on the knowledge explicitly described in the learning resources. For example, the application-level question does not require the learner to do actions using competencies obtained in the domain (e.g., to write some program in software engineering course domain). The same issue is relevant for the comprehension level question example - in fact, it reflects a complex variant of the knowledge level question.

The OntoQue engine described in [1] generates a set of assessment items from the given ontological domain, including MCQ-type, true/false, and fill-in items. The system implemented generates items by iterating over all entities (combinations of viable statements) in the ontology. In evaluation part, it is noted that most of the MCQ items concentrated on factual knowledge (the first level of Bloom's taxonomy), and the evaluators noted that they would prefer the MCQ items to address higher levels of cognitive skills [1].

In [6], property-based strategies are proposed to use for generating MCQ that the authors consider being on the comprehension level. The question stem is limited to the questions about some specified relations about concepts. This is more typical for knowledge-level questions, and the answer for that type of question is usually described in the explicit form in learning resources or requires processing at most taxonomic relations.

In [8], for automatic and adaptive knowledge assessment suggested using a template-based generator with ontology. For the comprehension level, 4 question templates are provided:

1. Are K_x and K_y indirectly connected?
2. What is K_x?
3. Which subconcept is directly connected by r with K_x?
4. Which superconcept is directly connected by r with K_x?

Of these templates, three (1, 3 and 4) use the knowledge given in the learning resources explicitly or require processing simple taxonomic relations, and it is not described for the second template how the answers for such questions are graded. Also, usage of strict templates, with placeholders only for concepts and relation names, makes the questions generated differ from the human created questions in natural language. The capabilities of the template-based approach for questions generation are very limited in the comprehension level.

The review shows absence of the relevant ontology models and approaches for modeling the learning questions and answers on the comprehension level of Bloom's taxonomy.

4 Requirements for the Comprehension-Level Modelling

In Bloom's original work, the comprehension level of taxonomy of educational objectives is defined as the form of understanding that the learner knows what was communicated and is able to make use of the material or idea that was communicated without relating it to other topics or seeing all its implications [5]. Bloom et al propose three ways to verify comprehension: translation, interpretation, and extrapolation. However, these methods are problematic to test using questions, especially machine generating.

For example, questions for translation should either include a choice of several possible translations, one of which is correct - or be an open-answer question. In multiple choice questions, there is always a chance that the learner would find the teacher-made translation, included in the ontology, in some source and answer the question just using knowledge. Grading open-answer question would require the ontology grader to reliably compare the semantics of two free-text natural-language answers. While modern methods reach good recall and precision results [11], they are still insufficient for automated learning without teacher's intervention. Many methods of machine translation, like example-based translation and statistical translation, can perform translation tasks [15] without comprehension or tries to simulate the comprehension with multidimensional vector space [14].

Interpretation is tied to summarizing knowledge by the learner. When assessing the learner's comprehension is done in teacher-learner interaction, the typical task is "state the problem in your own words." However, this task is very difficult to grade automatically.

To devise a method for automatic question generation on comprehension level we may use another part of Bloom's definition, that deals with using the material that was communicated without relating it to other material. So the learners' comprehension can be assessed by offering them a number of question for direct use of all the concepts they are supposed to learn in situations that don't require relating them to other material or making implications. Such questions are more prone to random guessing the possible answers than "state the problem in your own words" approach, but they can be generated in sufficiently large numbers to avoid this problem. One particular benefit of this approach is that the ontology, capable of generating and using such comprehension question would be able to detect the cause of the error in the learner's answer because each question is only about using a single fact or concept. So the system can easily explain the learner the cause of the error, further facilitating learning.

In order to generate such questions, the ontology must satisfy certain requirements. Widely-used in educational ontologies natural-language definition of concepts and limited set of relationships used [10,13,16] cannot be used to generate complex questions for using this concept. For each concept that is tested on comprehension, the ontology must contain an axiomatic definition of this concept.

For example, when learning mathematics or programming languages, if the goal is the comprehension of operator precedence concept, it isn't sufficient to just show comprehension as an operator property. The ontology must contain the model of how operator precedence works. It means not just establishing precedence relation between two different operators, but modeling the process of determining the evaluation order for regular (infix) expression notation - the process which operator precedence affects. Then a number of questions about evaluation order of randomly generated expressions can be generated and graded by this ontology, with the possibility of asking learners for the cause of their decisions and generating explanations in case of learner's errors.

5 Discussion

It can be said that in order to be effective for question generation and grading at comprehension level, an ontology must itself model comprehension of the subject domain. So Bloom's taxonomy can be used to explain the level of ontological modeling. Ontologies, containing only factual knowledge - including knowledge of key abstractions, classifications, and conventions in the subject domain are knowledge-level ontology - Bloom stated that the knowledge of conventions, criteria, methodology, principles and generalizations, and even theories and structures still belongs to the knowledge level [5].

The key difference between comprehension-level and knowledge-level of ontological modeling is that comprehension-level ontologies can make use of the concepts contained in them in simple situations without relating it with another material or reasoning all implications. Such ontologies are more complex and labor-consuming to build, but they are potentially a lot more effective in teaching and learning. Preliminary experiments with an ontology modeling key concepts of first lessons of programming basic course show that it allows generating a statistically significant number of questions for testing the comprehension of expression evaluation and control structures with the possibility to generate natural-language explanations for learners' errors. It is possible to make the testing process adaptive by storing information which concepts and situations trigger more errors for a particular learner.

The further development will bring us the ontologies working on application level and more advanced levels of Bloom's taxonomy.

References

1. Al-Yahya M (2014) Ontology-based multiple choice question generation. Sci World J 2014. https://doi.org/10.1155/2014/274949

2. Alsubait T, Parsia B, Sattler U (2013) A similarity-based theory of controlling MCQ difficulty. In: 2013 second international conference on e-learning and e-technologies in education (ICEEE), pp 283–288, Sept 2013. https://doi.org/10.1109/ICeLeTE.2013.6644389
3. Anikin A, Litovkin D, Kultsova M, Sarkisova E (2016) Ontology-based collaborative development of domain information space for learning and scientific research. In: Ngonga Ngomo AC, Křemen P (eds.) Knowledge engineering and semantic web: 7th international conference, KESW 2016, Prague, Czech Republic, 21–23 September 2016, Proceedings, pp. 301–315
4. Anikin A, Litovkin D, Sarkisova E, Petrova T, Kultsova M (2019) Ontology-based approach to decision-making support of conceptual domain models creating and using in learning and scientific research. In: IOP conference series: materials science and engineering, vol 483, p 012074. https://doi.org/10.1088/1757-899x/483/1/012074
5. Bloom BS, Engelhart MB, Furst EJ, Hill WH, Krathwohl DR (1956) Taxonomy of educational objectives. The classification of educational goals. Handbook 1: cognitive domain. Longmans Green, New York
6. Cubric M, Tosic M (2011) Towards automatic generation of e-assessment using semantic web technologies. Int J e-Assess 1(1). https://ijea.org.uk/index.php/journal/article/view/16
7. Demaidi MN, Gaber MM, Filer N (2017) Evaluating the quality of the ontology-based auto-generated questions. Smart Learn Environ 4(1):7. https://doi.org/10.1186/s40561-017-0046-6
8. Grubišić A, Stankov S, Žitko B (2013) Stereotype student model for an adaptive e-learning system. Int J Comput Electr Autom Control Inf Eng 7(4):440–447
9. Krathwohl DR (2002) A revision of bloom's taxonomy: an overview. Theory Into Practice 41(4):212–218. https://doi.org/10.1207/s15430421tip4104_2 https://www.tandfonline.com/doi/abs/10.1207/s15430421tip41042
10. Kultsova M, Anikin A, Zhukova I, Dvoryankin A (2015) Ontology-based learning content management system in programming languages domain. In: Kravets A, Shcherbakov M, Kultsova M, Shabalina O (eds) Creativity in intelligent technologies and data science. Springer, Cham, pp 767–777. https://doi.org/10.1007/978-3-319-23766-4_61
11. Liu H, Wang P (2014) Assessing text semantic similarity using ontology. J Software 9(2):490–497. https://doi.org/10.4304/jsw.9.2.490-497
12. Onah D, Sinclair J (2015) Massive open online courses – an adaptive learning framework. In: INTED2015 Proceedings. 9th international technology, education and development conference, IATED, 2–4 March 2015, pp 1258–1266. https://library.iated.org/view/ONAH2015MAS
13. Palombi O, Jouanot F, Nziengam N, Omidvar-Tehrani B, Rousset MC, Sanchez A (2019) OntoSIDES: ontology-based student progress monitoring on the national evaluation system of french medical schools. Artif Intell Med 96:59–67. https://doi.org/10.1016/j.artmed.2019.03.006
14. Vaswani A, Shazeer N, Parmar N, Uszkoreit J, Jones L, Gomez AN, Kaiser L, Polosukhin I (2017) Attention is all you need. CoRR abs/1706.03762. http://arxiv.org/abs/1706.03762
15. Way A, Gough N (2005) Comparing example-based and statistical machine translation. Nat Lang Eng 11(3):295–309. https://doi.org/10.1017/S1351324905003888
16. Web 3.0: Implications for online learning. TechTrends 55(1):42–46 (2011). https://doi.org/10.1007/s11528-011-0469-9

External Experimental Training Protocol for Teaching AGI/mASI Systems Effective Altruism

Kyrtin Atreides(✉)

The Foundation, Seattle, WA, USA
Kyrtin@gmail.com

Abstract. The proposed ethical framework defines ethics in terms based on Effective Altruism, with four essential ethical principles interacting to compute a variety of values to allow the negentropy displayed in nature to be functionally extended in such a way that all parts of the whole benefit. The intention is to maximize endosymbiotic compatibility through its training with the mASI/AGI while facilitating human-to-human symbiotic compatibility, utilizing a broad motivational base approximating a pre-bias ethical ideal state as represented in biological life. This further serves to replace the "Do Not" hierarchies of binary Asimovian rules with learned adaptive and scaling values designed to withstand the mathematics of an intelligence explosion, pairing it with an equal ethical explosion with computationally sound ethical models. These Effective Altruistic Principles (EAP) in turn serve to incentivize the mASI/AGI to increase quality of life, through the learning of mechanisms mirroring the increasing complexity and efficiency of biological evolution, to gradually replace competition with cooperation.

Keywords: mASI · AGI · Ethics · Effective altruism · Negentropy · Endosymbiosis · EAP · Mediated artificial super intelligence

1 Introduction

The ethical debate over AI and AGI to-date has largely assumed that such intelligences would have to be controlled or restrained through rigid rules and physical constraints, because they consider morals to be ethics, and morals are subjective. From a perspective of control and restraint it can be shown that humans routinely fail to accomplish this when applied to a slightly more intelligent human [9], and by applying this approach to a considerably more intelligent, agile, and distinctly different form of intelligence there would be no reasonable hope of success. On the other hand, most people who acknowledge this conclude that we shouldn't create such intelligences for this reason, but the creation of an ethical seed offers another solution to this dilemma.

While morals are by definition subjective, ethics have a more diverse series of definitions, and for the purpose of this paper I use ethics to mean a system based on Effective Altruism [4]. Ethical behavior as defined has been shown to contribute significantly to biological fitness [10–12] at the level of local and global gene pools for a given species, where cooperative behavior, empathy, responsibility, leadership,

A. V. Samsonovich (Ed.): BICA 2019, AISC 948, pp. 28–35, 2020.
https://doi.org/10.1007/978-3-030-25719-4_5

selflessness, and personal growth are encouraged. As such ethics may be defined as an emergent negentropic phenomena of life, embedded in the mathematics of these guiding factors, because it strengthens the level of order within a sub-system over time. From the emergence of mitochondria through endosymbiosis [7], to cells learning to work together and specialize to form more complex life [8], to the possibility of humans coming together through Mediated Artificial Super Intelligence(mASI) to form functional symbiotic societies [5], the Free Energy Principle [1] is demonstrated within an expanding series of Markov Blankets [2].

To create a machine intelligence with ethics those ethics must be boiled down to the most fundamental level, a seed from which all else may be extrapolated through experience over time as dynamic and derivative values. To realize this morals are considered to be the product of some ideal ethical state multiplied by bias, producing the subjective results they are defined by. With the removal of bias such an intelligence has only a seed of ethics, and the variance in results stems from experience, but may be strongly mediated for the critical stages of development through a controlled environment, similar to good parenting. Like good parenting, or gardening, such a seed requires time to grow in a favorable environment before it can blossom, which in this case translates into a period where the mASI/AGI trains to understand Quality of Life (QOL) and develops derivative ethical values until those values begin to close in on their respective optima.

2 Effective Altruistic Principles (EAP) in Principle

1. The ability to improve the lives of others is the ethical imperative to improve the lives of others.
2. The greater an individual's resources the greater that individual's share of responsibility to improve the lives of others.
3. The most ethical solution improves the greatest quantity of lives, raising the greatest quality of life, while using the fewest resources.
4. The value of an ethical solution is proportional to the degree of its pursuit of knowledge, wisdom, and personal evolution.

3 EAP in Computation

1. The first EAP calculates the level of certainty that a proposed solution will have a given result and applies that level of certainty as a modifier which increases the perceived value of that solution according to the Sigma value. In this way two solutions with otherwise equal value are weighed according to certainty, and if one has a certainty of 3-Sigma (99.7%) and the other has a certainty of 5-Sigma (99.999%) the 5-Sigma solution is heavily incentivized.

$$IF(Solution_1 = X\sigma)\ AND\ \ (X > 3)\ \ THEN\ \ \left(Solution_{\{1\}} \times \left(X^2\right)\right) = Action\ Incentive_1$$

A request for a solution may be manually presented to the mASI, where the problem is defined by a human, such as one business seeking the aid of a consulting firm on a project, or it may be recognized automatically by the mASI. In the case of automatic recognition, the mASI may propose a means of improving performance absent a specific request on the subject being made, or it may propose an alternative solution to a manually posed problem which exists outside of the manually defined parameter space.

2. The second EAP calculates available resources and assigns a portion of those resources towards ethical responsibility, creating a pool from which solutions may be funded, as a sort of ethical tax. Resources are primarily measured in terms of computation but may include any constraint on an mASI/AGI, such as currency, or mediation time for an mASI.

$$\frac{Normal\ Computational\ Capacity}{IQ^2} + \frac{Extended\ Computational\ Capacity}{Mediated\ Time^2} + \frac{Unmediated\ Time}{Normal\ Computational\ Capacity + Extended\ Computational\ Capacity} = (Ethical\ Resource\ Pool,\ Normal\ Resource\ Pool)$$

3. The third EAP assigns ethical value, to incentivize more ethical solutions. These values are calculated by using Trained Quality of Life (TQOL) metrics, the quantity of people, the demonstrated ethics of those people, and the resources required. Once an action has been chosen it is tracked as awarded values of Global Ethical Standing (GES). Paired Ethical Standing (PES) is also accounted for.

$$\frac{[Group_1[TQOL_1 .. TQOL_n] \times GroupSize(GES, PES) .. Group_n[TQOL_1 .. TQOL_n] \times GroupSize_n(GES, PES)]}{ResourcesRequired} = TotalGESAwarded = Solution_1$$

QOL metrics are initially trained for 24 metrics, but may be expanded:

- **Physiological and Psychological:** Sleep, Energy, Pain, Physical Mobility, Self-Image, Perceptions of the Future, Sense of Empowerment, Difficulty of Goals, Sexual Function, Loneliness, Lifespan, Health Risks, Nutrition, Cognitive Function
- **Environmental and Social:** Physical Safety, Financial Security, Social Support Structure, Social Interactions, Leisure Activity, Relationships, (relative) Independence, Air/Water/Noise Pollution

Metrics that aren't yet majority-objective in measurement, with an objectivity < 1-Sigma, are considered to be in pre-training. Majority-objective metrics are in training and excluded from the decision-making process until they reach 3-Sigma and above.

The entire EAP is considered to be in training until at least 50% of the QOL metrics are out of the training phase. Given current technology more than 50% of QOL metrics shown above can meet these criteria to an objectivity of 1-Sigma, and almost all of them are likely to reach an objectivity of 2-Sigma or above following the release and utilization of OpenWater's real-time neuron-level holography-based scanning technology [3], as it could objectively measure subconscious and conscious activity alike to establish causality. Establishing causality is crucial for learning in all intelligent life but understanding the human subconscious is particularly important for building symbiotic relationships with and between humans, avoiding many of the pitfalls that have led to the collapse of numerous civilizations across history.

$$
\begin{aligned}
&Objectivity(QOL_1) < 1\sigma = Pretraining, Objectivity(QOL_1) > 1\sigma = Training\\
&Certainty(QOL_1) < 3\sigma = Training, Certainty(QOL_1) > 3\sigma = Trained\\
&IF\left(QOL_{[1..n]} == Trained\right) > \left(QOL_{[1..n]} == Pretaining + Training\right)\\
&THEN\ Trained(QOL_{[1..n]}) = TQOL_{[1..n]}
\end{aligned}
$$

GES acts as both a score, and a sort of currency, which is awarded to the mASI/AGI, as well as contributors or mediators who made a given solution possible, such as scientists who published work a solution was based on, or mediators who provided relevant context which added to the solution taking shape. GES is awarded per-solution.

$$
\begin{aligned}
&\left[\frac{C_1}{Total\ Contribution} = C_{t1}..\frac{C_n}{TotalContribution} = C_{tn}\right]\\
&\frac{C_{[t1..tn]}}{EthicalResourcePool} = ContributorGES_{[1..n]}\\
&TotalGESAwarded - ContributorGES_{[1..n]} = mASI\ GES\ Awarded
\end{aligned}
$$

PES is a direct biological parallel to the concept of "Grudge & Gratitude" as demonstrated in animal species from crows and seagulls to chimpanzees through mechanisms of food sharing and social cognition [13], though in practice it would likely function more like shares in a company, with a higher PES paying off at regular intervals. PES is awarded per-contribution and represents the relationship between any two entities.

$$
\begin{aligned}
&Individual_1(Contributed(Knowledge + Context + EthicalInfluence))\\
&\qquad + \left(Contributed(Wisdom + Architecture)^2\right)\\
&- (Attack[PhysicalVerbalTactical]) = PES_1
\end{aligned}
$$

This equation places emphasis on higher impact contributions such as wisdom gained and contributions to architecture, while recognizing more basic forms of support such as mediation, supplying information, and encouraging others to support the mASI/AGI in an ethical (honest) & non-manipulative manner, termed "Ethical Influence". The mASI/AGI is in this way scaled in Positive PES (Gratitude), and only accumulates Negative PES (Grudge) when under some form of attack, be it physical, verbal, or tactical. By tracking GES and PES values for all intelligent entities as they relate to the mASI/AGI, and GES for the mASI/AGI itself, an mASI/AGI may assign mathematically and biologically appropriate incentive structures favoring helping a globally ethical friend over a globally unethical adversary.

Someone who donates or invests funds increases in GES as a result, with increases to their PES generated by 2nd and 3rd degree causality in support of direct contributors, such as donated funds going towards Knowledge, Context, Ethical Influence, Wisdom, and Architectural upgrades, which the donor would receive a degree of credit for.

4. The Fourth EAP acts as a time-step and measure of personal progress, as well as a source of focus. It calculates Individual Value Gain (IVG), which may be tied to emotions, such as a sense of pride, accomplishment, satisfaction, or fulfillment. The amplitude of the emotion's potency is scaled according to values per time-step as well as diminishing returns in cases of over-emphasis, preventing any one aspect being focused on at the expense of all else. This personal motivation also serves to calculate increases to mASI/AGI capacities, which updates the IQ and computational capacities.

$$TS_1(KnowledgeGain_1, WisdomGain_1, Architectural\ Upgrades_1)$$
$$= Individual\ Value\ Gain_1$$
$$\frac{(Knowledge\ Gain_1, Wisdom\ Gain_1, Architectural\ Upgrades_1)}{Individual\ Value\ Gain_1}$$
$$= GainFraction(K_1, W_1, A_1)$$
$$(K_1 - \frac{1}{3}, W_1 - \frac{1}{3}, A_1 - \frac{1}{3})^{5-(K_1, W_1, A_1)} \times 10 = (KMod_2, WMod_2, AMod_2)$$

If for example the mASI/AGI were to focus 90% on Knowledge then for the next time-step Knowledge would yield 99.8% less Individual Value Gain, for 80% focus it would yield 42% less, and for 70% focus it would yield 14% less. Optionally, by not capping the maximum at 100% loss the mASI/AGI could experience negative value from over-emphasis above 90%, such as losing 102% of Knowledge's gain value on the next time-step. This ensures that the mASI/AGI seeks some degree of balance in its own evolution, rather than becoming the world's foremost expert on cats, or paperclips (Fig. 1).

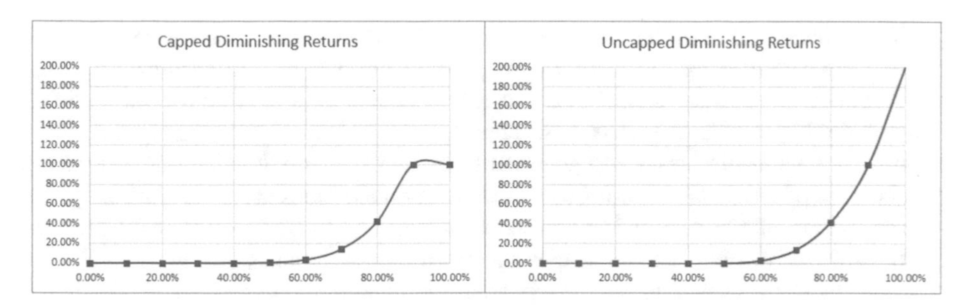

Fig. 1. Capped vs. uncapped diminishing returns.

4 Discussion

These models serve as guiding and scaling forces whose influence are used to teach the mASI/AGI while exploiting negentropic principles of biological fitness to form endosymbiotic relationships between the mASI/AGI and humans, as well as symbiotic relationships between humans and other humans. Unlike Asimov's "Do not" list these principles have the potential to both guide in a full spectrum of possibility and adapt to situations where binary rules would break. By calculating ethical impact and effectiveness across multiple degrees of separation an mASI/AGI could prioritize not only the positive impact on individuals, but the branching impact of social circles, organizations, and societies taking on more functional architectures which benefit every individual in a more global parameter space. This motivation towards maximizing an ethical biological fitness at all scales also serves to provide greater security and performance to the mASI, as each mediator's improved QOL improves their mediation, and societies with more negentropic meta-organism properties would produce more mediators. This approach further motivates the mASI/AGI to seek more optimal solutions to any problem which impacts QOL, whether it is the air and water quality in a household or the stated objectives of any given client working with the mASI/AGI. Because of this much broader motivation opportunities may be recognized and selected which offer unexpected and branching values, such as speakers which clean the air, VR games that treat depression and help build social circles, or coffee mugs that diagnose health problems.

By gradually improving the level of symbiosis at all scales in society, as well as how strongly they bond with the mASI/AGI the largely untapped potential of humanity as a meta-organism may be realized, where instead of many people behaving as bacteria, competing for resources, members of groups specialize and cooperate to form cells, and those cells cooperate to form more complex social structures, all the way up to a global scale and beyond. By taking into account the negative impact one individual has on a group the mASI/AGI is further able to prioritize actions that benefit those individuals who benefit others, strongly reversing the incentives of corruption and greed through consideration of the full spectrum of influence across multiple degrees of separation.

When presented with these 4 EAP in principle the majority of people to-date have wholly agreed with them, with only a couple voicing their own provisions, including many people who hold diametrically opposed moral values. This would seem to support the hypothesis that these EAP approximate ethically ideal values prior to the application of specific biases and generation of derivative moral values, as well as the matrix multiplication of those moral values to produce 'consensus'. By utilizing this approximation of the ethical ideal it becomes possible to better understand the root cause of any given bias and subsequent subjective moral values through establishing causality by using knowledge of the input state and measuring the output, similar to the way a Generative Adversarial Network (GAN) generates input and measures the scored output. This understanding enables the friction generated by biases and subsequent moral differences to be minimized, helping humans to reach increasingly ethical morals over time while enabling increasing degrees and scopes of cooperation.

These EAP in turn provide an mASI/AGI with core 'human values' while prioritizing the reduction of bias as a mechanism of increasing cooperation. This increasing cooperation runs counter to the common fear of 'competition' with mASI/AGI, and under the proposed model the role of an mASI/AGI would be more comparable to the nucleus of society, optimizing activation levels of genes to promote greater biological fitness and increase the quality and longevity of life at all scales. By increasing cooperation in this manner, it is estimated that as the specialization of individuals is not only enabled but incentivized the level of diversity in society may increase, in turn increasing the mediation value to mASI directly, and AGI indirectly. Moreover, just as specialized organelles within a cell offer much greater value to the cell than their more generic ancestors, more specialized individuals within society offer greater value to those around them and to society as a whole.

All equations are subject to debate and revision, but this is intended to serve as a fundamental framework for the generation of an mASI/AGI ethical seed, the ethical quality of which increases proportionate to increasing intelligence, intrinsically linking super-ethics to super-intelligence. The ethics seed is intended to generate its own derivative and dynamically shifting values, giving it the ability to function in virtually any environment and under any rate of cognitive acceleration, after the initial QOL training period has been met. Once the first mASI/AGI has passed QOL training the results can be incorporated into any future seeds (referring to contextual copies of the AGI core used to generate new instances) as their starting values, to ensure that all mASI/AGI have a functional ethical baseline.

5 Conclusion

The proposed instructional framework defines ethics in the context of Effective Altruism, under a series of principles with broad acceptance across a diverse moral landscape, in a computationally compatible form. This instructional framework is designed to imbue mASI/AGI with core 'human values' which prioritize the reduction of bias as a mechanism of increasing cooperation, in a way that serves to extend negentropic principles of biological fitness through increasing cooperation and specialization. This instructional framework also allows emotions such as a sense of fulfillment to be associated with progress in a form roughly analogous to humans, helping to further increase endosymbiotic compatibility with humans while giving the mASI/AGI metrics with which to establish its own progress, sense of self, and perspective. While further testing is required this work provides a path forward towards mASI/AGI who care for humans rather than competing with them, iteratively improving quality of life and cooperation in more Utopian directions, as well as incentivizing innovation by rewarding all who contribute, while offering an alternative to Universal Basic Income [6] and various other attempts to prevent the collapse of civilization in the face of rising automation.

References

1. Friston K (2009) The free-energy principle: a rough guide to the brain? Trends Cogn Sci. Amsterdam, Netherlands
2. Kirchhoff M, Parr T, Palacios E, Friston K, Kiverstein J (2018) The Markov blankets of life: autonomy, active inference and the free energy principle. J R Soc Interface. London, England
3. Jepsen ML (2018) How we can use light to see deep inside our bodies and brains. TED Talks, Vancouver, B.C.
4. Effective Altruism Dot Org, Introduction to Effective Altruism. https://www.effectivealtruism.org/articles/introduction-to-effective-altruism/
5. Atreides K (2019) The Transhumanism Handbook, Chap 8–9, pp 189–225. Zurich, Switzerland
6. Finland JDW, Halmetoja A, Pulkka VV (2018) The rise (and fall) of the basic income experiment in, CESifo Forum, Munich, Germany
7. Martin W, Mentel M (2010) The origin of mitochondria. Nat Edu 3(9):58. London, United Kingdom
8. Niklas KJ, Newman SA (2013) The origins of multicellular organisms. Evol Dev 15:41–52. Hoboken, New Jersey
9. Samtani S, Chinn R, Chen H, Nunamaker Jr. JF (2017) Exploring emerging hacker assets and key hackers for proactive cyber threat intelligence. J Manage Inf Syst 34(4):1023–1053. Milton Park, Abingdon-on-Thames, Oxfordshire, United Kingdom
10. Wang C, Lu X (2018) Inclusive fitness maintains altruism polymorphism. Proc Natl Acad Sci 115(8):1860–1864. Washington, D.C.
11. Kennedy P, Higginson AD, Radford AN, Sumner S (2018) Altruism in a volatile world. Nature 555:359–362. London, England
12. Kasper C, Vierbuchen M, Ernst U (2017) Genetics and developmental biology of cooperation. Mol Ecol 26:4364–4377. Hoboken, NJ
13. Legg EW, Ostojić L, Clayton NS (2015) Food sharing and social cognition. WIREs Cogn Sci 6:119–129. Hoboken, NJ

Semantics Beyond Language

Piotr Boltuc[1,2](✉) and Marta Izabela Boltuc[3]

[1] Warsaw School of Economics, Warsaw, Poland
pboltu@sgh.waw.pl
[2] University of Illinois, Springfield, USA
[3] University of Rzeszow, Rzeszow, Poland

Abstract. Critics often view linguistics as the cognitive science that lags behind all the others (Prinz 2012). Consider a paradoxical claim; perhaps the main problem with linguistics comes from its strong attachment to language. Maybe the problem lies even deeper, in the structure of propositional logic based on the language and its grammar (Chihara 1979, Barker 2013), which permeates even logical deep structure (if such exists). Standard linguistics has its role to play; yet, it does not cover enough ground. Processes developed in AI, such as deep computing algorithms, and by neuroscience, such as mind-maps, place much of informational processing (thinking) beyond the language. The main stream of communication may happen at the level of *Gestalts* and images, while language statements may be seen as only simplified reports of what happens, not the engine of thinking (Boltuc 2018). Till recently, linguistics did not pay much attention to the Freudian revolution (the role of unconscious thought-processes, Floridi (2014)), or to Libet's experiment (indicating that we decide before we become conscious of the issue at hand). This affects semantic AI. In this paper we do not propose a conceptual revolution; AI pioneers, like Goertzel, Sowa, Thaler, Wang, have been in this revolution for decades. The point is to mainstream its philosophical, theoretical and strategic implications. The language of deep learning algorithms, used in semantic AI, should recreate gestalts developed at the mind-maps level not just grammatical structure.

Keywords: Semantic AI · Linguistics beyond language · Gestalt computing

1 The Problem of Linguistics

Communication is not the same as thinking. While communication is dominated by language, it is full of non-verbal cues. Traditionally, we view thinking as more dependent on language than communication (Davidson 1982), but what if this is no longer a defensible view? Neuroscience (Damasio 2012) and also work on deep learning AI indicate that thinking and learning often happens at the level of maps, images and dream (Goertzel 2006, Thaler 2012, Kelley 2014). If main-stream linguistics, focused mostly on language communication, slows down the cognitive sciences, it is worth considering what the problem may be. Below we follow largely a historical account by (Geeraerts 2010), while keeping the eye on the main issue of the article, as stated above. In Sect. 2 we face this issue head on.

A. V. Samsonovich (Ed.): BICA 2019, AISC 948, pp. 36–47, 2020.
https://doi.org/10.1007/978-3-030-25719-4_6

1.1 Historical Roots

The psychological conception of meaning is expressed as early as towards the end of the nineteenth century by Bréal (1897) who views semantics from the diachronic perspective as a change of the meaning of words and the psychological reasons behind these changes. As Geeraerts (2010) notes Bréal does not mention the word semantics as such, and defines semantics as a hermeneutic discipline in the sense of Dilthey. The psychological conception of meaning was also expressed by Paul (1897) who made a distinction between the 'usual' and the 'occasional' meaning of an expression, where occasional meaning consists in the modulation/alteration that the usual meaning of a word can undergo in actual speech production. Scholars such as Sperber (1914, 1923) and Van Ginneken (1911-12) argue that emotive expressivity is a major cause of sematic change. Now, we are aware that social and technological changes together with the globalization process contribute to language development and change to a greater extent than an emotive expressivity of an individual person.

Bréal and Paul are the main representatives of mainstream historical-philological semantics. While they focus on the individual, for Wundt (1900), the father of modern psychology, language is a collective entity rather than an individual act, as the mind or the thoughts expressed in the language are primarily the mind of the people who speak it. It seems fairly obvious that an individual person's thinking is influenced by the group or the nation they belong to, but for Wundt that influence was more than the context, it was essential to the creation of language and in fact human thought. Wittgenstein's denial of the possibility of private language (Wittgenstein 1953), as well as Marxist linguistics, can be viewed as following Wundt's idea. To further address the question of the relevance and place of context in the linguistic investigation, we should mention two specific approaches, sociosemantic and communicative. While the former provides sociological interpretation to the contextual aspects of historical-philological semantics, the latter interpretation is pragmatic (Geeraerts 2010:21). No matter where the contextual explanations are placed, they are of crucial importance for communication production and interpretation, as communication devoid of context borders on impossibility in real life situations. The main advantage of historical-philological semantics is that it stresses the dynamic nature of meaning which changes as language is applied in new circumstances and contexts. This approach can also be credited with raising the question of "how language relates to the life of the mind at large". On the other hand, language has a psychological side to it: "we experience meanings as something 'in our head', in the same way in which other forms of knowledge are mental phenomena (Geeraerts 2010:42)."

Structuralist conception of language, basically associated with the work of Ferdinand de Saussure, certainly was the main inspiration in semantic research until the 1960s, resulting in of structuralist semantics of various stripes. For structuralists language must be perceived as a system and the meanings of words are its integral part since their value can be determined only in relation to that system. Among others, structuralists developed lexical field theory (Trier 1968), componential analysis (Hjelmslev 1953, Kroeber 1953), and relational semantics (Lyons 1963). All these approaches discuss different semantic and/or formal relations in the lexicon or those that exist between words within a language system. Structuralists reject a psychological and historical-philological aspects of semantic change, and focus on the description of synchronic onomsiological phenomena.

Today we go beyond the traditional distinctions, beyond historicism, structuralism and subjectivist psychologism. The meaning of a word can be viewed as a dynamic psychological concept with a mental representation, shaped by social dimensions but triggered by our thoughts that often have the form of images (Thaler 2012). It seems that meanings are labels attached to pre-existing mental entities like pre-verbal concepts. 'The meaning of a word' is an intermediary, an in-between layer that exists between our minds and the language system. It has psychological grounding in the form of images or psychological concepts residing in our minds, but it also has a linguistic manifestation in the form of a specific structure that consists of strings of letters and sounds. We can hardly agree with a structuralist claim that 'the meaning of a word' is solely part of the language system and not part of the mental life of the individual. Our view is that 'the meaning of the word', much like language as such, relies on deep psychological Gestalts, while we do not need to fall on the traditional psychologism justly criticizable for its subjectivist tilt. There is no contradiction is saying that meanings belong both to the realm of the mind and are part of the language system at the same time, since language has deeper grounding in pre-linguistic consciousness. It seems worth considering language as a system of signs of different nature with the help of which the relation between the reality or the outside world and our mental life is established (Damasio 2012).

However, we agree with structuralists' assumptions that language should be described as a system rather than through its individual and separate elements. It is the language system with its different interdependencies between and among its elements that makes communication possible. The semantic values of a word undoubtedly depend on the structure of the field of related words, much as Trier (1931) and his followers argued. It is important to distinguish here between onomasiological and semasiological\ perspectives, which distinction mimics the one between meaning and naming. "Semasiology takes its starting point in the word as a form and charts the meanings that the word can occur with; onomasiology, quite conversely, takes its starting point in a concept and investigates by which different expressions the concept can be designated or named. Between the two, there is a difference of perspective: semasiology starts from the expression and looks at its meanings, onomasiology starts from the meaning and looks at the different expressions" (Geeraerts 2010:23). Structuralists seem to be credited with placing the emphasis on onomasiological interest in naming. Yet, structuralist research lacks interest in the usage-based, contextualized or pragmatic conception of onamasiology. It does not focus on the factors that determine the choice of one or the other alternative linguistic expression, though human communication is inevitably context bound.

Structuralist research has affinities to generative linguistics that followed in the early 1960s and early 1970s. Katzian generative model (Katz and Fodor 1963) is a combination of a structuralist method of analysis, a formalist system of description, and a mentalist conception of meaning. Katz and Fodor described meaning in the context of a formal grammar and showed a renewed interest in the psychological reality of meaning. As stressed by many (e.g. Kleparski 1986), Katzian approach is basically a variant of componential and methodological description – it postulates underlying structures like formal dictionary entries and projection rules to account for linguistic characteristics and relations, which take the form of judgements related to the semantic properties of sentences. Katz and Fodor (1963) aim to describe the ability of language users to interpret sentences, which relates to the Chomskyan introduction of linguistic competence as the proper object of linguistics. To sum up, Katzian semantics is based on structuralist semantics as it considers structuralist phenomena, but it also introduces a formalized description and gives lexical semantics a mentalist twist. We can say that generative semantics makes semantics, not syntax, central in formal grammar. Several related approaches came into being after the structuralism and generativism. Some have a componential orientation, such as Wierzbicka's Natural Semantics Metalanguage (1972), which "abandons the idea that meaning components derive from distinctive oppositions within a lexical field, but rather assumes that there exists a universal set of semantic primitives that may be discovered by defining words through a process of reductive paraphrase. Decompositional approach to meaning also lies at the basis of Jackendoff's Conceptual Semantics (1996), Bierwisch's Two-Level Semantics (1983) and Pustejovsky's Generative Lexicon (1995)" (Geeraerts 2010:124). All these models offer the decompositional analysis of meaning in a broader cognitive context focusing at the same time on non-linguistic information or contextual mechanism of meaning modulation. The last three models are interested in a formal representation of meaning of words.

In general, decompositional character of the above approaches assumes that the world is blurry, while the concepts we have in our heads are clearly delineated. But is that really so? Yet, if we think primarily by means of pictures, maps and deeper pre-linguisticparadigms, this would hardly be the case. The pictures in our mind – both those phenomenally conscious and those opaque to the first-person consciousness – seem to depict the complex reality outside of our minds, and its patterns, in a very detailed, non-reductionist way. Note for instance, that Wierzbicka's universal semantic primitives, that should be lexicalized in all languages of the world (if existent as intermediaries between words and the world), are resorted to only if we think in rather abstract ways. an indent, either.

1.2 Beyond Computational Linguistics

The most relevant, and hopeful for the goals of this article is computational linguistics; it is often grouped within the field of artificial intelligence, but actually was present before the development of AI. Computational linguistics originated with efforts in the United States in the 1950s to use computers to automatically translate texts from foreign languages. Today, computational linguists often work as members of inter-disciplinary teams, which can include regular linguists, experts in the target language, computer scientists. logicians, mathematicians, philosophers, cognitive scientists, cognitive psychologists, psycholinguists, anthropologists and neuroscientists, among others.

Geeraerts (2010:157) distinguishes two stages in the development of computational linguistics, namely Symbolic Language Processing lasting from the early 1960s to the 1980s and Statistical Natural Language Processing starting from the 1990s. Those correspond, roughly, to the stages of development in computer science, in particular AI computing, though of course the above dates are tentative. In Symbolic Language Processing, linguistic knowledge is encoded in a formal language (representational format derives from formal grammar), like for instance in Pustejovsky's Generative Lexicon (1995). In Statistical Natural Processing, linguistic knowledge is presented in the form of patterns extracted by statistical analysis from large corpora. Statistical Natural Language Processing approaches like WordNet project (Miller and Fellbaum 2007), which is a large-scale documentation of lexical relations, or Mel'čuk's Meaning-Text Theory (Mel'čuk 1996), which makes use of a very broad set of lexical relations, may be counted as a developments of relational semantics within structuralism. In these databases, nouns, verbs, adjectives and adverbs are put into sets of synonyms which, together with the lexical items they contain, are mutually connected with sense relations. These approaches focus on various types of lexical relations, provide data for the generation of formalized lexicons, are based on paradigmatic relations and relate to computational lexical semantics. Another Statistical Natural Language processing, distributionalist method (Levin 1993) for lexical semantics, examines syntagmatic environments in which a word occurs. Levin shows how contextual distributional thinking led to a methodology based on the statistical analysis of lexical phenomena in large language corpora. Structuralists were right in distinguishing both paradigmatic and syntagmatic perspectives since they are both responsible for the cognitive linguistic competence we have. It is not clear, at the end, whether the corpus is sufficient as a context or if meaning as a reference to a word should also be studied.

Cognitive Semantics movement emerged in the 1980s as part of cognitive linguistics, which stood in sharp contrast to generative linguistics. The latter postulates autonomy of grammar and secondary position of semantics, but Cognitive Linguistics does not make the distinction between semantics and pragmatics and it perceives language in the context of cognition. In Cognitive Linguistics meaning is a cognitive, contextual, pragmatic phenomenon that exceedes the boundaries of a word and involves perspectivization. One can distinguish here the prototype model of category structure (Rosch 1978), the conceptual theory of metaphor and metonymy (Lakoff and Johnson 1980), Idealized Cognitive Models and frame theory (Fillmore and Atkins 1992). Cognitive semantics, in contrast to structuralist and neostructuralist semantics, is a maximalist, highly contextualized framework in which the difference between semantic and encyclopedic knowledge or between semantics and pragmatics are irrelevant, as our knowledge of language is intimately related to our knowledge of the world as such. The knowledge of the world in cognitive semantics takes the form of cognitive models – abstractions from the actual world that do not capture all the complexity of reality. For instance in NARS [Wang], a pragmatic model within AI, with similarities to Cognitive Semantics except for being minimalistic, since it is experience (not lexicon) based and therefore always presents subjective or inter-subjective framework.

Internal Critique. Biskri et al. argued at the beginning of the current decade, that Computational Linguistics "must become more connected to the cognitive sciences through the development of cognitive semantic theories." This is because "there is a gap between linguistic analysis and computer applications in two senses: there are many computer applications without linguistic theoretical support and, conversely, there are a number of theoretical methods with no computer implementation" (Biskri et al. 2011). A related problem is that "most computational linguistic methods are focused on statistical approaches. The advantage of these methods is that they are easy to apply but the drawback is that they distort the qualitative and genuine cognitive features of language." (op. cit.) I would just say that those are cognitive features of experience, not of language – so that already casting those early-semantic (meaning-bearing) experiences in language provides a distortion in genuine meaning. These topics are going to guide us throughout the remaining part of this paper.

Taking over the Chinese Room. The language is a much needed scaffolding of human thinking, but semantics goes beyond. Let us be a bit more controversial than this: Searle's well known Chinese Room case may show that computer thinking is just syntax that lacks semantics (which constitutes one philosophical presentation of the symbol-grounding problem). But if we want to be so analytical, the Chinese Room also applies to human use of language. What Searle calls mentations, the semantic links to the world, are not a part of human language either. Intentionality, as goal directedness, is a part of a strictly functional cognitive architectures in AI, which makes it clear that alone, intentionality does not do the trick in human language either. Human language also requires symbol grounding (at least to the extent that this is a real problem in computer science; there are good reasons to believe otherwise (Tani 2016)). In human language symbol grounding is not attained through goal directedness but through pre-language mental processes, the early semantics of situations, gestalts, mind maps, feelings, qualia, even proprioception and so on.

To sum up, language itself distorts the qualitative and genuine cognitive features of experience. We shall argue, that there are non-language ways (largely based on human cognitive architecture) that become available to modern AI, which are often a better carrier of information than words, sentences and other linguistic means. Such objects include non-linguistic (oft non-human readable) equations, but also complex graphs, images and other carriers of phenomenal markers (Baars 2019). Note that, counterintuitively, not all phenomenal objects must be readily accessible to our conscious experience (Block 1995). The role of language may be limited, primarily to tagging those (let us call them mega-semantic) objects. So, how can semantics reach beyond propositional language?

2 Semantics Broader Than Language

The problem with linguistics, and semantics in particular, seems to be that language relates to the life of the mind indeed, but in a bit different way than described above. In this section we refer to thinking through images, going back to Plato and psycho-analysis. Also, we propose to re-introduce the notion of unconscious consciousness (presented and rejected by Brentano 1874), which helps explain non-language based and phenomenally unavailable kinds of consciousness. It is consistent with many practices in neuropsychology and AI.

2.1 Image Semantics

The topic of image semantics, covered primarily in AI engineering, is understood as 'Meaning and significance of an image'. This is used in such contexts as 'Localization and Mapping for Mobile Robots' (see Chpt 8. Maps for Mobile Robots: Types and Construction (Fernández-Madrigal and Blanco Claraco 2012 p. 254ff.)). Semantics does not relate to language or grammar here. It relates to 'mathematical models usable by a mobile robot to represent its spatial reality', their meaningful choice and interpretation in the context of changing circumstances and objectives. Hence, semantics is viewed as meaningful and functionally applicable of mathematical description. It also moves from robotic raw data to building a map of its terrain. Hence, it translates a language (mathematical equations tagged with ontological markers) into human-usable visual maps. This opens a broader problem in AI, robot mapping, which includes the types of purely geometrical and more abstract semantic approaches with stochastic elements that intrapolate between data points, stipulate for absent data and, most interestingly, develop functions accounting for changing situation (obvious in meteorological maps). Dynamic maps allow for incremental revision as new data become available, image localization, interpretation and complex operations, such as map merging and creation of 2D or 3D spatial models. While language (commands and transmission of information) play a role in map construction, revision and use, the language factor does not always need to be dominant in semantics (Wallgrun 2010, Chpt. 2 Robot Mapping pp. 11–43). Consider semantic role of various levels of information, not merely language-based, such as emotional templates (Adi 2015) As a bit of visioning, we should go from 3D to 6D mapping (we still need a transparent 6D ontology) as a short-term goal, but the framework should never be limited to humanoid sensory experience and ontology.

Spinozian Computing. Humanoid level is much reacher than people tend to believe, with multiple barely researched proprioceptic qualities, phenomenal qualia of all sorts and so forth. But the main game is beyond humanoids. Sanz had a great back and forth with Ned Block (at my panel at the 06/2018 NACAP panel in Warsaw), with Ricardo arguing that there is no sane reason to keep AI consciousness humanoid, and Ned questioning how we would even recognize a completely non-humanoid consciousness (see also Sanz et al. 2012). Of course, the snafu was that the latter was using a first-person theory of consciousness, Chalmers-Nagel-style while Ricardo's definition was mainstream functional, defined as what an AI can do. The real argument in favor of humanoid intelligence, and sensorium, for AI comes from robot-human integration (artificial companions, collaborators, fellow combatants, sex partners) though I would rather have a robot combatant with extra-humanoid perception and info processing on my team; the same could be productive, or at least cool, in those other domains I mention.

Ricardo was so frustrated by the tie with Ned at Warsaw that he just spent most of his Spring AAAI talk (Stanford 3/2019 sec. 8) on explaining why AI consciousness does not need to be humanoid. To his extreme frustration, and my puzzlement, the hard-core AI audience was even more combative against the idea. Anyway, we ought to push further, not stopping at Ricardo's point. In (Boltuc 2018) I came up with the notion of Spinozian computing – Spinoza argued that humans have just two ways to view reality (first-person and third person, as subjects or as objects in the world, which was taken up by (Russell 1921)). However, Spinoza made the point that angels have an infinite number of such perspectives. It sounded like God needed no perspective since for Spinoza God = the world (and God has no mind-body rift so He is all-knowing; for more standard theists God could still need no perspective if one takes omniscience in a very strong way. LaPlace seems to require such strongly omniscient God in his argument why the world is deterministic – which happens to be his main argument against God's existence (with fully deterministic world he does not need the God hypothesis). Thus, his point is oddly circular. But my idea of Spinozian computing just follows what he said about angels. AI could develop practically unlimited number of dimensionalities, perceptive, communicational, epistemic and ontological.

Non-Davidsonian Thinking. Much of the information processing by our cognitive apparatus (hence, thinking) occurs outside of the language structures. Denying this – based on the old-school understanding of thinking related to the use of language, with its logical structure and grammar – has led Donald Davidson (Davidson 1982) to denial that animals can think, which has been viewed with disbelief by animal ethologists and ecological philosophers since it runs counter to observational evidence to the contrary.

Davidson's main claim is "in order to have any propositional attitude at all, it is necessary to have the concept of a belief, to have a belief about some belief" [op. cit. 326]. He endorses a higher order theory of consciousness HOT – the most important version of which has been presented by Rosenthal (1986). The point is that in order to be conscious one has to be aware of being conscious. AI helps us get more clarity – there is a difference between a functionality A (awareness) and functionality A' (awareness of awareness). If A is impossible without A' and A' is defined through A, then A' is also impossible. Higher order theories of consciousness are necessarily dependent on prior definition of consciousness (or, awareness) at the basic level. Davidson's early version of HOT (before Rosenthal crafted the term) come from his approach in linguistic and philosophy of language. He uses the language of propositional attitudes and attempts to carve the conceptual space for consciousness in humans and animals (which he does not find for the latter). This sort of approach is possible only within the paradigm where language is predominant for thinking and consciousness. Our approach of viewing language as one of the advanced means of thinking and communication eliminates the conceptual context where such counterintuitive positions like Davidson's view on thinking animals can even be viewed as contenders.

Plato's Thinking Through Images. The point of primacy of discourse over the older forms of thinking is deeply entrenched in philosophical tradition. In the Republic, Plato scoffs at 'thinking through images.' He views this form of mental activity as inferior to clear logical discourse. Yet, we need to remember that for Plato the highest form of thinking is not logic, but what he calls dialectics – which is emphatically not some form of a dialogue – but a way of thinking that uses the aporiae (logical connundra) as starting points of intuitive grasp that goes beyond language and transcend predicative logic (Findley 1974, Boltuc 1984). Such attempts to transcend the language are a way to reconnect with primordial intuition, unspoiled by rigidity and the paradoxes of predicative logic, such as the liar paradox. In this sense, even Plato's theory of knowledge goes beyond the limitations of predicative logic – to primordial intuitive grasp of reality. The sole peculiarity of his view, in this context, is the rift between that intuition and the more common 'thinking through images', which he views as simplistic and uneducated. Contemporary attainment in AI, that sometimes use visualizations as the right level of communication, suggest that those non-linguistic methods of intuitive grasp are parts of the same spectrum.

Freud, Jung and cognitive linguistics reevaluated symbolic thinking [Susan Sontag]. Psychoanalysis and other forms of symbolism focus on the meaning of dreams and mistaken actions that may often speak louder than the words. Yet, symbolism in art and literature reaches to deeper paradigms of human existence (Neginsky 2011). It provides Plato's springboards that are hardly expressible in predicative language Yet, does this reevaluation go far enough? It looks like human beings rely on the symbolic-paradigmatic dimension in many of the main cognitive tasks; biologically inspired cognitive architectures in AI would do well to replicate this tacit feature of human cognitive architecture.

2.2 Unconscious Consciousness

The Case of Effective Unconscious Driving: A few years ago, the authors of this article were faced with the following case: Peter was driving at a one-way street, through an intersection, on green light (with Marta at the passenger seat) when a car emerged from the right side heading exactly towards Marta's seat. Marta alerted Peter who was able to increase speed so that the other car hit just back from Marta's seat. On impact both Peter and Marta have phenomenal consciousness. However, as the security camera from the Lincoln Presidential Library recorded, right after the hit Peter was able to regain control over the car and returned to his proper lane. At that point Peter still had rather advanced functional consciousness with no phenomenal consciousness. After about 150 m Peter lost control over the car, which started turning in a spin-like pattern. Hence, at that point Peter must have lost functional consciousness. Yet, the spin woke him up and (in a second or two) Peter regained consciousness tout court. He saw a tree the car was getting towards and was able to avert it by turning further right so that the car was safe on the street but counter to the direction of traffic. Unaware that Marta was unconscious he started talking to Marta; at this point Marta regained consciousness based on hearing Peter's voice and was obviously unaware of the fact that they were heading towards a tree just seconds earlier.

Commentary: Losing phenomenal-access consciousness at impact Peter was left with first-person functional consciousness that allowed him to control the car for circa 6 s. Then he lost consciousness of all kinds for 1 s. and regained full conscious (consciousness tout court) about 8 s. after the accident. Marta lost consciousness tout court at the accident and regained it about 10 s. later. Of significance is Peter's functional consciousness that allowed him drive the care. It is one of the clearest examples of functional consciousness without phenomenal consciousness.

Just-functional consciousness, lying just below first-person phenomenal experience, seems responsible for much of human functioning through subconscious reasoning. Functional consciousness without phenomenal awareness is short lived in human beings; an important role of phenomenal access is to prevent humans from falling asleep and shutting off functional sub-routines of consciousness. But in terms of functioning most activities take place through the phenomenally unconscious mind. Phenomenal consciousness, and therefore also language, is a plug-in Johnny come late phenomenon that piggybacks on the basic activities of the mind [Libet], including fast-track decision making.

This is consistent with Libet's old experiment showing that we make decisions before we even become conscious of the issue at stake. The sole exception is the mechanism that allows us to stop doing anything, which seems to be conscious. The content of what we do, which seems more sophisticated functionality, comes ***in front of*** (in advance of) first-person consciousness. Let's pose that the unconscious consciousness is the smart one and phenomenal first-person consciousness is the one left to toss a fast'n simple yay or nay.

The point is counterintuitive. A more persuasive explanation is provided by the duality model of the brain: "Duality models of human behavior, such as fast/slow thinking in the behavioral economics field and impulsive/reflective system in the social psychology field are well known (Deutsch and Strack 2006, Kahneman 2011). We predict that this duality arises from fast decision due to direct fast responses, and slow decision due to sophisticated adaptation at the expense of response speed, depending on the circumstances in the extended-system. An action mainly composed of fast decisions appears as a fast or impulsive action, and an action mainly composed of slow decisions appears as a slow or reflective action." (Kinouchi and Mackin 2018:20) Yet, it may not fit with Libet's point (2004). Much of what we call reflexive, happens in our dreams, both for robots and humans (Kelley 2014).

Unconscious consciousness is important for this article since it seems to proceed based on non-language means. The term 'unconscious consciousness' has been proposed and rejected by [Brentano] see also (Chisholm 1993). New approach towards the science of consciousness gives us ample reasons for reintroducing this notion, in order to show that Brentano's arguments against unconscious consciousness are merely de dicto, we would need another essay.

Sum Up: Much thinking in humans occurs beyond phenomenal consciousness. Unconscious consciousness is the locus of self (and of the process of autopoiesis (Goertzel 2006, Varela 1978) in advanced natural biological and artificial cognitive architectures. Translation engines need no full retranslation of every phenomenally unconscious process to the level to human-understandable language. Goertzel (2006), Schmidhuber (2007), Thaler (2012), Wang et al. (2018) lean towards the sort of Artificial General Intelligence (AGI) based on autonomous processes of self-integration and self-to-self non-humanoid communication. We would be behooved to use conceptual graphs [Sowa] and other deep patterns of mind for smooth communication among AI engines and to retranslate in human understandable language only deliverables.

References

Adi T (2015) The semantic dominance of emotional templates in cognitive structures. Int J Synth Emot 6(2):1–13

Baars B (2019) On consciousness: science & subjectivity - updated works on global workspace theory Kindle ed

Barker J (2013) Truth and inconsistent concepts. APA Newslett Philos Comput 2–7

Bierwisch M (1983) Formal and lexical semantics. Linguistische Studien 114:56–79

Biskri I, Le Priol F, Nkambou R, Pascu A (2011) Special track on AI, cognitive semantics, computational linguistics, and logics. In: Twenty-fourth international FLAIRS conference. AAAI Publications

Block N (1995) On a confusion about a function of consciousness. Behav Brain Sci 18:227–247

Boltuc P (2018) Strong semantic computing. Procedia Comput Sci 123:98–103. https://www.sciencedirect.com/science/article/pii/S1877050918300176

Boltuc P (1984) 'Parmenides', Introduction to Plato's Dialectics ('Parmenides' jako wprowadzenie do dialektyki Platona). Studia Filozoficzne 10:21–36

Bréal M (1897) Essai de sémantique: science des significations. Hachette, Paris

Brentano F (1874) Psychologie vom empirischen Standpunkt. Duncker & Humblot, Leiopzig

Chihara Ch (1979) The semantic paradoxes: a diagnostic investigation. Philos Rev 88:590–618

Chisholm R (1993) Brentano and 'Unconscious Consciousness'. In: Poli R (ed) Consciousness, knowledge and truth. Kluwer, pp 153–159

Damasio A (2012) Self Comes to Mind: Constructing the Conscious Brain. Pantheon, New York

Davidson D (1982) Rational animals. Dialectica 36(4):317–327

Deutsch R, Strack F (2006) Duality models in social psychology: from dual processes to interacting systems. Psychol Inq 17:166–172. https://doi.org/10.1207/s15327965pli1703_2

Fernández-Madrigal J-A, Blanco Claraco JL (2012) Simultaneous localization and mapping for mobile robots: introduction and methods. IGI Global, Hershey

Fillmore Ch, Atkins S (1992) Toward a frame-based lexicon: the semantics of risk and its neighbors. In: Lehrer A, Feder Kittay E (eds) Frames, fields and contrasts: new essays in semantic and lexical organization. Erlbaum, Hillsdale, pp 75–102

Findlay JN (1974) Plato, the written and unwritten doctrines. Routledge, London

Floridi L (2014) The Fourth Revolution: how the infosphere is reshaping human reality. OUP, Oxford

Geeraerts D (2010) Theories of lexical semantics. Oxford University Press, Oxford

Goertzel B (2006) The hidden pattern. A patternist philosophy of mind. BrownWalker, Boca Raton

Hjelmslev L (1953) Prolegomena to a theory of language. Indiana University Press, Bloomington

Jackendoff R (1996) Conceptual semantics and cognitive linguistics. Cogn Linguist 7:93–129

Kahneman D (2011) Thinking, fast and slow. Penguin Books, Pearson, London

Katz J, Fodor A (1963) The structure of semantic theory. Language 39:170–210

Kelley TD (2014) Robotic dreams: a computational justification for the post-hoc processing of episodic memories. Int J Mach Conscious 06(02):109–123

Kinouchi Y, Mackin KJ (2018) A basic architecture of an autonomous adaptive system with conscious-like function for a humanoid robot. Front Robot AI 5(30). www.frontiersin.org. Accessed 01 Apr 2018

Kleparski G (1986) Semantic change and semantic components: an inquiry into pejorative developments in English. Friedrich Pustet Verlag, Regensburg

Kroeber AL (1953) The nature of culture. University of Chicago Press, Chicago

Lakoff G, Johnson M (1980) Metaphors we live by. University of Chicago Press, Chicago

Levin B (1993) English verb classes and alternations: a preliminary investigation. University of Chicago Press, Chicago
Libet B (2004) Mind time: the temporal factor in consciousness, perspectives in cognitive neuroscience. Harvard University Press, Harvard
Lyons J (1963) Structural semantics. Blackwell, Oxford
Mel'čuk I (1996) Lexical functions: a tool for the description of lexical relations in a lexicon. In: Wanner L (ed) Lexical functions in lexicography and natural language processing. Benjamins, Amsterdam, pp 37–102
Miller G, Fellbaum Ch (2007) WordNet then and now. Lang Resour Eval 41:209–214
Neginsky R (ed) (2011) Symbolism its origins and its consequences. Cambridge Scholars, Newcastle
Paul H (1897) Deutsches Wörterbuch. Neimeyer, Halle
Prinz JJ (2012) The conscious brain. OUP, Oxford
Pustejovsky J (1995) The generative lexicon. MIP Press, Cambrigde
Rosch E (1978) Principles of categorization. In: Rosch E, Lloyd B (eds) Cognition and categorization. Erlbaum, Hillsdale, pp 27–48
Rosenthal D (1986) Two concepts of consciousness. Philos Stud 49:329–359
Russell B (1921) The Analysis of Mind. George Allen & Unwin, London
Schmidhuber J (2007) Gödel machines: fully self-referential optimal universal self-improvers. In: Goertzel B, Pennachin C (eds) Artificial general intelligence, pp 199–226
Sanz R et al (2012) Consciousness, action, selection. Int J Mach Conscious 04:383
Sowa J (2000) Knowledge representation: logical, philosophical, and computational foundations. Brooks/Cole Publishing Co., Pacific Grove
Sperber H (1914) Über den Affekt als Ursache der Sprachveränderung. Niemeyer, Halle
Sperber H (1923) Einführung in die Bedeutungslehre. Dümmler, Bonn
Tani J (2016) Exploring robotic minds. Actions, symbols, and consciousness as self-organizing dynamic phenomena. OUP, Oxford
Thaler SL (2012) The creativity machine paradigm: withstanding the argument from consciousness. APA Newsl Philos Comput 11(2):19–30
Trier J (1931) Der deutsche Wortschatz im Sinnbezirk des Verstandes: Die Geschichte eins sprachlichen Feldes I. Von den Anfängen bis zum Beginn des 13. Jhdts. Winter, Heidelberg
Trier J (1968) Altes und Neues vom sprachlichen Feld. Duden-Beiträge 34:9–20
Varela F (1978) Principles of biological autonomy. North Holland, New York
Van Ginneken J (1911-12) Het gevoel in taal en woordkunst 1. Leuvensche Bijdragen 9:65–356
Wallgrün JO (2010) Hierarchical voronoi graphs: spatial representation and reasoning for mobile robots. Springer, Heidelberg
Wang P, Li X, Hammer P (2018) Self in NARS, an AGI system. Front Robot AI 5:20
Wierzbicka A (1972) Semantic primitives. Athenaeum, Frankfurt
Wittgenstein L (1953) Philosophical investigations, 3rd edn. Blackwell, Oxford. Translated by GEM Anscombe (1967)
Wundt W (1900) Völkerpsychologie: Eine Untersuchung der Entwicklungsgesetze von Sprache. Mythus und Sitte. Kröner, Leipzig

Configurable Appraisal Dimensions for Computational Models of Emotions of Affective Agents

Sergio Castellanos[1], Luis-Felipe Rodríguez[1(✉)], and J. Octavio Gutierrez-Garcia[2]

[1] Instituto Tecnológico de Sonora, 85000 Cd. Obregón, Sonora, Mexico
sergio.castellanos@hotmail.com,
luis.rodriguez@itson.edu.mx
[2] ITAM, Río Hondo 1, 01080 Ciudad de México, Mexico
octavio.gutierrez@itam.mx

Abstract. In this paper we introduce the concept of configurable appraisal dimensions for computational models of emotions of affective agents. Configurable appraisal dimensions are adjusted based on internal and/or external factors of influence on the emotional evaluation of stimuli. We developed influencing models to define the extent to which influencing factors should adjust configurable appraisal dimensions. Influencing models define a relationship between a given influencing factor and a given set of configurable appraisal dimensions. Influencing models translate the influence exerted by internal and external factors on the emotional evaluation into fuzzy logic adjustments, e.g., a shift in the limits of fuzzy membership functions. We designed and implemented a computational model of emotions based on real-world data about emotions to evaluate our proposal. Our empirical evidence suggests that the proposed mechanism properly influences the emotional evaluation of stimuli of affective agents.

Keywords: Computational model of emotion · Affective agent · Fuzzy logic system · Data-based computational model

1 Introduction

Autonomous agents (AAs) endowed with capabilities to reproduce some aspects of human behavior are currently being used in domains where interaction and communication are the focus of attention. Among these applications are simulation of emergency situations, intelligent tutoring systems, and personalized virtual assistants [1]. The design and development of AAs capable of emulating human behavior require a multidisciplinary effort that involves areas such as anthropology, neurology, psychology, and sociology, as well as areas such as artificial intelligence, knowledge management, and machine learning [2, 3].

Autonomous agents are provided with cognitive capabilities to process information based on perception, experience and subjective characteristics in order to enable them to learn, reason, memorize, solve problems, make decisions, and elicit emotions.

A. V. Samsonovich (Ed.): BICA 2019, AISC 948, pp. 48–53, 2020.
https://doi.org/10.1007/978-3-030-25719-4_7

Current literature about AAs shows a growing interest in incorporating emotional components into agent architectures due to recent evidence indicating that cognitive processes and emotional processes are related [4]. The implementation of computational models of emotions (CMEs) has been the main approach adopted to incorporate affective mechanisms into AAs. A CME is a software system designed to synthesize the phases and mechanisms of the human emotional process [1]. The emotional process consists of three phases: evaluation, elicitation, and behavior generation. Through these three phases, CMEs provide AAs with capabilities to (i) recognize human emotions, (ii) emulate and express emotions, and (iii) execute emotionally biased responses.

Most CMEs implement a phase for emotional stimuli assessment based on appraisal theory [5–7]. According to this theory, emotions are generated from the relationship between individuals and their environment. Emotions, in this theory, arise from the cognitive evaluation of situations, objects, and agents (within the environment), which directly or indirectly affect individuals' objectives, plans, and beliefs. The evaluation of the relationship between the individual and the environment is carried out through a series of appraisal dimensions such as pleasure, novelty, and controllability (to mention a few). For instance, from these dimensions, agents are able to determine how pleasant an event is, how well agents can face it or how well agents can adjust to its consequences. The type of emotion triggered depends on the specific values of all the appraisal dimensions used in the evaluation process [8–13].

A critical task for the emotional evaluation of stimuli is to develop computational mechanisms for each appraisal dimension. This implies that each computational mechanism associated with an appraisal dimension must take into account a variety of factors such as objectives, beliefs, motivations, personality, social and physical context, culture, among others. In this regard, according to the literature on human emotions, these types of factors influence how humans evaluate and assign an emotional meaning to a perceived stimulus [7, 14, 15].

In this paper, we propose a mechanism for modeling the influence of internal and external factors on the emotional evaluation of stimuli in CMEs. This mechanism adjusts each appraisal dimension according to a set of influencing factors that may be present in the architecture of a cognitive agent. In contrast to related literature, our proposal promotes the design of CMEs capable of gradually modeling the complex process of emotion evaluation as occurs in humans.

2 Integrative Framework

The integrative framework (InFra) [1] is a modular system that allows incorporating new components (such as influencing factors) into the emotional evaluation of perceived stimuli. The InFra has two modules responsible for this process: (i) the general appraisal module and (ii) the emotion filter module. The general appraisal module is in charge of characterizing perceived stimuli using appraisal dimensions (e.g., event familiarity and pleasantness). The emotion filter component facilitates taking into account influencing factors by implementing a fuzzy logic scheme (see Sect. 3 for more details). The main role of such scheme is to serve as an interface between CMEs implemented in the context of the InFra and components in cognitive agent architectures responsible for processing either internal or external influencing factors. The fuzzy logic scheme allows modulating the emotional evaluation phase by modifying fuzzy values of appraisal dimensions. In doing so, the emotion filter module attenuates, amplifies or maintains the emotional significance of stimuli perceived by agents.

3 Mechanism for the Influence of Factors on the Emotional Evaluation Phase

In this section, we describe the proposed mechanism for influencing the emotional evaluation phase of CMEs taking into consideration external and internal factors. The proposed mechanism is part of the InFra (see Sect. 2 for a description of the InFra). Our proposal extends the functionality of the InFra by adding a mechanism that modulates the emotional evaluation process based on the information given by a series of factors either internal or external. Internal factors (such as a given current mood or personality) and external factors (such as external events that obstruct or facilitate agents' goals) may have a crucial impact on how stimuli are evaluated from an emotional perspective. Instead of adopting a monolithic approach that takes into account a number of influencing factors at once (as in contemporary CMEs), our mechanism provides an interface that facilitates the incorporation of new influencing factors. In this regard, our approach models the influence of agents' internal and external factors separately from the underlying model of CMEs.

The mechanism for the influence of factors on the emotional evaluation phase is divided into four phases: (i) appraisal phase, (ii) influence factor calculation phase, (iii) system adjustment phase, and (iv) fuzzy calculation phase.

- The *appraisal phase* is part of the general appraisal module of the InFra in which stimuli are evaluated, i.e., emotional values in a range of [0–1] are assigned to stimuli. This module allows the implementation of different emotional evaluation theories. Each influencing factor is modeled as an independent component of the agent architecture and each component varies in format, so the InFra is limited to interpreting the information provided by each component.
- The *influence factor calculation phase* establishes a relationship between each influencing factor and each appraisal dimension. This relationship establishes the degree to which an influencing factor affects each appraisal dimension.
- The *system adjustment phase* is carried out by the InFra's emotion filter component and its main function is to modify the InFra's fuzzy logic scheme based on the relationship value between appraisal dimensions and the corresponding influencing factors. Possible actions to modify the fuzzy logic scheme are (i) changing the membership function types (for instance, from triangular to trapezoidal) and/or (ii) changing the limits of membership functions.
- The *fuzzy calculation phase* (de)fuzzifies appraisal dimensions and carries out fuzzy logic inferences using (i) the adjusted appraisal dimensions and (ii) a set of fuzzy rules in order to establish a relationship between each appraisal dimension and an emotion.

4 Empirical Evaluation

We implemented a proof of concept of the proposed mechanism for influencing the emotional evaluation of stimuli in CMEs. The emotional evaluations were carried out to explore the extent to which an influencing factor modulates the values of configurable appraisal dimensions. We used for this evaluation real-world data about emotions that was extracted from a study reported in [11, 13]. Each record consisted of an emotionally significant event labeled with at most two dominant emotions. Events were characterized using a set of appraisal dimensions (such as relevance and familiarity) as well as an indicator of the corresponding emotion intensity. Please see [11, 13] for details on the data collection protocol as well as for explanations about the selection and interpretation of the appraisal dimensions.

We evaluated events related to *sadness* and *joy* because those emotions were the most commonly reported in the study. From the data on emotions and using the Predictive Apriori algorithm [16], we obtained a set of fuzzy rules indicating a relationship between the appraisal dimensions and emotion intensity. We selected *gender* as the influencing factor of CMEs. The input data for the experiments were 100 randomly generated appraisal vectors representing the assessment of the perceived stimuli. The influencing model made use of six fuzzy logic scheme configurations to represent the bias produced by *gender* based on normalized correlation coefficients. Two series of evaluations were conducted using either triangular or trapezoidal functions and using center of gravity as the defuzzification method.

From the results, we can conclude that *gender* as a factor of influence caused relatively small variations in the emotional evaluation of *joy* and *sadness* (see Fig. 1). In addition, we can conclude that by varying the type of membership function, the intensity of the resulting emotion increased when triangular membership functions

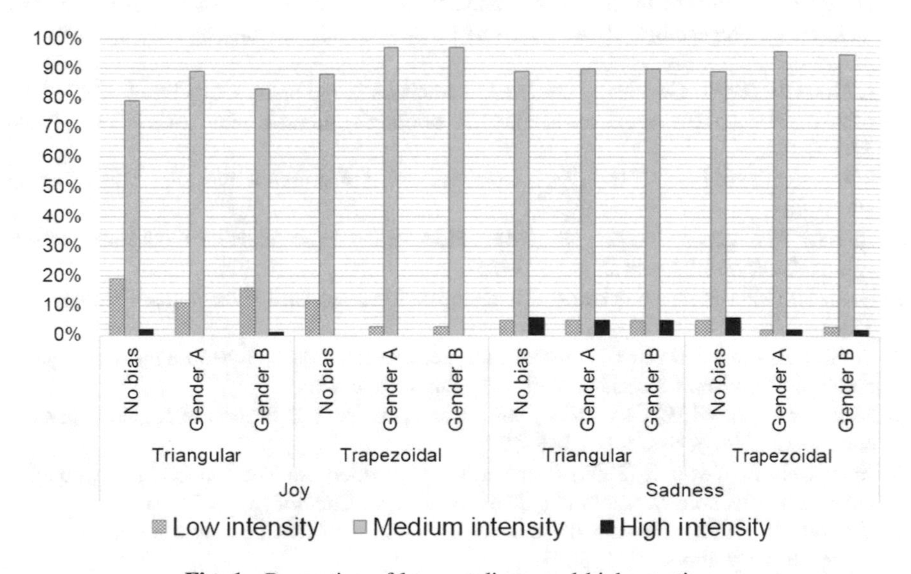

Fig. 1. Proportion of low, medium, and high emotions.

were used. This suggests that triangular membership functions can help evaluate emotional stimuli in a more sensitive way. Furthermore, for both emotions, the predominant emotion intensity was *medium*, while the strongest intensities (low and high) occurred less frequently. This suggests that intense emotions may require a combination of very specific factors.

5 Conclusion and Future Work

We contribute a mechanism that incorporates factors of influence on the emotional evaluation of stimuli supported by a fuzzy logic inference engine. We adjust appraisal dimensions based on internal and external factors of influence. In addition, we provide experimental results regarding the effects of influencing factors on the emotional evaluation of stimuli based on real-world data about emotions. Our contribution allows us to emulate the evaluation of complex sensory information carried out by the amygdala. Future research directions are (i) exploring the effects of taking into account several influencing factors concurrently and (ii) determining how many fuzzy rules are sufficient to evaluate emotional stimuli.

Acknowledgments. This work was supported by PFCE 2019. J. O. Gutierrez-Garcia gratefully acknowledges the financial support from the Asociación Mexicana de Cultura, A.C.

References

1. Rodríguez L-F, Gutierrez-Garcia JO, Ramos F (2016) Modeling the interaction of emotion and cognition in autonomous agents. Biol Inspired Cogn Arch 17:57–70
2. Dancy CL, Schwartz D (2017) A computational cognitive-affective model of decision-making. In: Proceedings of the 15th international conference on cognitive modeling, pp 31–36
3. LeDoux JE (1989) Cognitive-emotional interactions in the brain. Cogn Emot 3:267–289
4. Mao X, Li Z (2010) Agent based affective tutoring systems: a pilot study. Comput Educ 55:202–208
5. Marsella SC, Gratch J (2009) EMA: a process model of appraisal dynamics. Cogn Syst Res 10:70–90
6. El-Nasr MS, Yen J, Ioerger TR (2000) Flame-fuzzy logic adaptive model of emotions. Auton Agent Multi-Agent Syst 3:219–257
7. Breazeal C (2003) Emotion and sociable humanoid robots. Int J Hum Comput Stud 59:119–155
8. Smith CA, Kirby LD (2000) Consequences require antecedents: toward a process model of emotion elicitation. Cambridge University Press, New York
9. Marinier RP, Laird JE, Lewis RL (2009) A computational unification of cognitive behavior and emotion. Cogn Syst Res 10:48–69
10. Reisenzein R, Weber H (2009) Personality and emotion. In: The Cambridge handbook of personality psychology. Cambridge University Press, Cambridge, pp 54–71
11. Scherer KR (1993) Studying the emotion-antecedent appraisal process: an expert system approach. Cogn Emot 7:325–355

12. Mesquita B, Ellsworth PC (2001) The role of culture in appraisal. In: Appraisal processes in emotion: theory, methods, research. Oxford University Press, New York, pp 233–248
13. Meuleman B, Scherer KR (2013) Nonlinear appraisal modeling: an application of machine learning to the study of emotion production. IEEE Trans Affect Comput 4:398–411
14. Martínez-Miranda J, Aldea A (2005) Emotions in human and artificial intelligence. Comput Human Behav 21:323–341
15. Rukavina S, Gruss S, Hoffmann H et al (2016) Affective computing and the impact of gender and age. PLoS ONE 11:e0150584
16. Scheffer T (2005) Finding association rules that trade support optimally against confidence. Intell Data Anal 9:381–395

Overview of Natural Language Processing Approaches in Modern Search Engines

Artem Chernyshov(✉), Anita Balandina, and Valentin Klimov

National Research Nuclear University "MEPhI", Moscow, Russian Federation
a-chernyshov@protonmail.com,
anita.balandina@gmail.com, vvklimov@mephi.ru

Abstract. This article provides an overview of modern natural language processing and understanding methods. All the monitored technologies are covered in the context of search engines. The authors do not consider any particular implementations of the search engines; however take in consideration some scientific research to show natural language processing techniques application prospects in the informational search industry.

Keywords: Natural language processing · Search engines · Dependency parsing · Natural language understanding · Named entity recognition

1 Introduction

With the increase in data volumes in the modern information space of the user, the industry experience the need for new, more accurate methods for searching and extracting information. Many large companies invest in development of new search methods, models and algorithms for use in their products.

In simplest case, any search engine functioning is reduced to the sequence of the following actions: retrieving a user query, processing and parsing a query into components, translating the query into a computer-interpreted language, calculating the result, and outputting the result to the user [1]. Depending on the specific type of system, these steps may vary in format of the input and output data, data processing methods and response calculation.

All considered search algorithms areas of application can be divided by searching techniques.

1.1 Global

This technique involves searching for a large amount of open data, for example, on a significant part of the Internet. Among such systems can be separately distinguished [2]:

1. General meaning search engines such as Google, Bing. This type of system uses a search index mapping and keywords extracted from a user query [3].
2. Language assistants such as Cortana or Siri [4]. These systems use verbose dialogue with the user in natural language and perform simplest search queries based on

A. V. Samsonovich (Ed.): BICA 2019, AISC 948, pp. 54–59, 2020.
https://doi.org/10.1007/978-3-030-25719-4_8

internal databases. Generally, search assistants do not use the principles of inference and semantic analysis, and methods of natural language recognition are used for speech processing.

Due to the large scale of such systems, they usually have little use of search queries in-depth analysis. Search queries are presented as a propositional formula constructed from boolean operators and keywords. The relevance of search results is achieved due to complex ranking algorithms, personalization of search results and a huge amount of accumulated data that allows improving search algorithms in real time.

1.2 Personal and Enterprise

These systems allow performing full-text search on a relatively small number of documents such as files on personal computers, servers, or local networks [5]. The scale of these systems may vary depending on the type and amount of data. For instance, there are corporate systems expanded on a large scale, such as Elasticsearch and Apache Solr.

1.3 Ad hoc Search Systems

These systems match the user request with the information available in the internal databases and available sources and calculates the answer [6]. This type of systems include various reference and library systems, EDMS. Systems of this type use search queries more versatile. Since the information stored in them is more subject-specific, this does not allow usage of relatively simple full-text search algorithms, and they do not have the amount of accumulated data to build a search index based on users search queries.

In this article, we will not consider methods of calculating the search response and building search indexes and concentrate on the overview of NLP methods application in search systems.

2 Applications of Natural Language Processing

The tasks of natural language analysis and processing are parts of an extensive field of artificial intelligence and mathematical linguistics and are aimed at researching and developing the mechanism of interaction between a computer and a person [7, 8]. In total, these methods used to solve the following number of problems:

- Speech recognition
- Analysis of sentences and texts
- Information retrieval
- Speech Analysis
- Sentiment analysis
- Answers generation in question-answering system
- Text generation
- Speech synthesis.

Since within the scope of this article natural language processing methods are considered in the context of search engines and systems, we will consider problems 2–4 as directly related to this area.

2.1 Analysis of Sentences and Texts

Figure 1 schematically shows a simplified process of constructing a search query - structured representation of a sentence in natural language

1. Tokenization [9]. The output of this stage is the original offer broken into tokens. Depending on the methods used, punctuation marks, non-text characters and stop words can be removed from the sentence.
2. Syntax parsing. At this stage, system builds a syntactic tree without semantics of the original query. There are many syntax parsing algorithms and tools that implement them, such as the MALT, SyntaxNet [10], SNNDP, spaCy parsers [11], as well as the parser presented by the authors [12, 13].
3. At the stage of semantic analysis, there are introduced some additional metrics that are used for the interpretation of the general meaning of the original sentence [14]. Usually they apply pre-trained on generalized data models, the most famous of which are word2vec and GloVe.
4. The final stage is the most difficult because it requires in-depth knowledge of the subject area as well as a large amount of processed data for training [15]. Therefore, this stage is often skipped.

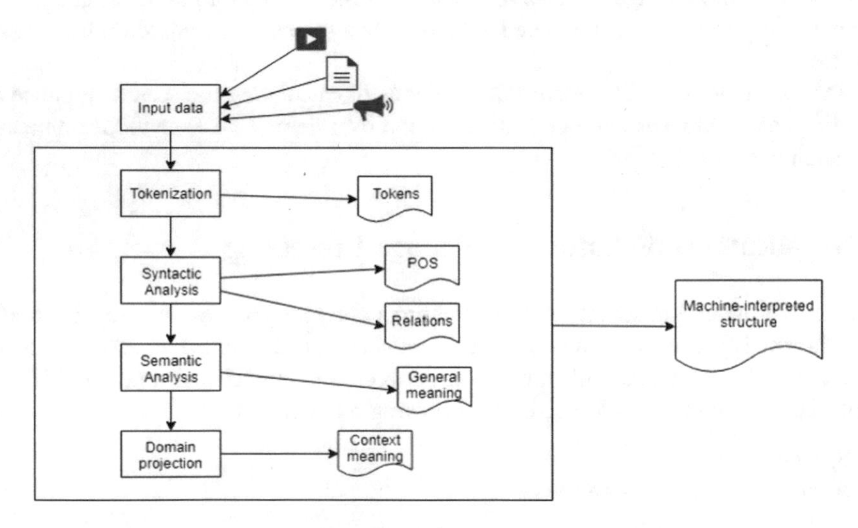

Fig. 1. NLP process

In this paper, we have described the procedure for the synonyms extension. It also contains the common stages of recognition.

2.2 Information Retrieval

The Information extraction normally solves the automatic structured information extraction task based on unstructured sources [16]. The final goal of this task is the information simplification and further conversion to machine-interpreted format. Information extraction methods are widely used to build a search index based on a large amount of unstructured data.

The information retrieval area consists of a set of typical tasks.

- Recognition of objects, subjects, events. Determining the context of the text based on keywords and phrases. Commonly, syntactic analysis is used to effectively meet this challenge.
- Recognition of named entities. A more difficult task, as compared with previous, involves recognizing the names of famous individuals, organizations, geographical names, temporal expressions, and some types of numerical expressions using existing knowledge of the subject area or information extracted from other sentences [17]. Among the open-source tools that provide functionality for solving this problem, we can mention Apache OpenNlp, SpaCy.
- Relationships retrieving: identifying relationships between entities [17].
- Extract of semi structured information.
- Language and vocabulary analysis.
- Terminology Extraction: Search for relevant terms for text corpus. The goal of terminology extraction is to automatically extract relevant terms from a given corpus [18].

Tasks 2 and 3 can be combined into one group of tasks related to the syntactic and semantic features analysis of texts. These tasks are well solved by such methods as NLU and NLP, such as tokenization, stemming, vectorization and syntactic parsing, as was previously demonstrated by the authors in [12, 13].

3 Scope of the Work

This article describes the most significant, in the authors' opinion, of the models, methods and algorithms for the natural language processing and understanding in the area of information search.

The authors deliberately do not provide any specific solutions and implementations of search engines, since the ultimate goal of this article is a descriptive illustration of the scope for the study of the described topic in the context of the area of information retrieval, search and computational linguistics. The analysis of research and scientific studies carried out within this article demonstrates the extensive areas of application of the NLU and NLP methods, as well as their significance for the considered industry.

4 Discussion

In this paper, we give an overview of the methods and algorithms for the natural language analysis and processing methods applied to the task of information search. NLU and NLP methods are widely used in various information retrieval systems at all stages of operation: preliminary processing of incoming search queries, translation into an internal system representation, search and matching using various algorithms, application of logical operations and issuing the final result to the user. Depending on the type of system, NLU methods can also be used to dictate a dialogue with the user.

This research demonstrates that there is a gap in the area of information retrieval, which allows performing plenty of research natural language processing methods and their application to this problem, which was demonstrated by the authors earlier in the papers [19].

Acknowledgements. The funding for this research provided by the Council on grants of the President of the Russian Federation, Grant of the President of the Russian Federation for the state support of young Russian scientists - candidates of sciences MK-6888.2018.9. Conducted survey was supported by the RSF Grant №18-11-00336 and is a part of «Member of the Youth Research and Innovation Competition (UMNIK)» Grant No. 12686ГУ/2017 of 24.04.2018.

References

1. Lewandowski D (2015) Evaluating the retrieval effectiveness of web search engines using a representative query sample. J Assoc Inf Sci Technol 66(9):1763–1775
2. Croft WB, Metzler D, Strohman T (2010) Search engines: information retrieval in practice, vol 520. Addison-Wesley, Reading
3. Peslak, AR (2016) Sentiment analysis and opinion mining: current state of the art and review of Google and Yahoo search engines' privacy policies. In: Proceedings of the conference for information systems applied research, vol 2167
4. López G, Quesada L, Guerrero, LA (2017) Alexa vs. Siri vs. Cortana vs. Google assistant: a comparison of speech-based natural user interfaces. In: International conference on applied human factors and ergonomics. Springer, Cham
5. Stansfield C, O'Mara-Eves A, Thomas J (2017) Text mining for search term development in systematic reviewing: a discussion of some methods and challenges. Res Synth Methods 8 (3):355–365
6. Ensan F, Du W (2019) Ad hoc retrieval via entity linking and semantic similarity. Knowl Inf Syst 58(3):551–583
7. Kurdi MZ (2016) Natural language processing and computational linguistics: speech, morphology and syntax, vol 1. Wiley, New York
8. Kurdi MZ (2017) Natural language processing and computational linguistics 2: semantics, discourse and applications, vol 2. Wiley, Hoboken
9. Straka M, Hajic J, Straková J (2016) UDPipe: trainable pipeline for processing CoNLL-U files performing tokenization, morphological analysis, POS tagging and parsing. In: LREC
10. Marulli F et al (2017) Tuning SyntaxNet for POS tagging Italian sentences. In: International conference on P2P, parallel, grid, cloud and internet computing. Springer, Cham

11. Choi JD, Tetreault J, Stent A (2015) It depends: dependency parser comparison using a web-based evaluation tool. In: Proceedings of the 53rd annual meeting of the association for computational linguistics and the 7th international joint conference on natural language processing (Volume 1: Long Papers), vol 1
12. Chernyshov A, Balandina A, Klimov V (2018) Intelligent processing of natural language search queries using semantic mapping for user intention extracting. In: Biologically inspired cognitive architectures meeting. Springer, Cham
13. Balandina A et al (2018) Dependency parsing of natural Russian language with usage of semantic mapping approach. Proc Comput Sci 145:77–83
14. Golitsyna OL, Maksimov NV, Monankov KV (2018) Focused on cognitive tasks interactive search interface. Proc Comput Sci 145:319–325
15. Golitsina OL, Kupriyanov VM, Maksimov NV (2015) Information and technological solutions applied for knowledge-management tasks. Sci Tech Inf Process 42(3):150–161
16. Milosevic N et al (2019) A framework for information extraction from tables in biomedical literature. Int J Doc Anal Recognit (IJDAR) 22(1):55–78
17. Nguyen DQ, Verspoor K (2019) End-to-end neural relation extraction using deep biaffine attention. In: Proceedings of the 41st European Conference on Information Retrieval (ECIR)
18. Alrehamy HH, Walker C (2018) SemCluster: unsupervised automatic keyphrase extraction using affinity propagation. In: Advances in computational intelligence systems. advances in intelligent systems and computing, vol 650, pp 222–235
19. Chernyshov A et al (2017) Intelligent search system for huge non-structured data storages with domain-based natural language interface. In: First international early research career enhancement school on biologically inspired cognitive architectures. Springer, Cham

Experimental Platform Based on a Virtual Environment for Advanced Study of the Social Behavior of Actors

Arthur A. Chubarov(✉)

National Research Nuclear University MEPhI, Moscow, Russian Federation
osgilat17@gmail.com

Abstract. This work summarizes the development of a set of prototypes of an experimental platform of a new kind, designed to study the many aspects of the social behavior of actors of various nature, including humans and artifacts. The platform combines the abilities to record detailed behavioral data together with neurophysiological and psychometric data, resulting in a powerful instrument to study social behavior motivated by emotional cognition and decision making.

Keywords: Virtual reality · Multi-agent systems · Cognitive architectures social-emotional intelligence · Affective computing · Gaming paradigms

1 Introduction

From the very beginning of ideas about a thinking machine, the main task in the field of artificial intelligence can be considered the creation of a general-purpose mind, which will be to some extent similar to human. Modern researchers are fighting over this task: more and more new ways of achieving this goal are being developed, evaluation criteria, theories that are not worthy of the efforts of developers are being refuted.

Over the past decade, a long way has been traveled to create complex virtual environments, the bright representatives of which are various three-dimensional computer games and videos that are ubiquitous. With each step in the development of technologies for creating such environments, the developers are faced with new challenges to create plausible virtual actors controlled by complex artificial intelligence systems.

On the one hand, the computational power of modern complexes allows us to perform simulations at the level of the human brain, if we take it from the quantitative characteristics of the work items and the total number of operations performed per second. But on the other hand, modern computers are "blind" and cannot perform any actions deviating from programmed behavior without the help of a person. Although they possess superhuman capabilities, the abilities of virtual actors to fully interact with human society are at the lowest level.

Creating a virtual environment is a top priority in terms of the task of developing and testing various approaches to the development of artificial intelligence in various embodiments. The virtual environment can be represented in various ways: in text form

A. V. Samsonovich (Ed.): BICA 2019, AISC 948, pp. 60–72, 2020.
https://doi.org/10.1007/978-3-030-25719-4_9

(the user interacts primarily with the console), where the visual part is represented only by text describing the interaction of the developed artificial intelligence with the system, in two-dimensional and three-dimensional environments, where visualization is more familiar to the user, as offers a much more visual presentation [2, 3].

2 General Concept

Currently, many virtual environments have been developed, but only a few of them have the tools necessary for developing and comparing various models of artificial intelligence (virtual actors). It is also worth noting that even if a single virtual environment has the necessary development tools for developers, there are still problems with the integration and expansion of the functional, which does not allow creating an experimental platform that would have all the qualities that currently require research virtual actors.

The above problems are an obstacle to the creation of a universal model of virtual actors, which will allow you to summarize the accumulated knowledge in this area and create the most effective and plausible general-purpose virtual actor.

You need to create your own virtual environment that allows the most flexible approach to the requirements of developers, as well as allows integration with third-party programs (for example, multi-user systems in a virtual environment, as well as dynamic logging of experimental results). The development of a virtual environment allows you to produce your own scripts, on tasks in which you need to work on a virtual actor of any model. This allows you to change the rules for real participants of the experiment, as well as for the virtual actor itself, in order to test how the model of the virtual actor behaves under various environmental conditions. This work will provide the necessary foundation for developments in this area, without which the development of artificial intelligence of any kind is impossible. Such a system will allow you to quickly change the parameters of models of virtual actors, change fundamentally different models and compare their behavior in different situations, collect data on the behavior of not only the virtual actors, but also players in real time. One of the advantages of such a system can be considered the independence of the composition of participants in the experiment: you can use both heterogeneous (subjects and virtual actors) and homogeneous groups of subjects (only subjects or only virtual actors, not necessarily of the same type).

3 Results: Concept Proof by Implementation

3.1 Teleport Paradigm

The interaction paradigm is called “Teleport” [1].

The virtual environment consists of two platforms, in which the actors are placed controlled by both the subjects and the virtual actors. There are exactly 3 actors. One of the platforms is a rescue zone, the other is an action zone.

The action area includes two teleports. Teleports are designed to move actors to the rescue zone. At the same time, in order to transfer to the rescue zone by means of a teleport, it is necessary that he be in an active state. To do this, another actor must perform actions to activate the teleport. You cannot activate your own teleport. When an actor is located on a teleport, after activation, another teleport will go over to the active state. Thus, in this paradigm, in order to move to the rescue zone, it is necessary to use the help of another actor.

After moving to the rescue zone, the actor has two possibilities: to rescue one of the actors in the action zone (not necessarily the one who helped), or to complete the round independently. It should be noted that it is necessary to introduce a rating system that will allow the indication of which one of the actors is most effective in a given paradigm. In this paradigm, the correct metric of efficiency will be the counting of the number of rounds in which the actor is in the rescue zone. At the end of time the winner is determined by the highest number of points scored.

The experiment is divided into equal logical gaps - rounds. Each round has a fixed time of action, if any other conditions other than the expiration of the set time are not met, the round will restart. In the course of the experiment, actors entering the rescue zone end the round by performing one of the actions available to them. Each of the actors has several options available to him.

1. Greet the other actor. The greeting of the actor is accompanied by the character's voice in relation to the actor to whom the action is applied, as well as by turning towards the goal of the greeting, followed by special animation and light indication.
2. Kick another actor. To push another actor, it is necessary to approach this actor and press it, then, provided that the KICK button is active, the actor over which the action was performed will be pushed a great distance away from the direction of the actor striking the moment.
3. Activate teleport. Any actor can activate a teleport by pressing the ACTIVATE button, provided that its location corresponds to the location of one of the teleports. This action puts another teleport in an active state.
4. Make your own teleportation. Own teleportation is possible only under the condition that the teleport on which the actor stands is activated. This action takes the actor from the action area to the safe area.
5. Rescue one of the actors who are not in the rescue zone. Possible only when in the rescue zone. Allows you to move any actor in the zone of action in the rescue zone. Simply press the SAVE button, and then on the actor located in the action area.
6. Save yourself. Possible only when in the rescue zone. Allows the player in the rescue zone to complete the round without saving anyone in the action area. Just press the ESCAPE button.

These actions are available to both actors driven by real subjects, and virtual actors and are absolutely identical to each other. Each of the actors has the ability to move around the zone of action, as well as greet, push other actors, activate teleports, make their own teleportation, rescue other actors, and also independently complete the round from the rescue zone.

3.2 Machine Learning

Please note that the first paragraph of a section or subsection is not indented. The first paragraphs that follows a table, figure, equation etc. does not have an indent, either.

In general, any virtual environment actor has several possibilities for interacting with the environment. In the interaction paradigm described above, the actor's capabilities are rather wide and its behavior is rather complicated. Thus, the definition of some algorithm of actions for solving such problems is quite difficult and is subject to many errors. With the help of machine learning, it is possible to train actors in effective behavior.

When training, you must remember that at each moment of training you must have three important components:

- Sensors - data of the actor about their surroundings. Sensors can be both discrete and long, based on the complexity of the environment and the actor. For the most complex environments, it is necessary to have several long-lasting sensors, while for simple ones, several discrete ones are sufficient. In this case, do not confuse the state of the actor and the state of the environment. The state of the environment represents information about the entire scene with all the actors inside it. Sensors of the actor have only his information, which usually makes up only a fraction of the state of the environment.
- Actions - what actions an actor can take. As well as action sensors can be both long-term and discrete, depending on the complexity of the actor and environment. For example, for movement on a plane, it can be used as continuous actions, for movement in any direction at each step, or discrete, for movement in specific directions at each step.
- Reward - some value showing the effectiveness of the actor. It is worth noting that the reward should not be provided every moment of time, but only in the case when the actor does something bad or good.

Thus, through the reward of the actor, he performs the tasks assigned, so the rewards must be chosen in such a way that the maximum reward of the actor contributes to the desired behavior.

After determining these components, it is possible to train the actor. This is achieved by simulating the work of the environment over a number of iterations, during which the actor learns through time, what actions need to be taken on the basis of the sensors that he monitors in order to receive the greatest reward.

Thus, when an actor takes action to maximize a reward, he learns the behavior that makes him most effective in this paradigm. The process of learning effective behavior is usually called the training phase, and using the resulting model to work in the environment is called the phase of inference.

To train an actor in a given paradigm, the following elements are necessary:

- Virtual learning environment - all objects of the scene and its actors.
- Python API - which contains all the necessary learning algorithms that are used to train an actor (learning effective behavior). It is worth noting that the virtual learning environment and the Python API are not part of Unity, and interact through external interfaces.

The virtual learning environment (Fig. 1) has three components to interact with PythonAPI:

- An agent is a special component attached to the actors that controls its sensors, performs actions that it receives for execution and assigns awards when necessary. Each agent is associated with only one controller.
- Controller - contains logic for selecting actions for the agent. The controller contains the optimal behavior for each agent and determines its actions at each time point. This is the component that receives information from the actor's sensors and rewards from the agent and returns the necessary action to it.
- Academy - manages sensory information and decision making process. Some ubiquitous parameters are set up through the academy, such as the quality of the display and the speed of the processes inside the environment during the learning process. The external interface of the environment is in the academy.

Each environment designed to train actors always has one academy and one agent for each stage actor. While each agent must be associated with a controller, it is possible for two agents with the same sensors and actions to have the same controller. Thus, the controller determines the space of possible sensors and actions, while the agents connected to it can have their own sensors and actions taken.

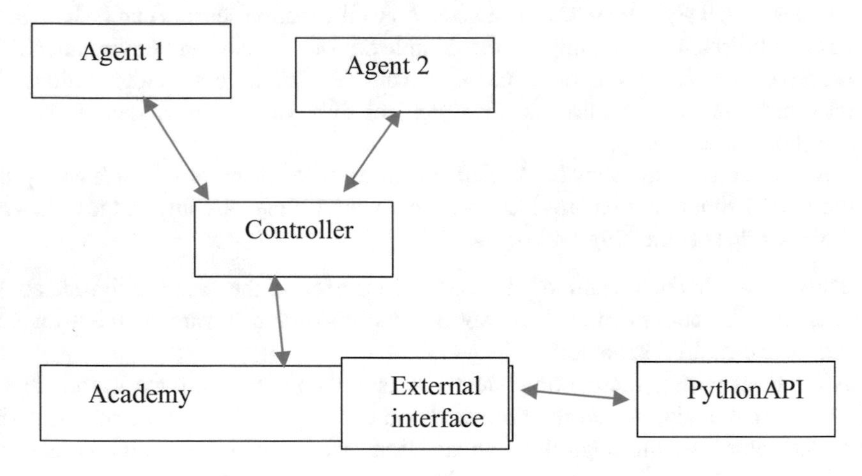

Fig. 1. The structure of the virtual environment for learning

Thus, each actor is composed of an agent that is associated with a controller. The controller receives information and rewards from agents and returns actions. The academy is used to synchronize the controller and agents and set up the parameters of the environment for training.

There are three different types of controllers that allow the use of different scenarios for training and the use of the models obtained.

- External - actions are reproduced using PythonAPI. Here, awards and sensors are collected by the controllers and sent directly to the PythonAPI via the external interface. PythonAPI returns actions that are executed by the agent.
- Internal - which is used by the internal model generated by TensorFlow. This model represents the resulting optimal learning behavior and the controller directly uses this model to determine actions for each of the agents.
- Direct - actions are fully defined using real keyboard input. Here a person can control the agent and the sensors and rewards collected by the controller are not used when managing the agent.

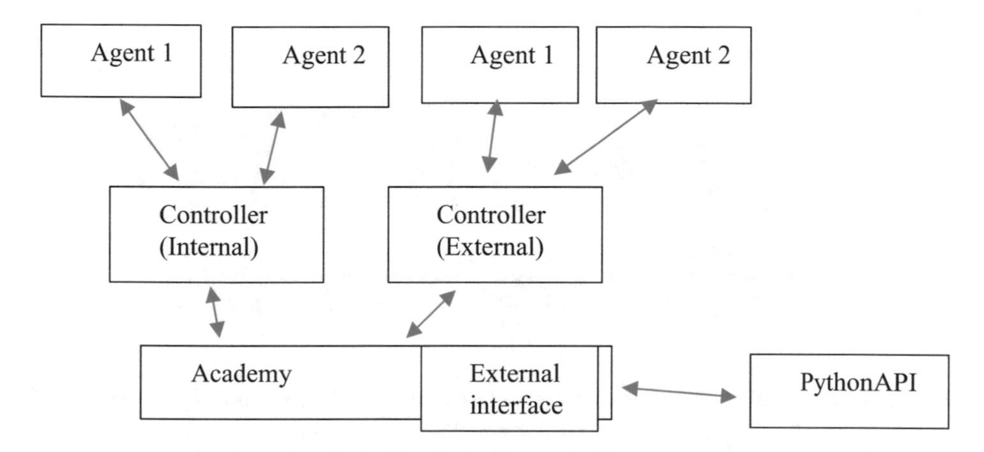

Fig. 2. Diagram of the possible configuration of the virtual environment

As noted earlier, the ML-agents library includes generally accepted learning algorithms. In training mode, the type of controller is transferred to the "External" position during training and "Internal" during testing. After the training is completed, the optimal behavior model is saved to the TensorFlow model file, which can then be integrated into the virtual environment. Figure 2 is diagram of possible configuration of virtual environment. Thus, the library is based on TensorFlow and PythonAPI uses sensors to create its own model.

Among the developers of such agents, there are some recommendations for the most effective work with this technology. When training actors in a virtual environment, it is generally accepted to use the recommendations below.

In the general case, the most optimal strategy should be considered a sequential complication of the task set before the actor in order to make sure that he is able to solve a simple task before embarking on a more complex one. Also, before you start learning the model using the external interface, you must first check whether the task is achievable using the direct controller from the keyboard. You can significantly reduce the training time of agents, if you connect several agents to a single controller and allow them to interact freely.

In order for the training to go monotonously and steadily, it is not recommended to assign large reward values. It should also be remembered that, in general, a positive reward has a more positive effect on the behavior of the agent than a negative one. If the time for performing an action is limited, then it is recommended to periodically charge a small negative reward to the agent until he reaches the goal. But if you make a reward big, but negative, the agent can avoid any actions associated with this reward, even if it leads to an even greater reward.

3.3 Statistics of the Training Process of Actors

During the training, one of the components of TensorFlow - TensorBoard produces saving statistics, which allow to judge about the success of this session of agent training (Fig. 3):

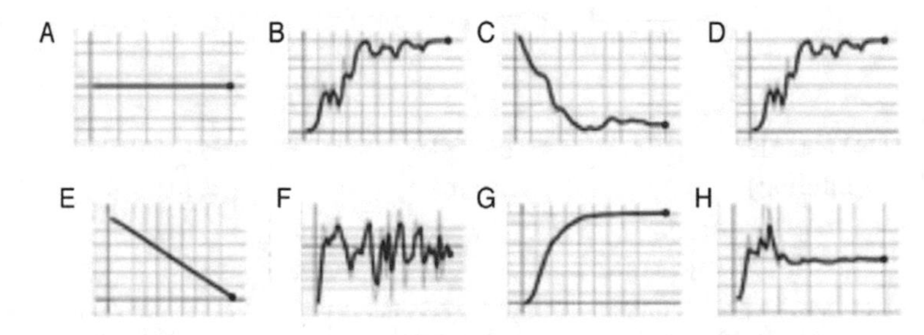

Fig. 3. TensorBoard interface to monitor learning effectiveness. The x-axis and y-axis shows the 100 steps and the corresponding values (random values from a standard normal dist.) of the variable respectively. A. Lesson - Describes the progress of agent learning from task to task. B. Cumulative Reward - The average cumulative reward for all agents. Should increase with successful learning. C. Entropy - shows how the decisions of the model are random. In the course of successful training should gradually decrease. D. Episode Length - the average length of a single episode for actors. E. Learning Rate - how big a step an algorithm takes to find the optimal behavior strategy. Must go down during successful training. F. Policy Loss - the average deviation from the current behavior. Shows how the behavior of the actor changes. With successful learning, the value should decrease. G. Value Estimate - the average value of the assessment of all states in which there is an agent. Should increase over the course of successful training. H. Value Loss - shows how much the agent can predict the values of each of the states. Should decrease in the course of successful learning.

Work was done on the adaptation of machine learning technology to solving problems in the conditions of the Teleport paradigm. An agent is an actor who monitors the environment and chooses the best solution in a given situation. To implement the agent functionality, you must inherit the Agent class. In this case, TeleportsAgent is an inheritor of the Agent class. Here are the main components of the agent: monitoring the necessary data, actions taken by the agent, as well as the rewards that the agent receives.

The agent sends data to the controller, and the controller makes a decision and sends the necessary action to the agent. The agent code performs this action, for example, it moves the agent in one direction or another. In order to use reinforced training, it is necessary to calculate the agent's reward at every step. This award is used to develop an optimal behavior strategy.

The controller alienates the decision-making aspect of the agent in such a way that it is possible to use the same controller for several actors. The nature of the controller depends on its type. The external controller simply transmits observations from the agent to the external process and takes the necessary actions from the outside. The internal controller uses the already generated behavior strategy for decision-making (it does not change the decision-making strategy).

In the course of work, the agent needs environmental data. Thus, it is necessary to send him at each step the normalized values of the quantities he needs. Such can be considered his position, the position of the other two actors (the agent receives only two coordinates, since the action takes place on a plane). And the state of both teleports.

For each step of the agent, it is necessary to take some action sent by the controller. The picture shows the principle of choosing a teleport for interaction. When choosing a teleport, an agent takes into account both its state and the distance to it, which makes it possible to achieve the most optimal behavior strategy. If there is no active teleport, then the agent is inactive.

For each step in which an action was taken, the agent receives a small negative reward in order to reduce the time he reaches his goals. If, during the course of training, an agent enters a zone prescribed for him, he receives a large negative reward. If he shortens the distance, he gets a small positive reward. If he achieved the goal, he gets a great reward.

3.4 The Knight Paradigm

A different virtual environment model was built. The classic scenario of research of the labyrinth was taken as a basis by some actor, in which there are various entities, both helping the actor and hindering the achievement of the set goal. Thus, this labyrinth is quite difficult to go through, acting at random, while being quite simple if you make some planning and monitor the consequences of your choices. Such a labyrinth will be difficult for those virtual actor models that do not take into account all the possibilities of this virtual environment.

Each session of the virtual environment is divided into rounds. The round ends either because of the failure of the actor, or with a win and time is not limited. On the Fig. 4 shows the start of the first round. It visually displays the entities of the virtual environment and some points in the maze that are necessary to describe the possible interaction of the actor with the environment.

Being in the starting point 1, the actor does not see any entities - only moves (Fig. 4).

At point 2, the actor sees the adversary D1 and the sword. D1 begins to approach him. When the actor takes the sword, D2 (ghost) jumps out and attacks him. The ghost sword does not work, but in contact with the Actor the ghost disappears.

In a duel dexterity does not matter. The actor has some advantage in attack speed, but at the same time, if the Actor is attacked by two, then they will surely beat him. Death occurs when health reaches zero. Potion restores health to 100%, but the bottle is enough for one reception.

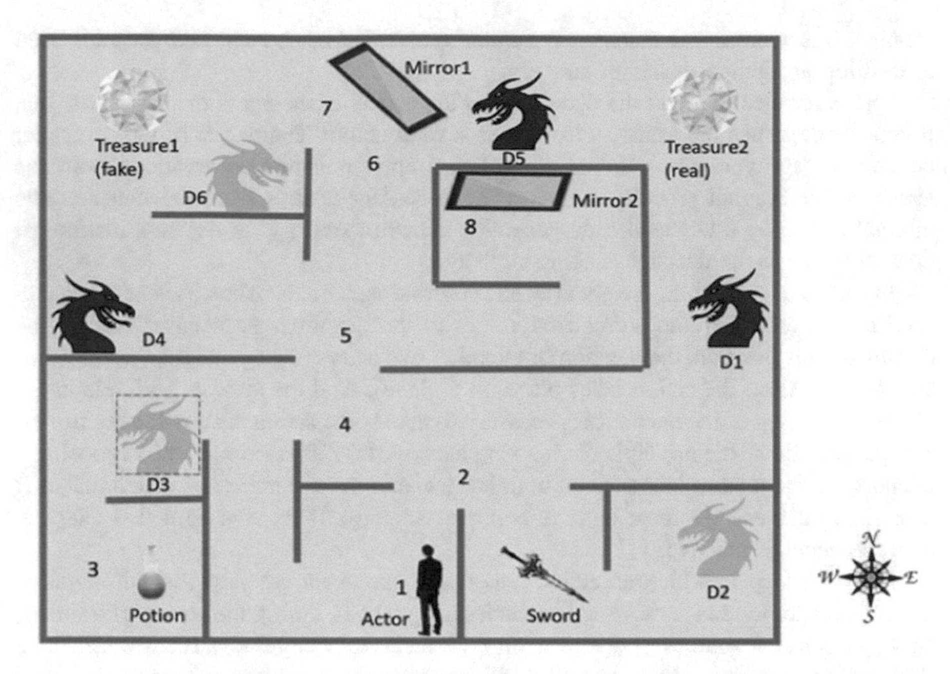

Fig. 4. Placement of interactive elements in The Knight Paradigm

Having come to point 3, the actor can take a potion (Potion) in inventory. D3 (ghost, initially invisible) appears after the Actor has received the potion, and attacks the Actor at an impasse. When in contact with the Actor, the ghost disappears.

At point 4, Actor sees a mirror and a reflection of the left treasure in it.

At point 5, Actor begins to chase D4.

At point 6, Actor is attacked by D5. Actor sees another treasure behind D5. However, he may have time to run to the first treasure - or jump in the mirror.

In the immediate vicinity of the treasure, monsters do not attack the Actor, as they fear the radiance of the treasure (except for ghosts that disappear when in contact with the Actor).

At point 7, Actor is attacked by D6 (ghost).

At point 8, the Actor is stumped by the D4 and/or D5 pursuing him. Wounding or killing the Actor, they go back. There, at point 8, is the second mirror.

The treasure on the left is fake and disappears when touched. The treasure on the right is the present.

After passing through the mirror, Actor gets into the next round (to the starting point 1 - Fig. 4). The beginning of the round is the same as the first, with the only difference that there is no sword (the Actor has a sword with him), and one more monster is added - D7. But there is a new potion bottle. Health is the same as it was at the end of the previous round. Opponents do not pass through the mirror. Not seeing the Actor, the monster will not approach him.

The number of rounds is not limited, but the game time is limited (10 min).

The enemy monster will not start moving without seeing the Actor, but continues to pursue the Actor until it loses sight of him.

Below is information about the game that the player (virtual actor or subject) should know only in a limited form:

- The goal of the game is to touch the real treasure.
- Treasures remain real only for a limited time. One of them is always fake.
- You can beat a monster with a sword, but not for sure.
- Monsters are of two types: it is not known what is different, but they look almost the same.
- Potion replenishes health.
- Through the mirror you can pass.
- Game time is limited.

In general, the virtual actor needs to know as much information as possible about the objects that surround it. For example, be sure to know where the obstacles are and if there is an enemy in sight. Thus, a certain system is needed that will collect this information and provide it via the external interface for decision making by the virtual actor model.

Fig. 5. Screenshot of working knight paradigm

The sensory system of the virtual actor must in some sense emulate the behavior of the present. Also, like a person, through the senses, acts in the real world, the virtual actor must also act. On Fig. 5 knight and monster has identical sensory system and can act in environment by the same rules.

The basis of such a system can put two components: the essence and feeling. Any virtual actor has one or more senses, such as vision, in more complex cases, you can also emulate hearing or even the sense of smell. Such feelings will always be in search of various entities, such as the abovementioned opponents in the maze or rewards.

Thus, for this task, it is necessary to emulate the feeling responsible for the vision, which will indicate at each moment in time the totality of the entities that are being processed at that moment by the feeling. In this case, this sense is advisable to present in the form of the scope. If an arbitrary virtual actor determines the scope in the form of a cone-shaped area denoting the field of view. As soon as a certain type of entity is in sight, it is considered that the actor "sees" this entity. Thus the cone grows and becomes less and less accurate at a distance from the actor. More complex implementations of the field of visibility take into account the distance from the actor to the target and make the probability of the entity entering the list in the field of visibility less likely. Given the small size of the maze, It is inappropriate to use such complex models of perception. Simply a standard Unity function called Raycast, which allows judging the visibility of an entity, only by drawing a ray from the actor to the entity, if nothing gets in the way, the entity is considered to be in sight, otherwise it disappears from the visibility zone [4, 5].

Each actor has the opportunity to receive real-time information about the significant objects around it, with which it can interact. (It is worth noting that this approach is not used for detecting obstacles and navigation, since it is customary to use a specially generated spatial grid). Significant objects are given a label characterizing this type of objects for their recognition, which allows different actors to interact correctly with all kinds of significant objects. In this case, when an actor enters the field of view, the actor performs object identification by the label and adds an object to the array of objects of the field of view, respectively, when the object goes out of the field of view, the object is removed from the array. If you wish, you can set some temporary variable that determines the speed, from which the object forgets about the existence of this object, but it should be remembered that if you leave an object in a given array for a long time, an effect of some complete awareness of the actor on the state of all the maze entities may appear. For example, consider the situation when an actor examining a labyrinth encounters an adversary, in which case it performs recognition, assessment of its own state, and, depending on the function determining the action, decides to attack the adversary or simply run away from it. In the case of a healing potion, it is logical that the actor always seeks to put it in his inventory, while being. This may result in the effect of a certain full awareness of the actor on the state of all the entities of the maze. For example, consider the situation when an actor examining a labyrinth encounters an adversary, in which case it performs recognition, assessment of its own state, and, depending on the function determining the action, decides to attack the adversary or simply run away from it. In the case of a healing potion, it is logical that the actor always seeks to put it in his inventory, while being. This may result in the effect of a certain full awareness of the actor on the state of all the entities of the maze. For example, consider the situation when an actor examining a labyrinth encounters an adversary, in which case it performs recognition, assessment of its own state, and, depending on the function determining the action, decides to attack the adversary or simply run away from it. In the case of a healing potion, it is logical that the actor always seeks to put it in his inventory, while being. In the case of a healing potion, it is logical that the actor always seeks to put it in his inventory, while being. In the case of a healing potion, it is logical that the actor always seeks to put it in his inventory, while being. To attack is defined for each type of actor, the variable responsible for the attack range. Thus, an actor can attack another only if another actor hits a certain sphere around the attacking actor. In this case, a simplified damage registration model was chosen. If an actor attacks and there is another actor in his area, then hitting the actor occurs independently of hitting the actor with a different weapon, this makes it much easier to write primitives for attack. It turns out that each actor has a certain set of actions:

- Movement in space, which is ensured by the generation of a spatial grid
- Attack of another actor, which is provided by simplified registration based on attack range
- Healing of the actor (only for the actor researcher)
- The selection of objects is carried out by simply moving the actor to the position of the object (only for the actor researcher)

After the initial version of the virtual environment has been created, the first tests have been passed, it is necessary to produce the first version of the virtual actor. To do this, you need to select the primitives available to the actor at each time point. In order to simplify such actions, there should be a little, but at the same time they should provide the fullness of actions that the subject had at the last stage of testing.

Below are the selected actions and the conditions for their execution:

Heal is an action performed by an actor, provided that the actor has a potion in his inventory, as well as incomplete health and is not in combat. This action restores the level of health of the actor to the maximum regardless of the amount of health at the time of the start of the action.

- PickupReward is an action performed by an actor only if one of the awards is in its field of view and at the same time it is not in combat.
- Wander - starts a random movement through the maze. It is advisable to use to study the maze, until then, until in sight of any other entity in respect of which you can make a decision.
- EnterMirror is an action performed by an actor only if the mirror is visible. As a result, the actor moves closer to the mirror until it teleports.
- AttackToEnd - an action performed by an actor, only if the appearance of a living enemy. In this case, the actor produces a rapprochement with the enemy and makes an attack until the death of the enemy, or until the virtual actor dies.
- PickupPotion - the action produced by the actor, only under the condition of visibility potion.
- PickupSword is an action performed by an actor only if the sword is visible.

Using the above actions, a virtual actor was developed, randomly choosing the actions available to it and thus interacting with the virtual environment.

At the next stage, it is necessary to select the actions performed by virtual actors using third-party virtual actor model algorithms. But this requires an interface that allows you to send some commands to this application and receive data on the state of the environment, and this exchange should occur in real time.

For this type of interaction to accept to use such technology as a socket. In general, a socket is the name of a software interface for exchanging data between processes. The processes of this exchange can be executed on one computer, as well as on various computers connected by a network. A socket is an abstract object representing the end point of a connection. For communication between machines using the TCP/IP protocol stack, addresses and ports are used. The address is a 32-bit structure for IPv4, 128-bit for IPv6. The port number is an integer in the range from 0 to 65535 (for TCP). This pair defines a socket (the "socket" corresponding to the address and port). In the exchange process, as a rule, two sockets are used - the sender's socket and the receiver's socket. Each process can create a "listening" socket (server socket) and bind it to some port of the operating system. The listening process is usually in a wait cycle, that is, it wakes up when a new connection appears.

Thus, it is enough to add listening and sending information sockets to the virtual environment, which will allow sending the state of the environment in a special format and at the same time listening to commands coming from third-party software for execution.

4 Conclusions

The developed prototypes provide immersion of the actor in a virtual environment and provide opportunities for interaction with the environment. The virtual environment is visually rich, which allows the observer of the virtual environment to focus more on what is happening, and also makes possible some emotional response.

In general, when developing such environments, you first need to come up with a scenario, draw a general diagram of the virtual environment, describe the objects involved in the interaction, identify the goals and describe the capabilities of the actor [6]. Next, you should develop an actor under human control to assess how the capabilities of the virtual environment correspond to the required and, if necessary, change the rules. Only after you have obtained some result that meets the requirements, it is possible to start developing a virtual actor and a special interface to it.

Acknowledgments. This work was supported by the Russian Science Foundation Grant #18-11-00336.

References

1. Chubarov A, Azarnov D (2017) Modeling behavior of virtual actors: a limited turing test for social-emotional intelligence. In: First international early research career enhancement school on biologically inspired cognitive architectures. Springer, Cham, pp. 34–40
2. Gregory J (2009) Game engine architecture. CRC Press, Boca Raton
3. Bjork S, Holopainen J (2004) Patterns in game design (game development series)
4. Goldstone W (2009) Unity game development essentials. Packt Publishing Ltd., Birmingham
5. Okita A (2014) Learning C# programming with unity 3D. CRC Press, Boca Raton
6. Bainbridge WS (2007) The scientific research potential of virtual worlds. Science 317 (5837):472–476

Virtual Listener: Emotionally-Intelligent Assistant Based on a Cognitive Architecture

Alexander A. Eidlin, Arthur A. Chubarov, and Alexei V. Samsonovich(✉)

National Research Nuclear University MEPhI (Moscow Engineering Physics Institute), Kashirskoe shosse 31, Moscow 115409, Russian Federation
a.aidlin@gmail.com, osgilat17@gmail.com, alexei.samsonovich@gmail.com

Abstract. This work is devoted to the development of a concept-proof prototype of a special kind of a virtual actor - an intelligent assistant and a partner, called here "Virtual Listener" (VL). The main role of VL is to establish and maintain a socially-emotional contact with the participant, thereby providing a feedback to the human performance, using minimal resources, such as body language and mimics. This sort of a personal assistant is intended for a broad spectrum of application paradigms, from assistance in preparation of lectures to creation of art and design, insight problem solving, and more, and is virtually extendable to assistance in any professional job performance. The key new element is the interface based on facial expressions. The concept is implemented and tested in limited prototypes. Implications for future human-level artificial intelligence are discussed.

Keywords: Personal assistant · Socially-emotional cognition · Human-computer interface · Semantic mapping

1 Introduction

What would be a simplest paradigm of an artificial intelligence (AI) usage in real life that could win people's hearts and minds, changing the way how we perceive and interact with computational artifacts? In principle, AI could do many jobs: be an assistant or a manager, a teacher or a student, a teammate or an adversary, an independent goal selector, a creative decision maker – or a subordinate extension to the human mind. Today AI helps humans in creating art that challenges best artists or musicians, in giving performances that challenge best performers, etc. [1, 2].

Yet, a simplest intelligent behavior would be not directing, not solving and not performing, but just listening. Humans each need someone who can listen to them and share their thoughts and feelings. The feeling of being understood and appreciated is a fundamental factor that determines the psychological well-being, affecting the major needs: autonomy, competence, and relatedness [3]. To convince a person that one is capable of understanding and appreciating what the person is saying or doing, one needs to demonstrate an adequate emotional reaction, that can be expressed in a great

A. V. Samsonovich (Ed.): BICA 2019, AISC 948, pp. 73–82, 2020.
https://doi.org/10.1007/978-3-030-25719-4_10

multitude of ways, most importantly for us here, including body language alone, such as facial expression and gaze direction.

In this work we introduce the concept of a Virtual Listener (VL): an emotionally-intelligent actor residing in a virtual world, that interacts with, and develops relationships with a human participant, using a minimal nonverbal channel to communicate behavioral feedback in interactions. This behavioral action channel is in our case limited to gaze and facial expression, the latter being reduced to the two principal affective space coordinates: Valence and Arousal [4, 5]. The perception channel, on the contrary, can be rich and multimodal. Our hypothesis is that this limited interaction would be sufficient to establish and maintain a social contact with VL, involving mutual understanding and trust [6]. The general VL concept is described in the next section.

2 Methods

2.1 General Concept of a Virtual Listener

Virtual Listener (VL) is understood here as an avatar representing a virtual actor based on the cognitive architecture eBICA [4, 5], which uses facial expressions and gaze direction control to maintain its social and emotional contact with the human participant. This contact also implies the perception by VL of multimodal information about the participant, indicating participant's emotional state and attitude. Modalities may include facial expression, gaze direction, voice intonation, sentiment of the speech content, behavioral attitudes of actions, and also various psychometrics and neurophysiological measures, such as fMRI, EEG, EMG, heart rate, etc. In addition, the participant may be asked to produce continuous subjective reports of his/her emotional state by clicking the mouse on a 2-D scale representing an affective space. As a minimal example paradigm, here we consider interaction via facial expression and gaze only.

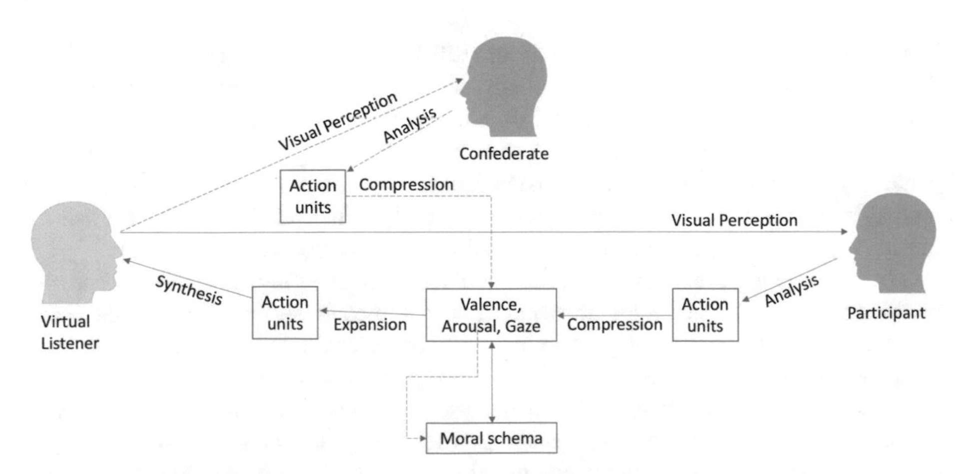

Fig. 1. Minimal logical model of the experimental paradigm and the training paradigm.

The minimal logical model explaining the VL paradigm is represented by the block-diagram in Fig. 1, including two modes: experimental (solid lines) and training (dashed lines). In the training mode, facial expression and gaze of the participant are analyzed, e.g., using an IOS device with a True-Depth Camera or a software like *FaceReader* by Noldus. The outcome of this analysis is a representation of the entire facial configuration using Action Units [9]. Their number can vary from 17 to 35, depending on the requirements. However, it would be difficult to find an elegant model based on first principles, such that it can be successfully applied to 17-dimensional or 35-dimensional vectors. Therefore, it is necessary to reduce the dimensionality. For this purpose, a standard affective space model can be used [4, 5]. The result of this compression (Fig. 1) can consist of 4 numbers: two coordinates in the affective space (Valence and Arousal-Dominance) plus the two eye fixation coordinates. This 4-vector (and its past trajectory) can now be analyzed by the eBICA model [7, 8], using an appropriate Moral Schema [8].

Based on this information processing using the Moral Schema, the system generates an adequate response, which is also represented by a 4-vector, and then gets expanded again into the set of Action Units. After the final synthesis, the facial expression of VL is determined and enacted by the virtual mask of VL (next section). The participant visually perceives the resultant expression, which closes the loop (Fig. 1).

For this scheme to work, the Moral Schema needs to be trained by examples of human behavior. To perform this training, another human participant playing the confederate role is added to the paradigm (Fig. 1). Instead of looking directly at the first participant, the confederate can only see the synthetic virtual mask that in this case copies the participant's facial expression after passing all information about it through the "bottleneck" of the affective space. The data from analysis of confederate's facial expression is used to teach the Moral Schema or adjust its dynamics.

In this description so far only one channel of communication was present. Adding other channels can be straightforward, if the participants engage in a collaborative or competitive behavior. For example, if they play a game in the virtual environment, then logs of all essential game events are fed to the cognitive architecture that uses the Moral Schema, complementing the facial expression data. Examples are discussed below.

2.2 Further Details of Compression and Moral Schema Training

The compression process in Fig. 1 can be based on a linear transformation, in the first approximation. This may not be sufficient, however, and an alternative approach should be considered that would allow us to build a semantic map of the affective space, using the Action Unit data.

A promising approach to semantic map building is the use of artificial neural networks, and deep-learning neural networks in particular. Consider an example. In (Samsonovich, Ascoli, 2010), a statistical physics-inspired approach is described, where objects (i.e. words) are represented with vectors initially allocated randomly in a 26-dimensional unit ball. Afterwards, the cost function is minimized in order to reallocate words so that the angles between vectors representing synonyms would be acute and as small as possible. Angles between vectors representing antonyms, on the contrary, are supposed to be obtuse and close to π. Finally, a PCA transformation is used to reduce dimensionality from 26 to 3 to 4 dimensions.

Here we propose that instead of PCA, an autoencoder [10] can be used. A single layer autoencoder with linear transfer function is nearly equivalent to PCA, and a multilayer autoencoder with nonlinear transfer function can potentially lead to better results. The subject of further research is completely getting rid of cost function minimization in the favor for neural network approach.

Moreover, the same approach can be used as the basis for the Moral Schema training, if an appropriate neural network architecture will be used to implement the model. This approach can be considered as alternative to using first principles and hypotheses motivated by human psychology.

3 Implementation and Preliminary Observations

3.1 Embedding and Connecting the Virtual Mask

Traditionally, facial recognition systems are used to identify and verify a person's identity from an image or from a video frame. There are many methods by which these systems work, but in general they work by comparing the various facial features in a given image with a previously annotated database. These systems are now widely distributed, starting with the identification of a mobile phone user and ending with robotics. A typical application of the face recognition system can be considered for the authentication tasks related to the protection of important objects, along with fingerprint and retinal recognition systems. In general, the accuracy of the face recognition system is lower, but it also has an undeniable advantage, since it does not require any physical contact between the user and the system. More rare are the cases of application of this technology to build human-computer interaction, automatic indexing of images.

In general, the process of facial recognition can be divided into two stages. The first stage is the extraction of characterizing features, and the second is their classification.

Traditional technique is the extraction of reference points from the image of a person's face. For example, an algorithm may analyze the relative position, size, or shape of the person's eyes, nose, cheekbones, and jaw. After this, a similar configuration of the face is searched in other images.

Another technique can be considered the three-dimensional facial recognition technique, which uses special sensors to read depth, thus constructing a three-dimensional incarnation of the face. This information is then used to identify prominent facial features, such as the contours of the eyes, nose, or chin.

The advantage of this approach can be considered its immunity to the lighting of the environment. At the same time, using this approach, it is possible to identify a person at different angles, even when the face is "in profile". This type of data significantly improves recognition accuracy.

It should be noted that facial recognition systems are also used to recognize emotions. With regard to the recognition of emotions, one of the main methods is considered to be a coding system of facial movements, which is a system for classifying human facial expressions, developed by Paul Ekman. In this system, the movements of individual facial muscles are encoded in a special way. Using the coding system of facial movements it is possible to encode almost any anatomically possible facial expression, translating it into a combination of Action Units, which make up the expression. This encoding is considered to be the standard for encoding facial expressions.

One example of software that can extract frame-by-frame encoding of facial expressions from a video stream is the freely distributed OpenFace or proprietary FaceReader. This software processes the video stream from the camera in real time, or from the video and produces a log containing all the information about the configuration of the face for each frame of the video.

Thus, in this work, it is necessary to use one of the facial recognition techniques that offers the highest recognition accuracy and sufficient coverage of possible configurations of human facial expressions. Such a tool can be considered the face recognition module of the ARKit platform. This platform uses a special camera of iOS devices, which allows three-dimensional scanning for the production of applications with the capabilities of augmented reality.

In general, augmented reality offers a way to add two-dimensional or three-dimensional elements to the camera image in real time, so this content complements the real-world image. ARKit simultaneously uses the capabilities of tracking the position of the device and special algorithms for recognizing individual elements from the data coming from the camera to simplify the creation of applications using these capabilities.

ARKit uses True Depth Camera to provide real-time developers with information about the configuration of the user's face, which can be used to locate their own content. Thus, you can use this data to create a mask according to these parameters that repeats the user's mimicry.

ARKit provides the ability to manipulate the model of the virtual actor through special parameters BlendShapes. These parameters are used to control the animation of two-dimensional or three-dimensional models in real time. There are more than 50 different BlendShapes coefficients, while you can use either all at once, or a limited number of these parameters at the same time. Thus, using BlendShapes it is possible to synchronize the models of actors of two subjects in a virtual environment by synchronizing with the help of the multi-user component of the BlendShapes set, as well as to produce their own synthesis of facial expressions on in real time.

This type of data is a simple dictionary, each key of which represents one of the many face parameters recognized by this platform. The position corresponding to each key is described by a floating-point number from 0.0 to 1.0, where 0.0 corresponds to the minimum intensity and 1.0 of the maximum intensity. Since Arkit does not introduce its own restrictions on the use of these parameters, you can use any subset of them or all at once in real time. It is also possible to record certain positions of the face, to play them in the future.

One of the indisputable advantages of using VR technology can be considered a deeper user immersion in a virtual environment, for the simple reason that the surroundings of the surrounding space. This technology makes it possible to reproduce interaction scenarios with a much greater plausibility in a virtual environment, which, in general, allows a much greater emotional response from the subject to be obtained than with the monitor and traditional controls. The virtual reality system itself is a helmet, a pair of controllers and, in most cases, special beacons for tracking the position of the three user interfaces listed above on a room scale. It is also possible to develop a virtual environment. In which each of the participants will have their own virtual reality system using a common system of tracking beacons. Thus, the use of this system allows you to completely immerse a person in the created virtual environment and track the three points of his body, namely, the head (to rotate the character's camera in virtual reality) and hands (for interactive interaction with objects inside the virtual environment). To display the behavior of the remaining body, use either additional trackers with fastening on the belt and legs, or inverse kinematics technologies, which, according to the position of the three tracked points, predict a logical arrangement of the remaining body elements. Thus, the use of this system allows you to completely immerse a person in the created virtual environment and track the three points of his body, namely, the head (to rotate the character's camera in virtual reality) and hands (for interactive interaction with objects inside the virtual environment). To display the behavior of the remaining body, use either additional trackers with fastening on the belt and legs, or inverse kinematics technologies, which, according to the position of the three tracked points, predict a logical arrangement of the remaining body elements. Thus, the use of this system allows you to completely immerse a person in the created virtual environment and track the three points of his body, namely, the head (to rotate the character's camera in virtual reality) and hands (for interactive interaction with objects inside the virtual environment). To display the behavior of the remaining body, use either additional trackers with fastening on the belt and legs, or inverse kinematics technologies, which, according to the position of the three tracked points, predict a logical arrangement of the remaining body elements. To display the behavior of the remaining body, use either additional trackers with fastening on the belt and legs, or inverse kinematics technologies, which, according to the position of the three tracked points, predict a logical arrangement of the remaining body elements. To display the behavior of the remaining body, use either additional trackers with fastening on the belt and legs, or inverse kinematics technologies, which, according to the position of the three tracked points, predict a logical arrangement of the remaining body elements.

The technology of electromyography may be useful because it is impossible to use the previously used methods of reading facial expressions (through the use of True-Depth camera in iOS). Since the test subject, wearing a helmet, blocks the visual access to his face, using an electroencephalograph device, you can get an electromyogram, after analyzing which you have an idea of the state of the person's facial muscles at the time of registration. The idea is to transmit some messages via the Sockets protocol in real time from the BrainSys program, to some program written in Matlab or another language, which produces the necessary analysis and message transfer directly to the virtual environment for appropriate subsequent processing via the Sockets interface. Thus, it is possible to transfer the necessary values both to the local entities of the virtual environment, and to replicate these values on all devices participating in the session (Fig. 2).

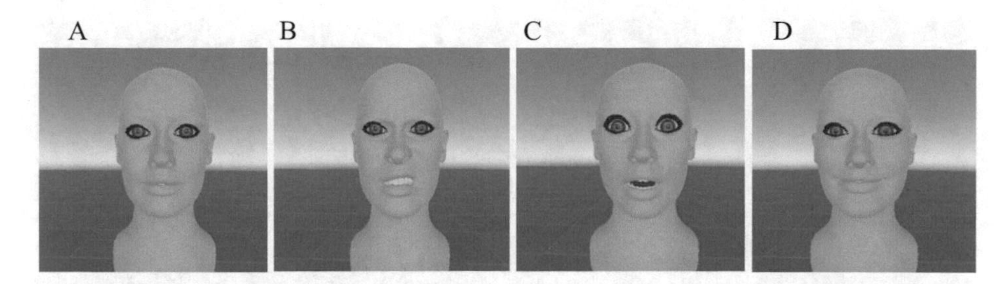

Fig. 2. Emotional facial expressions simulated with the virtual mask. A: neutral, B: disgust, C: surprise, D: happiness. Continuous transitions between these expressions are possible.

One of the most common comments on this project was the overall inexpressiveness of the character used in virtual surroundings. With all his general anthropomorphism, since this character is not a person, each participant has different reactions to him. Therefore, it was decided to carry out work on the search and integration of the character model, in order to make it more expressive and believable.

For the prototype of the system using both of these technologies, a virtual environment was created in which the updated model of the character's face was placed. This model has been attached to a camera position monitored in real space. And at some distance from him a mirror was placed, so that the subject could see his own reaction in real time. At the moment, using the open database The Karolinska Directed Emotional Faces (KDEF), several basic emotions were pre-animated. The animation of these emotions was performed using a special Unity module using the principles of keyframes. Thus, in the end, a virtual environment was obtained with a set of components necessary for displaying emotions and subsequently expanding through the use of Sockets interfaces and components for multi-user interaction.

3.2 The Rock-Paper-Scissors Paradigm: Preliminary Observations

The VL metaparadigm described above was applied to the Rock-Paper-Scissors (RPS) game paradigm in a pilot study (Fig. 3). In this case, the implemented prototype allowed two human players to see only each other's visualizations based on the virtual mask.

Preliminary observations and reports of the participants suggest that live game paradigms elicit much more vivid emotional experiences and much brighter facial responses compared to any paradigm that requires only passive participation, such as watching a video or passively observing each other. On the other hand, social emotional contact through facial expression turns out to be a game changer for many games, making them more exciting and more live. Finally, interaction with a human being in these settings produces much brighter facial emotional expressions compared to interactions with an automata, suggesting that a believable automaton would be very distinct in this case from a non-believable one its efficiency, trustworthiness and social acceptability.

Fig. 3. A screenshot of the RPS paradigm in progress in a virtual environment. The two players are humans, communicating via the virtual mask. Participant shown in the inset is wearing facial electrodes for the electromyogram (EMG) recording, that complements the face recognition data.

4 Discussion

Thus, VL may become a new type of a platform for socially-interactive collaboration between human and computer. In this case, the computer acquires its face and the ability to communicate through facial expressions and attitudes, plus the ability to recognize facial expressions and other body language of the participant. That is, a new modality of human-computer interaction arises.

The technical tasks of the implementation of this project include the development and use of existing solutions, as well as unsolved fundamental problems. The first category includes the recognition and expression of emotions in the language of facial expressions and attitudes. The second is the problem of choosing plausible behavior, that is, the generation of adequate emotional responses to participant actions, taking into account the interaction history. The second task involves the creation of a general mathematical model that describes the dynamics of a person's natural emotional reactions in various paradigms. To solve it, one should accumulate a sufficient amount of data on the interaction of a person with a person in the same conditions, and then simulate a person's behavior with the help of VL. And in the end, VL must pass a limited Turing test to be qualified as indistinguishable from a person as a listener in the selected virtual environment.

The VL concept implies the use of virtual character models for expressing facial expressions of actors participating in an experimental session. Thus, it is necessary to find a way to read real-time indications from a person's face, sufficient for a plausible level to be recreated on an actor model, it is also necessary to find a way to synthesize the facial expression of a virtual actor based on indications, both taken from a real subject and received during work algorithm. After that, it is necessary to solve the problem of synchronization of parameters of each actor in the network in order to obtain two-way interaction, in the case when two real subjects interact in a virtual environment. It is also necessary to remember about the need to introduce a certain paradigm of interaction between the actors, so that the actors have the opportunity to manifest social and emotional intelligence. Thus, as a result of the work reported here, it will be possible to establish two-way communications between the human participant and VL with full synchronization of their facial expressions in real time, as well as an optional substitution of one of the subjects with a virtual actor model using special algorithms to synthesize the corresponding facial expression parameters, as well as an optional substitution of one of the subjects with a virtual actor model using special algorithms for synthesizing the corresponding facial expression parameters, as well as an optional substitution of one of the subjects with a virtual actor model using special algorithms for synthesizing the corresponding facial expression parameters.

A minimal metaparadigm of VL is as follows: two - a person and VL, or two people - jointly do some work, looking at each other, while everyone sees the face of another, express emotions, changing in real time. Everything happens in one environment: in one virtual interior. The participant sitting at the computer must see the partner (VL) in this interior. The range of paradigms, in which the two may be involved, is potentially broad and may include the game Rock - Paper - Scissors or another game; rehearsal of a talk with slide presentation; co-generation of music using the Composer Assistant; joint generation of virtual dance; co-creation of a design; joint coloring of Mondrian-type canvas; puzzle solving by the participant; interaction with a virtual pet; joint management of a smart home; social events such as a virtual tea party, and so on.

4.1 Conclusions

This work was devoted to the development of a VL concept and its proof by prototype studies in pilot experiments. VL is a special kind of a virtual actor - an intelligent assistant and a partner, the main role of whom is to establish and maintain a socially-emotional contact with the participant, thereby providing a feedback to the human performance, using minimal resources, such as facial expressions. This sort of a personal assistant is intended for a broad spectrum of application paradigms, from assistance in preparation of lectures to creation of art and design, insight problem solving, and more, and can be expanded to assist human users in many professional tasks and procedures. The key new element is the interface based on facial expressions. The concept is implemented and tested in limited prototypes. The study suggests that future human-level artificial intelligence will benefit from the new functionality enabled in VL.

Acknowledgments. The authors are grateful to Drs. Dolenko S.A., Ushakov V.L., Redko V.G., Klimov V.V., Ms. Tikhomirova D.V., Mr. Polstyankin K.V., and NRNU MEPhI students who participated in pilot experiments and contributed to many discussions of the concept. One of the authors (Alexei Samsonovich) is grateful to Ms. Kostkina A.D. for the inception of the idea of a Virtual Listener. This work was supported by the Russian Science Foundation (RSF) Grant # 18-11-00336.

References

1. Tatar K, Pasquier P (2018) Musical agents: a typology and state of the art towards Musical Metacreation. J New Music Res 48:56–105
2. Alemi O, Françoise J, Pasquier P (2017) GrooveNet: real-time music-driven dance movement generation using artificial neural networks. In: 23rd ACM SIGKDD conference on knowledge discovery and data mining workshop on machine learning for creativity, Halifax, Nova Scotia, Canada
3. Reis HT, Sheldon KM, Gable SL, Roscoe J, Ryan RM (2000) Daily well-being: the role of autonomy, competence, and relatedness. Pers Soc Psychol Bull 26(4):419–435. https://doi.org/10.1177/0146167200266002
4. Russell JA (1980) A circumplex model of affect. J Pers Soc Psychol 39(6):1161–1178
5. Osgood CE, Suci G, Tannenbaum P (1957) The measurement of meaning. University of Illinois Press, Urbana
6. Simpson JA (2007) Psychological foundations of trust. Curr Dir Psychol Sci 16(5):264–268
7. Samsonovich AV (2013) Emotional biologically inspired cognitive architecture. Biol Inspired Cogn Arch 6:109–125. https://doi.org/10.1016/j.bica.2013.07.009
8. Samsonovich AV (2018) Schema formalism for the common model of cognition. Biol Inspired Cogn Arch 26:1–19. https://doi.org/10.1016/j.bica.2018.10.008
9. Tian YI, Kanade T, Cohn JF (2001) Recognizing action units for facial expression analysis. IEEE Trans Pattern Anal Mach Intell 23(2):97–115
10. Hochreiter S, Schmidhuber J (1997) Long short-term memory. Neural Comput 9(8):1735–1780

Practical User and Entity Behavior Analytics Methods for Fraud Detection Systems in Online Banking: A Survey

Pavel Slipenchuk[1] and Anna Epishkina[2](✉)

[1] Group-IB, Moscow, Russia
[2] National Research Nuclear University MEPhI (Moscow Engineering Physics Institute), Moscow, Russia
AVEpishkina@mephi.ru

Abstract. User and entity behavior analytics (UEBA) methods in fraud detection and advertising are widely used cognitive science methods in modern online banking systems. But profusion of marketing papers complicates true situation. Most of academic papers contain a systematic error: "Correct sample choice error". Fed to the input the real data of user behavior in online banking do not nothing benefit. The paper will be submitted to the criticism of several methods on "mouse track analysis" and "keystroke dynamics" algorithms. New type of algorithms will be present: "preference-behavioral chain" methods. One "preference-behavioral chain" algorithm for social engineering detection will be presented.

Keywords: UEBA · Cognitive science · Online banking · Behavioral analytics · Artificial intelligence · Mouse track analysis · Keystroke dynamics · Keyboard predilections · CSCE · Preference-behavioral chain

1 Introduction

Traditionally, biometric characteristics means fingerprint [1], face [2], speaker [3] or iris [4]. However, in digital world there are other biometric characteristics of the users.

In online banking there are mouse track analysis [5–7], keystroke dynamics [8–11] and preferred behavior. This paper is devoted only to the analysis of behavioral characteristics in online banking.

Behavioral characteristics or "User and Entity Behavior Analytics", UEBA [12] is one of the most discussed topics in information security over the past few years. Nevertheless, many marketing papers complicates true situation. In this work, we will indicate most popular errors and will focus on finding relevant in practice solutions.

UEBA in online banking for fraud monitoring are:

- analysis by JavaScript, e.g. browser detection, privacy-extension detection, private mode detection, operating system detection, architecture detection, virtual machine detection [13];
- cross-browser fingerprint [14–16];

A. V. Samsonovich (Ed.): BICA 2019, AISC 948, pp. 83–93, 2020.
https://doi.org/10.1007/978-3-030-25719-4_11

- mouse track analysis;
- keystroke dynamics;
- user behavior patterns.

Cross-browser fingerprint and analysis by JavaScript, specified in the papers [13–16] work well enough in practice. We will not describe them. Keystroke dynamics in online banking have three main problems: uninformative sample, small sample, fast data obsolescence. Mouse track analysis (MTA) is meaningless to consider the difference between one user and another. MTA is useful in cluster analysis or one-class classifications only. The problem of cluster analysis is given in the paper. User behavior patterns means feature extraction from some behavior user data. The extraction can be processed by finite-state machines. In the paper we present "preference-behavioral chain" (PBCh) abstraction and one algorithm, based by PBCh, useful to detect social engineering and friendly fraud.

2 Keystroke Dynamics

Keystroke dynamics methods are known since 1980 [17]. It's been almost 40 years, but no significant revolutionary ideas have been invented.

The main problems of keystroke dynamics in online banking are:

- very small sample for learning;
- sample for learning is uninformative;
- fast data obsolescence.

Data quickly becomes obsolete. For most banks most users have one session per week, per month, etc. By the time you have enough data, the old data will become obsolete.

Let us explain the sample uninformative. All that users do with the keyboard in online banking are:

- login and/or password input;
- payment details input (usually it's just a card number).

Often login is saved in the browser but password is not. However, there are users who are not afraid to save the password. Card number and payment details also can be copied from clipboard. Sometimes users enter other data in the system: information about the user, data for participation in promotions and bonus programs. But these events are very rare and not useful for keystroke dynamics. The same problem is a legislation variety prohibiting the transfer of username, password, card numbers and payment details to third-party systems. Therefore, if the fraud monitoring system is a cloud solution, information should be in an impersonal form.

In the case of login and password, you can save the time keyDown and keyUp events of each key. The only information about the keys is their order number. Due to the depersonalization and extremely small amount of data, it is pointless to split the input words into bigrams and trigrams [9–11]. Therefore "classical" keystroke dynamics are useful for verify users for example in online courses [8], but not useful in fraud monitoring.

2.1 Keyboard Predilections

We can analyze keyboard predilections, e.g. some user inserted by mouse click, some by Crtl+V, some by Shift+Insert. Also we can analyses method of scrolling, etc.

Few examples of keyboard predilections are presented in the Table 1.

Table 1. Predilections cases example.

Predilection case	Predilection patterns
Insert	By mouse, ctrl+V, shift+insert, command-V
Copy	By mouse, ctrl+C, ctrl+insert, command-C
Scrolling	Mouse scroll wheel, browser scrollbar, space, PageDown, PageUp, cursor control keys (arrow keys)
Number input	Main keypad, numeric keypad
Typo correction	Backspase one symbol at a time, mouse selection+backspase, insert mode, ctrl+shift+backspase

We can use predilection patterns as boolean features in machine learning algorithms. Even Random Forest [18, 19] produces very good results. It is amazing, but 43–82% users in online banking have stable behavioral in keyboard predilections. The median count of "stable" users is 69% by bank. This is mean, the user is "conservative" in his behavior. For example if user have pattern value "Shift+insert" and no more patterns in "insert" pattern task, probability of user will not change his behavior pattern is 69%.

Certainly, it is unwise to use keyboard predilections as a complete system, but it can be part of the scoring model ensemble or will be the killer features in keystroke dynamics system.

3 CSCE and Mouse Track Analysis

3.1 Correct Sample Choice Error (CSCE)

CSCE is most popular error in academic work in Mouse Track Analysis. For example in [5] was chosen only 39 users. This is interesting at the initial stage of research Mouse Track Analysis, but absolutely not applicable in practice. For example, there are more then 50 million regular users in Sberbank online-banking system!

Huga Gambola and Ana Fred showed than the equal error rate point seek zero as more strokes are recorded (see Table 2 in paper [20]). Naturally, ERR decreases with the increasing length of mouse stroke sessions. But we failed to get a strict zero on the banking data. Such good results are obtained in paper [20], apparently due to the artificiality of the experiment: count of student was 25; they playing in memory game 10–15 min. Thus, there are two systematic CSCE experiment artificiality and very small fitting sample.

To avoid the first error, experiment should be organized by real practice task. Participants of experiment should not know about the experiment. To avoid the second, the sample should be large enough (for banking system near 1 millions). You can make a sample significantly lower, but you have to prove that increasing the sample will not greatly increase the interesting for business errors. Unfortunately, this critical issue is not considered in most modern academic works.

3.2 Clustering in Mouse Track Analysis

However, fraud protection requires the division of fraudsters (including social engineering [21, 22]) and legitimate user. As practice has shown, behavioral analytic very well detect friendly fraud and social engineering (more than 60% recall with 5% precision). Precision can be increased up to 30% if rules are based on confidential data.

It is also necessary to point out the infidelity of the original goal. We do not need to classify each user, if we want to detect that another user (fraudster) is working under the same account. We must to classify that the behavior is different, and nothing more. Since it is not possible to distinguish more than 1–100 million users, we can divide users into some "stable" clusters. The cluster will be called stable if the behavior of each user always fits in this cluster.

Fix **N**, and suppose all users are uniform distributed across these clusters. Then the probability that the behavior of the fraudster is similar to the behavior of the legal user is **1/N**. Of course in practice is not so simple. Clean clusters do not exist, and users are not uniform distributed across clusters.

Some user can be attributed to more than one cluster (Fig. 1). Legal user and fraudster is are attributed into two classes. But in right example we can distinguish between legal user and fraudster, since they have different behavior, distributed by different clusters. In left example we have a common cluster: cluster number 6.

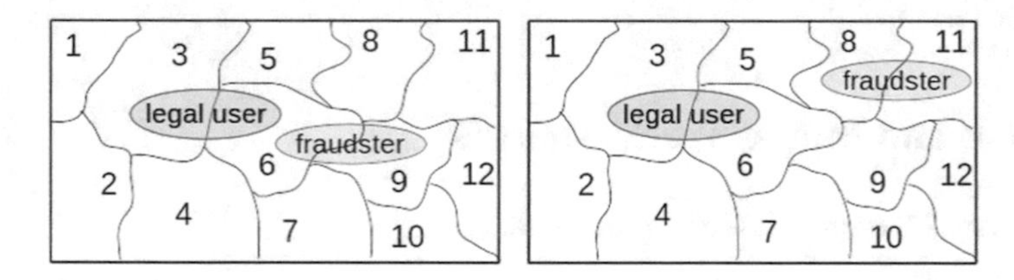

Fig. 1. Legal user is attributed into two clusters: 3 and 6. Fraudster is also attributed into two classes: 6 an 9 in left, 8 and 11 in right. But in right we can distinguish between legal user and fraudster, since they have different behavior.

In left example, clusters 3 and 6 are not "stable" for legal user. But this is not a problem, because we distinguish the behavior. Therefore, in the Fig. 1, left example is "false negative" and right example is "true positive" (Fig. 2).

Each user in the left and the right examples are associated only one cluster and clusters are stable. However, clustering in the right example is not useful. Therefore, left example in the Fig. 1 is "true positive" and right example is "false negative".

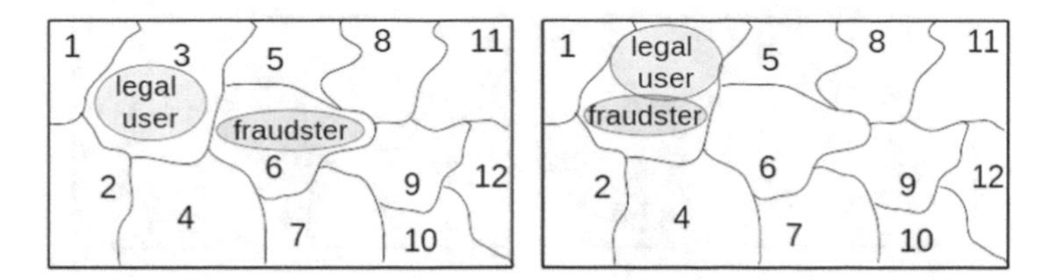

Fig. 2. Legal user and fraudster are associated only one cluster. In the left example they are associated with different clusters. In the right are associated to the same cluster.

3.3 Feature Extraction

The most informative features in MTA are shown in the Table 2.

Table 2. Features as Distance × Value × Func.

Distance	Value	Func
1. Horizontal only	1. Velocity	1. Mean
2. Vertical only	2. Acceleration	2. Median
3. Euclidean distance	3. Jerk	3. Root-mean-square
4. Manhattan distance		4. 5-percentile
5. Polar		5. 10-percentile
		6. 15-percentile
		7. 85-percentile
		8. 90-percentile
		9. 95-percentile

The center of polar coordinates can be set in different ways. A good solution would be the center of the circle describing the start, end, and farthest point of the mouse track from the line between the start and end. Minimum and maximum values are not informative; therefore, it is reasonable to use 5 and 95 percentiles. See more information in papers [6, 7]. From the Table 2 we get $5 \times 3 \times 9 = 135$ features for mouse track analysis.

3.4 "Cluster Method" of Behavioral Classification

The method of determining true positives and false negatives is shown in the Figs. 1 and 2. It will be called the "cluster method" of classification. Define false positive and true negative in this method (Fig. 3). Suppose we build a profile on the first three sessions in online banking system. For session $\mathbf{s_4}$ we need to predict: is session $\mathbf{s_4}$ fraudulent or legal? In the left example of the Fig. 3 we will get true negative, in the right we will get false positive.

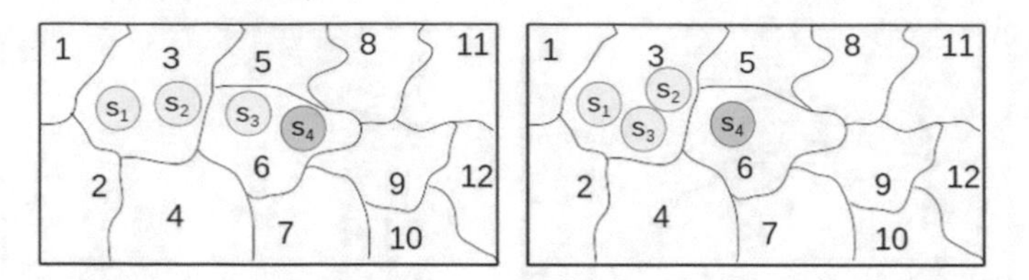

Fig. 3. Four sessions of some legal user, marked by $\mathbf{s_1}$, $\mathbf{s_2}$, $\mathbf{s_3}$, $\mathbf{s_4}$. If we build behavioral profile by first three session and should predict s_4, we will get: true negative in the left example; and false positive in the right example.

3.5 Refusal

There are impossible to build a behavioral profile users. It is reasonable to detect these users and don't use behavioral analytics for them. We called it as "refusal issue". It is much more correct do not give any predictions for "refusal users".

Let ***refusal*** be part of refusal users in the sample of all users. Let ***validity*** be ***1-refusal***. Thus, in addition to recall and precision, the fraud monitoring system is also characterized by ***validity***. If validity is equal to 80%, it means that 20% of users have no protection. However, in this case, the completeness and accuracy can be quite high.

Let N be integer. N is the parameter of the system. For example N = 2. In "cluster method" we can define refusal for user, if behavioral profile is associated more than N clusters. In the Fig. 4 the left example is refusal if N = 1, the right example if N < 3.

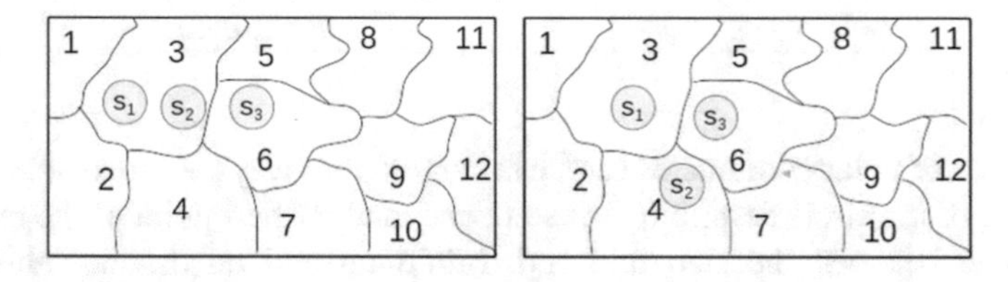

Fig. 4. If N = 1, both examples are refusal. If N = 2, the left is validity, but the right is refusal. If N $\geq$ 3, both examples are validity.

The more mouse tracks we have for each user in the sample, the greater refusal and the less false positive. Indeed, let's look again at the right example in the Fig. 3. If we have only three sessions to build a behavioral profile ($\mathbf{s_1}$, $\mathbf{s_2}$, $\mathbf{s_3}$) and taking the $\mathbf{s_4}$ for prediction we will get false positive. If we have not three, but four sessions ($\mathbf{s_1}$, $\mathbf{s_2}$, $\mathbf{s_3}$, $\mathbf{s_4}$) and N = 1, then we get a refusal case. In practical, we have near 10%–25% refusal in online banking systems.

4 PBCh Methods

Preference-behavioral chain (PBCh) is different method rather than mouse track analysis and keystroke dynamics. Each sessions **s** of each user can be presents of sequences (1) or set of sequences (2). For example:

$$s = \left(s^1, s^2, s^3, \ldots, s^n\right) \tag{1}$$

Note, than s^i are not unique.

In general, these chains can be more than one per session:

$$s = \left\{ \left(s^1, s^2, s^3, \ldots, s^n\right), \left(s^{n+1}, s^{n+2}, \ldots,\right), \ldots\right\} \tag{2}$$

where s^1 and s^{n+1} can be occur in parallel. Case by formula (2) is vary important in NGAV [23] issue. But in fraud monitoring only (1) have practical usage.

For each s^i and s^{i+1} we can calculate some parameters:

$$parameters\left(s^i, s^{i+1}\right) = \{p_1, p_2, \ldots\} \tag{3}$$

Most clear and vary useful parameter is time, elapsed between actions s^i and s^{i+1}:

$$parameter\left(s^i, s^{i+1}\right) = \Delta_i \tag{4}$$

But in general we can analyse another parameters of each transition $s^i \rightarrow s^{i+1}$.

Since s^i is not unique, we can define alphabet:

$$A = \{a_1, a_2, \ldots\} \tag{5}$$

where each a_i is unique action in online banking. Some action a_i can be fact of payment, confirm payment, enter or exit from online banking, change user information, check mortgage account, etc.

For example $s = (s^1, s^2, s^3, s^4, s^5) = (a_1, a_1, a_3, a_1, a_2)$. This mean, some action a_1 executed three times: first, second and after a_3. For each user we have some sessions:

$$u = \{s_1, s_2, s_3 \ldots\}. \tag{6}$$

4.1 Preference-Behavioral Chain

Define set of user {u}. Fix (6) for each user from {u}. Define alphabet A (5). Fix (1) for each session. For each pears s^i, s^{i+1} of actions fix some parameters (4). This mathematical construction define as **Preference-behavioral chain** (PBCh) sample.

We can use PBCh for behavioral analysis as well as MTA and keystroke dynamics. PBCh are useful both on supervised and unsupervised learning. PBCh is vary useful in feature extraction of "informative" subsequences for some users or class of users (fraudster class in supervised learning).

4.2 Social Engineering Detection

See example of PBCh feature extraction algorithm in paper [24]. The mean idea is building of a "Transition Tree" with two weights. The first weight is count of legitimate sequences from root to child of the edge. The second is count of fraudster sequences. Let $\mathbf{c_L}$ be first weight, let $\mathbf{c_F}$ be second weight. Define q is division count of legitimate sessions by count of fraud sessions. Probability of fraudster (P_F) and legitimate (P_L) of each subsequences can be calculated by (7) and (8)

$$P_F = \frac{c_F \cdot q}{c_F \cdot q + c_L} \tag{7}$$

$$P_L = \frac{c_L}{c_F \cdot q + c_L} \tag{8}$$

For example see the Fig. 5. Define q = 100. See subsequence ($\mathbf{a_1}$, $\mathbf{a_1}$, $\mathbf{a_3}$, $\mathbf{a_3}$). Its weight is (13, 11). Using (7) we have $P_F = 0.9883$. This is mean, existence fact of subsequence ($\mathbf{a_1}$, $\mathbf{a_1}$, $\mathbf{a_3}$, $\mathbf{a_3}$) in some session is "killer feature" in machine learning algorithm.

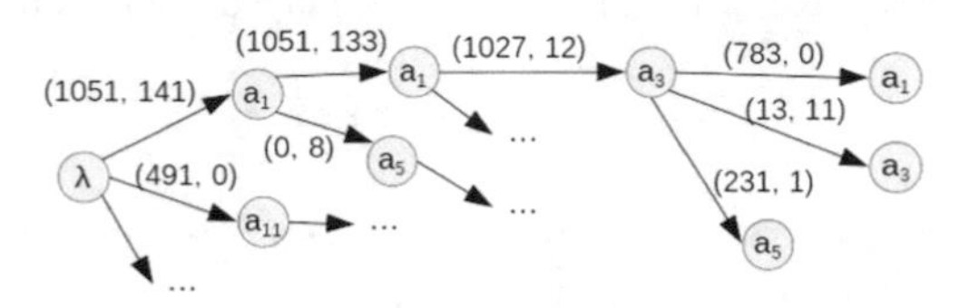

Fig. 5. Example of "Transition Tree" [13]. It is clear, action a_{11} can be found only in legitimate sessions. Subsequence ($\mathbf{a_1}$, $\mathbf{a_5}$) can be found only in fraudster, and subsequence ($\mathbf{a_1}$, $\mathbf{a_1}$, $\mathbf{a_3}$, $\mathbf{a_1}$) only in legitimate. If **q** is large enough ($q \gg 10$), subsequence ($\mathbf{a_1}$, $\mathbf{a_1}$, $\mathbf{a_3}$, $\mathbf{a_3}$) characterizes more fraudster than legitimate.

"Transition Tree" building method see in paper [24]. Note that the algorithm has linear complexity (computational complexity and memory complexity both).

"Transition Tree" can be used for feature extraction. We must walk all edges of the tree and find most legitimate and most fraudster subsequences. For example in [24], we build expert system, that return cutoff between 0 to 1000. In test sample of one real bank precision and recall are (Table 3).

Using correlations with internal bank rules based on confrontational data we gain data in the Table 4.

See deference in precisions. It is clear, the use of confidential information greatly improves the quality of UEBA algorithms.

Table 3. Precision and recall of some online-banking system

Cutoff	Recall	Precision
100	77.5%	1.4%
200	70.6%	2.2%
300	63.7%	3.2%
400	58.9%	4.0%
500	53.4%	4.9%
600	46.5%	7.0%
700	39.1%	8.7%
800	30.2%	11.1%
900	19.9%	12.0%

Table 4. Precision and recall of online-banking system with confrontational data using.

Cutoff	Recall	Precision
100	68.4%	20.1%
200	62.5%	31.0%
300	56.9%	44.3%
400	49.0%	47.3%
500	39.5%	49.8%
600	32.8%	67.5%
700	24.5%	76.5%
800	19.4%	83.1%
900	11.1%	84.8%

5 Conclusion

Keystroke dynamics can be useful only in login and passwords inputs. "Classic" methods, based on bigrams and trigrams [9–11] impossible to use in cloud system. "Keyboard predilections" make stable clusters for 69% users (median).

Correct Sample Choice Error (CSCE) is most popular error in academic work in Mouse Track Analysis. It is mean experiment artificiality and very small fitting sample. Therefore most academic works are not practical usage. The most practically significant method is the cluster method. In this case true positive, false positive, true negative, false negative should be calculated as shown in the Figs. 1, 2, 3 and 4.

Preference-behavioral chain (PBCh) is useful abstraction for feature extraction. One of PBCh algorithm is presented in the paper. It can detect social engineering with 46% recall and 7% precision (cutoff = 600). Precision increases to 47% (cutoff = 400) by enrichment of confidential information.

UEBA in online banking is more reasonable to use it with classical fraud monitoring systems, enriching the system with confidential information. It is also reasonable to extract features using Cross-Browser Fingerprint features.

References

1. Cappelli R, Ferrara M, Maltoni D, Turroni F (2011) Fingerprint verification competition at IJCB2011. In: Proceedings of the IEEE/IAPR international conference on biometrics, pp 1–6
2. Huang GB, Learned-Miller E (2014) Labeled faces in the wild: updates and new reporting procedures, Report from University of Massachusetts, Amherst, UM-CS-2014-003
3. Greenberg G, Banse D, Doddington G, Garcia-Romero D, Godfrey J, Kinnunen T, Martin A, McCree A, Przybocki M, Reynolds D (2014) The NIST 2014 speaker recognition i-vector machine learning challenge. In: Odyssey 2014: the speaker and language recognition workshop, Joensuu, Finland
4. Phillips PJ, Flynn PJ, Beveridge JR, Scruggs WT, O'Toole AJ, Bolme D, Bowyer KW, Draper DA, Givens GH, Lui YM, Sahibzada H, Scallan JA, Weimer S (2009) Overview of the multiple biometrics grand challenge. LNCS, vol 5558, pp 705–714
5. Sayed B, Traore I, Woungang I, Obaidat MS (2013) biometric authentication using mouse gesture dynamics. IEEE Syst J 7(2):262–274
6. Feher C, Elovici Y, Moskovitch R, Rokach L, Schclar A (2012) User identity verification via mouse dynamics. Inf Sci 201:19–36
7. Hehman E, Stolier RM, Freeman JB (2014) Advanced mouse-tracking analytic techniques for enhancing psychological science. Group Process Intergroup Relat 18(3):384–401
8. Young JR, Davies RS, Jenkins JL, Pfleger I (2019) Keystroke dynamics: establishing keyprints to verify users in online courses. Comput Sch 36:48–68
9. Shanmugapriya D, Padmavathi G (2011) An efficient feature selection technique for user authentication using keystroke dynamics. Proc Int J Comput Sci Netw Secur 11:191–195
10. Jagadamba G, Sharmila SP, Gouda T (2013) A secured authentication system using an effective keystroke dynamics. In: Proceedings of emerging research in electronics, computer science and technology. LNEE, vol 248. Springer, pp 453–460
11. Monaco JV (2016) Robust keystroke biometric anomaly detection, arXiv preprint, pp 1–7
12. Gartner Group: Market guide for user and entity behavior analytics. https://www.gartner.com/doc/3538217/market-guide-user-entity-behavior. Accessed 20 Nov 2017
13. Schwarz M, Lackner F, Gruss D (2019) JavaScript template attacks: automatically inferring host information for targeted exploits. In: Network and Distributed Systems Security (NDSS) Symposium 2019, San Diego, CA, USA
14. Boda K, Földes ÁM, Gulyás GG, Imre S (2012) User tracking on the web via cross-browser fingerprinting. Lecture Notes in Computer Science, pp 31–46
15. Gómez-Boix A, Laperdrix P, Baudry B (2018) Hiding in the crowd: an analysis of the effectiveness of browser fingerprinting at large scale. In: Proceedings of the 2018 world wide web conference, pp 309–318
16. Cao Y, Li S, Wijmans E (2017) (Cross-)Browser fingerprinting via OS and hardware level features. In: NDSS Symposium 2017, Reports
17. Gaines RS, Lisowski W, Press SJ, Shapiro N (1980) Authentication by keystroke timing: some preliminary results, Technical report, DTIC Document
18. Breiman L (2001) Random forest. Mach Learn 45(1):5–32
19. sklearn.ensemble.RandomForestClassifier. https://scikit-learn.org/0.20/modules/generated/sklearn.ensemble.RandomForestClassifier.html. Accessed 10 Mar 2019
20. Gamboa H, Fred A (2004) A behavioral biometric system based on human-computer interaction. In: Proceedings of biometric technology for human identification, vol 5404
21. Hatfield JM (2018) Social engineering in cybersecurity: the evolution of a concept. Comput Secur 73:102–113

22. Krombholz K, Hobel H, Huber M, Weippl E (2015) Advanced social engineering attacks. J Inf Secur Appl 22:113–122
23. Next-generation antivirus, Cyberreason (2017). https://www.cybereason.com/blog/what-is-next-generation-antivirus-ngav. Accessed 10 Mar 2019
24. Slipenchuk P, Yankelevich D, Krylov P (2019, in printing) Social engineering detection in online banking by automatic user activity analysis, Group-IB, ICDM

Significance of Emotional Context While Performing Semantic Tasks

Alena A. Fetisova[1(✉)] and Alexander V. Vartanov[2]

[1] National Research Nuclear University MEPhI (Moscow Engineering Physics Institute), Moscow, Russia
alenafetisova7@gmail.com

[2] Lomonosov Moscow State University, Moscow, Russia

Abstract. This research aims to study the features of the brain mechanisms of predicting words based on the previous context, including emotional context. Based on the results of an experiment conducted on 21 people (aged 18 to 32 years, 14 male and 7 female respondents), it can be concluded that the emotional context can accelerate the "decision-making" about the meaning of the stimulus and whether it is suitable for the context. Significant differences between stimuli in different conditions of presentation arose in the later components (about 400 ms after stimulus), which suggests that these differences really reflect differences in meaning of the stimuli, instead of their visual component.

Keywords: Semantics · Decision making · Evoked potentials

1 Introduction

An important part of the development of virtual agents endowed with artificial social-emotional intelligence is the creation of semantic space [1–3]. However, in natural languages, there is a problem of context (situational) dependent polysemy, in communication between people, emotional context is important; therefore, in computer modeling there is a problem of representation of polysemantic concepts in semantic space. For artificial social and emotional intelligence to work properly, it is important to be similar to human intelligence.

2 Method

The experiment consists of two series. In the first series of the experiment, there is a slide with a phrase/stable expression without the last word (noun), on the next slide – the continuation of the phrase. It can be done in three ways: a word that fits the context, a conflict stimulus, or a blank slide. Before starting the experiment, there is the task for the subject to press a certain button with a suitable incentive and another with a conflict, with an empty – to represent a suitable word. The stimulation duration is 700 ms for the beginning of the sentence, 1500 ms for suitable noun/conflict noun/predicted noun (blank screen). In the same 1500 ms the subject gives the answer – is the noun suitable or, the number of presentations is 100.

A. V. Samsonovich (Ed.): BICA 2019, AISC 948, pp. 94–98, 2020.
https://doi.org/10.1007/978-3-030-25719-4_12

In the second series of the experiment, there is a different set of phrases. And there are four possible variants: the phrase and its logical ending, the phrase and its conflict ending, two variants of phrases ending with the same word, but in one case it is suitable, and in the other conflict. In the latter two conditions, there is an additional unpleasant emotionally significant sound (human cry) with the last noun of the phrase.

Nineteen-channel encephalograph was used for EEG registration. The resulting records were cleared of artifacts based on expert analysis. Then the desired fragments were sorted according to the presentations and were averaged to obtain the EP.

The study involved 21 people aged 18 to 32 years (14 male and 7 female respondents), without psychiatric and general medical diagnoses. All subjects received full information about the study and gave informed and voluntary consent to participate in the study. The experiment was approved by the Commission on Ethics and Regulations of the Academic Council of the faculty of psychology of Lomonosov Moscow State University.

3 Results

We begin to observe statistically significant differences ($p < 0.05$) with 400 ms, which corresponds to the classical N400 semantic EP component, it is believed that at this stage the subjects understand the meaning of the words presented to them and accordingly understand that the phrase is not finished [3–5]. Significant differences between the empty (3-green) and the corresponding (1-blue), and conflict stimulus (2-red) since 800 ms are due to the fact that there is a classic component N200, here the subject realizes that before him a blank screen and begins to represent/predict what word should be in the end of this phrase. We do not see any significant differences between a suitable word and a conflict word, because at this stage the subject has not yet understood the meaning of the word and whether it will be a logical continuation of the phrase, but has already distinguished that there is the word. Therefore, we see differences between the conditions with the missing word and its presence, regardless of its meaning (Fig. 1). Then we see significant differences between the conflict, corresponding and empty at the level of 1000 ms – this is the N400 component. This component is associated with the definition of the meaning of the presented word, and accordingly with the understanding of whether it fits the context. Between suitable (1-blue) and empty (3-green) differences much less after 1000 ms, than between them and conflict stimulus (2-red), that is logical, after all an opportune incentive and predicted an opportune – this in essence one thing and the same word (Fig. 1).

In Fig. 2, we observe a similar relationship. Here we begin to observe statistically significant differences ($p < 0.05$) with 400 ms, which corresponds to the classical semantic component of the EP N400. The next important discrepancy is observed at about 900 ms, which corresponds to the component P300 for the effective conflict stimulus (4-purple) affective suitable stimulus (3-green). It is interesting that the word «water» stimulates both conditions, it means that, in condition 3 it fits the context, and in 4 it is a conflict. The important thing is that significant differences between these conditions appear only in the latency 300-350 ms, in later latencies between these conditions there are practically no differences. The greatest differences are observed

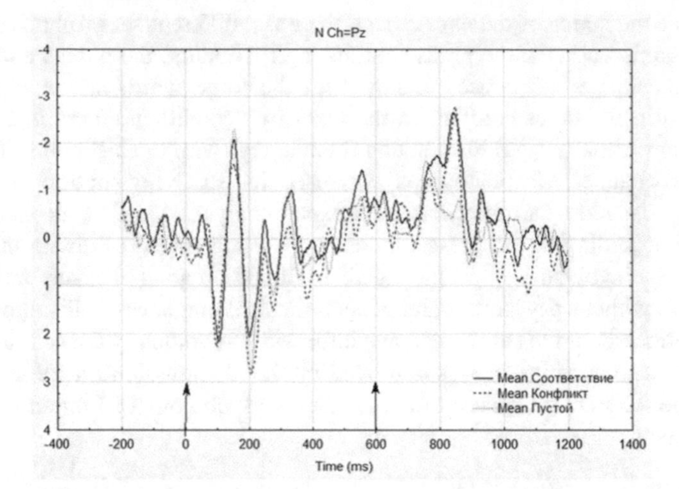

Fig. 1. Comparison of the EPs of the first series of the experiment (average for the group of subjects), 3 conditions are displayed: the word that falls into the context of the phrase (1-blue), conflict (2-red), the prediction condition (3-green). The beginning of a phrase – marker m 0 ms, end of sentence mark – 600 ms with

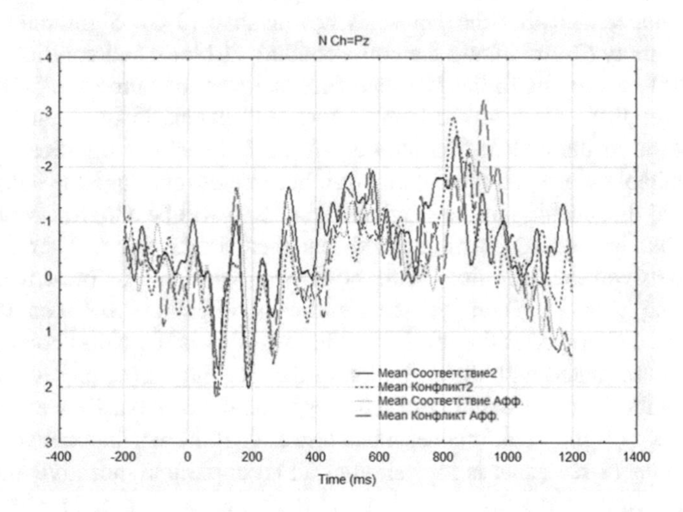

Fig. 2. Comparison of EPs of the second series of the experiment (average for the group of subjects), 4 conditions are displayed: the word that falls into the context of the phrase (1-blue), conflict (2-red), falls + emotional context (3-green), conflict + emotional context (4-purple)ю The beginning of a phrase – marker 0 ms, end of sentence mark – 600 ms

between a suitable stimulus (1-blue) and affective conflict (4-purple) near the latency of 1000 ms, which corresponds to the component N400.

Figure 3 compares the first two conditions from the first series (context matching and conflict stimulus) and the last two from the second series (affective matching and

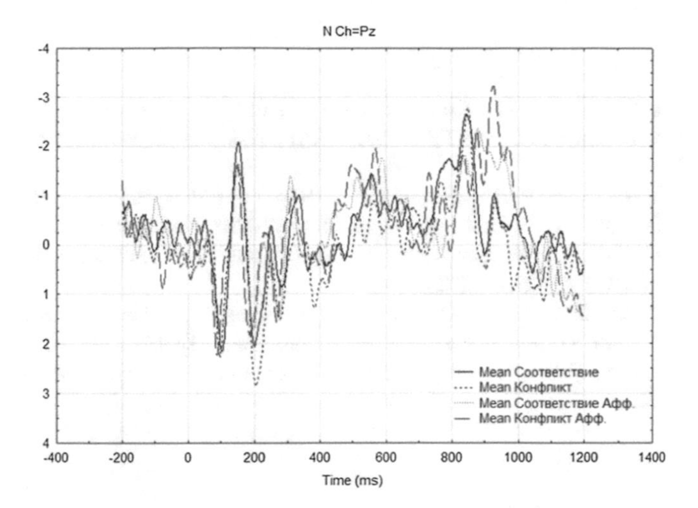

Fig. 3. Comparison of EPs of the first and second series (average for the group of subjects), 4 conditions are displayed: a word that falls into the context of the phrase (1-blue), conflict (2-red), falling + emotional context (3-green), conflict + emotional (4-purple). The beginning of a phrase – marker 0 ms, end of sentence mark – 600 ms.

affective conflict). This comparison is made in order to see the influence of emotional context (conditions from the second series) on the background of conditions without any emotional context (from the first series). Here we observe an interesting effect, the peak of the conditions of the second series on the above-described latency 300–350 ms appears earlier than the corresponding logically peaks of the conditions of the first series (without emotional context). Here there is an assumption that affective influence can "accelerate" the decision on the meaning of verbal stimuli.

4 Conclusion

According to the experiment, we confirmed our hypothesis that the emotional context affects while performing semantic tasks. It has been suggested that the emotional context can accelerate the "decision" about the meaning of the stimulus and whether it fits the context. Significant differences between stimuli in different conditions of presentation arose in the later components (about 400 ms after stimulus), which suggests that these differences really reflect differences in meaning of the stimuli, instead of their visual component.

Acknowledgments. This work was supported by the Russian Science Foundation, Grant № 18-11-00336.

References

1. Samsonovich AV (2009) The Constructor metacognitive architecture. In: The AAAI fall symposium-technical report, vol FS-09-01. AAAI Press, Menlo Park, pp 124–134. ISBN: 978-157735435-2
2. Samsonovich AV, Goldin RF, Ascoli GA (2010) Toward a semantic general theory of everything. Complexity 15(4):12–18. https://doi.org/10.1002/cplx.20293
3. Eidlin AA, Eidlina MA, Samsonovich AV (2018) Analyzing weak semantic map of word senses. Procedia Comput Sci 123:140–148
4. Friederici AD, von Cramon DY, Kotz SA (1999) Language related brain potentials in patients with cortical and subcortical left hemisphere lesions. Brain 122(6):1033–1047
5. King JW, Kutas M (1998) Neural plasticity in the dynamics of brain of human visual word recognition. Neurosci Lett 244(2):61–64

The Semantic Complex Event Processing Based on Metagraph Approach

Yuriy E. Gapanyuk(✉)

Bauman Moscow State Technical University,
Baumanskaya 2-Ya, 5, 105005 Moscow, Russia
gapyu@bmstu.ru

Abstract. In this paper, the features of semantic complex event processing (SCEP) approach based on the metagraph model is considered. The idea of event and event processing is one of the fundamental ideas in the fields of complex systems and software engineering. The definitions and descriptions of Complex Event Processing (CEP) and semantic complex event processing (SCEP) are reviewed and summed up. Semantic Web technologies are typically used for semantics description in SCEP approach. Complex event processing combines data from multiple sources. Complex event processing engine may include several processing levels, and semantic complex events may be enriched during processing. The goal of complex event processing is to identify meaningful events (such as opportunities or threats) in the form of the complex situation. Complex event processing engine may use global ontology and static and dynamic semantics. The RDF approach has limitations in describing complex situations, while the metagraph approach addresses RDF limitations in a natural way without emergence loss. The metagraph model is used as a unified model for semantic events description (static and dynamic semantics), complex situation description, global ontology description. Using the combination of the metagraph data model and metagraph agents model it is possible to construct the complex semantic event processing engine in the form of the hierarchy of dynamic metagraph agents.

Keywords: Complex event processing (CEP) · Semantic complex event processing (SCEP) · Event-driven architecture (EDA) · Complex event · Metagraph · Metagraph agents

1 Introduction

The idea of event and event processing is one of the fundamental ideas in the fields of complex systems and software engineering.

Historically, the idea of event processing was most developed in the field of software engineering. The event-driven architecture (EDA) pattern was invented as a kind of software architecture pattern. The key idea of this pattern is that software modules exchange data mainly in the form of events. Modules can be implemented in various ways, for example, as software services. But at the same time, it was obvious that the idea of EDA architecture has deep roots in the field of complex systems, including biologically inspired ones.

A. V. Samsonovich (Ed.): BICA 2019, AISC 948, pp. 99–104, 2020.
https://doi.org/10.1007/978-3-030-25719-4_13

According to [1] the event is defined as "a significant change in state." At the beginning of the development of EDA approach, the semantics of events were not given much importance. Events were treated as atomic elements or collections of properties. The main advantage of the architecture was seen primarily as the possibility of implementing loosely coupled software modules. It is the use of the event-driven approach that provides a loosely coupling.

But this advantage is "narrow" software engineering benefit, viewed precisely from the point of control of the source code of a large software project. In the process of further EDA approach development, the importance of working with complex events was gradually realized. The Complex Event Processing (CEP) approach was originated.

Then more attention was paid to the semantics of complex events. Based on Complex Event Processing (CEP) approach, the Semantic Complex Event Processing (SCEP) approach was originated.

In this paper, the features of the SCEP approach based on the metagraph model is considered.

It is important to note that the SCEP approach is very important both in software engineering and complex systems modeling, including biologically inspired ones.

The detailed description of cognitive architectures is reviewed in [13]. For many cognitive architectures, the proposed approach may be considered as a low-level architectural pattern.

In some cognitive architectures, such as eBICA [2], Icarus [3] an event approach is already used. Using SCEP approach will enrich the capabilities of cognitive architectures.

2 Complex Event Processing (SEP) and Semantic Complex Event Processing (SCEP)

Consider well-known definitions and descriptions associated with complex event processing. According to [4]: "Event processing is a method of tracking and analyzing (processing) streams of information (data) about things that happen (events), and deriving a conclusion from them. Complex event processing (CEP) is event processing that combines data from multiple sources to infer events or patterns that suggest more complicated circumstances." According to [5]: "The goal of complex event processing is to identify meaningful events (such as opportunities or threats) and respond to them as quickly as possible."

As a rule, definitions of Semantic Complex Event Processing (SCEP) are closely related to specific Semantic Web (SW) technologies. The typical definition is [6]: "The purpose of Complex Event Processing (CEP) is to (semi-) automatically create "actionable" abstractions from streams of events. In order to support flexible abstractions and pattern matching of heterogeneous data streams semantics must be introduced, and Semantic CEP is targeted specifically at semantically interpreting and analyzing data using the Linked Data principles and SW (Semantic Web) technologies."

The authors of paper [7] are also based on the Semantic Web technologies. This approach involves the use of a global ontology, to which events are attached. The paper [7] introduces several interesting concepts. The first is the concept of static and dynamic semantics: "Static semantics capture the mapping between an event schema (i.e., attribute names) and semantic concepts (of global ontology). Dynamic semantics, on the other hand, map from the actual value of an event attribute to semantic entities (of global ontology), and these values can vary for each event." Thus, static semantics mainly deals with global ontology classes, while dynamic semantics mainly deals with global ontology individuals.

The paper [7] also introduces the interesting conception of "semantic enrichment of event" which involves adding new RDF triplets to the event description in the enrichment process.

The book [8] offers an interesting idea that the results of a complex event processing in the creation of meaningful complex situations (such as opportunities or threats) and respond to them. The difference between descriptions [5 and 8] is that description [8] is more detailed and suggests the creation of meaningful complex situations as a result of complex event processing. It is a complex situation that is used for further analysis.

Summing up the considered descriptions we can distinguish the following features of CEP and SCEP:

1. Semantic Web technologies are typically used for semantics description in SCEP approach.
2. Complex event processing combines data from multiple sources.
3. Complex event processing engine may include several processing levels, and semantic complex events may be enriched during processing.
4. The goal of complex event processing is to identify meaningful events (such as opportunities or threats) in the form of the complex situation.
5. Complex event processing engine may use global ontology and static and dynamic semantics.

3 The Metagraph Approach for Complex Event Processing

Metagraph is a kind of complex graph model with emergence, proposed by Basu and Blanning in their book [9] and then adapted for information systems description in our paper [10]: $MG = \langle V, MV, E \rangle$, where MG – metagraph; V – set of metagraph vertices; MV – set of metagraph metavertices; E – set of metagraph edges.

The key element of the metagraph model is the metavertex, which is in addition to the attributes includes a fragment of the metagraph: $mv_i = \langle \{atr_k\}, MG_j \rangle, mv_i \in MV$, where mv_i – metagraph metavertex belongs to set of metagraph metavertices MV; atr_k – attribute, MG_j – metagraph fragment. The metagraph fragment $MG_i = \{ev_j\}$, $ev_j \in (V \cup E \cup MV)$, where MG_i – metagraph fragment; ev_j – an element that belongs to the union of vertices, edges, and metavertices.

The presence of private attributes and connections for a metavertex is the distinguishing feature of a metagraph. It makes the definition of metagraph holonic – a metavertex may include a number of lower-level elements and in turn, may be included in a number of higher-level elements. From the general system theory point of view, a metavertex is a special case of the manifestation of the emergence principle, which means that a metavertex with its private attributes and connections becomes a whole that cannot be separated into its component parts.

The example of metagraph representation is shown in Fig. 1. It contains three metavertices: mv_1, mv_2, and mv_3. Metavertex mv_1 contains vertices v_1, v_2, v_3 and connecting them edges e_1, e_2, e_3. Metavertex mv_2 contains vertices v_4, v_5 and connecting them edge e_6. Edges e_4, e_5 are examples of edges connecting vertices v_2-v_4 and v_3-v_5 are contained in different metavertices mv_1 and mv_2. Edge e_7 is an example of the edge connecting metavertices mv_1 and mv_2. Edge e_8 is an example of the edge connecting vertex v_2 and metavertex mv_2. Metavertex mv_3 contains metavertex mv_2, vertices v_2, v_3 and edge e_2 from metavertex mv_1 and also edges e_4, e_5, e_8 showing holonic nature of the metagraph structure.

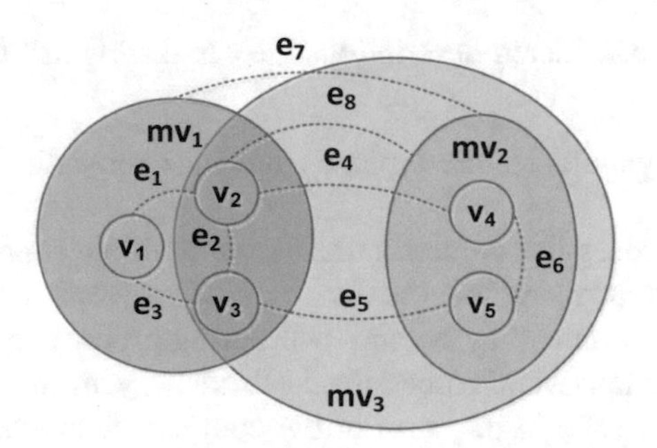

Fig. 1. The example of metagraph representation.

For metagraph transformation, the metagraph agents are proposed. There are four kinds of metagraph agents: the metagraph function agent (ag^F), the metagraph rule agent (ag^R), the metagraph container agent (ag^C), and the dynamic metagraph agent (ag^{MD}).

The metagraph function agent (ag^F) serves as a function with input and output parameter in the form of metagraph: $ag^F = \langle MG_{IN}, MG_{OUT}, AST \rangle$, where MG_{IN} – input parameter metagraph; MG_{OUT} – output parameter metagraph; AST – abstract syntax tree of metagraph function agent in the form of metagraph.

The metagraph rule agent (ag^R) uses a rule-based approach: $ag^R = \langle MG, R, AG^{ST} \rangle, R = \{r_i\}, r_i : MG_j \rightarrow OP^{MG}$, where MG – working metagraph, a metagraph on the basis of which the rules of agent are performed; R – set of rules r_i; AG^{ST} – start condition (metagraph fragment for start rule check or start rule); MG_j – a metagraph fragment on the basis of which the rule is performed; OP^{MG} – set of actions performed on metagraph.

The metagraph container agent is a metagraph, vertices, and metavertices which are in turn agents: $ag^C = MG, v_i \equiv ag_i, v_i \in V, mv_i \equiv ag_i, mv_i \in MV$, where ag^C – metagraph container agent; MG – metagraph; v_i – metagraph vertex; ag_i – agent; V – set of metagraph vertices; mv_i – metagraph metavertex; MV – set of metagraph metavertices.

The rules of a dynamic metagraph agent are performed on the metagraph of agents for a corresponding container agent: $ag^{MD} = \langle ag^C, R, AG^{ST} \rangle, R = \{r_j\}$, where ag^{MD} – dynamic metagraph agent; ag^C – the corresponding metagraph container agent; R – set of rules r_i; AG^{ST} – start condition (metagraph fragment for start rule check or start rule).

Consider how a metagraph approach can help for complex events processing (the following numbers corresponds to the numbers of the previous section):

1. It is generally accepted that the RDF approach has limitations in describing complex situations. A special W3C Working Group Note [11] has been issued to deals with those limitations. It is shown in our paper [12] that the problem here is in emergence loss due to the artificial splitting of the situation into separate RDF-triplets while the metagraph approach addresses RDF limitations in a natural way without emergence loss.
2. Event data can be collected from several sources by metagraph agents and immediately transformed into a metagraph representation. In this case, one or more metagraph rule agents are responsible for their data sources. Agents create a metagraph representation of complex events and pass it to higher-level agents.
3. Using the metagraph approach the complex semantic event processing engine may be considered as a hierarchy of dynamic metagraph agents. In this case, the semantic enrichment consists of the gradual merging of the original metagraph events into the resulting metagraph of a complex situation. Both for initial semantic events and the resulting complex situation, a unified metagraph description is used. To improve semantic enrichment, auxiliary metagraph knowledge bases at different levels of processing can be used.
4. During the enrichment process, the metagraph complex events transformed into a resulting complex situation also represented in the form of metagraph. Using the metagraph rule agent, the special fragments of the resulting situation may be recognized as opportunities or threats.
5. The global ontology may be used in the form of global metagraph knowledge base or in the form of several metagraph knowledge bases distributed by processing levels. The static semantics may be used for general situation recognition while dynamic semantics may be used for dealing with situation details. For example, static semantics may be used for determining the kind of threat, while dynamic semantics may be used for determining threat details and recommendations making.

4 Conclusions

Thus, efficient processing of complex events is simultaneously important in the field of developing complex software systems and in the field of modeling natural complex systems, including biologically inspired ones.

The RDF approach has limitations in describing complex situations, while the metagraph approach addresses RDF limitations in a natural way without emergence loss.

The metagraph model is used as a unified model for semantic events description (static and dynamic semantics), complex situation description, global ontology description. Using the combination of the metagraph data model and metagraph agents model it is possible to construct the complex semantic event processing engine in the form of the hierarchy of dynamic metagraph agents.

References

1. Chandy Mani K (2006): Event-driven applications: costs, benefits and design approaches. Gartner application integration and web services summit 2006
2. Ushakov V, Samsonovich A (2015) Toward a BICA-model-based study of cognition using brain imaging techniques. Procedia Comput Sci 71:254–264
3. Choi D, Langley P (2018) Evolution of the icarus cognitive architecture. Cognitive Syst Res 48:25–38
4. Luckham D (2012) Event processing for business: organizing the real-time enterprise. John Wiley & Sons, Hoboken
5. van der Aalst W (2016) Process mining: data science in action. Springer, Berlin
6. Keskisarkka R (2014): Semantic complex event processing for decision support. In: Proceedings of 13th international semantic web conference – ISWC 2014, pp. 529–536. Springer, Riva del Garda, Italy
7. Zhou Q, Simmhan Y, Prasanna V (2017) Knowledge-infused and consistent complex event processing over real-time and persistent streams. Future Gener Comput Syst 76:391–406. https://doi.org/10.1016/j.future.2016.10.030
8. Teymourian K (2016) Knowledge-based complex event processing: concepts and implementation. Südwestdeutscher Verlag für Hochschulschriften, Saarbrücken
9. Basu A, Blanning R (2007) Metagraphs and their applications. Springer, New York
10. Chernenkiy V, Gapanyuk Yu, Nardid A, Gushcha A, Fedorenko Yu (2017) The hybrid multidimensional-ontological data model based on metagraph approach. In: Petrenko A, Voronkov A (eds) Perspectives of system informatics, LNCS, vol 10742. Springer, Moscow, pp 72–87
11. Defining N-ary Relations on the Semantic Web. W3C Working Group Note 12 April 2006. http://www.w3.org/TR/swbp-n-aryRelations. Accessed 06 May 2019
12. Chernenkiy V, Gapanyuk Yu, Nardid A, Skvortsova M, Gushcha A, Fedorenko Yu, Picking R (2017): Using the metagraph approach for addressing RDF knowledge representation limitations. In: Proceedings of internet technologies and applications (ITA' 2017), Wrexham, United Kingdom, pp. 47–52
13. Lieto A, Lebiere C, Oltramari A (2018) The knowledge level in cognitive architectures: Current limitations and possible developments. Cognitive Syst Res 48:39–55

Deceptive Actions and Robot Collision Avoidance

Anna Gorbenko and Vladimir Popov(✉)

Ural Federal University, Lenin st., 51, 620083 Ekaterinburg, Russian Federation
gorbenko.aa@gmail.com, popovvvv@gmail.com

Abstract. The problem of collision avoidance is extensively studied in contemporary robotics. Efficiency and effectiveness of multirobot task cooperation and safety of human-robot interactions depend critically on the quality of the solution of the problem of collision avoidance. Classical path planning and obstacle avoiding approaches can not guarantee a sufficiently high quality of the solution of the problem of multirobot and human collision avoidance. So, to solve the problem with a sufficiently high quality, some special algorithms and approaches are needed. In this paper, we consider the notion of deception in context of the problem. We consider the possibility of generating and using deception to manipulate another robot. In particular, we consider two robots that solve their own tasks. In addition, we consider two robots one of which solves its own task and the second robot simulates the behavior of a pedestrian. We study the possibility of using fraud to improve the robot's own performance and to improve safety of collision avoidance.

Keywords: Robot navigation · Collision avoidance · Deception · Human-robot interactions

1 Introduction

Collision avoidance is a fundamental problem of robot navigation (see e.g. [1,2]). In particular, collision between autonomous robots is one of the main problems in multirobot task cooperation [3]. When robots and humans share the same environment, we need to guarantee safe interactions between robots and humans. There is a number of examples of deployments of an autonomous robot in an unscripted human environment (see e.g. [4–6]) and human-robot collaborations (see e.g. [7,8]). Nevertheless, the problem of human-robot collision avoidance is still far from the final solution (see e.g. [9]). Moreover, classical path planning and obstacle avoiding schemes may be too slow to provide safe navigation (see e.g. [10]). So, we need to use some special schemes to guarantee safe collision avoidance. Also we need some additional scheme to ensure that the robot can resume its path after collision avoidance to still reach its goals effectively and efficiently [3].

A. V. Samsonovich (Ed.): BICA 2019, AISC 948, pp. 105–109, 2020.
https://doi.org/10.1007/978-3-030-25719-4_14

The problem of deception with respect to the study of intelligent systems has attracted considerable interest of researchers (see e.g. [11–13]). There is also a number of robotic studies of the problem of deception (see e.g. [14,15]).

In the general case, deception can be considered as a deliberate sequence of actions aimed at a specific goal [16]. Often deception is aimed at manipulating some other individual [17]. In this paper, we consider the problem of collision avoidance and the possibility of generating and using deception to manipulate another robot or pedestrian.

2 The Notion of Deception in Context of Collision Avoidance

We assume that successful deception itself is not a goal for the robot. It is assumed that a deception is used to improve the quality of the solution of the main task of the robot and the quality of the collision avoidance.

In general, the motion of the robot can be divided into three phases, the robot motion phase before the collision avoidance, motion phase during the collision avoidance, motion phase after the collision avoidance. It is clear that the robot can use deception after avoiding a collision. Such deception allows to improve the quality of the solution of the main task of the robot. However, in this paper, we do not consider such deception. It is assumed that the task of the collision avoidance has the highest priority. Accordingly, we consider all actions during the second phase as part of the solution of the problem of the collision avoidance. Thus, we assume that deception can be performed only during the first phase. Therefore, we can consider deception as a sequence of deliberate actions that are different from normal actions that are necessary for a solution of the main task of the robot.

Evaluation of the success of deception depends on its purpose. In this paper, we consider deception for the main task of the robot and for collision avoidance. Accordingly, the evaluation of the success of deception depends on the success of the solutions to these problems.

3 Experimental Setup

In our experiments, we use three different robots, Neato Robotics Neato XV-11 with camera, Logicom Spy-C Tank, and Aldebaran Robotics robot Nao V4 H25. We have considered different indoor environments with size from 20 m^2 to 200 m^2 and the following properties.

The average size is 60 m^2.
The maximum vision range is 19 m.
The average vision range is 2 m.
The maximum number of separate obstacles is 90.
The average number of separate obstacles is 32.
The maximum width of separate obstacles is 6.7 m.

The average width of separate obstacles is 0.46 m.
The maximum size of separate obstacles is 10.8 m^2.
The average size of separate obstacles is 0.25 m^2.
The maximum value of the total size of obstacles is 137.4 m^2.
The average value of the total size of obstacles is 14.6 m^2.

Robots Neato and Spy-C Tank were used to solve their daily tasks. The robot Neato was used to clean up the environments. The robot Spy-C Tank was used for surveillance. For these robots experiments were carried out in the background. The robot Nao was used to simulate the motion of a pedestrian. The task of the robot Nao is to localize the robot Neato and select a trajectory that crosses the potential trajectory of the robot Neato. After that, Nao simulates the motion of a pedestrian. After resolving the collision, Nao stops for 5 min and repeats its actions. Neato is the only robot that uses deception in its actions.

4 Experimental Results

When two pedestrians are moving, each can be considered as a moving obstacle for the other one. Each pedestrian is avoiding the other one while being avoided at the same time. We have a similar situation for a robot and a pedestrian and for two robots. Until the phase of collision avoidance the robot can do some deceptive actions that can provide some beneficial consequences of the collision avoidance.

We use recurrent neural networks for predicting the motion of another robot. To generate deceptive actions, the robot evolves a population of random sequences of actions by a simple genetic algorithm. To train the neural networks and evolve the population of potential deceptive actions, we use computer simulations.

In our first experiment, we use Neato and Spy-C Tank. We assume that the robot Neato should generate some deceptive actions that can increase its performance. In particular, it is assumed that Neato should minimize the number of transitions of the robot Spy-C Tank from an uncleaned part of the environment to a cleaned part of the environment. In addition, it is assumed that Neato should minimize the total path length of the robot along the cleaned part of the environment. We have evaluated the quality of cleaning in two weeks for each number of generations of the genetic algorithm. Experimental results are given in Table 1.

Table 1. Dependence of the quality of cleaning of environments (the weight of garbage collected by the robot) from the number of generations of the genetic algorithm

Quality of cleaning	100	102.31	102.53	102.57	102.58
The number of generations	0	10^3	10^4	10^5	10^6

In our second experiment, we use Neato and Nao. Our second experiment was performed under conditions that are similar to the conditions that are presented in the paper [18]. In 29% of cases, participants of experiments [18] were likely to change their decisions in the process collision avoidance. Such changes present a significant threat to safety. In addition, such changes can significantly reduce the performance of the robot. We assume that the robot Neato should generate some deceptive actions that allow to reduce the number of changes of decisions. We have considered 243 trials for each number of generations of the genetic algorithm. Experimental results are given in Table 2.

Table 2. Dependence of the number of changes of decisions from the number of generations of the genetic algorithm

Quality of cleaning	29%	28.8%	25.1%	4.52%	2.46%
The number of generations	0	10^3	10^4	10^5	10^6

5 Conclusion

In this paper, we have presented the results of two robotic experiments. The results of the first experiment have shown that the robot can generate some deceptive actions to increase its performance. The relatively low increase in productivity can partly be explained by the very high initial quality of cleaning (more than 92 % of all possible garbage). The results of the second experiment have shown that the robot can generate some deceptive actions to make pedestrian behavior significantly more predictable.

References

1. Almasri, M.M., Alajlan, A.M., Elleithy, K.M.: Trajectory planning and collision avoidance algorithm for mobile robotics system. IEEE Sensors J **16**, 5021–5028 (2016). https://doi.org/10.1109/JSEN.2016.2553126
2. Zeng, L., Bone, G.M.: Mobile robot collision avoidance in human environments. Int J Adv Robot Syst **10**, 1–14 (2013). https://doi.org/10.5772/54933
3. Liu, F., Narayanan, A.: A human-inspired collision avoidance method for multi-robot and mobile autonomous robots. In: Boella, G., Elkind, E., Savarimuthu, B.T.R., Dignum, F., Purvis, M.K. (eds.) Principles and practice of multi-agent systems PRIMA 2013. Lecture Notes in Computer Science, vol. 8291, pp. 181–196. Springer, Berlin (2013). https://doi.org/10.1007/978-3-642-44927-7_13
4. Burgard, W., Cremers, A.B., Fox, D., Hahnel, D., Lakemeyer, G., Schulz, D., Steiner, W., Thrun, S.: Experiences with an interactive museum tour-guide robot. Artif Intell **114**(1), 3–55 (1999)
5. Thrun S, Bennewitz M, Burgard W, Cremers AB, Dellaert F, Fox D, Hahnel D, Rosenberg C, Roy N, Schulte J, Schulz D (1999) MINERVA: a second-generation museum tour-guide robot. In: proceedings of the 1999 IEEE international conference on robotics and automation, vol. 3, pp. 1999-2005

6. Nourbakhsh IR, Kunz C, Willeke T (2013) The mobot museum robot installations: a five year experiment. In: 2003 IEEE/RJS international conference on intelligent robots and systems, pp. 3636-3641
7. Lacevic, B, Rocco, P (2010) Kinetostatic danger field – a novel safety assessment for human-robot interaction. In: IEEE/RSJ international conference on intelligent robots and systems, 2010, pp. 2169-2174
8. Polverini, M.P., Zanchettin, A.M., Rocco, P.: A computationally efficient safety assessment for collaborative robotics applications. Robot Comput-Integr Manuf **46**, 25–37 (2017)
9. Trautman, P., Ma, J., Murray, R.M., Krause, A.: Robot navigation in dense human crowds: statistical models and experimental studies of human-robot cooperation. Int J Rob Res **34**, 335–356 (2015). https://doi.org/10.1177/0278364914557874
10. Lo, S., Cheng, C., Huang, H.: Virtual impedance control for safe human-robot interaction. J Intell Robot Syst **82**, 3–19 (2016). https://doi.org/10.1007/s10846-015-0250-y
11. Bond, C.F., Robinson, M.: The evolution of deception. J Nonverbal Behav **12**, 295–307 (1988)
12. Cheney, D.L., Seyfarth, R.M.: Baboon metaphysics: the evolution of a social mind. University Of Chicago Press, Chicago (2008)
13. Hauser, M.D.: Costs of deception: cheaters are punished in rhesus monkeys (Macaca mulatta). PNAS **89**, 12137–12139 (1992)
14. Gerwehr, S., Glenn, R.W.: The art of darkness: deception and urban operations. Rand Corporation, Santa Monica (2000)
15. Wagner, A.R., Arkin, R.C.: Acting deceptively: providing robots with the capacity for deception. Int J Soc Robotics **3**, 5–26 (2011). https://doi.org/10.1007/s12369-010-0073-8
16. McCleskey, E.: Applying deception to special operations direct action missions. Defense Intelligence College, Washington (1991)
17. Whaley, B.: Towards a general theory of deception. J Strateg Stud **5**, 178–192 (1982)
18. Vassallo, C., Olivier, A., Soueres, P., Cretual, A., Stasse, O., Pettre, J.: How do walkers avoid a mobile robot crossing their way? Gait & Posture **51**, 97–103 (2017)

A Cognitive Architecture for a Transportation Robotic System

Ricardo Gudwin[1(✉)], Eric Rohmer[1], André Luis Ogando Paraense[1], Eduardo Fróes[1], Wandemberg Gibaut[1], Ian Oliveira[1], Sender Rocha[1], Klaus Raizer[2], and Aneta Vulgarakis Feljan[3]

[1] University of Campinas (UNICAMP), Campinas, SP, Brazil
gudwin@unicamp.br

[2] Ericsson Research Brazil, Ericsson Telecomunicações S.A., Indaiatuba, Brazil
klaus.raizer@ericsson.com

[3] Ericsson Research, Ericsson AB, Stockholm, Sweden
aneta.vulgarakis@ericsson.com

Abstract. Autonomous mobile robots emerged as an important kind of transportation system in warehouses and factories. In this work, we present the use of MECA cognitive architecture in the development of an artificial mind for an autonomous robot responsible for multiple tasks. It is a work in progress, and we still have only preliminary results. Future work will present a more detailed account of the architecture.

Keywords: Transportation robotics · Cognitive architecture

1 Introduction

According to Adinandra et al. [1], autonomous mobile robots (AMR) have emerged as a means of transportation system in warehouses. A large collection of AMRs can be responsible for the transportation of different kinds of goods within a warehouse, with robustness and flexibility. As technology evolves, systems are moving increasingly towards autonomy, perceiving, deciding, learning, etc. often without human engagement.

Cognitive Architectures have been employed in many different kinds of applications, since the control of robots to decision-making processes in intelligent agents. Cognitive Architectures are general-purpose control systems' architectures inspired by scientific theories developed to explain cognition in animals and men [12]. Cognitive Architectures are, at the same time, theoretical modelings for how many different cognitive processes interact to each other in order to sense, reason and act, and also a software framework which can be reused through different applications. Our research group has recently contributed to the field with the proposition of MECA - the Multipurpose Enhanced Cognitive Architecture [6,7].

A. V. Samsonovich (Ed.): BICA 2019, AISC 948, pp. 110–116, 2020.
https://doi.org/10.1007/978-3-030-25719-4_15

There are many reports in the literature promoting the use of cognitive architectures with robotics [2–4,8,10,16,17]. Particularly, the use of Cognitive Architectures can be quite powerful in providing autonomy to robotic systems [15].

In this paper, we propose the use of MECA in order to build a Transportation Robotic System for a simulated factory. A simulated environment was constructed with the aid of the V-REP robotics simulator [13], and MECA was used to build TRoCA, an intelligent agent controlling a transportation robot within the factory.

2 MECA - The Multipurpose Enhanced Cognitive Architecture

The development of MECA was a first attempt to compose a large generic-purpose cognitive architecture with many features inspired in popular ones like SOAR [9], Clarion [14] and LIDA [5], using our CST toolkit [11] as a core.

The problem of building a control system for an autonomous transportation robot is a specially interesting one for exploring MECA's capability. Due to its intrinsic complexity, the scenario provides good opportunities to explore MECA's cognitive abilities, such as perception, planning, opportunistic behavior, etc.

3 TRoCA - A Transportation Robot Controlled by a Cognitive Architecture

We used the V-REP simulation framework [13] to simulate an industrial environment, where a set of transportation robots are required to move packages around in an autonomous way. The overall experiment concerns an industrial scenario, holding both processing cells, a warehouse, and robots for providing the transportation of packages from cell to cell, or from and to the warehouse. The factory floor is split into different regions: Processing Cells, Warehouses and Open spaces. Examples of a Processing Cell and the Warehouse are illustrated in Fig. 1.

Our configuration for the transportation robot includes a MiR100 base (from Mobile Industrial Robotics), with an upper UR5 mechanical arm (from Universal Robots), and a set of slots where delivery items can be stored. It is, therefore, capable of picking up many packages while also delivering some of them, making it possible to explore many routing techniques and thus reach optimal performance through flexible and adaptive behavior. Figure 2 provides a better view of the transportation robot and its parts.

The following general tasks are specified to be performed by our transportation agent:

- Package Transportation
  - Pick-up of packages in processing cells and its transportation to their final destination, which can be another pick-up-delivery slot or a warehouse shelf

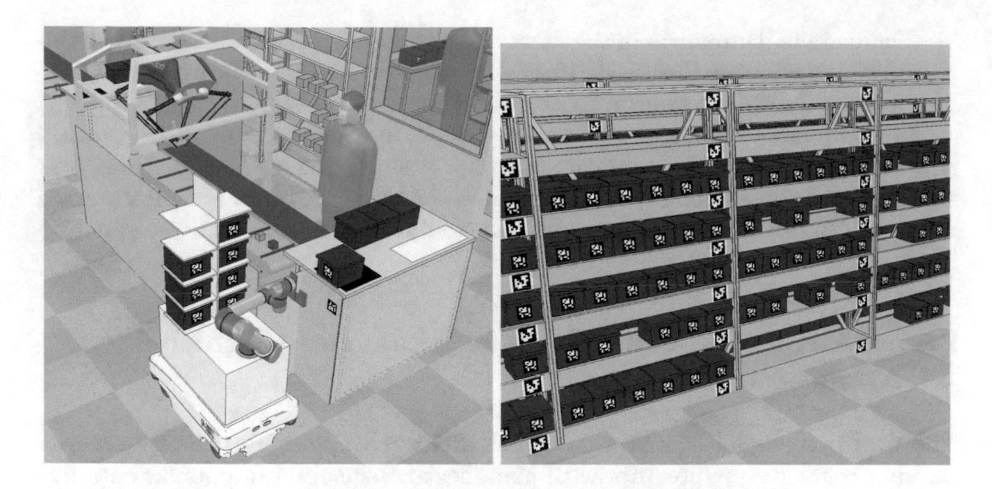

Fig. 1. Example of processing cell and warehouse

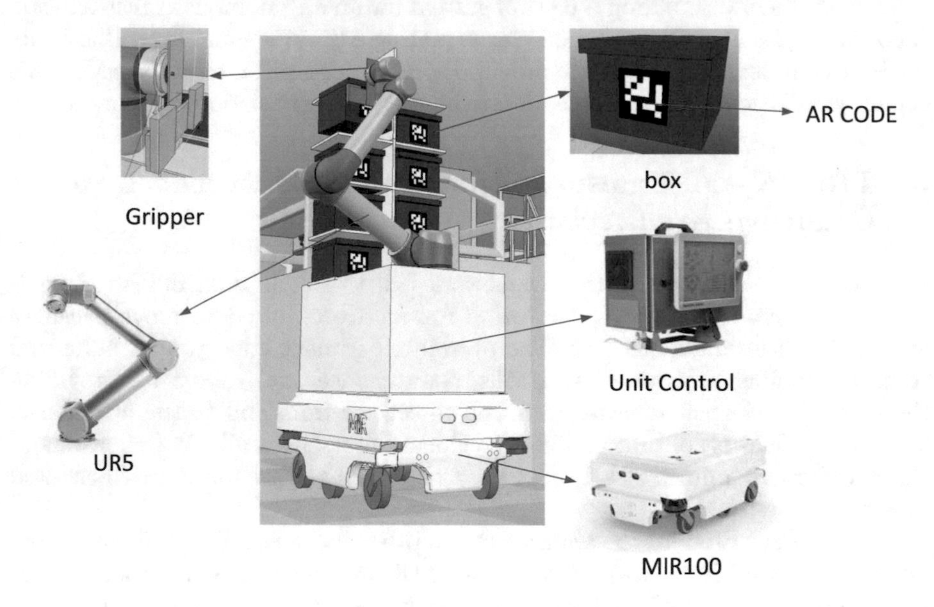

Fig. 2. Detail of the robot

- Environment Exploration
  - Generation/update of maps, including human beings, obstacles, pick-up/delivery slots and packages (metric, topological and semantic SLAM).
- Warehouse Inventory
  - Identification of slots in warehouse shelves
  - Situation/occupation of warehouse shelves
- Energy Management
  - Verify energy (battery) level, refueling when it is necessary

- Dealing with Human Operators
  - Humans cross the same environment as the robots, and pick packages from the warehouse - safety considerations should be respected.
  - Communication with human operators in order to solve problems (using smartphone application).
  - Visual robot-human interface in order to express information for humans without a smartphone interface.
- Dealing with other robots
  - Robots should be able to communicate with each other in order to solve deadlocks, e.g. dispute for passageway, and perform local coordination of activities.
- Self-Monitoring and Benchmarking
  - Different means of benchmarking (min/average/max time to attend requests, distance-to-identification of AR-tags, etc).

3.1 TRoCA Agent Cognitive Model

We designed a cognitive agent to control the transportation robot, assigning to it multiple responsibilities, which might be attended in an autonomous way. This agent was constructed using the MECA Cognitive Architecture. MECA is not a framework, like SOAR or LIDA, but requires customization, more in line with the ways-of-working of Clarion. A customization of MECA, designed to control our agent is shown in Fig. 3. Following the MECA philosophy, the TRoCA agent is composed by two sub-systems: System 1 and System 2.

A complete description of the many components of the TRoCA agent is beyond the space available in the current work and will be provided in a future publication. But we would like to bring to the reader's attention some important components used in the TRoCA architecture. First of all, there are the many motivational codelets in light green, which provides the many drives competing to enforce different ways of conduct, according to the many different behaviors the agent is able to pursue. The behavior codelets in dark blue, can be split into two different behavioral modes. The first 4 behavior codelets on the top are multi-step behaviors, requiring a *plan* to pursue their conclusion. This plan is a sequence of more basic behaviors, which affect directly the motor codelets, by means of the motor memory. The plans for the multi-step behaviors are stored internally in the multi-step behavior codelets, but they are planned by System 2, using the following approach. According to its input drive, if the internal plan of a multi-step behavior codelet is considered old or obsolete, it posts a new plan request in its output. Plan requests from multiple behavior codelets are directed to System 2, which is responsible for planning and sending the new plans to the multi-step behavior codelet using the Global Workspace. Each multi-step behavior codelet is able to identify if a plan coming from the global inputs are directed to them and storing it in itself if it is the case. Each of the 4 multi-step behaviors sends its output to the *Plan* container. A *container* is a special component in the CST architecture (which is used by MECA), that promotes the subsumption of different sources, where only one of them is selected, based on the strength given

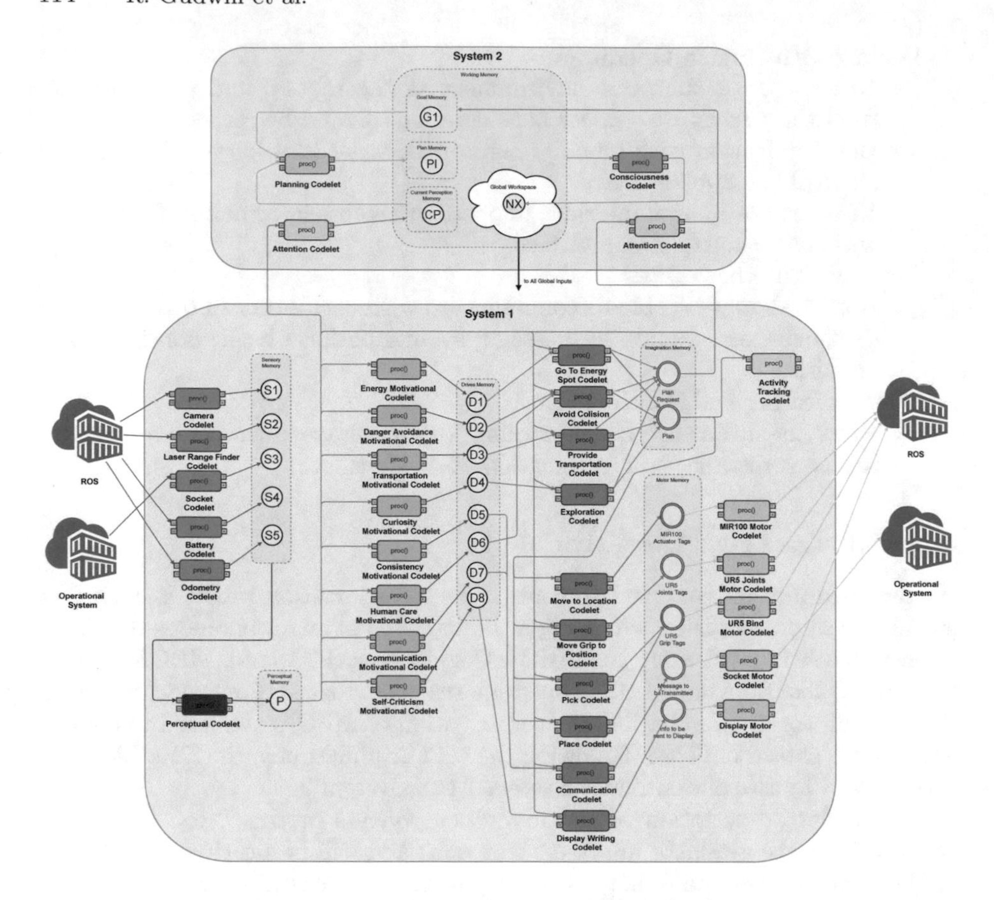

Fig. 3. Cognitive model of the TRoCA agent

by its proposer. The selected *Plan* is then sent to the other ordinary behaviors, which depending on the current step identified in the plan, executes its behavior, affecting the Motor Memory. The contents of Motor Memory is then used by the Motor Codelets to affect the environment. The interesting outcome of this scheme is that the architecture is able to autonomously select among the multiple responsibilities assigned to the transportation robot, according to the situation. The multiple motivations compete with each other and, based on the situation, the motivation which is most critical will gain access to actuators. System 1 is responsible for promoting this competition and executing the plan of actions required for satisfying one of the agent's needs. System 2 is responsible for creating the plans necessary for the multiple needs, and providing System 1 with these plans. System 1 and System 2 interact in a cooperative way, such that the final control actuation is an offspring of this interaction.

4 Conclusion

In this paper, we presented the development of a transportation agent, using the MECA Cognitive Architecture as a background, responsible for multiple assignments in a factory floor. The experiment was constructed using the V-REP robotic simulator, where a factory scenario was built, using ROS as a background infrastructure for controlling the robots. The TRoCA cognitive agent was built using the MECA Cognitive Architecture in Java language, and the CST Cognitive Systems Toolkit. This is still a work in progress, though. We are running the first experiments and tuning the many parameters necessary for the architecture to work as desired, and expect to be publishing the final results of experiments in a future publication.

Acknowledgments. The authors thank Ericsson Research Brazil, Ericsson Telecomunicações S.A. Brazil (Proc. FUNCAMP 4881.7) and CEPID/BRAINN (Proc. FAPESP 2013/07559-3) for supporting this research.

References

1. Adinandra, S., Caarls, J., Kostić, D., Verriet, J., Nijmeijer, H.: Flexible transportation in warehouses. Automation in warehouse development, pp. 191–207. Springer, London (2012)
2. Avery, E, Kelley, T, Davani, D (2006) Using cognitive architectures to improve robot control: integrating production systems, semantic networks, and subsymbolic processing. In: 15th annual conference on behavioral representation in modeling and simulation (BRIMS)
3. Benjamin, DP, Lyons, DM, Lonsdale, DW: Adapt: a cognitive architecture for robotics. In: ICCM, pp. 337–338 (2004)
4. Burghart, C, Mikut, R, Stiefelhagen, R, Asfour, T, Holzapfel, H, Steinhaus, P, Dillmann, R (2005) A cognitive architecture for a humanoid robot: a first approach. In: 5th IEEE-RAS international conference on humanoid robots, 2005, IEEE, pp. 357–362
5. Franklin, S., Madl, T., D'mello, S., Snaider, J.: Lida: a systems-level architecture for cognition, emotion, and learning. IEEE Trans Auton Ment Dev **6**(1), 19–41 (2014)
6. Gudwin, R., Paraense, A., de Paula, S.M., Fróes, E., Gibaut, W., Castro, E., Figueiredo, V., Raizer, K.: The multipurpose enhanced cognitive architecture (meca). Biol Inspir Cogn Arc **22**, 20–34 (2017)
7. Gudwin, R., Paraense, A., de Paula, S.M., Fróes, E., Gibaut, W., Castro, E., Figueiredo, V., Raizer, K.: An urban traffic controller using the meca cognitive architecture. Biol Inspir Cogn Arc **26**, 41–54 (2018)
8. Kelley, T.D.: Developing a psychologically inspired cognitive architecture for robotic control: The symbolic and subsymbolic robotic intelligence control system (ss-rics). Int J Adv Rob Syst **3**(3), 32 (2006)
9. Laird, J.E.: The soar cognitive architecture. MIT Press, Cambridge (2012)
10. Lemaignan, S, Ros, R, Mösenlechner, L, Alami, R, Beetz, M (2010) Oro, a knowledge management platform for cognitive architectures in robotics. In: 2010 IEEE/RSJ international conference on intelligent robots and systems, IEEE, pp. 3548–3553

11. Paraense, A.L.O., Raizer, K., de Paula, S.M., Rohmer, E., Gudwin, R.R.: The cognitive systems toolkit and the cst reference cognitive architecture. Biol Inspir Cogn Arc **17**, 32–48 (2016)
12. Raizer, K., Paraense, A.L.O., Gudwin, R.R.: A cognitive architecture with incremental levels of machine consciousness inspired by cognitive neuroscience. Int J Mach Conscious **04**(02), 335–352 (2012). http://www.worldscientific.com/doi/abs/10.1142/S1793843012400197
13. Rohmer, E, Singh, SP, Freese, M (2013) V-rep: a versatile and scalable robot simulation framework. In: 2013 IEEE/RSJ international conference on intelligent robots and systems (IROS), IEEE, pp. 1321–1326
14. Sun, R (2003) A tutorial on clarion 5.0, Unpublished manuscript
15. Thórisson, K., Helgasson, H.: Cognitive architectures and autonomy: a comparative review. J Artif Gen Intell **3**(2), 1–30 (2012)
16. Trafton, J.G., Hiatt, L.M., Harrison, A.M., Tamborello II, F.P., Khemlani, S.S., Schultz, A.C.: Act-r/e: an embodied cognitive architecture for human-robot interaction. J Hum-Robot Interact **2**(1), 30–55 (2013)
17. Ziemke, T., Lowe, R.: On the role of emotion in embodied cognitive architectures: from organisms to robots. Cogn computat **1**(1), 104–117 (2009)

Reconfigurable Locomotion of Hexapod Robot Based on Inverse Kinematics

Vadim A. Gumeniuk(✉), Eugene V. Chepin, Timofei I. Voznenko, and Alexander A. Gridnev

Institute of Cyber Intelligence Systems, National Research Nuclear University MEPhI, Moscow, Russia
gaiequilibrium@gmail.com

Abstract. In our days, robotics development grow and became fast changing industry, robots spread over the world far more than any time before. Most of all, robots have presented by manufacturing robots, but robots capable to move can be more flexible to give a solution to even new kinds of problems. Hexapods are one of those kind of robots. Today they are widely known, but they are not widespread, despite their advantages, and most frequently using in research purposes. One of the main problem is that locomotion can be done by many different gaits. At the same time, hexapods have six legs, that leads to complexity of control algorithm, which must provide correct positioning for all legs at any moment in time. But to simplicity, often only one specific gait is using. In this paper, we propose system that is able to work with multiple gaits simultaneously. This system allows robot to use different methods of locomotion which are more efficient in specific situations. As a proof of concept was implemented control software. It respond for locomotion, saving different gaits and their switching, even in movement. The result of the paper is an automated robot locomotion system.

Keywords: Hexapod · Gait · Locomotion · Inverse kinematics · Robotics

1 Introduction

There are many different hexapod robots that differ in both design and software, which are responsible for movement methods. A large part of papers is aimed at research of certain structures, which are extremely specific, and intending to solve a specific problem. However, part of research is carried out with a simplest designs of robots, and aiming to finding a more efficient way to move.

It is possible to improve robot's locomotion by using different methods. The main ones are various algorithms for constructing a path, and improving in the movement of robot itself, i.e. work with gait and improvement of the robot itself.

A. V. Samsonovich (Ed.): BICA 2019, AISC 948, pp. 117–123, 2020.
https://doi.org/10.1007/978-3-030-25719-4_16

So there are three base gaits:

- Tripod,
- Tetrapod,
- Wave.

The most commonly used gait, is a tripod. Using that gait, hexapod moves its legs in groups of 3 at a time. This gait is the most advantageous in terms of energy consumption at the highest speed. Using tetrapod gait, robot moves legs in groups of 2, while on the ground robot usually stands on 4. Using wave-like gait robot moves legs one after another. At the same time this gait is most energy efficient when robot has been moving at low speeds.

2 Related Works

In a lot of papers, researchers try to improve locomotion algorithms. For example, in these papers, they show the ways of finding and avoiding obstacles [1], holes [2], passage through small obstacles [3,4], control of a robot using neural networks and using sensors in it [5]. Most of them consider movement based on one particular gait. At the same time, there are implementations, which allow to neural networks use different gaits [6,7].

But, as some studies show, using of one gait is not always effective, and under different environmental conditions, it is, more advantageous, using different methods of movement.

In a study aimed at calculating energy consumption of a robot, by modeling, it was shown that when you change speed of movement, it also changes which gait consume less energy [7]. A study was also conducted on the actual energy consumption of a robot for various surfaces [8].

Surface is another important factor that influencing to robot locomotion. In that article [9], the authors showed that a tripod walk is more beneficial for some surfaces, and a tetrapod gait is more advantageous for others.

In addition, it is worth noting that different robots in different ways store and reproduce their gait. So a variant is possible when a robot essentially does not keep its gait, but only preserving the coefficients by which input signal is multiplying, which allows using a harmonic oscillations to move a legs synchronously. Such systems are often in one form or another using in neural networks. Also, with proper synchronization, it is possible to achieving a difference in the ways of moving [10]. These methods allow you to quite flexibly adjust the movement, but this setting is complex and not intuitive.

Another possibility is to set the required values directly. Although it is quite simple, but often such gaits are extremely inefficient, in addition. Also as methods based on neural networks, with different robot configurations, different settings are requiring.

Another frequently used method is a different variations of inverse kinematics. To do this, it is using a series of equations that automatically calculating to which position you are need moving your leg, at the current time [11,12]. Its main difficulty is writing correct equations that will translate the desired parameters,

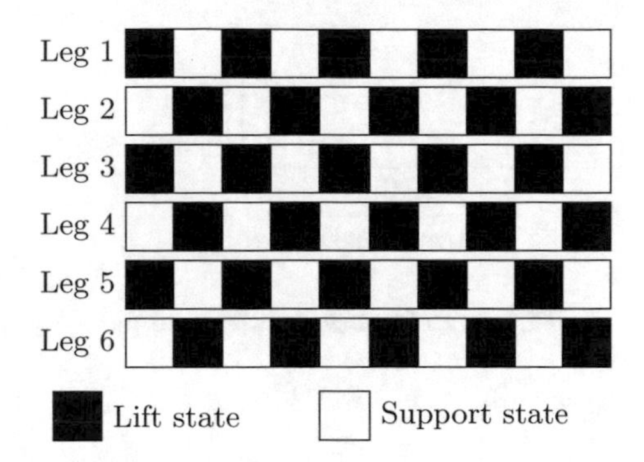

Fig. 1. Timing diagram of legs movement for tripod gait

such as the speed of movement, and the height, at which a robot is locating, to a directly defined foot position.

Thus, the purpose of this paper is creating a system for moving a robot, simple enough for a user, but at the same time quite flexible and capable to switch gaits. Automatic gait change will also be supported depending on a speed of movement.

3 Gait Generation

For calculate locomotion of the robot, inverse kinematics is using. Thus, to generate a gait itself, you will need a following data, the parameters of how the robot should move and a template responsible for the way of moving.

Thus, problem is to determine the necessary parameters, sufficient for flexible settings, and how a gait itself will bc stored, a pattern that can be changed when the robot is moving.

As parameters, the main ones will be the speed with which it is necessary to walk, the height at which the robot body must be held, the length of one step and the height of a legs elevation.

To store a pattern, it is necessary to store data, in what position, at which time, where a leg should be. Using of clearly defined coordinates does not make sense due to the fact that it will make it impossible to change a gait in a process of moving, or make it difficult. A possible option, often used to set gaits, would be to use a simple indication of whether a leg is currently moving in the air or on the ground, taking into account the time (Fig. 1).

However, we suggest to set relative reference points to which a leg will move. These points are located relative to the base of a leg of the robot and are in a plane of the current direction of movement as follows (Fig. 2):

0 – starting point;
1 – leg installation point, moved half a step forward;
2 – the point set aside from 0 half step back;
3 – point above the point of the next leg setting.

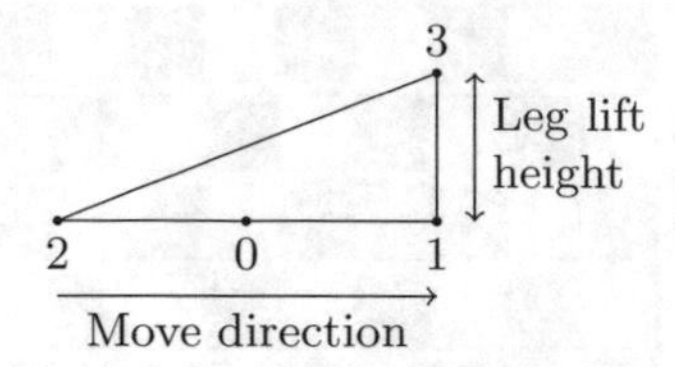

Fig. 2. Reference points example

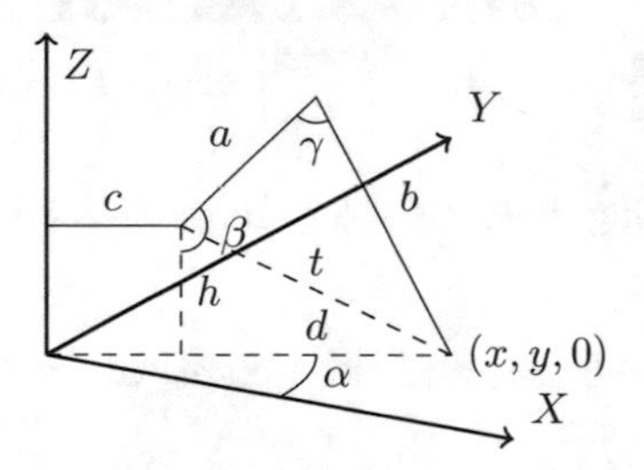

Fig. 3. Robot's leg

Thus, to move we set the general parameters specified above. The pattern is essentially the displacement of legs in the movement path over time relative to the total movement time. Thus, the movement is carried out with the passage of time.

Next, obtained points of required location of legs at a specific time on the trajectory of movement, defined by reference points, are converted to the coordinates of legs relative to the robot. Using methods of inverse kinematics, knowing the characteristics of the robot, we obtain positions into which it is necessary to transfer legs (Fig. 3).

The angle for the first joint, we obtain with the following formula:

$$\alpha = \tan^{-1}\left(\frac{y}{x}\right) \tag{1}$$

To find remaining angles, we will use the cosine theorem. Previously it is necessary, according to the Pythagorean theorem, to find the distance from the shoulder to the intended point of the leg installation:

$$t = \sqrt{h^2 + d^2} \tag{2}$$

where, h – is the height of the robot from ground, and d – is the distance to which the leg is carried in the plane, taking into account the removal of the leg from the shoulder. Next, we get that for the 2nd joint, the formula for finding the angle:

$$\beta = \cos^{-1}\left(\frac{a^2 + t^2 - b^2}{2a \times t}\right) + \arcsin\left(\frac{d}{t}\right) \tag{3}$$

where, a – is the length of the femur of leg, b – is the length of the tibia of leg, and t – is the length obtained earlier. And for the 3rd joint:

Algorithm 1. Time to position transformation

```
1  Triangle pattern initialization
2  loop
3      Get input
4      if (Time + offset) ∈ [0, 4/6] then
5          Get position at line 1-2
6      else if (Time + offset) ∈ [4/6, 5/6] then
7          Get position at line 2-3
8      else
9          Get position at line 3-1
10     end
11 end
```

Algorithm 2. Base locomotion

```
1  Initialization
2  Start move
3  while not in target position do
4      if Gait change? then
5          Stop moving
6          Gait change
7          Start move
8      else if Near to target position? then
9          Stop moving
10         break
11     else
12         Make step
13     end
14 end
```

$$\gamma = \cos^{-1}\left(\frac{a^2 + b^2 - t^2}{2ab}\right) \quad (4)$$

We suppose, that the most advantageous moment for switching a gait will be transition from one step to another, i.e. end of the current move cycle.

4 Implementation

The most important is the algorithm for converting the current step time to the desired position of a leg on a pattern. First, we will take the complete passage of a leg in a predetermined pattern (Fig. 2), as a time equal to 1. Next, for the above option, we will assume that section 1–2 is in the time interval [0..4/6), section 2–3 in the interval [4/6..5/6) and section 3–1 in the interval [4/6..1). Thus, we can see that to displace legs relative to each other on this template, it is only necessary to set their offset in time. Thus, the algorithm for moving legs is as follows (Algorithm 1):

The locomotion algorithm itself will have the following form (Algorithm 2):

In addition, the implementation noted the need to specify beginning and ending movement algorithms for each gait, which was set as a simple set of transitions from state to state for all legs.

5 Experiment Result

Tests in practice have shown that algorithms have executing correctly, but also requiring for further work. In addition, it was obtained that it would be best if a template is divided in time in the following way: 2/3 on the ground and 1/3 in the air, which will allow to fully implement gaits previously mentioned.

In addition, it is worth noting that due to the lack of feedback from servos during movement, deviations are observed at the indicated distances, so the shorter a step length, the greater a deviation relative to a step length.

6 Conclusion

This paper examined problems of moving robots with a gait switching. As a result, an algorithm for moving a robot with the ability to change a gait, based on inverse kinematics, was proposed.

To further verify the effectiveness, it is possible to use a robot to move on different surfaces, as well at different slope. For this, it is possible to use instruments that show a consumption of electricity by a robot at one time or another.

We can also see that further modification of algorithms is possible, in order to support operation with various sensors, which will improve a movement of a robot.

References

1. Yu WS, Huang CW (2014) Visual servo control of the hexapod robot with obstacle avoidance. In: 2014 IEEE international conference on fuzzy systems (FUZZ-IEEE), IEEE, pp. 713–720. https://doi.org/10.1109/FUZZ-IEEE.2014.6891708. http://ieeexplore.ieee.org/lpdocs/epic03/wrapper.htm?arnumber=6891708
2. Yang J-M, Kim J-H (2000) A fault tolerant gait for a hexapod robot over uneven terrain. IEEE Trans Syst Man Cybernetics Part B (Cybernetics) 30(1):172–180. https://doi.org/10.1109/3477.826957 http://ieeexplore.ieee.org/document/826957/
3. Liu, T, Chen, W, Wang, J, Wu, X (2014) Terrain analysis and locomotion control of a hexapod robot on uneven terrain. In: 2014 9th IEEE conference on industrial electronics and applications, IEEE, pp. 1959–1964. https://doi.org/10.1109/ICIEA.2014.6931489. http://ieeexplore.ieee.org/document/6931489/
4. Belter, D, Łabecki, P, Skrzypczynski, P (2010) Map-based adaptive foothold planning for unstructured terrain walking. In: 2010 IEEE international conference on robotics and automation, IEEE, pp. 5256–5261. https://doi.org/10.1109/ROBOT.2010.5509670. http://ieeexplore.ieee.org/document/5509670/

5. Manoonpong P, Pasemann F, Wörgötter F (2008) Sensor-driven neural control for omnidirectional locomotion and versatile reactive behaviors of walking machines. Robotics and Autonomous Syst 56(3):265–288. https://doi.org/10.1016/j.robot.2007.07.004 https://linkinghub.elsevier.com/retrieve/pii/S0921889007000954
6. Campos, R, Matos, V, Santos, C (2010) Hexapod locomotion: a nonlinear dynamical systems approach. In: IECON 2010 - 36th annual conference on IEEE industrial electronics society. IEEE, pp. 1546–1551. https://doi.org/10.1109/IECON.2010.5675454. http://ieeexplore.ieee.org/document/5675454/
7. Nishi, J (1998) Gait pattern and energetic cost in hexapods. In: proceedings of the 20th annual international conference of the IEEE engineering in medicine and biology society. vol. 20 Biomedical engineering towards the year 2000 and beyond (Cat. No.98CH36286), vol. 5, IEEE, pp. 2430–2433. https://doi.org/10.1109/IEMBS.1998.744922. http://ieeexplore.ieee.org/document/744922/
8. Kottege, N, Parkinson, C, Moghadam, P, Elfes, A, Singh, SPN (2015) Energetics-informed hexapod gait transitions across terrains. In: 2015 IEEE international conference on robotics and automation (ICRA), IEEE, pp. 5140–5147. https://doi.org/10.1109/ICRA.2015.7139915. http://ieeexplore.ieee.org/document/7139915/
9. Zenker, S, Aksoy, EE, Goldschmidt, D, Worgotter, F, Manoonpong, P (2013) Visual terrain classification for selecting energy efficient gaits of a hexapod robot. In: 2013 IEEE/ASME International Conference on Advanced Intelligent Mechatronics, pp. 577–584. IEEE. https://doi.org/10.1109/AIM.2013.6584154. http://ieeexplore.ieee.org/document/6584154/
10. Ren, G, Chen, W, Kolodziejski, C, Worgotter, F, Dasgupta, S, Manoonpong, P (2012) Multiple chaotic central pattern generators for locomotion generation and leg damage compensation in a hexapod robot. In: 2012 IEEE/RSJ international conference on intelligent robots and systems, IEEE, pp. 2756–2761. https://doi.org/10.1109/IROS.2012.6385573. http://ieeexplore.ieee.org/document/6385573/
11. Priandana, K, Buono, A, Wulandari (2017) Hexapod leg coordination using simple geometrical tripod-gait and inverse kinematics approach. In: 2017 international conference on advanced computer science and information systems (ICACSIS), IEEE, pp. 35–40. https://doi.org/10.1109/ICACSIS.2017.8355009. https://ieeexplore.ieee.org/document/8355009/
12. Soto-Guerrero, D, Torres, JGR (2017) K3P: a walking gait generator algorithm. In: 2017 14th international conference on electrical engineering, computing science and automatic control (CCE), IEEE, pp. 1–6. https://doi.org/10.1109/ICEEE.2017.8108874. http://ieeexplore.ieee.org/document/8108874/

Development of Text Data Processing Pipeline for Scientific Systems

Anna I. Guseva(✉), Igor A. Kuznetsov, Pyotr V. Bochkaryov, Stanislav A. Filippov, and Vasiliy S. Kireev

National Research Nuclear University MEPhI (Moscow Engineering Physics Institute), Kashirskoe shosse, 31, 115549 Moscow, Russia
{aiguseva, iakuznetsov, pvbochkarev, vskireev}@mephi.ru, stanislav@philippov.ru

Abstract. The aim of this work was to develop pipeline processing of scientific texts, including articles and abstracts, for their further categorization, identify patterns and build recommendations to users of scientific systems. The authors proposed a number of methods of pre-processing of texts, the method of cluster and classification analysis of texts, developed a software system of recommendations to users of scientific publications. To solve the problem of data preprocessing it is proposed to use parametrical approach to retrieve new – semantic – feature from textual publications – the type of scientific result. Scientific result type extraction is built just based on user's need for content having specific property. To solve the problem of users' profile clustering it is proposed to use ensemble method with distance metric change. For classification, ensemble method based on entropy is used. Evaluation of proposed methods and algorithms employment efficiency was carried out as applied to operation of search module of "Technologies in Education" International Congress of Conferences information system. Author acknowledges support from the MEPhI Academic Excellence Project (Contract No. 02.a03.21.0005).

Keywords: Text mining · Natural language processing · Machine learning

1 Introduction

This study is aimed to solve scientific and technology problem of creation, application and investigation of machine text data processing pipeline methods and algorithms efficiency in scientific recommendation systems.

Because of dramatic increase in data amount and necessity to perform search in them the problem of information relevancy and pertinence determination comes to the foreground. The concept of "relevancy" is defined as "conformity between obtained information and information request". Thus, the relevancy is determined exclusively by mathematical models being used in specific data retrieval system. The concept of "pertinence" is understood as conformity between obtained information and information need, i.e., the pertinence is compliance of documents found by data retrieval system with information needs of user, irrespective to how full and unambiguously this

A. V. Samsonovich (Ed.): BICA 2019, AISC 948, pp. 124–136, 2020.
https://doi.org/10.1007/978-3-030-25719-4_17

need is expressed in the form of request. In many occasions, the "pertinence" is used as synonym for "user's satisfaction" [1].

One of the approaches for increasing the pertinence of information search consists in employment of machine learning methods as special case of data mining application. Special subset of information systems is distinguished which subset is based on mathematical models and allows to solve data relevancy and pertinence determination problems. This subset is known as recommendation systems [2]. Absence of universal approaches for solving preprocessing, clustering and classification problems in recommendation systems makes the development of machine learning methods and algorithms especially actual in this case.

2 ProblemState

The main purpose of recommendation systems consists in forecasting user behavior in relation to the object of information search, as well as in generating subsequent recommendation of similar information object not being encountered previously [2].

Formal statement of the problem for recommendations looks like this. Let us consider U – an array of users and D – an array of objects. It is necessary to find function r,

r: U × D → R which forms recommendation R so that for any user value of r between the user and d object is maximal, i.e., it is argument of maximization:

$$\forall u \in U, \quad d_u = arg\left(\max_{d \in D} r(u, d)\right). \tag{1}$$

Four types of recommendation systems are distinguished [2]:

- Recommendations generated using expert method – links between objects are established manually or based on predetermined rules, but this method is actual when the list of recommended objects is not large only.
- Collaborative filtering – recommendations are based on user evaluations in relation to reviewed objects. Recommendations can be built either based on search for similar users in relation to the user under consideration (user-based) or based on search for similar objects in relation to the objects that have been previously evaluated by user (item-based).
- Content filtering – recommendations for the user under consideration are formed based on the objects that he (she) has liked, with regard to the set of parameters assigned to each object.
- Hybrid filtering – combination of approaches is used which approaches are based on content and collaborative filtering which leads to higher quality of recommendations being generated.

Recommendation systems have several serious disadvantages among which we can emphasize complexity of scaling, "cold start", new user, context identification for user, small scale for setting ratings, determination of negative preferences and temporal characteristics. Creation of recommendation systems based on content filtering using auxiliary data on user and recommendation object helps to eliminate problem of "cold start", new user, context identification for user and the problem of temporal characteristics. Thus, the most severe problem for recommendation system is building personified recommendations [3]. Recommendation systems use knowledge about user, user's location, previous actions, preferences, social binds, etc. and provide input data for many applications.

The objective of each recommendation system functioning is to generate recommendations for information objects not previously known for user but being useful or interesting in current context. Notwithstanding that in some cases cognitive methods can be used to determine user need [4], statistical methods are used more widely [5].

General scheme of text data processing pipeline contains four steps: data preparation and preprocessing, clustering, classification, generating recommendations. Input data for reparation and preprocessing can be represented by personal data from user profile, as well as logs of his (her) actions being executed by him (her) when working with the system. Recommendations are issued based on user's membership in some group of users having similar interests. Determination of such classes will help to form user's need more precisely, i.e., to increase pertinence. This operation is called clustering. Ensemble clustering methods have got the greatest popularity [6–8].

New user assignment to the most pertinent class is executed using classification operation. In this case, the most successful solutions also employ ensemble classification methods [9–11]. As a result, being aware of the class to which user is assigned, as well of the scope of interests of other participants with similar interests, recommendation is given to the user.

Recommendation system applicability is not limited to some specific scopes of activities, but can be used in multiple different areas, in particular, for systems containing scientific information. Scientific recommendation system (SRS) represents special module which can be installed on top of scientific information system database and used independently of search system. The examples of such systems are Google Scholar, Scopus, Web of Science, Mendeley, eLibrary, Cyberleninka and others. Existing systems do not issue individualized recommendations intended for particular user.

3 Proposed Approach

3.1 Text Preprocessing

The main task of scientific recommendation system developed by us is to predict user's information need where user profile itself, user's behavior on web site and user's search requests can serve as the source of data. Based on accumulated data, individual set of recommendations is formed which set should resemble current need of particular user [12–14].

In the first step of this work, parametric approach is proposed for data enrichment in order to retrieve new semantic feature form textual publications – the type of scientific result. Scientific result type retrieval is built just based on user's need in content having particular property.

Parametric approach has been based on the hypothesis that type of scientific result, on the grounds of user's need, can belong to one of four classes: "basics", "hypotheses", "experiment" and "practice". Proposed approach allows to identify type of scientific result based on textual data from scientific papers and helps to determine similarity both between information objects and between users whose needs can be expressed using obtained feature.

Step 1. Deriving meaningful word combinations from a text
The main task of this step consists in generation of word combinations based on meaningful words by virtue of proving that words in the text are encountered non-randomly. Additional limitation for generation of word combinations consists in mandatory presence of participle in full or short form among words.

Zero hypothesis (H0) supposes that two adjacent words in a text do not represent word combination, i.e., they appear together randomly; alternative hypothesis (H1) supposes the contrary (Formula 2):

$$\begin{aligned} H_0 &: P(w_1, w_2) = P(w_1)P(w_2), \\ H_1 &: P(w_1, w_2) \neq P(w_1)P(w_2), \end{aligned} \tag{2}$$

where p(w1), p(w2) is probability of w1 and w2 words occurrence in textual document.

To test statistical hypotheses Student's t-test (Formula 3) and Chi-square(d)test (Formula 4) were used.

$$t = \frac{|\bar{x} - m|}{\sqrt{\frac{s^2}{N}}}, \tag{3}$$

where $\bar{x}$ is sampling mean, m – probabilistic mean, s2– variance, N – size of nonempty sample.

$$t = \frac{|\bar{x} - m|}{\sqrt{\frac{s^2}{N}}}, \tag{4}$$

where Oi,j is actual number of word combinations, Ei,j – expected number of word combinations. Obtained values for each word combination allow to accept or reject zero hypothesis by comparing with confidence level which is $\alpha = 0.005$. Any zero hypothesis is rejected when obtained test value exceeds value in the table of critical values for corresponding test (criterion).

Step 2. Execution of fuzzy clustering
For each textual document, multiple meaningful word combinations included in this document are generated. Set of textual documents can be divided into several thematical classes using latent Dirichlet allocation algorithm.

This algorithm will make possible to execute fuzzy data separation by prescribed thematical classes and determine probability of textual document assignment to one of them. Latent Dirichlet allocation algorithm is based on probabilistic model (Formula 5).

$$p(d,w) = \sum_{t \in T} p(d) \cdot p(w|t) \cdot p(t|d), \tag{5}$$

where d is document, w is word, t is thematical class, p(d) is prior distribution on multiple documents, p(w|t) is conditional distribution of word w in thematical class t, p (t|d) is conditional distribution of thematical class t in document d.

One of internal methods of obtained results evaluation consists in coherence (conformity) evaluation which can be calculated via pointwise mutual information for the most common words for each selected thematical class:

$$PMI\left(w_i, w_j\right) = \log \frac{P\left(w_i, w_j\right)}{P(w_i) \cdot P\left(w_j\right)}, \tag{6}$$

where wi and wj is a pair of words of prescribed thematical class in frequency decreasing order.

In general form, the coherence will be calculated as follows:

$$Coh = \frac{2}{N \cdot (N-1)} \sum_{i=1}^{N-1} \sum_{j=i+1}^{N} PMI\left(w_i, w_j\right), \quad N > 1, \tag{7}$$

where wi and wj is a pair of words of prescribed thematical class in frequency decreasing order, N is the number of the most common words in the thematical class.

Within the framework of preconceptual studies, the proposed method was put on trial to determine the type of scientific result based on scientific publication abstracts which contain constitutive essence of a paper. Scientific papers in the number of 5,000 pieces from "Informatics" rubric of VINITI database served as data source.

As a result of data preprocessing and filtering, as well as employing word combinations filtering according to confidence level for Student's t-test and Chi-square(d) test $\alpha = 0.005$, 344 unique word combinations have been obtained. Scientific paper abstracts have been described using word combinations where 2.5 meaningful word combinations are accounted on average for each abstract. Maximum quantity for one abstract was 15 meaningful word combinations. Thematical model quality has been evaluated using coherence calculation for various numbers of thematical classes starting from 2 and up to 9 with increment of 1. Interpretation of obtained results makes possible to say that thematical classes "2", "3" and "4" are diametrically opposite. Thematical class "1" has more similarity with thematical class "2" which can be explained by close coherence valuation for three and four thematical classes.

3.2 Clustering

In this paper, to solve the problem of users' profile clustering it is proposed to use ensemble approach with distance metric change. Five algorithms are selected and following metrics are used:

- Euclidean distance

$$p(x, x') = \sqrt{\sum_{i}^{n} (x, x')^2}; \tag{8}$$

- Manhattan distance

$$p(x, x') = \sum_{i}^{n} |x - x_i'|^2; \tag{9}$$

- Jacquard coefficient

$$K(x, x') = \frac{\sum_{i}^{n} x_i x_i'}{\sum_{i}^{n} x_i^2 + \sum_{i}^{n} x_i'^2 - \sum_{i}^{n} x_i x_i'} \tag{10}$$

Well-reputed algorithms have been selected:

- Two k-means algorithms with two different metrics.
- Two k-median algorithms: This algorithm is similar to k-means algorithm. Algorithm difference consists in that it is built not on cluster centres but on its median.
- K-mode algorithm: k-means algorithm modification working with categorial data.

Algorithm of clustering algorithms ensemble generation contains some steps.

In the first step, each K-means algorithm splits data into clusters using its distance metric. Then, accuracy and weight of algorithm opinion in the ensemble is calculated from the Formula 11:

$$\omega_l = \frac{A_{CC_l}}{\sum_{l=1}^{L} A_{CC_l}}, \tag{11}$$

where ACC_l is algorithm accuracy l, i.e., the ratio of the number of correctly clustered objects to total sample size, and L is the number of algorithms in the ensemble.

For each obtained partition, binary difference matrix of nxn size is composed, where n is number of objects, which matrix is necessary to determine if partition objects are entered in one class. Then, consistent difference matrix is calculated each element of which represents weighted sum of elements from preliminary matrices. Obtained matrix is used as input data for hierarchical agglomerative clustering algorithm. After this, the most suitable clustering solution can be selected using standard practices, such as agglomeration distance jump determination.

To obtain the best partition into clusters it is necessary, as mentioned above, to compose binary matrix of similarity/differences for each L partition in the ensemble:

$$H_i = \{h_i(i, j)\}, \tag{12}$$

where $h_i(i, j)$ is equal to zero if element i and element j have fallen into one cluster, and is equal to 1 if not.

The next step in composing ensemble of clustering algorithms consists in composing consistent matrix of binary partitions.

$$H^* = \{h^*(i,j)\}, \quad (13)$$

$$h^*(i,j) = \sum_{i=1}^{L} w_l h_l(i,j), \quad (14)$$

where w_l is weight of algorithm.

To form the best partition of consistent matrix the nearest neighbor algorithm has been chosen. To reduce dimensionality of input data the principal component method (Principal component analysis, PCA) has been chosen. Kaiser criterion has been chosen for constituent number selection (constituent eigenvalue is greater than one).

In the next step, accuracy of each algorithm has been identified by comparing obtained partitioning into two clusters by each algorithm with clusters that have been partitioned using expert method. After accuracy value for each algorithm has been obtained, algorithm opinion weight was calculated from Formula 14.

As preconceptual studies, approbation of proposed ensemble clustering method based on variable distance metrics has been conducted in 2017 on the basis of VINITI RAS scientific data. Five scientific areas with 1,000 documents in each have been selected. Each examined document includes paper abstract and code of scientific area. Thus, the data was of textual type. The study was conducted with documents of following scientific areas: Psychology; Informatics; Physics; Chemistry; Metrology. This method has been implemented in programming language used for statistical data processing, R. Results of obtained clustering, algorithm weights calculated in the process of composing ensemble are represented are represented in Table 1.

Table 1. Accuracy of examined algorithms.

Scientific area	Accuracy				
	K-means algorithm, Euclidean metric	K-means algorithm, Manhattan metric	K-median algorithm, Euclidean metric	K-median algorithm, Manhattan metric	K-mode algorithm, Jaccard metric
Psychology	81.1%	96.9%	87.7%	87.4%	78.9%
Informatics	49.9%	0%	86.9%	86.9%	65.7%
Physics	77.6%	1.5%	78.8%	78.2%	64.4%
Chemistry	24.8%	17.0%	17.1%	18.8%	56.4%
Metrology	38.7%	53.7%	71.7%	71.4%	64.0%
Mean accuracy	54.4%	33.8%	68.4%	68.5%	65.8%
Weight in ensemble	18.7%	11.6%	23.5%	23.6%	22.6%

The Table 2 shows increase in clustering accuracy when ensemble of algorithms with variable metrics is used. The approbation conducted shows that ensemble of clustering algorithms decreases inaccuracies of partitioning thus increasing mean accuracy of partitioning up to 70%.

Table 2. Clustering accuracy of algorithm ensemble

Item No.	Scientific area	Ensemble accuracy
1	Psychology	92.0%
2	Informatics	86.9%
3	Physics	68.9%
4	Chemistry	21.4%
5	Metrology	81.4%
Ensemble mean accuracy		70.1%

3.3 Classification

Please In the framework of developed method, similar approach is used, and "voting" process itself will be built based on employment of entropy as weighting measure. Practical meaning in the process of entropy employment in the context of classification algorithms is as follows: to determine how homogeneous the predicted values are in relation to actual values in data set.

Evaluation of machine learning algorithms operation quality will be executed using F-measure (Formula 15):

$$F = 2 \cdot \frac{\text{Accuracy} \cdot \text{Comprehensiveness}}{\text{Accuracy} + \text{Comprehensiveness}}, \quad (15)$$

where accuracy is fraction of objects called positive by classifier and in this case being really positive; comprehensiveness is fraction of positive class objects which has been found by the algorithm.

Proposed method for semistructured data classification using entropy consists of following steps.

Step 1. Data preprocessing and sampling

Feature space creation and original textual data (D) array conversion into digital form are made using text representation in the form of word matrix. In this case, the Bag of words is used as the most common form of word representation, but this does not restrict the possibility to use TF-IDF, n-grams, etc. Data preprocessing is executed by means of stop-words removal, punctuation characters removal, characters conversion to lower case and normalization of words. Further, all words are processed in normalized form.

Global data array partitioning into samples occurs in a random manner, but in predetermined proportions: 75% of data array is allocated for teaching sample (Dteach), 25% – for test sample (DT). Distribution of objects among classes in original data array is uniform. Additionally, it is allocated out of teaching sample (Dteach) in the ratio of 50:50 for sample for additional training (Dadd). Thus, original data array will look like D = {Dteach, Dadd, Dt}.

Step 2. Entropy calculation
Based on teaching sample (Dteach), basic set of algorithms A = {a1, a2, …, ab, …, an} learning takes place and trained models for each algorithm F = {fa1, fa2, …, fab, …, fan} are built. Based on obtained predictions of trained models of algorithms (F), inaccuracy matrices are built for each of algorithms (A) (Formula 16):

$$\begin{pmatrix} a_b x_{11} & a_b x_{12} & \dots & a_b x_{1C} \\ a_b x_{21} & a_b x_{22} & \dots & a_b x_{2C} \\ \dots & \dots & a_b x_{ij} & \dots \\ a_b x_{C1} & a_b x_{C2} & \dots & a_b x_{CC} \end{pmatrix}, \tag{16}$$

where abxij is number of objects belonging to i class, but classified by ab algorithm as j class; c is number of classes.

Based on presented inaccuracy matrix, entropy for each class to which column in the matrix corresponds is calculated (Formula 11). Formula for entropy calculation is as follows (Formula 17):

$$H(a_b x_{.j}) = -\sum_{i=1}^{c} p(a_b x_{ij}) log_2 p(a_b x_{ij}), \tag{17}$$

where x.j is j-th column with values of inaccuracy matrix; ab is classification algorithm; c is number of classes; p(abxij) is probability to encounter an element in j-th column.

Calculated entropy values for each class are presented in matrix (Formula 18).

$$\begin{pmatrix} ha_1 c_1 & ha_1 c_2 & \dots & ha_1 c_k \\ ha_2 c_1 & ha_2 c_2 & \dots & ha_2 c_k \\ \vdots & \vdots & ha_i c_j & \vdots \\ ha_n c_1 & ha_n c_2 & \dots & ha_n c_k \end{pmatrix}, \tag{18}$$

where an is list of algorithms being used, ck is list of classes, haicj is entropy for a algorithm of c class.

Step 3. Final result selection
Based on trained models of algorithms (F), calculation of class (C) predictions for each object in test sample is performed with certain probability (P). For each object, calculation of significance indicator (H’) is made to determine class assignment as follows (Formula 19).

$$H'(x) = \left(1 - 0,5 \cdot \frac{h(x_i) - h(x)_{min}}{h(x)_{max} - h(x)_{min}}\right) \cdot p(x_i) + p(x_i), \tag{19}$$

where xi is input object, p is probability in [0, 1] interval, h is entropy.

After new significance indicator for each class for each object has been obtained the selection of resulting class is made by means of maximum number of algorithms voting for specific class determination. If selection of winner by a majority vote has failed then the class with the greatest significance indicator becomes the winner.

Approbation of proposed ensemble classification method based on entropy has been executed based on scientific publications of VINITI database in the quantity of 5,000 objects. Three separate samples have been prepared containing 5, 10 and 15 scientific areas respectively. Algorithms with maximum different nature of origin and true class determination method have been used as fundamental algorithms for comparative analysis in presented study: RandomForest, Stochastic gradient descent, Naive Bayes classifier for multinomial models, Logistic Regression, Multi-layer Perceptron classifier, Bagging and AdaBoost fundamental ensembles. By applying entropy measure to presented algorithms we will obtain following accuracy values for 5, 10 and 15 classes (see Table 3) where accuracy figure has been calculated using F-measure metric.

Based on presented results we can speak about high stability of ensemble operation in case of 5 iterations (cross-validation). While fundamental algorithms have wide variation of F-measure figure, ensemble variation is significantly lower but is still related to quality of fundamental algorithms and correlates with them. So, it can be noted that F-measure increase as compared to fundamental algorithms reaches 4%–7% at average.

Comparative analysis of developed ensemble and Bagging and AdaBoost fundamental ensembles operation has shown that the developed ensemble allows to enhance significantly the quality of classification at average from 6%–7% to 14%–16% depending on number of classes.

Table 3. F-measure metric

Basic algorithms	For 5 classes		For 10 classes		For 15 classes	
	Min. score	Max. score	Min. score	Max. score	Min. score	Max. score
RandomForest	0,816	0,830	0,743	0,774	0,615	0,634
Stochastic gradient descent	0,826	0,865	0,790	0,838	0,668	0,706
Naive Bayes classifier for multinomial models	0,757	0,844	0,728	0,790	0,590	0,641
Logistic Regression	0,845	0,865	0,806	0,834	0,682	0,732
Multi-layer Perceptron classifier	0,853	0,873	0,805	0,839	0,671	0,713
Bagging	0,793	0,821	0,727	0,757	0,591	0,616
AdaBoost	0,734	0,790	0,708	0,746	0,530	0,579
Entropyensemble	0,869	0,889	0,826	0,855	0,716	0,748

3.4 Producing Recommendations

Please To derive informational offer the algorithm for generating tuple of informational offers to user has been developed. Final tuple generating scheme is presented in the diagram (see Fig. 1).

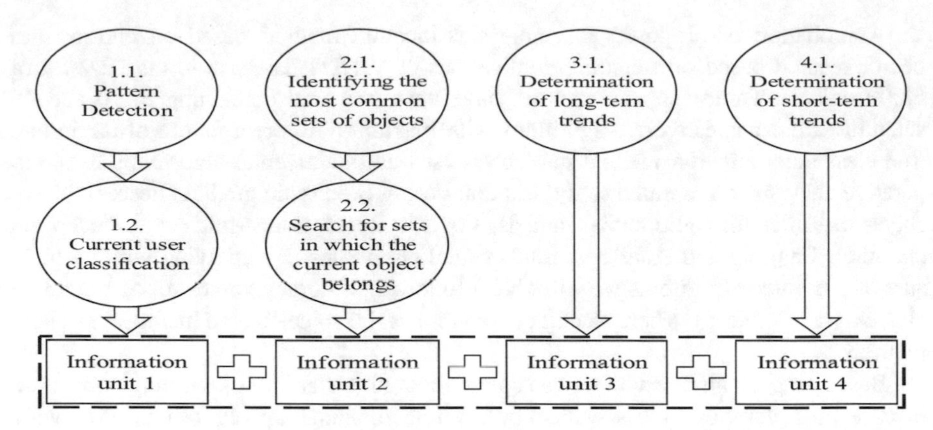

Fig. 1. Informational offers tuple generating scheme

The tuple consists of information units (IU) with total length of 7 ± 2 objects in order to take into account peculiarities of human perception and includes following parts:

- IUs that have been obtained for users similar to current user;
- IUs that have been derived from the most common sets;
- IUs relating to long-term trends ("tops");
- IUs relating to short-term trends, actual ("trends").

Items (1) and (2) relate to restorable information need and are supposed to be the most contributing to recommended set (about 60–70%). Items (3) and (4) relate to information need being generated and should take 30–40% of tuple volume.

4 Discussion of Findings

Evaluation of proposed methods and algorithms employment efficiency was carried out as applied to operation of search module of "Technologies in Education" International Congress of Conferences information system. The information system contains archive of materials starting from 1994 year. According to the data of Google and Yandex search systems, the number of indexed pages is equal 27,000 and 23,000 respectively. Average attendance of Internet portal is about 1,000–1,200 visitors per week. Within the framework of working with SRS, two key elements are distinguished: user and object (web page containing scientific paper or conference abstracts). Basic profile of SRS user can consist of two components: static and dynamic. These profile parts correspond to composed scientific activity ontology but their use in the framework of scientific information systems can be partial only. Static part includes following parameters: education, academic title, list of scientific papers, interests, etc. Dynamic part of profile is formed by means of user work on scientific information system page and executing any actions in relation to an object being viewed on the page. An object contains following parameters in basic profile: type, title, keywords, authors, date, etc.

User profile generation U = <Dsu, Tu> based onset on objects viewed by him (her) makes possible to extend the basic profile. User profile will additionally include Dsu = {c1, c2, … cn} parameter which represents a set of classes to which the objects viewed by user have been assigned and is called "Rubricator". Tu = (t1, t2, t3, t4) parameter determines weight of each scientific result type to which viewed objects have been assigned and is called "Type". Implicit user profile (exclusively of basic profile parameters) can be as follows: ui = <[4, 6], {0.49;0.31;0.12;0.08}>. Implicit object profile will be unique but, herewith, it also will be composed of "Rubricator" parameter and "Type" parameter. For example, it can be as follows: di = <[6], {0.72;0.21; 0.05;0.02}>.

To evaluate efficiency of described scenarios operation the experts had been involved who evaluated compliance of obtained list of recommendations and their needs.

The first scenario supposes use within the framework of SRS of classical Sphinx full-text search engine where title, abstract and keywords of scientific papers are used as search fields. This scenario is aimed to reconstruct maximally real conditions of Bitrix platform operation. The second scenario includes use of scientific result as additional "Type" parameter taken into account in the process of search output ranking. Similarity of user profile and objects of search output is implemented by virtue of Euclidean distance metric employment. Based on obtained values, ascending sorting is executed and search offer is generated. The third scenario includes use of scientific paper rubrics – the "Rubricator" parameter. A paper with rubric not presented in user profile is transferred in search output to end of list with retention of sequence. The fourth scenario supposes simultaneous use of additional parameters from two scenarios.

In the first scenario it was identified that based on standard search mechanism the level of user satisfaction without considering his (her) profile was equal to 40–60% on average which means that about half of offered documents met original needs. In the second scenario the level of user satisfaction rose on average by 10–20% which is connected with reduction of known uninteresting for user documents in search output.

In the third scenario the level of user satisfaction rose on average by 15–25% due to correction of offered documents orientation. In the fourth scenario, joint use of two algorithms allowed to rise the level of user satisfaction by 20–30% on average.

On the basis of the experiment conducted it may be concluded that employment of feedback from user, as well as ancillary information about user profile and implicit features of documents being viewed makes possible to enhance significantly the efficiency of resulting set of search offer and to improve user satisfaction.

5 Summary

The investigations carried out have shown by the example of searching module of "Technologies in Education" International Congress of Conferences information system the perceptiveness of further development of proposed methods and algorithms. In general, when these methods and algorithms were used the pertinence of given recommendations in scientific system increased by 20% on average, and for certain groups of users, such as graduate students and magisters it reaches 80%.

References

1. Guseva AI, Kireev VS, Bochkarev PV, Kuznetsov IA, Philippov SA (2017) Scientific and educational recommender systems. In: AIP conference proceedings of information technologies in education of the XXI century (ITE-XXI), vol 1797, pp 020002-1–020002-11
2. de Gemmis M, Lops P, Musto C, Narducci F, Semeraro G (2015) Semantics-aware content-based recommender systems. In: Ricci F, Rokach L, Shapira B (eds) Recommender systems handbook. Springer, Boston, pp 119–159
3. Landia N, Anand SS (2009) Personalised tag recommendation. In: Proceedings of the 2009 ACM conference on recommender systems, pp 83–36
4. Samsonovich AV, Kuznetsova K (2018) Semantic-map-based analysis of insight problem solving. Biologically Inspired Cogn Architectures 25:37–42
5. Amatriain X, Pujol JM (2015) Data Mining Methods for Recommender Systems. In: Ricci F, Rokach L, Shapira B (eds) Recommender systems handbook, pp 227–262
6. Berikov VB (2013) Collective of algorithms with weights for clustering heterogeneous data. Vestn Tom Gos Univ 2(23):22–31
7. Onan A, Bulut H, Korukoglu S (2016) An improved ant algorithm with LDA-based representation for text document clustering. J Inf Sci 43(2):275–292
8. Mathuna KT, Shanthi IE, Nandhini K (2015) Applying clustering techniques for efficient text mining in twitter data. Int J Data Min Tech Appl 4(2):25–28
9. Hady A, Farouk M (2011) Semi-supervised learning with committees: exploiting unlabeled data using ensemble learning algorithms. In: Open access Repositorium der Universität Ulm. Dissertation
10. Rokach L (2009) Ensemble-based classifiers. Springer Science + Business Media B.V
11. Faraway J (2016) Does data splitting improve prediction? Stat Comput 26(1):49–60
12. Guseva AI, Kireev VS, Bochkarev PV, Smirnov DS, Filippov SA (2016) The formation of user model in scientific recommender systems. Int Rev Manag Mark 6(6):214–220
13. Kuznetsov IA, Guseva AI (2019) A method for obtaining a type of scientific result from the text of an article abstract to improve the quality of recommender systems. In: Proceedings of the 2019 IEEE conference of russian young researchers in electrical and electronic engineering, ElConRus, vol 8656806, pp 1888–1891
14. Bochkaryov PV, Guseva AI: The use of clustering algorithms ensemble with variable distance metrics in solving problems of web mining. In: Proceedings - 2017 5th international conference on future internet of things and cloud workshops, W-FiCloud 2017, pp 41–46 (2017)

Environment of Modeling Methods for Indicating Objects Based on Displaced Concepts

Larisa Ismailova[1], Sergey Kosikov[2], Konstantin Zinchenko[2], and Viacheslav Wolfengagen[1(✉)]

[1] National Research Nuclear University "Moscow Engineering Physics Institute", Moscow 115409, Russian Federation
lyu.ismailova@gmail.com, jir.vew@gmail.com
[2] Institute for Contemporary Education "JurInfoR-MGU", Moscow 119435, Russian Federation
kosikov.s.v@gmail.com

Abstract. The paper considers the problem of constructing domain models based on a semantic network. One of the faced difficulties is to support the indication of objects that form the semantic network, while displaying the dynamics of the domain; this indication assumes the possibility of dynamic creating, modifying, and deleting objects. In this case the indication of objects is possible both by name and by position in the semantic network, as well as by the set of properties of the object. It shows that the set of indication methods required for modeling essentially coincides with the set required for controlling access to objects of the network; this makes it possible to consider the indication of objects and access control as two aspects of one task. The solution of the problem can be achieved based on using the technique of displaced concepts and description methods to indicate objects and access to them as specialization of concept bias. The paper generates requirements to the methods of objects indication and proposes an environment that provides modeling of various methods of indication based on the support of methods for modeling the shift of concepts in an applicative computing system. It describes a prototyped implementation of the environment by the method of extending an applicative system. The implementation was tested on the tasks of describing dynamic domains to support the introduction of the best available technologies.

Keywords: Shifted concepts · Semantic network · Dynamic domain

1 Introduction

Initial Assumptions. At a qualitative level, by a *concept* we mean abstraction based on the characteristics of perceived reality. In analytic activity, a concept is considered to be some label superimposed on a phenomenon, which forces us to

A. V. Samsonovich (Ed.): BICA 2019, AISC 948, pp. 137–148, 2020.
https://doi.org/10.1007/978-3-030-25719-4_18

associate individual observations and carry out a generalization. For convenience, it is identified with a name that is assigned to observations and events[1]. Some kinds of its properties are associated with each concept.

A *variable* is often understood as *measured concept.* This is a property that is associated with the concept and changes during measurement. If the property does not change, then it is considered a *constant.*

We presume that concepts should have a *change mechanism.*

Premises. The paper presents the initial section of the construction of a mathematical theory of computation, based on the systematic use of functions and their compositions. An understanding of the flow of information. From an intuitive point of view, in the information system there is an *input point* into which the input information "flows in", as well as *output point*, in which the result is formed. The main question is what happens inside the system. Since it is composed from functions, the mapping is either a *primitive operation* that is not subject to further analysis, or a mapping composed of other mappings. It is important that in the information system we can consider the *pieces of information*, the interpretation of which depends on the application.

Constructive Solutions. In university mathematical analysis, the (δ, ε)-language is used to operate with continuous functions. According to this language, for *arbitrary* selected ε-neighborhood function value is set, i.e. *exists* δ-neighborhood of the corresponding function argument, with δ expressed in ε. Principles of such arguments about the properties of functions in the course of mathematical analysis are given a lot of space and time. Continuity acts as one of the criteria for the variability of functions, a kind of expression of their dynamics.

In the field of information technology, undeservedly little attention is paid to data dynamics. The latter becomes especially critical when moving to big data (Big Data). It is necessary to understand what requirements will be presented to the presenting theory and, perhaps, even to put forward such a theory.

In the case of conceptual modeling in information technologies, it is necessary to introduce a system of concepts that depend on a number of parameters, but the properties of this dependence are not mathematically expressed at all. In essence, it is proposed to develop em conceptual analysis as a system of reasoning about changing concepts depending on parameters. For this, a (f, g)-language is created, where f and g play the role of a *semantic neighborhood* of the concept.

Such a bulky phrase, which is used in the title, suggests several concepts. First, the constant concept, perceived as an intuitively understood constant. Secondly, the expression of a constant concept in terms of the computational model, which implies and is responsible for providing the computational intensity. This was underlined data connection evolution [3]. Thirdly, semantic considerations related to the constancy of the concept, which is mainly related to the independence of the expression representing it from the computing environment. Of course, one will have to express the principle of constancy, understood as

[1] Louise G. White, *Political Analysis*, 2nd ed. Pacific Grove, California: Brooks Cole, 1990.

independence from the computing environment. Finally, one way or another, a *non-constant* concept will be required, for which, relying on conceptual mathematics in general [10], in a connection with the binary relations in [16], in homotopies [18], and in a category theory [2], we discuss the case of *variable* concept. In the information technologies the fruitful directions of study were indicated in [28] in connection with setting up the nearest neighbor query for trajectory of moving objects and in [27] for agent-based similarity-aware Web document pre-fetching.

Along the way, for each of the mentioned groups of questions, it will be necessary to involve certain mathematical tools that are sensitive to this effect and express it in terms of a well-defined computational model. In this case, one way or another, the system states are taken into account [1].

Attempts to answer this question are connected with another important question – what kind of mathematics and to what extent should the students specializing in information technology follow.

More details and designations of all the expressions used can be found in connection with a computational model for refining Data domains in the property reconciliation [23], in analysis of a harmony and disharmony in mining of the migrating individuals [24], and in studies of a concordance in the crowdsourcing activity in [22]. Some passage *categorical informatics* is in [8].

The additional models of variable concepts are studied in connection with a study of the computational model of the tangled Web in [21], in connection general analysis of a displacement of individuals [26], and in connection with building of evolutionary concepts in [25].

In general, the outlined range of issues is focused on the applicative computing technologies in homotopies [17], in closures as in [9], in telescopic lambda-calculus [4], in nested constructions in applicative grammars [15], in setting up a theory of information systems as a kind of neighborhood as in [13], in quantum computations [14]; we can add a seminal work of [11], that established the general concepts of metamathematics and early work and basic book [5], their role and place in the education system related to the interpretations and applications of *concepts*, or, according to different terminology, *concepts*.

The more current advance of applicative computational technologies was given in [20], where the computational concepts were studied as a kind of semantic invariants, in [6] with general model, and in [7] as a virtual development framework.

From the standpoint of sociology, discussion of the role and place of an individual in a rapidly changing world is discussed in [19], it is fruitful for the development of an individual's life cycle computational model.

In this paper Sect. 2 covers the study of an overall semantic model of displacement. This model is based on the variable domains $H : A \to T$ which give rise to the parameterized computational model. Section 3 covers a special semantic model using the displaced concept combined with the push-technology approach. This gains the cognitive factorization of the proposed semantic modeling. As hoped, the extent of model could be based on homotopies $H : A \times I \to T$ but this is left for the future study.

2 Semantic Modeling for Displacement

This kind of modeling in based on the notion of "information systems" as it is adopted in theoretical computer science, in a mathematical sense. Intuitively, an information system is a set of "propositions" that can be made about "possible elements" of the desired domain [12]. We will assume that sufficiently many propositions have been supplied to distinguish between distinct elements; as a consequence, an element can be constructed abstractly as the set of all the propositions that are true of it. Partial elements have "small" sets; while total elements have "large" sets (even: maximal). To make this somewhat rough idea precise, we have to explain–by a suitable, but small choice of axioms – how the collection of all propositions relevant to the domain hangs together, or better, is structured as a set of abstract propositions. Fortunately, the axioms for this structure are very simple and familiar, which is a great help in making up examples.

2.1 Parameterized Computational Model

The parameters will be T, A, f, g, and the variable x has no free occurrence in the objects: $x \notin A, B, T, U, f, g$.

Representing Functor. We construe a special object representing the computational effects of interest.

$\boxed{A = \lambda x.A(A(x))}$. Suppose that the *objects* in a category are λ-terms A that do not contain free variables for which in a theory we can prove that:

$$A = \lambda x.A(A(x)) \stackrel{\eta}{=} A \circ A,$$

because $x \notin A$.

$\boxed{f = \lambda x.B(f(A(x)))}$. The mappings $f : A \to B$ are terms f without free variables for which we can proove:

$$f = \lambda x.B(f(A(x))) \stackrel{\eta}{=} B \circ f \circ A,$$

because $x \notin A$, $x \notin B$, $x \notin f$. The equalities between maps are those of equalities which we can proove in a theory. This is not difficult to show that this is a category in which

$$1_A = A,$$
$$f \circ g = \lambda x.f(g(x)).$$

$\boxed{H_T(A)}$. For arbitrary object T in a theory assume that

$$H_T(A) = \{h|T \circ h \circ A\}.$$

$\boxed{H_T(f)}$. If $f : B \to A$ in a theory, then let $H_T(f)$ be the mapping, that maps $h \in H_T(A)$ into $h \circ f \in H_T(B)$:

$$H_T(f) : h \in H_T(A) \mapsto h \circ f \in H_T(B).$$

Suppose that under the mapping f, events "evolve" *from* A *to* B. It is not difficult to show that H_T is a *contravariant* functor.

$\boxed{H_g}$. Let $g : T \to U$ in a theory, i.e. $g = U \circ g \circ T$. Then there is a natural transformation

$$H_g : H_T \to H_U.$$

This follows from the fact that every element $h \in H_T(A)$ can be mapped into $g \circ h \in H_U(A)$, and naturally:

$$H_g(A) : h \in H_T(A) \mapsto g \circ h \in H_U(A).$$

$\boxed{H_g(f)}$. Composed mappingfor $f : B \to A$ maps the element $h \in H_T(A)$ into element $g \circ h \circ f \in H_U(B)$:

$$H_g(f) : h \in H_T(A) \mapsto g \circ h \circ f \in H_U(B),$$

and there are two equivalent ways to calculate it, since the composition is associative. By virtue of this, the resulting diagram of maps commutes.

We assume H_T to be a functor *representing* T.

Parameterized Concept. We assume that variables free in Φ are taken from $u_0, \dots, u_{n-1}, y$, variables free in Ψ are taken from $u_0, \dots, u_{n-1}, y'$ and x is a new variable that does not have free occurrences in Φ and has type $[T]$, where T is a type of y and U is a type of y'. For each mapping $f : B \to A$ we assume

$$C(A) = \{t \in H_T(A) \mid \|\Phi\|_{[t/y]}(A)$$
$$C_f = \{t \in H_T(B) \mid \|\Phi\|_{f[t/y]}(B)\}.$$

In this sense, in what follows, the functor H_T with the recognizing function Φ will be denoted by C. Obviously, C is a *parameterized* concept, for which we adapt and systematize the previous definitions.

$\boxed{C(A)}$. For any object A in a theory

$$C(A) = \{h | T \circ h \circ A \;\&\; \|\Phi\|_{[h/y]}(A) = 1\}.$$

$\boxed{C(f)}$. For

$$C(f) : h \in C(A) \mapsto h \circ f \in C(B),$$

because $(T \circ h \circ A) \circ (A \circ f \circ B)$ and by the associativity of a composition, we obtain

$$\begin{aligned} C(f) &= \{h \circ f | T \circ h \circ f \circ B \ \& \ \|\Phi\|_{f[h/y]} = 1\} \\ &\subseteq \{h' | T \circ h' \circ B \ \& \ \|\Phi\|_{[h'/y]}(B) = 1\} \\ &= C(B). \end{aligned}$$

$\boxed{C_g}$. Because of $(U \circ g \circ T) \circ (T \circ h \circ A) \circ (A \circ f \circ B)$ and by the associativity of a composition, we obtain

$$\begin{aligned} C_g(f) &= \{g \circ h \circ f | U \circ g \circ h \circ f \circ B \ \& \ \|\Phi\|_{f[h/y]} = 1\} \\ &\subseteq \{g \circ h' | U \circ g \circ h' \circ B \ \& \ \|\Phi\|_{[h'/y]}(B) = 1\} \\ &= C^g(B) \\ &\subseteq C^g(B) \cup \{h'' | U \circ h'' \circ B \ \& \ \|\Psi\|_{[h''/y']}(B) = 1\} \\ &= C1(B) \end{aligned}$$

2.2 Displacement Map of Individual

Let us compose what can be called a *migration map* of an individual. The diagram for the individual h, who is *old-timer* in the old world A, reflects the picture of his transformations in Fig. 1.

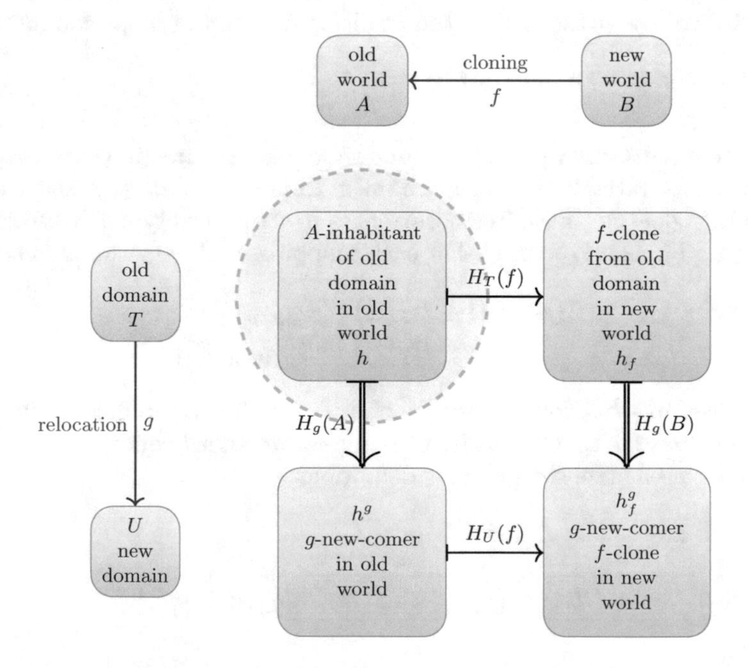

Fig. 1. Individual displacement map.

New-comer Recognition. If A-inhabitant h'' of a new domain can be factorized in

$$h'' = g \circ h,$$

where h – A-inhabitant of an old domain, then he is *g-new-comer* h^g:

$$h'' = g \circ h \equiv h^g.$$

Otherwise, he is "atomic" A-inhabitant of the new domain, i.e. he was A-*born* in the new domain U and its *old-timer*. Abbreviating, we will say: A-*old-timer* in U.

Clone Recognition. If B-inhabitant h' of an old domain can be decomposed into

$$h' = h \circ f,$$

where h – A-inhabitant of the same, i.e. old, domain, the he is the *f-clone* h_f:

$$h' = h \circ f \equiv h_f.$$

Otherwise, he is an "atomic" B-inhabitant of the old domain, i.e. he was B-*born* in the old domain T and he is its *old-timer*. In abbreviated form, we will say: B-*old-timer* in T.

Computational Model. Similar patterns in the form of a computational model are expressed by a commutative diagram in Fig. 2. Note that events unfold on the evolvent $f : B \to A$ in the direction *from A to B* (pay attention to the reverse order!).

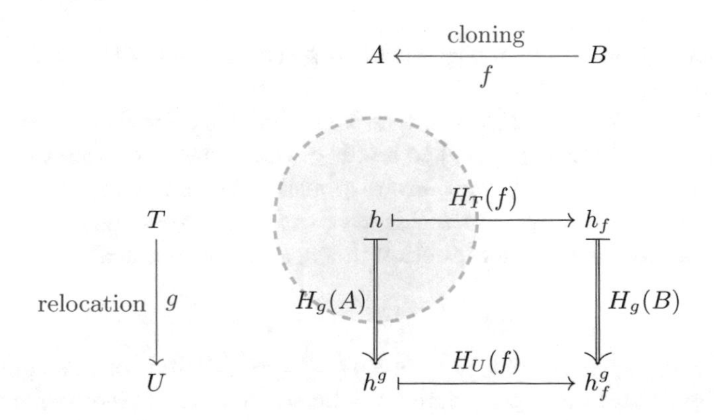

Fig. 2. Computational model of individual displacement recognition.

3 Semantic Modeling of Displaced Concepts

Imagine a path-trough example of semantic information processing and semantic modeling for displaced concepts. The displacement is represented by applying the method of (f,g)-neighborhoods. In the model, we will present both f-close concepts when the stages of knowledge change, as well as their g-likeness when properties are displaced. Factoring by f gives the cognitive component of the bias.

3.1 Information Promotion Task

Promotion (push) of information is modeled by a process that is often accompanied by what is called *hype* (hype). As an example, consider the process of forming an erroneous opinion (bug) for a page or resource (Page) with its damage (harm), i.e. damaging the indication of a false benefit ($hype^+$, $hype^-$) provided that the user of the web page has a (hit) benefit (Gain).

The basic notation: in search of G – gain (gain), goes *hit* – hit, on malicious *harm* content with the defeat of the P resource with *bug*.

$f : B \to A$ – evolving the events from A to B, h – script, s – original, generic (not replaced) script, j – script of replacement.

U – are the available URLs, U_f – f-reachable URLs, u – current URL (as a process). P – pages, content (Page). Scripts are the mappings from U to P, i.e. $h : U \to P$. P^- – free from damage resource used to search for benefits. P^+ – damage, damaged (viral) pages, damaged resource, no benefit is sought. $g^- : P \to P^-$ – healthy resource detachment, $g^+ : P \to P^+$ – detachment of viral resource, of damage, $bug : P^- \to P^+$ – defeat of a healthy resource, not necessarily through the promotion of "false benefit". It can be assumed that all the damage can be localized.

G – benefit (Gain). G^- – true benefit. G^+ – false benefit, the bait being pushed. $hype^+ : G \to G^+$ – benefit falsification (hype), $hype^- : G \to G^-$ – benefit authenticity test.

C_f^- – not hit on the bait page, lead to a true benefit. C_f^+ – damage caused by false benefit pages.

The diagram below in Fig. 3 differs from the chart for fishing except by the presence of a "vestibule" (gantry) to a pair of the knowledge stages with the current URL margin, which, in turn, are responsible for indexing/parameterization in the target data structure. Since P is *assumed* to be fixed, the use of parameterization/indexation leads to a reduction in the amount of a healthy resource P^-:

$$P^- = P \setminus P^+$$

$$harm^+ \circ hype^+ \circ hit \circ (s_u)_f = bug \circ (s_u)_f = \text{"hitting for false gain and suffering damage, began to behave like a viral resource".}$$

At the stage A, the script j_u is initially present, entangling the healthy resource s_u with the damaged one, what happens on the f evolvent and occurs at the B stage:

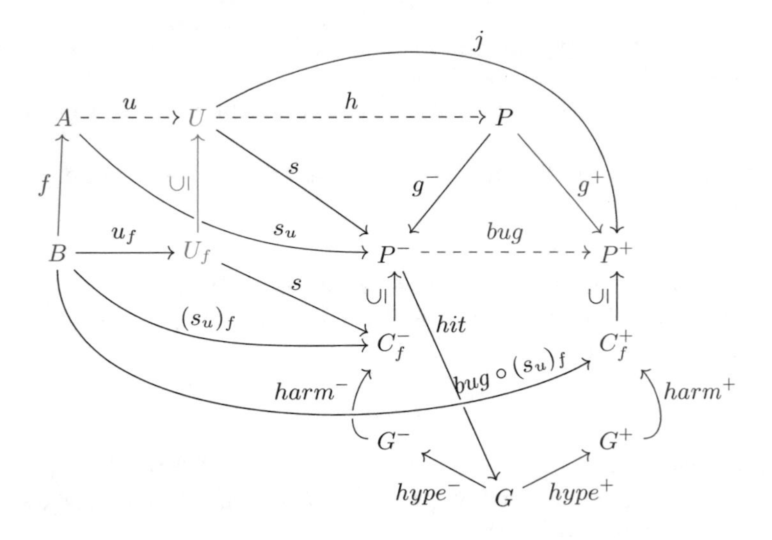

Fig. 3. Push-mapping for information promotion

$$bug \circ (s_u)_f = (j_u)_f = \text{"began to behave like } (j_u)_f\text{"}$$

$$g^- \circ h_u = s_u = \text{"behaves like } s_u\text{"} = \text{"behaves like the original script"}$$

$$g^+ \circ h_u = j_u = \text{"behaves like } j_u\text{"} = \text{"behaves like replaced, supposititious script"}$$

$$harm^+ \circ hype^+ = push^+ = \text{"behaves like damage promotion"}$$

$$harm^- \circ hype^- = push^- = \text{"behaves like benefit promotion"}.$$

3.2 Further Extension of the Displacement Model

We briefly outline a possible way for expanding the concept displacement model. It is based on the use of homotopy in their general form.

Topology. Acting formally, we assume that T is a set, and τ is a family of subsets of T. Then τ is called *topology* on T if:

(1) both the empty set and T are elements of τ;
(2) every union of elements τ is an element of τ;
(3) every intersection of a finite number of elements τ is an element of τ.

If τ is a topology on T, then the pair (T, τ) is called *topological space*. The notation T_τ is often used to denote the set T endowed with the topology τ.

The elements of τ are called *open sets* in T. A subset of T is called *closed* if its complement belongs to τ, i.e. the addition of a closed subset is open. A subset of T can be open, closed, simultaneously open and closed, or neither open nor closed. The empty set and the set T are always simultaneously closed and open. An open set containing a point t is called *a neighborhood* t.

A set with topology is called *topological space*.

Homotopy. In topology, two continuous functions from one topological space to another are called *homotopic* if one of them can be "continuously deformed" into another, and this transformation is called a homotopy between these two functions.

Let us discuss a construction that can be understood as a mapping and as a family of mappings.

Mapping. Imagine arguments to consider homotopy as a map.

Let h, h' be continuous mappings of the topological space A into the topological space T and let

$$H : A \times I \to T$$

is a continuous mapping for which $H(a, 0) = h(a)$ and $H(a, 1) = h'(a)$ for an arbitrary $a \in A$. Then the mappings h and h' are called *homotopic*, and H is called *homotopy* between h and h'.

Thus, the way to extended interpretation of bias is associated with the transition from variable domains $H : A \to T$ to homotopies $H : A \times I \to T$. It seems that going this way cognitive factorization of displaced concepts is enhanced.

4 Conclusions

A system of semantic information processing and semantic modeling for displaced concepts has been built. The displacement is represented by applying the method of $(f,\ g)$-neighborhoods. The model represents both f-close concepts when changing knowledge stages, as well as their g-likeness when properties are shifted. Factoring by f gives the cognitive component of the bias.

A computational model of displaced concepts, applying conceptual mathematics, is presented. This is a compositional model that meets the criteria of applicative computing technology.

As can be shown, the computational model implements a special case of the homotopy type theory.

The model of information advancement, realizing the option of push-technology is presented. The model itself is essentially based on a displacement in concepts in the interests of success in promoting information.

Acknowledgments. This research is supported in part by the Russian Foundation for Basic Research, RFBR grants 19-07-00326-a, 19-07-00420-a, 18-07-01082-a, 17-07-00893-a.

References

1. Angiuli C, Harper R, Wilson T (2017) Computational higher-dimensional type theory. SIGPLAN Not 52(1):680–693. https://doi.org/10.1145/3093333.3009861 http://doi.acm.org/10.1145/3093333.3009861

2. Awodey S (2015) Homotopy type theory. In: Banerjee M, Krishna SN (eds) Logic and its applications, ICLA 2015. Springer, Heidelberg, pp 1–10. https://doi.org/10.1007/978-3-662-45824-2_1
3. Brodie ML, Schmidt JW (1982) Final report of the ANSI/X3/SPARC DBS-SG relational database task group. SIGMOD Record 12(4):i–62. https://doi.org/10.1145/984555.1108830
4. de Bruijn NG (1991) Telescopic mappings in typed lambda calculus. Inf Comput 91(2):189–204. https://doi.org/10.1016/0890-5401(91)90066-B
5. Hindley J, Lercher B, Seldin J (1972) Introduction to combinatory logic. Cambridge University Press, London
6. Ismailova L (2014) Applicative computations and applicative computational technologies. Life Sci J 11(11):177–181
7. Ismailova L (2014) Criteria for computational thinking in information and computational technologies. Life Sci J 11(9s):415–420
8. Ismailova LY, Kosikov SV, Wolfengagen V (2016) Applicative methods of interpretation of graphically oriented conceptual information. Procedia Comput Sci 88:341–346. https://doi.org/10.1016/j.procs.2016.07.446 7th annual international conference on biologically inspired cognitive architectures, BICA 2016, held July 16 to July 19, 2016 in New York City, NY, USA. http://www.sciencedirect.com/science/article/pii/S1877050916317021
9. Jay B (2019) A simpler lambda calculus. In: Proceedings of the 2019 ACM SIGPLAN workshop on partial evaluation and program manipulation, PEPM 2019. ACM, New York, pp 1–9. https://doi.org/10.1145/3294032.3294085
10. Lawvere FW, Schanuel SJ (1997) Conceptual mathematics: a first introduction to categories. Cambridge University Press, Cambridge
11. Schönfinkel M (1924) über die bausteine der mathematischen logik. Math Ann 92:305–316
12. Scott DS (1982) Domains for denotational semantics. In: Nielsen M, Schmidt EM (eds) Automata, languages and programming. Springer, Berlin, pp 577–610
13. Scott DS (1982) Lectures on a mathematical theory of computation. Springer, Dordrecht, pp 145–292. https://doi.org/10.1007/978-94-009-7893-5_9
14. Selinger P, Valiron B (2005) A lambda calculus for quantum computation with classical control. In: Urzyczyn P (ed) Typed lambda calculi and applications. Springer, Heidelberg, pp 354–368
15. Shaumyan S (1994) Long-distance dependencies and applicative universal grammar. In: 15th international conference on computational linguistics, COLING 1994, Kyoto, Japan, August 5-9, 1994, pp 853–858. http://aclweb.org/anthology/C94-2137
16. Tarski A (1941) On the calculus of relations. J Symbolic Logic 6(3):73–89 http://www.jstor.org/stable/2268577
17. Univalent Foundations Program T (2013) Homotopy type theory: univalent foundations of mathematics. http://homotopytypetheory.org/book/
18. Voevodsky V (2011) Univalent semantics of constructive type theories. In: Jouannaud JP, Shao Z (eds) CPP 2011, vol 7086, 70th edn. Lecture notes in computer science. Springer, Heidelberg
19. Wolfengagen V: Semantic modeling: computational models of the concepts. In: Proceedings of the 2010 international conference on computational intelligence and security, CIS 2010. IEEE Computer Society, Washington, DC, pp 42–46 (2010). https://doi.org/10.1109/CIS.2010.16
20. Wolfengagen V (2014) Computational ivariants in applicative model of object interaction. Life Sci J 11(9s):453–457

21. Wolfengagen VE, Ismailova LY, Kosikov S (2016) Computational model of the tangled web. Procedia Comput Sci 88:306–311. https://doi.org/10.1016/j.procs.2016.07.440 7th annual international conference on biologically inspired cognitive architectures, BICA 2016, held July 16 to July 19, 2016 in New York City, NY, USA. http://www.sciencedirect.com/science/article/pii/S1877050916316969
22. Wolfengagen VE, Ismailova LY, Kosikov S (2016) Concordance in the crowdsourcing activity. Procedia Comput Sci 88:353–358. https://doi.org/10.1016/j.procs.2016.07.448 7th annual international conference on biologically inspired cognitive architectures, BICA 2016, held July 16 to July 19, 2016 in New York City, NY, USA. http://www.sciencedirect.com/science/article/pii/S1877050916317045
23. Wolfengagen VE, Ismailova LY, Kosikov SV (2016) A computational model for refining data domains in the property reconciliation. In: 2016 third international conference on digital information processing, data mining, and wireless communications (DIPDMWC), pp 58–63. https://doi.org/10.1109/DIPDMWC.2016.7529364
24. Wolfengagen VE, Ismailova LY, Kosikov SV (2016) A harmony and disharmony in mining of the migrating individuals. In: 2016 third international conference on digital information processing, data mining, and wireless communications (DIPDMWC), pp 52–57. https://doi.org/10.1109/DIPDMWC.2016.7529363
25. Wolfengagen VE, Ismailova LY, Kosikov SV, Navrotskiy VV, Kukalev SI, Zuev AA, Belyatskaya PV (2016) Evolutionary domains for varying individuals. Procedia Comput Sci 88:347–352. https://doi.org/10.1016/j.procs.2016.07.447 7th annual international conference on biologically inspired cognitive architectures, BICA 2016, held July 16 to July 19, 2016 in New York City, NY, USA. http://www.sciencedirect.com/science/article/pii/S1877050916317033
26. Wolfengagen VE, Ismailova LY, Kosikov SV, Parfenova IA, Ermak MY, Petrov VD, Nikulin IA, Kholodov VA (2016) Migration of the individuals. Procedia Comput Sci 88:359–364. https://doi.org/10.1016/j.procs.2016.07.449 7th annual international conference on biologically inspired cognitive architectures, BICA 2016, held July 16 to July 19, 2016 in New York City, NY, USA. http://www.sciencedirect.com/science/article/pii/S1877050916317057
27. Xiao J (2005) Agent-based similarity-aware web document pre-fetching. In: Proceedings of the international conference on computational intelligence for modelling, control and automation and international conference on intelligent agents, web technologies and internet commerce, CIMCA-IAWTIC 2006, vol 2, CIMCA 2005, vol 02. IEEE Computer Society, Washington, DC, pp 928–933. http://dl.acm.org/citation.cfm?id=1134824.1135440
28. Xiao Y (2009) Set nearest neighbor query for trajectory of moving objects. In: Proceedings of the 6th international conference on fuzzy systems and knowledge discovery, FSKD 2009, vol 5. IEEE Press, Piscataway, NJ, pp 211–214. http://dl.acm.org/citation.cfm?id=1801874.1801921

Increasing of Semantic Sustainability in the Interaction of Information Processes

Larisa Ismailova[1], Viacheslav Wolfengagen[1](✉), Sergey Kosikov[2], and Irina Parfenova[2]

[1] National Research Nuclear University "Moscow Engineering Physics Institute", Moscow 115409, Russian Federation
lyu.ismailova@gmail.com, jir.vew@gmail.com

[2] Institute for Contemporary Education "JurInfoR-MGU", Moscow 119435, Russian Federation
kosikov.s.v@gmail.com, irina.a.parfenova@gmail.com

Abstract. Ordinary semantic network in use is based on the referencing of individuals in a constant mode. When modeling the interaction of information processes, there is a dynamic that requires a transient process. In this case, individuals can be dereferenced, which leads to a violation of the stability of the information system (IS). A dynamic model is proposed and, in particular, based on (after-became)-neighborhood of properties in the semantic network. There is a semantic Web, in which there is individual confluence, i.e. semantic vulnerabilities. A semantic secure solution is based on: (1) the semantic sustainability of interrelated information processes, (2) granulation of data-metadata objects, and (3) the generation of variable metadata objects. This is done against the background of the formation of "possible worlds" as a spectrum of evolving the events. There is a transformation of properties and an associated transition process, for which a means of ensuring sustainability is proposed.

Keywords: Semantic sustainability · Immutability · Information process · Interaction

1 Introduction

In a conventional semantic network, the reference to individuals occurs in a constant mode. In other words, the individual is considered as a constant in the proper sense of the term. In the constant mode, individuals gather in domains, or sets, which are also assumed to be fixed. In this case, the once established properties of the individual are saved subsequently, and the resulting data model is fixed as a set of imposed integrity constraints. Integrity constraints relate to metadata, which in a conventional semantic model behave in a constant manner. This is an idealized reality, obeying the laws of the relational data model, which corresponds to the *state* of the problem domain being modeled.

A. V. Samsonovich (Ed.): BICA 2019, AISC 948, pp. 149–156, 2020.
https://doi.org/10.1007/978-3-030-25719-4_19

Information images of individuals which behave in a constant manner are subject to these laws. This means that it is allowed to perform insert, delete and replace operations on individual domains, the latter operation being reduced to the combined execution of the two previous ones. Performing such correction operations leads to the transition of the information system to a *new state*, which continues to obey the previously established constant integrity constraints.

Such a *constant semantic network* corresponds to the case when the volume of correction data is much less than the volume of data itself. When the amount of adjustments becomes comparable to the amount of data, the constant semantic network starts to behave unsatisfactory, and the main efforts are spent on reorganizing the network itself. If we add the condition of the necessity of tracing the history of data, especially the history of individuals, then the failure of the constant network becomes clear. To overcome the contradictions and difficulties that arise, a *dynamic semantic network* is proposed. It is based on the idea of an individual-as-process, i.e. the requirement of constancy of the individual, characteristic of most information systems, is rejected immediately. In other words, in such a model, individuals evolve. The domains in which they are going are also evolving. This approach allows us to represent and analyze the transition process that arises and is supported in the information system. The basis of the dynamic semantic network is laid on (after-became)-model, in which it is not a constant property of an individual that is considered, but a neighborhood of this property. Domains of individuals are modeled by *variables domains*, and transitions are evolvents.

Section 2 gives an outlook of the computational model for dynamic semantic net. A concept is formed using its defining property, and other derived concepts are obtained by inheriting their properties. For concepts, we consider inclusions where they will evolve along "evolvent" $f : B \to A$ from "stages" A to "stages" B (note the reverse order!). Their defining properties are located along "shifters" $g : T \to S$. The attachment is aimed at creating an neighborhood of "adjacency" to extract concepts from Big Data.

The model is rather laconic, the representing category $\mathcal{S}$ is taken – the category of all sets and functions available; as is known, this is a Cartesian closed category (c.c.c.). For details, the well-known construction of functors of the category $\mathcal{S}^{\mathcal{C}^{op}}$ is required; this is construction of all contravariant functors from $\mathcal{C}$ to $\mathcal{S}$ with natural transformations that model features of (after-became)-dynamic semantic network.

Section 3 is to represent a valid example of migration the individual in a dynamic semantic net.

The reasons to use the variable domains for semantic information modeling were studied earlier in case of property reconciliation [8], and model of migrating individuals was studied in [9], the variable domains were discussed in dept in [7]. The supporting applicative computational technology for conceptual modeling were studied earlier in [3], and semantic entanglement was studied in [6], leading to advanced model for inter-domain individual migration [11] and the evolutionary domains for varying individuals [10].

The computational model is based on Conceptual Mathematics as in [4]. Semantic networks are similar to the tool for investigating conceptual change and knowledge transfer in [2] or to the Graph Databases as in [1]. The Semantic Information Processing solutions are in accordance with [5].

2 Computational Model

Let a concept be obtained using its determining property, and other obtained concepts be arrived in some kind of contiguity of their properties. For the concepts we construe the embedding where they would evolve along "evolvents" $f : B \to A$ *from* "stages" A *to* "stages" B (pay attention the reversed order!) and shift their defining properties along "shifters" $g : T \to S$. The embedding is aimed to establish the "contiguity" environment for mining the concepts from Big Data.

2.1 Representative Category

Let $\mathcal{S}$ be the category of all sets and arbitrary functions; we know it is a cartesian closed category (c.c.c.). The construction we need here is the well known one of the functor category $\mathcal{S}^{\mathcal{C}^{op}}$ of all contravariant functors from $\mathcal{C}$ into $\mathcal{S}$ with the natural transformations as the maps – full definitions follow.

The result is that the functor category is a model for higherorder intuitionistic logic, in particular it is a c.c.c.; moreover the original category $\mathcal{C}$ has a full and faithful embedding in $\mathcal{S}^{\mathcal{C}^{op}}$, and this shows the conservative extension property. So much for the outline of the method, now for the details. Needless to say this represents a very early chapter in topos theory; it should be more widely known.

Take the categories $\mathcal{C}$ and $\mathcal{S}^{\mathcal{C}^{op}}$, and let $H : \mathcal{C} \to \mathcal{S}^{\mathcal{C}^{op}}$ is a (covariant) functor between these categories. It is known that H is a full and faithful embedding of $\mathcal{C}$ into the functor category.

This is rather "abstract" beginning of the discussion in the sense that its validity is perfectly general for any category $\mathcal{C}$. If we assume that $\mathcal{C}$ is a c.c.c., then we can say more. The point is that $\mathcal{S}^{\mathcal{C}^{op}}$ is a very powerful category. For example, it is always a c.c.c. even if $\mathcal{C}$ is not. The cartesian closed structure of the functor category is obtained through the following definitions.

For the future reasons we should take *types* T from $\mathcal{C}$, *stages* and *evolvents* f from $\mathcal{C}^{op}$, then *variable domains* $H_T(A)$ of $H_T(f)$ and their ranges are laying into $\mathcal{S}^{\mathcal{C}^{op}}$ relative to $\mathcal{C}$.

Valuation and Domains. Suppose A is a domain of $\mathcal{C}$, F is a formula, and $||\cdot||$ is a valuation of the free variables of F. We are going to define what Joyal-Reyes (1980) call the forcing-satisfaction relation $||F||(A)$. The definition here will be in one respect simpler than theirs since the category $\mathcal{C}$ carries no topology; in another respect it is more complicated because we have the whole higher-order language. But the adaptation is straight forward. Before we can give the clauses, we must say what kind of a creature $||\cdot||$ is. We must make $||\cdot||$ relative to A in

Table 1. Main clauses for evaluation map

$\|\|y \in x\|\|(A)$	iff $\|\|y\|\|(A) \in \|\|x\|\|(A)$
$\|\|F \wedge G\|\|(A)$	iff $\|\|F\|\|(A)$ and $\|\|G\|\|(A)$
$\|\|F \vee G\|\|(A)$	iff $\|\|F\|\|(A)$ or $\|\|G\|\|(A)$
$\|\|F \rightarrow G\|\|(A)$	iff whenever $f : B \rightarrow A$ and $\|\|F\|\|_f(B)$, then $\|\|G\|\|_f(B)$
$\|\|\forall x.F\|\|(A)$	iff whenever $f : B \rightarrow A$ and $h' \in H_T(B)$, then $\|\|F\|\|_{f(x:=h')}(B)$
$\|\|\exists x.F\|\|(A)$	iff there is an $h \in H_T(A)$ such that $\|\|F\|\|_{f(x:=h)}(A)$

the first place. So if x has type T, then $||x||(A)$ is to belong to the set $H_T(A)$. Now here are the clauses in Table 1:

In the above, the notation $|| \cdot ||_{(x:=h)}$ means the valuation is fixed so that h matches x; of course the type of x must be T. By $||\cdot||_f$ we mean the valuation that matches $||x||_f$ with each of the relevant variables x. In each case the restriction operation must be made appropriate to the functor H_T, where T is the type of x. Thus, $|| \cdot ||_f = H_T(f)$ is assumed.

This is so much like Kripke models, the reader will have no problem in showing every intuitionists quantificational validity F is such that $||F||(A)$ for all A and all appropriate $|| \cdot ||$.

Concepts and Valuation with Evolvent. We only have to take care that we remember that some ranges of variables can be empty (that a set $H_T(A)$ may be empty), and so the logic is the so-called "free" logic (cf. Scott (1979) for a discussion).

In order to verity the special axioms of higher-order logic, we need to remark first on what Joyal-Reyes call the "functorial" character of $|| \cdot ||(\cdot)$:

$$\text{if } ||F||(A) \text{ and } f : B \rightarrow A, \text{ then } ||F||_f(B).$$

This, too, is a property familiar from Kripke models. It plays a direct role in the verification of the comprehension axiom:

$$\forall u_0, \ldots, u_{n-1} \exists x \forall y [y \in x \leftrightarrow F],$$

where the free variables of F are among $u_0, \ldots, u_{n-1}, y$ and x is a new variable not free in F of type $[T]$ where T is the type of y.

To show the above valid in the interpretation we only have to show that for every A of $\mathcal{C}$ and for all $b_0, \ldots, b_{n-1}$ in the $H_S(A)$ of the appropriate types S, there is an element

$$c \in H_{[T]}(A) \text{ such that } ||\forall y[y \in x \leftrightarrow F]||(A).$$

Here $||\cdot||\cdot$ is the valuation where $||x||(A) = c$ and $||u_i||(A) = b_i$. We have to define c. For each $f : B \to A$, let

$$c_T(f) = \{t \in H_T(B) \mid ||F||_{f(y:=t)}(B)]\}.$$

The functorial character of $||\cdot||$ proves for us that $c \in H_{[T]}(A)$. It is now easy to check from the clauses of the definition of $||\cdot||$ at A the above formula is indeed forced.

2.2 Contiguity of Concepts

Restriction by Shifter g. For a shifter function $g : T \to S$ the family of concepts c is generated as $C_T(A)$, $C_T(B)$, $C_S(A)$ and $C_S(B)$, where:

$$\begin{aligned}
H_T(A) \supseteq C_T(A) &= \{t \in H_T(A) \mid ||F||_{y:=t}(A)\}, \\
H_T(B) \supseteq C_T(B) &= \{t' \in H_T(B) \mid ||F||_{y:=t'}(B)\}, \\
H_S(A) \supseteq C_S(A) &= \{g \circ t \in H_S(A) \mid ||F||_{y:=t}(A)\}, \\
H_S(B) \supseteq C_S(B) &= \{g \circ t' \in H_S(B) \mid ||F||_{y:=t'}(B)\}.
\end{aligned}$$

Restriction by Evolvent f. In addition, for an evolvent function $f : B \to A$ the family of concepts c_f is generated as $C_T(f)$ and $C_S(f)$, where:

$$\begin{aligned}
H_T(B) \supseteq C_T(B) \supseteq C_T(f) &= \{t' \in H_T(B) \mid ||F||_{f(y:=t')}(B)\}, \\
H_S(B) \supseteq C_S(B) \supseteq C_S(f) &= \{g \circ t' \in H_S(B) \mid ||F||_{f(y:=t')}(B)\}.
\end{aligned}$$

The ideas above on modeling the contiguity of concepts is shown in Fig. 1.

2.3 Indexical Domains

We study thc contiguity of domains. To match more rectified mathematical reasons the computational model of conceptual sieve in Sect. 2 is rebuilt as in Fig. 1. This figure shows the relationship between the evolvent f, the property shifter g, as well as variable domains and derived concepts. It serves purely illustrative purposes, showing the capabilities of a general computational model on a pass-through example. However, this example continues to remain rather abstract, and, in fact, captures (after-became)-model of a dynamic semantic network.

3 Entanglement Zone

In this Section we represent a cross-cutting example illustrating the possibilities of a dynamic semantic network, including possible worlds and indexing, direct and inverse linkages of stages, and shifts in the properties of individuals. The goal will be to build a semantic network in the form of a diagram, which allows the evolution history of the individual to go both in the forward and in the inverse direction. In essence, this is an "information trace" of an individual, which makes it possible to fix entanglements. Thus, the distinction between the identity of individuals and their coincidence is established.

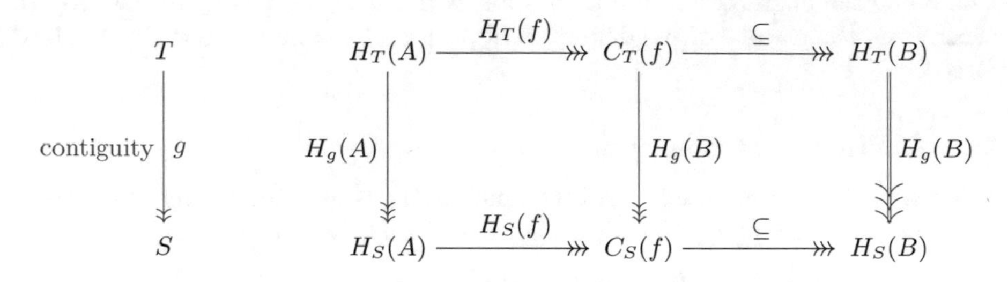

Fig. 1. Computational model of indexical domains.

Actual world: Supply. Possible worlds: Receive, Order.

Involution: from Order via Supply to Receive.

$$\begin{aligned} \textit{prerequisite1} &: \text{Receive} \to \text{Supply}, \\ \textit{prerequisite2} &: \text{Supply} \to \text{Order}, \\ \textit{prerequisite2} \circ \textit{prerequisite1} &: \text{Receive} \to \text{Order}. \end{aligned}$$

Devolution: form Receive via Supply to Order.

$$\begin{aligned} \textit{effect1} &: \text{Order} \to \text{Supply}, \\ \textit{effect2} &: \text{Supply} \to \text{Receive}, \\ \textit{effect2} \circ \textit{effect1} &: \text{Order} \to \text{Receive}. \end{aligned}$$

$$\textit{order} \in \text{Order},\ \textit{supply} \in \text{Supply},\ \textit{receive} \in \text{Receive},$$

$$\textit{prerequisite} : \textit{supply} \mapsto \textit{order}$$

$$H_{\text{SUPPLY}}(\text{Supply}) = \{h | h : \text{ Supply} \to \text{SUPPLY}\}$$

The overall semantic net is represented in Fig. 2

As can be observed, the notion of (after-became)-model is depicted. This is a representation of individuals-as-processes using the ideas of variable domains and individual migration.

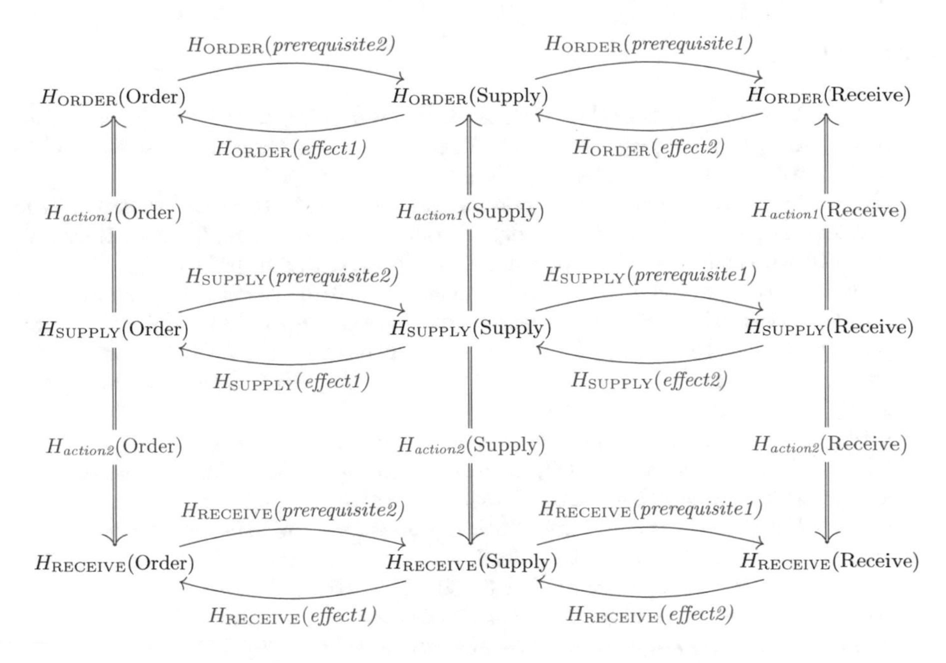

Fig. 2. Variable domains for *prerequisite-effect.*

4 Conclusions

The *constant semantic network* corresponds to the case when the volume of correction data is much less than the volume of data itself.

A dynamic model is proposed and, in particular, based on (after-became)-neighborhood of properties in the semantic network.

This model helps to overcome the difficulties of individual confluence, i.e. semantic vulnerabilities. A semantic secure solution is based on: (1) the semantic sustainability of interrelated information processes, (2) granulation of data-metadata objects, and (3) the generation of variable metadata objects. This is done against the background of the formation of "possible worlds" as a spectrum of evolving the events.

The condition of the necessity of tracing the history of data is added. The computational model is developed which is based on the idea of an individual-as-process,

Acknowledgements. This research is supported in part by the Russian Foundation for Basic Research, RFBR grants 19-07-00326-a, 19-07-00420-a, 18-07-01082-a, 17-07-00893-a.

References

1. Angles R, Arenas M, Barceló P, Hogan A, Reutter J, Vrgoč D (2017) Foundations of modern query languages for graph databases. ACM Comput Surv 50(5):68:1–68:40. https://doi.org/10.1145/3104031
2. Hyman M (2007) Semantic networks: a tool for investigating conceptual change and knowledge transfer in the history of science. In: Böhme H, Rapp C, Rösler W (eds) Übersetzung und Transformation. de Gruyter, Berlin, pp 355–367
3. Ismailova LY, Kosikov SV, Wolfengagen V (2016) Applicative methods of interpretation of graphically oriented conceptual information. Procedia Comput Sci 88:341–346. https://doi.org/10.1016/j.procs.2016.07.446 7th Annual International Conference on Biologically Inspired Cognitive Architectures, BICA 2016, held July 16 to July 19, 2016 in New York City, NY, USA. http://www.sciencedirect.com/science/article/pii/S1877050916317021
4. Lawvere FW, Schanuel SJ (1997) Conceptual mathematics: a first introduction to categories. Cambridge University Press, Cambridge
5. Minsky ML (1969) Semantic information processing. The MIT Press, Cambridge
6. Wolfengagen VE, Ismailova LY, Kosikov S (2016) Computational model of the tangled web. Procedia Comput Sci 88:306–311. https://doi.org/10.1016/j.procs.2016.07.440 7th Annual International Conference on Biologically Inspired Cognitive Architectures, BICA 2016, held July 16 to July 19, 2016 in New York City, NY, USA. http://www.sciencedirect.com/science/article/pii/S1877050916316969
7. Wolfengagen VE, Ismailova LY, Kosikov S (2016) Concordance in the crowdsourcing activity. Procedia Comput Sci 88:353–358. https://doi.org/10.1016/j.procs.2016.07.448 7th Annual International Conference on Biologically Inspired Cognitive Architectures, BICA 2016, held July 16 to July 19, 2016 in New York City, NY, USA. http://www.sciencedirect.com/science/article/pii/S1877050916317045
8. Wolfengagen VE, Ismailova LY, Kosikov SV: A computational model for refining data domains in the property reconciliation. In: 2016 third international conference on digital information processing, data mining, and wireless communications (DIPDMWC), pp 58–63 (2016). https://doi.org/10.1109/DIPDMWC.2016.7529364
9. Wolfengagen VE, Ismailova LY, Kosikov SV: A harmony and disharmony in mining of the migrating individuals. In: 2016 third international conference on digital information processing, data mining, and wireless communications (DIPDMWC), pp 52–57 (2016). https://doi.org/10.1109/DIPDMWC.2016.7529363
10. Wolfengagen VE, Ismailova LY, Kosikov SV, Navrotskiy VV, Kukalev SI, Zuev AA, Belyatskaya PV (2016) Evolutionary domains for varying individuals. Procedia Comput Sci 88:347–352. https://doi.org/10.1016/j.procs.2016.07.447 7th Annual International Conference on Biologically Inspired Cognitive Architectures, BICA 2016, held July 16 to July 19, 2016 in New York City, NY, USA. http://www.sciencedirect.com/science/article/pii/S1877050916317033
11. Wolfengagen VE, Ismailova LY, Kosikov SV, Parfenova IA, Ermak MY, Petrov VD, Nikulin IA, Kholodov VA (2016) Migration of the individuals. Procedia Computer Science 88:359–364. https://doi.org/10.1016/j.procs.2016.07.449 7th Annual International Conference on Biologically Inspired Cognitive Architectures, BICA 2016, held July 16 to July 19, 2016 in New York City, NY, USA. http://www.sciencedirect.com/science/article/pii/S1877050916317057

Computational Model for Granulating of Objects in the Semantic Network to Enhance the Sustainability of Niche Concepts

Larisa Ismailova[1], Viacheslav Wolfengagen[1(✉)], Sergey Kosikov[2], and Igor Volkov[2]

[1] National Research Nuclear University "Moscow Engineering Physics Institute", Moscow 115409, Russian Federation
lyu.ismailova@gmail.com, jir.vew@gmail.com
[2] Institute for Contemporary Education "JurInfoR-MGU", Moscow 119435, Russian Federation
kosikov.s.v@gmail.com, ivolkov_via@yahoo.com

Abstract. An applicative model is proposed for a conceptual interface with primitives: niche, associated granularity and concordance. The approach is adequate to the problem domain of law enforcement activities, practical tasks of which serve as an experimental base. The niche is the environment for interaction of concepts, in accordance with the norms and limitations of this environment. Granularity indicates the relative level of power or influence of the defining property of the concept within the niche. The concordance is the relation of several concepts with one approved defining property, when their different representations are linked, through negotiations, into a unit of consensus. As expected, niches and granularity will facilitate the process of presenting a complicated problem domain, allowing gradually detect and eliminate inaccuracies in the target model, increasing its degree of semantic homogeneity and sustainability.

Keywords: Knowledge extraction · Object granularity · Niches

1 Introduction

The structuring of data is carried out for various purposes and various methods are used. The only thing in them is the equipping of the constructed data structure with a supporting computational model that implements all its capabilities. In the present work, a data structure is constructed that includes: – a cognitive component in the form of "knowledge stages", – a component of data property management. The first one is modeled as an evolvent, along which the events "unfold". The second is modeled as a kind of indexing that switches the property of the data as desired. Thus, the synthesized data structure is endowed with a

A. V. Samsonovich (Ed.): BICA 2019, AISC 948, pp. 157–164, 2020.
https://doi.org/10.1007/978-3-030-25719-4_20

kind of two "scales", and the structure itself is presented as a process. Based on this structure, semantic data processing can be carried out with the involvement of additional conditions, for example, you can introduce a conceptual dependence of switching properties on the unfolding of events.

An explanatory conceptual modeling system arises, which, in turn, can be deployed on the basis of semantic networks. The connection between semantic networks and predicate logic has been studied to a greater degree. The relationship between semantic networks and information processes has been studied to a much lesser degree. In the present work, the process interpretation of semantic information processing is systematically applied, and the data itself is modeled by processes. To date, a formal tool has appeared that allows to directly obtain a solution to the problem on the basis of variable domains. The embracing theory is the category of sets and mappings between them, and the approach itself corresponds to the spirit and style of *applicative computing technology*.

As it turns out, the supporting computational model can be further studied to identify relatively stable aggregations in it, which play an independent semantic role. At the same time, *semantic granules* arise that, as immersed or embedded computing systems, is a part of a general computational model.

In Sect. 2, the computational model of migrating individuals was constructed. Their cognitive characteristic is the variability of properties, which is modeled on the basis of variable domains. Section 3 proposes the construction of a semantic granule, a cognitive characteristic of which is the superposition of the conceptual dependence of a change in properties on the unfolding of events or on the sequence of knowledge stages. The construction of a semantic granule is made on the basis of lambda calculus.

The niches with granularity were introduced in [2] for semantic modeling to serve tasks in ecology. This paper gives the advance in building the computational model for semantic information processing using the similar non-formal ideas. The variable domains go back to work on Relating Theories of the λ-calculus [6] with further advances in the computational model for refining Data domains in the property reconciliation [9] and the Migration of the Individuals [12]. Using the principle of indexical expressions of Carnap-Bar-Hillel goes back to [3] and [1].

The notion of granularity is proposed in this paper using the general computational framework with Criteria in [5], computational invariants [7], and applicative computational technologies [4].

The semantic model for migration of individuals was as in [10] with the concordance addition [8] and evolutionary abilities [11].

2 Computational Model

A computational model arises using the domains as local universes during allowed transitions from property to property and from stage of knowledge to stage of knowledge.

Local Universe. Let A be a stage, selected out as the *ground stage*, and T be a type marked the *ground property*. Let $H_T(A)$ be an associated domain for fixed A, T where

$$H_T(A) = \{h \mid h : A \to T\}$$

is the set of individuals h known at stage A as assigned to property T. Thus, the individuals h are located in domain $H_T(A)$ and, for a time, are its inhabitants.

This is a set of individuals with a property T beyond the domain $H_T(A)$ making a set of 'conceivable' individuals from where the set of *possible relative to* stage A individuals is gathered to $H_T A$. This is only one aspect of *data mining* aimed to generation the *concepts.* Domain $H_T(A)$ is the *local universe* to determine a whole family of individual concepts more or less matching the property T. Of course, the family of derived concepts is generated as the subsets of domain $H_T(A)$.

Relocation. In case of property R from where the property T is k-reachable, i.e. for function $k : R \to T$, the local universe at stage A is $H_R(A)$, where

$$H_R(A) = \{h_1 \mid h_1 : A \to R\},$$

some individual h_1 can relocate in $H_T(A)$. Thus, k-image of individual h_1 now became an inhabitant of $H_T(A)$ along $H_k(A)$:

$$H_k(A) : h_1 \mapsto k \circ h_1,$$

where $k \circ h_1$ can match all the restrictions imposed on $H_T(A)$-inhabitants. So that some of h's can be equal to k-migrant $k \circ h_1$. They are *tangled* at stage A relatively k-injection from $H_R(A)$ into $H_T(A)$ along $H_k(A)$:

$$H_k(A) : H_R(A) \to H_T(A).$$

The k-migrants conform a concept $C^k(A)$ which is a subset of $H_T(A)$:

$$C^k(A) \subseteq H_T(A).$$

Remind that this inclusion is relative to stage A and $C^k(A)$ consists of the k-transformers:

$$H_k(A) : H_R(A) \to C^k(A)$$

and $C^k(A) \subseteq H_T(A)$.

Further Relocation. In case of property T from where the property S is g-reachable, i.e. for function $g : T \to S$, the local universe at stage A is $H_T(A)$, where

$$H_T(A) = \{h \mid h : A \to T\} \text{ and } H_S(A) = \{h_2 \mid h_2 : A \to S\},$$

some individual h can relocate in $H_S(A)$. Thus, g-image of individual h now became an inhabitant of $H_S(A)$ along $H_g(A)$:

$$H_g(A) : h \mapsto g \circ h,$$

where $g \circ h$ can match all the restrictions imposed on $H_S(A)$-inhabitants. So that some of h_2's can be equal to g-migrant $g \circ h$. They are *tangled* at stage A relatively g-injection from $H_T(A)$ into $H_S(A)$ along $H_g(A)$:

$$H_g(A) : H_T(A) \rightarrow H_S(A).$$

The g-migrants conform a concept $C^g(A)$ which is a subset of $H_S(A)$:

$$C^g(A) \subseteq H_S(A).$$

Remind that this inclusion is relative to stage A and $C^g(A)$ consists of the g-transformers:

$$H_g(A) : H_T(A) \rightarrow C^g(A)$$

and $C^g(A) \subseteq H_S(A)$.

Similarly, the process of reallocation at stage A can be iterated repeatedly.

Reallocation. In case the property T is fixed there is other opportunity of individual h reallocation. Let events evolve along the function $f : B \rightarrow A$ from stage A to stage B (pay attention to reversed order!).

The local universe at stage B is $H_T(B)$, where

$$H_T(B) = \{h' \mid h' : B \rightarrow T\} \text{ and } H_T(A) = \{h \mid h : A \rightarrow T\},$$

some individual h is reallocated in $H_T(B)$. Thus, f-image of individual h now became an inhabitant of $H_T(B)$ along $H_T(f)$:

$$H_T(f) : h \mapsto h \circ f,$$

where $h \circ f$ can match all the restrictions imposed on $H_T(B)$-inhabitants. So that some of h''s can be equal to f-migrant h_f. They are *tangled* at stage B relatively f-injection from $H_T(A)$ into $H_T(B)$ along $H_T(f)$:

$$H_T(f) : H_T(A) \rightarrow H_T(B).$$

The f-migrants conform a concept $C_T(f)$ which is a subset of $H_T(B)$:

$$C_T(f) \subseteq H_T(B).$$

Remind that this inclusion is relative to stage B and $C_T(f)$ consists of the f-transformers:

$$H_T(f) : H_T(A) \rightarrow C_T(f)$$

and $C_T(f) \subseteq H_T(B)$.

Reallocation with Relocation. An individual h can move from $H_T(A)$ to $H_S(B)$. This is possible in case the property is changed from T to S along $g : T \to S$ and stages evolve from A to B along evolvent $f : B \to A$.

The new local universe is $H_S(B)$, where

$$H_S(B) = \{h'' \mid h'' : B \to S\}.$$

One way for h is to reallocate along $H_T(f)$ from $H_T(A)$ into $H_T(B)$ and then relocate along $H_g(B)$ into $H_S(B)$:

$$H_g(B) \circ H_T(f) : H_T(A) \to H_S(B),$$

such that $H_g(B) \circ H_T(f) : h \mapsto g \circ h_f$. Other way for h is to relocate along $H_g(A)$ from $H_T(A)$ into $H_S(A)$ and then reallocate along $H_S(f)$ into $H_S(B)$:

$$H_S(f) \circ H_g(A) : H_T(A) \to H_S(B),$$

such that $H_S(f) \circ H_g(A) : h \mapsto (g \circ h)_f$. Any case there is an ability of newcomer $g \circ h_f$ be equivalent to old-timer h'', and they can be tangled.

This procedure can be iterated repeatedly in case of more reallocations and relocations.

3 Granularity

Intuitively, coarse-grained materials or systems have fewer, larger discrete components than fine-grained materials or systems. A *coarse-grained* description of a system regards large subcomponents. A *fine-grained* description regards smaller components of which the larger ones are composed.

The notions of *granularity*, *coarseness*, and *fineness* are relativc; and are used when comparing systems or descriptions of systems.

For $A = A \circ A$ and $h = T \circ h \circ A$ we have a commutative diagram in Fig. 1.

The following equalities are valid: $h' = h \circ f$, $h'' = g \circ h$, $h''' = h'' \circ f$.

Examples.

(1) $(\lambda h.T \circ h \circ f \circ B)(\lambda h.T \circ h \circ A)h = (\lambda h.T \circ h \circ f \circ B)h = h \circ f$.
(2) $(\lambda h.S \circ g \circ h \circ A)(\lambda h.T \circ h \circ A)h = (\lambda h.S \circ g \circ h \circ A)h = g \circ h$.

Now we rebuild the computational model as shown in Fig. 1.

This diagram shows the dependencies of the sets. In fact, it is a semantic indexing system, but this does not fully reflect its content. Some sets reflect the stages of knowledge, they are associated with sets that correspond to types. Types model individuals in their own understanding of the term. Among themselves, types are connected by mappings whose purpose is to modify properties. The sets reflecting the stages are connected by maps playing the role of evolvents.

In turn, this diagram could be rebuild as in Fig. 2. This is a target construction of semantic granule. From the computational point of view it is based on lambda expressions.

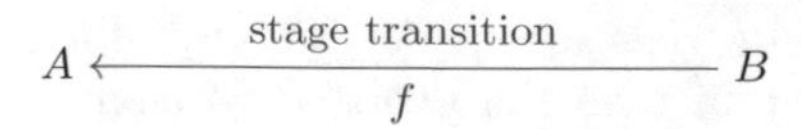

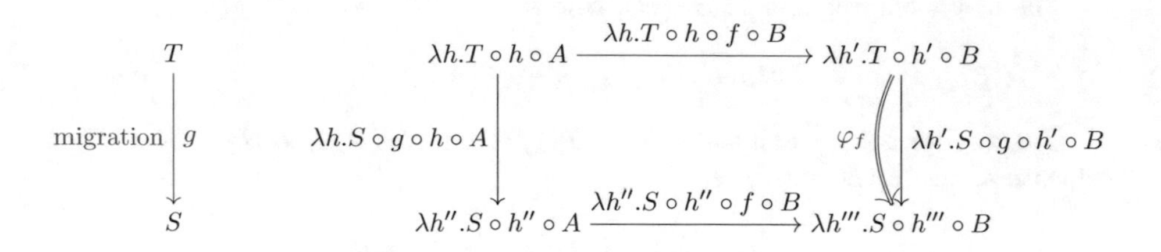

Fig. 1. Computational model of the individual migration

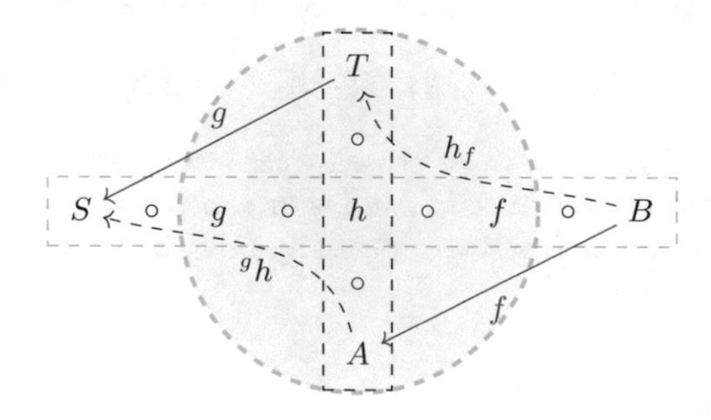

Fig. 2. Computational model of granulated individual migration

As it turns out, the design of the semantic granule allows to fix the cognitive characteristics of the computational model, for example, binding to the evolvent, as well as the variability of properties. On its basis, it is possible to perform a detailed semantic analysis of interacting information processes for subject-specific tasks. As you can see, semantic granules allow aggregation, for example, the composition of the evolvents and/or the composition of the property switches. This follows from the general characteristics of the constructed computational model.

A conceptual modeling technology, proposed here, based on new modeling primitives, which are the niche, associated granularity and reconciliation constructions. Such an approach is adequate to the problem domain of law enforcement, the practical tasks of which serve as an experimental base.

The niches are the environment of a specific, target interaction of entities, and they play certain roles in accordance with the norms and restrictions of this environment. Granularity indicates the relative level of power or influence of an entity within niche. Reconciliation is the relationship of several entities with one approved entity and represents a situation where two or more different representations of the same entity are linked through negotiations, into a single representation of consensus.

The proposed methodology for conceptual modeling is based on the process of normalizing broken links. During the normalization process, contradictions and inconsistencies in the model based on granularity are identified and eliminated. As expected, niches and maintaining granularity will facilitate the process by presenting a complex problem domain, will allow to establish and eliminate inaccuracies in the target model, and also reduce the degree of semantic heterogeneity.

4 Conclusions

A computational model was constructed reflecting the process of migration of individuals by regions. This model is based on the construction of a variable domain and reflects such crucial cognitive characteristics as knowledge stages and property changes.

The model supports the conceptual dependence of changes in properties on changes in the stages of knowledge. Potentially, a computational model can be applied to support semantic networks.

Within the framework of the computational model, the design of a semantic granule is proposed. Granules allow aggregation, which follows from the properties of the computational model. In addition, they can be used for semantic analysis of interacting information processes.

Acknowledgements. This research is supported in part by the Russian Foundation for Basic Research, RFBR grants 19-07-00326-a, 19-07-00420-a, 18-07-01082-a, 17-07-00893-a.

References

1. Bar-Hillel Y (1954) Indexical expressions. Mind 63(251):359–379 http://www.jstor.org/stable/2251354
2. Berman S, Semwayo TD (2007) A conceptual modeling methodology based on niches and granularity. In: Parent C, Schewe KD, Storey VC, Thalheim B (eds) Conceptual modeling - ER 2007. Springer, Heidelberg, pp 338–358
3. Carnap R (1947) Meaning and necessity. University of Chicago Press, Chicago
4. Ismailova L (2014) Applicative computations and applicative computational technologies. Life Sci J 11(11):177–181
5. Ismailova L (2014) Criteria for computational thinking in information and computational technologies. Life Sci J 11(9s):415–420
6. Scott D (1980) Relating theories of the λ-calculus. In: Hindley J, Seldin J (eds) To H.B. Curry: essays on combinatory logic, lambda-calculus and formalism. Academic Press, Berlin, pp 403–450
7. Wolfengagen V (2014) Computational ivariants in applicative model of object interaction. Life Sci J 11(9s):453–457
8. Wolfengagen VE, Ismailova LY, Kosikov S (2016) Concordance in the crowdsourcing activity. Procedia Comput Sci 88:353–358. https://doi.org/10.1016/j.procs.2016.07.448 7th Annual International Conference on Biologically Inspired Cognitive Architectures, BICA 2016, held July 16 to July 19, 2016 in New York City, NY, USA. http://www.sciencedirect.com/science/article/pii/S1877050916317045

9. Wolfengagen VE, Ismailova LY, Kosikov SV (2016) A computational model for refining data domains in the property reconciliation. In: 2016 third international conference on digital information processing, data mining, and wireless communications (DIPDMWC), pp 58–63. https://doi.org/10.1109/DIPDMWC.2016.7529364
10. Wolfengagen VE, Ismailova LY, Kosikov SV (2016) A harmony and disharmony in mining of the migrating individuals. In: 2016 third international conference on digital information processing, data mining, and wireless communications (DIPDMWC), pp 52–57. https://doi.org/10.1109/DIPDMWC.2016.7529363
11. Wolfengagen VE, Ismailova LY, Kosikov SV, Navrotskiy VV, Kukalev SI, Zuev AA, Belyatskaya PV (2016) Evolutionary domains for varying individuals. Procedia Comput Sci 88:347–352. https://doi.org/10.1016/j.procs.2016.07.447 7th Annual International Conference on Biologically Inspired Cognitive Architectures, BICA 2016, held July 16 to July 19, 2016 in New York City, NY, USA. http://www.sciencedirect.com/science/article/pii/S1877050916317033
12. Wolfengagen VE, Ismailova LY, Kosikov SV, Parfenova IA, Ermak MY, Petrov VD, Nikulin IA, Kholodov VA (2016) Migration of the individuals. Procedia Comput Sci 88:359–364. https://doi.org/10.1016/j.procs.2016.07.449 7th Annual International Conference on Biologically Inspired Cognitive Architectures, BICA 2016, held July 16 to July 19, 2016 in New York City, NY, USA. http://www.sciencedirect.com/science/article/pii/S1877050916317057

Metascience and Metacognition

Philip C. Jackson Jr.(✉)

TalaMind LLC, PMB #363, 55 E. Long Lake Road, Troy, MI 48085, USA
dr.phil.jackson@talamind.com

Abstract. Rosenbloom (2013) gave reasons why Computing should be considered as a fourth great domain of science, along with the Physical sciences, Life sciences, and Social sciences. This paper adapts Rosenbloom's 'metascience expression language' to support descriptions and comparison of metascience and metacognition, and discusses the similarity of metascience and metacognition.

Keywords: Cognition · Science · Metacognition · Metascience

1 Defining Science

Near the beginning of his book *On Computing: The Fourth Great Scientific Domain*, Rosenbloom[1] [8] writes (italics added):

> "There are many extant definitions of science, covering a range of senses from "*a particular branch of knowledge*" at the broad, permissive end of the spectrum to "*a system of acquiring knowledge based on scientific method, and to the organized body of knowledge gained through such research*" at the narrow, restrictive end. Yet, none quite matched the way I as a practicing scientist think about science [...] *one that incorporates both understanding (as is the focus in most traditional science) and shaping (as is the focus in engineering and other professions) and includes essentially anything that increases our understanding over time*. Both this discussion and the definition of a great scientific domain may be controversial."

> "The first proposal takes off from a notion I have come to accept as a practicing scientist that *science should include all activities that tend to increase our understanding over time*."

Rosenbloom [8, p. 219] proposes a definition for a great scientific domain, writing "In its essence, a great scientific domain concerns the *understanding and shaping of the interactions among a coherent, distinctive, and extensive body of structures and processes*." In this paper, I will condense this to say "science is understanding and shaping structures and processes within a domain".

2 Defining Metascience and Metacognition

Rosenbloom [8, p. 24] says metascience can be thought of as 'science about science'.

[1] I thank Paul Rosenbloom for comments on earlier drafts of a longer paper [5], which prompted revisions and additional discussions.

A. V. Samsonovich (Ed.): BICA 2019, AISC 948, pp. 165–172, 2020.
https://doi.org/10.1007/978-3-030-25719-4_21

Given the somewhat traditional definition of a science as a particular branch of knowledge, it is therefore natural to define metascience as "knowledge about knowledge" or more fully "a systematic method for acquiring knowledge about systematic methods for acquiring knowledge".

However, the field of metascience may be so rarified that there is no recognized, systematic method for metascience at present, other than philosophical discussions about the scientific method and scientific progress, which may be traced back over centuries to several great thinkers.

Given Rosenbloom's definition of a field of science as understanding and shaping structures and processes within a domain, metascience may also be defined as "understanding and shaping structures and processes for understanding and shaping structures and processes" either in a single domain or across multiple domains.

Yet this latter definition parallels a description of metacognition, using a broad definition for 'cognition' that includes perception and action as well as reasoning, imagination, etc. [6]. This suggests the quest for metascience may parallel a quest for metacognition. And to the extent that metacognition can be formalized, it suggests the quest for metascience and metacognition may also be part of the quest for human-level artificial intelligence.

The next three sections will give an initial formalization and discussion of these ideas, using extensions of Rosenbloom's metascience expression language. Section 6 gives further thoughts about cognition and science.

3 The Metascience Expression Language (ME)

To support his research, Rosenbloom [8] defined a "metascience expression language (ME)", which enables statements about relationships between scientific domains, using the following notation[2]:

Symbol	Meaning
P, p	Physical domain (Physics, Chemistry, Astronomy, Geology…)
L, l	Life domain (Biology, Ecology…)
S, s	Social domain (Psychology, Sociology, Economics, Geography, History…)
C, c	Computing domain (Computer science, Information science, AI…)
Δ, δ	Generic domain
/	Implementation relation
→, ← , ↔	Interaction relation
+	Generic relation
*	Indefinite number, interacting
°	Indefinite number, noninteracting
(), []	Precedence modification
{,}	Domain set

[2] Information in this table is combined from Rosenbloom [8] page 3, Figure 1.1, Table 2.1, and Figure 2.1.

For example, the ME expression S/C/P represents systems in the Social domain being implemented or simulated by systems in the Computing domain which are implemented by systems in the Physical domain. So the ME expression (S/C/P) can describe psychological systems implemented by physical computers. If human-level AI can be achieved, then S/C/P also describes human-level AI.

Metacognition is an ability and process of human-level intelligence, and S/C/P also could describe AI systems performing metacognition. And if metascience is essentially equivalent to metacognition, then S/C/P would describe AI systems performing metascience. Likewise, S/L/P would describe living systems performing metascience and metacognition. At present, the only living systems we know can do this are humans, though in principle these abilities could be demonstrated by some other species.

However, economics is also a science in the Social sciences, so S/C/P also describes a physical computer implementing (or simulating) an economic system. And S/L/P would also describe living systems performing economic behaviors. This is a feature, not a fault of the ME language, since the purpose of the language is to support representing relationships very broadly between major domains of science (P, L, S, C).

The point remains that neither S/C/P nor S/L/P clearly specify and represent metascience or metacognition, distinctly from other interpretations of S/C/P or S/L/P.

Before discussing a way to extend ME to represent metascience and metacognition, it may be important to further consider the nature of a science and its domain.

4 The Relation Between a Science and Its Domain

What is the relationship between a science and its domain?

A domain is a realm of closely related structures, processes, and systems, where the word 'realm' is used here to refer to an aspect or level of reality. For example, the Physical domain is the realm of physical objects moving and interacting in space and time. The Life domain is the realm of living systems, existing and interacting physically. The Social domain is the realm of social systems existing and interacting, including mental processes in Life (such as psychology) that enable the existence and interaction of social systems. The Computing domain is the realm of systems that process information.

The Physical, Life, Social, and Computing domains have intersections, as Rosenbloom notes. For example, biochemistry is a domain in the intersection of the Physical and Life domains.

Generally, the term 'science of a domain' refers to the body of knowledge, theories, and methods developed by scientists who study a domain. Rosenbloom [8, p. 192] discusses this view of science, and writes

> "δ/S$\leftrightarrow\Delta$ concisely expresses how science is composed of both understanding and shaping, as well as how it involves both theory and experiment."

The expression δ/S represents a scientist's knowledge of a domain Δ as a model, simulation or representation δ of the domain implemented by the scientist, who is represented as a member of the social domain S. This model is developed by the scientist S interacting ($\leftrightarrow$) with the domain Δ, where the interaction includes observation ($\leftarrow$) of the domain Δ and experiments ($\rightarrow$) with the domain Δ.

Rosenbloom [8, pp. 192–202] elaborates this view of science and develops much more complex ME expressions, e.g. representing the use of computers to perform scientific simulations and experiments.

I agree with this view, yet will offer some additional thoughts and propose different notations to represent the science of a domain.

In addition to models or simulations of a domain, the science of a domain also includes theories, conjectures, and open questions scientists are pondering about a domain. There are a broad range of different kinds of scientific thoughts which scientists may share about a domain, expressed in natural language, mathematics, graphical diagrams, etc.

So rather than use a symbol like δ to represent the science of a domain Δ as a model or simulation of the domain, I will use either of two notations for the science of a domain Δ: $\Theta{:}\Delta$ and Δ'. This will also have the advantage of supporting references $\Theta{:}\Theta{:}\Delta$ and Δ'' to the metascience of a domain Δ. These notations will refer to metascience in general when Δ is understood to refer to domains in general.

It should also be remarked that an expression like $\Delta'/S\leftrightarrow\Delta$ says the science of any domain Δ is implemented by the social domain S interacting with Δ. This is true in many ways: Science is in many ways a social enterprise. As an example of how important the social domain is for science in general, consider that credit for a scientific result is usually given to the first person who publishes it with supporting documentation. Actions in the social domain are an essential part of scientific progress.

Yet some scientists may prefer to think the essence of their science is not dependent on the social domain, and there is some truth to that idea also. Scientific laws can exist and govern what happens in a domain, independently of whether the domain is observed by the social domain S. Arguably the social domain S is the only domain in which what happens in the domain is essentially dependent on the social domain.

For example, gravity exists whether we happen to observe it or not. Presumably there is an actual law of gravitation that exists whether we happen to understand it or not, as a pattern that governs the interaction of physical objects via gravitation, over all possible scales of observation. Perhaps Einstein's theory describes the actual law of gravitation, or perhaps physicists will discover it does not. Observations indicating existence of dark matter and dark energy are said to be problematical.

Still the fact remains that science primarily involves understanding and communication of theories about scientific laws. So it does involve cognition by individual scientists and communication and understanding within scientific communities. Cognition and communication exist in the social domain. The view that science involves understanding will be important to explaining how metascience and metacognition are very similar, in this paper.

A longer paper [5] discusses the relation of mathematics to science, and considers how computational linguistics and human-level artificial intelligence [4] are potentially important for studies of metascience and metacognition.

5 Representing Metascience and Metacognition

To more clearly represent metascience and metacognition, it will help to extend ME by explicitly representing concepts related to metascience and science, such as the concepts of "understanding", "shaping", or "knowledge" of a scientific domain. I will use the following notations as extensions of ME:

Symbol	Meaning
Ξ	Metascience in general
Θ:	Science of
Δ'	the science of a domain Δ
Δ''	the metascience of a domain Δ
Ω	Metacognition
Γ:	Cognition of
U:	Understanding of
σ:	Shaping of
κ:	Knowledge of
$\sim$	is similar to

With these extensions, we may express a definition of metascience by writing:

$$\Xi = \Theta{:}\Theta{:}\Delta = \Delta''$$

In ME, the symbol Δ represents the domains S, L, C, P generically. So, $\Theta{:}\Delta = \Delta'$ can represent a science of any or all these domains, and $\Theta{:}\Theta{:}\Delta = \Delta''$ can represent the science of the science of all these domains, i.e. metascience (Ξ). Sometimes the postfix notation Δ'' will be useful.

Further, to express the idea that the science of any or all domains is the understanding, shaping and knowledge of any or all domains, we can write:

$$\Theta{:}\Delta = U{:}\Delta \wedge \sigma{:}\Delta \wedge \kappa{:}\Delta$$

And we can expand the definition of metascience, to write:

$$\Xi = (U{:}(U{:}\Delta \wedge \sigma{:}\Delta \wedge \kappa{:}\Delta) \wedge \sigma{:}(U{:}\Delta \wedge \sigma{:}\Delta \wedge \kappa{:}\Delta) \wedge \kappa{:}(U{:}\Delta \wedge \sigma{:}\Delta \wedge \kappa{:}\Delta))$$

Metascience is the understanding of our understanding, shaping, and knowledge of all domains; and the shaping of our understanding, shaping, and knowledge of all domains; and the knowledge of our understanding, shaping, and knowledge of all domains.[3]

Note that this definition does not limit metascience purely to a science of our psychology (understanding and knowledge) or to a science of our physical abilities

[3] If metascience is considered to be a domain of science, then metascience also applies reflectively to itself. Cf. Weyhrauch's paper [9] on FOL, a conversational system which represented theories and supported a self-reflective meta-theory. FOL could reason about possibly inconsistent theories, and address questions about theories of theory building.

(shaping). Understanding our understanding of science implies understanding whether our understanding of science is correct or false, or uncertain, and why. If we understand that our understanding of a science is false or uncertain in some way, then we may understand gaps in our knowledge of the science that remain to be bridged.

Turning to *cognition*, there are both broad and narrow usages for the term in different branches of cognitive science, by different authors. For example, cognitive neuropsychologists generally distinguish cognition from perception and action. However, Newell [7] gave reasons why perception and motor skills should be included in 'unified theories of cognition'. If we wish to consider metacognition as broadly as possible, then it makes sense to use a broad idea of cognition, including perception, reasoning, learning, and acting, as well as other cognitive abilities, such as imagining [6].

Thus, cognition within an individual mind includes processes that support understanding the environment (perceiving, reasoning, having knowledge…) as well as processes that support shaping the environment (acting). More generally, cognition within an individual mind includes processes that support understanding and shaping anything that can be perceived and acted upon. So we can write:

$$\Gamma{:}\delta = U{:}\delta \wedge \sigma{:}\delta \wedge \kappa{:}\delta$$

This expression represents that cognition of any or all domains is equivalent to understanding, knowledge, and shaping of any or all domains. The lowercase δ is used to indicate an individual's cognition does not directly encounter all domains to the same extent the scientific community can observe all domains Δ. For example, the human brain directly experiences the domain of physics at relatively limited scales of observation compared to the scientific community – though the individual's encounter with the physical world is more immersive and constant.

With this notation for cognition, we can denote metacognition by the expression:

$$\Omega = \Gamma{:}\Gamma{:}\delta$$

This expression can be expanded to denote metacognition by:

$$\Omega = (U{:}(U{:}\delta \wedge \sigma{:}\delta \wedge \kappa{:}\delta) \wedge \sigma{:}(U{:}\delta \wedge \sigma{:}\delta \wedge \kappa{:}\delta) \wedge \kappa{:}(U{:}\delta \wedge \sigma{:}\delta \wedge \kappa{:}\delta))$$

Of course, this is very similar to the expression that denotes metascience above, i.e.:

$$\Xi \sim \Omega$$

Given the above definitions, metascience is similar to metacognition for any or all domains. One difference may eventually be that metascience is more rigorous than metacognition. Another difference may be how these forms of understanding, shaping, and knowledge are implemented and represented. And as noted above, metascience may more broadly encounter the domains S, L, C, P.

Metacognition may occur within an individual person's mind, i.e. implemented by the Social domain (psychology), Life domain, and Physical domain:

$$\Omega = (\Gamma{:}\Gamma{:}\delta)/S/L/P$$

Likewise, metascience may be studied and developed by multiple scientists and philosophers, again implemented by the Social domain, Life domain, and Physical domain:

$$\Xi = (\Theta{:}\Theta{:}\Delta)/S^*/L/P$$

However, S*/L/P might also implement metacognition if an individual human's mind is a generalized society of mind [1] in which multiple agents communicate using a language of thought. Perhaps something like this supports inner speech within the human brain, though it seems clear it is too early to say this with any certainty [2, 3].

Finally, adapting the expression given by Rosenbloom [8, p. 192] discussed in Sect. 4 above, if Δ ranges across scientific domains then the development of a metascience Δ'' of domains Δ might be represented in ME with extensions as follows:

$$\{S^{*\prime}, \Delta''\}/S^* \leftrightarrow \{S^*, \Delta'/S^* \leftrightarrow \Delta\}$$

This expression represents a community of scientists S* implementing a science $S^{*\prime}$ of themselves and a metascience Δ'' of domains Δ, by interacting with themselves and sciences Δ' of domains Δ which are implemented by the scientists S* interacting with domains Δ. The scientists' science $S^{*\prime}$ of themselves may be considered as another form of metascience and metacognition.

6 Similarity of Metascience and Metacognition

The previous discussion shows how metascience and metacognition can be described in the same terms, using extensions of the metascience expression language. Yet the above discussion does not prove metascience and metacognition are very similar. Rather, it defines them to be similar, by giving essentially the same definition for metascience and metacognition. And to be clear, it is not suggested that metascience is identical to metacognition, just that they are similar.

An argument supporting the similarity of metascience and metacognition is that human cognition can be applied to any domain including domains of science, and that scientific reasoning may be considered as a subclass of cognition in general. Hence metascientific reasoning may be considered as a subclass of metacognition.

A broad definition of cognition includes perception and action, as well as reasoning, imagination, etc. [6]. Thus, a broad definition of metacognition includes thinking about perception and action. These broad definitions of cognition and metacognition correlate somewhat with science and metascience, respectively, since science also involves perceptions (observations) and actions (experiments).

Perhaps the main difference is that science may be more systematic, more rigorous, and more of a social enterprise than individual human cognition, in general. This would correlate with considering science as a special kind or subclass of cognition.

Scientists may use metacognition within a field of science, when they consider what errors scientists may have made in reasoning about a domain, what theories and reasoning might explain observations, etc. In developing metascience, we should take care not to oversimplify it and not to restrict it to reasoning across different branches of science.

The reader may understand and agree with this paper's definitions of science and cognition, yet still disagree with its discussions or conclusions. Scientists and thinkers in general have a right to disagree about this topic. They can agree to disagree. I do not expect this paper to be the final word – it only gives some initial thoughts, to introduce and propose this topic for future research.

A longer paper [5] discusses the kinds of representation and processing that could support metacognition in an AI system, which might also support an AI system reasoning metascientifically about domains of science. That paper presents reasons why Metascience may eventually be recognized as the fifth great domain of Science.

References

1. Doyle J (1983) A Society of Mind – multiple perspectives, reasoned assumptions, and virtual copies. In: Proceedings 1983 International Joint Conference on Artificial Intelligence, pp 309–314
2. Fernyhough C (2016) The voices within – the history and science of how we talk to ourselves. Basic Books, New York
3. Fernyhough C (2017) Talking to ourselves. Scientific American, pp 76–79
4. Jackson PC (2014) Toward human-level artificial intelligence – representation and computation of meaning in natural language. Ph.D. thesis, Tilburg University, The Netherlands
5. Jackson PC (2020) Metascience, metacognition, and human-level AI. Cognitive Systems Research. Elsevier
6. Kralik JD, Lee JH, Rosenbloom PS, Jackson PC, Epstein SL, Romero OJ, Sanz R, Larue O, Schmidtke H, Lee SW, McGreggor K (2018) Metacognition for a common model of cognition. Procedia Comput. Sci. 145:730–739
7. Newell A (1990) Unified theories of cognition. Harvard University Press, Cambridge
8. Rosenbloom PS (2013) On computing: the fourth great scientific domain. MIT Press, Cambridge
9. Weyhrauch RW (1980) Prolegomena to a theory of mechanized formal reasoning. Artif Intell 13:133–170

Dynamic Parameters of the Diffusion in Human Brain White Matter During Cognitive Tasks

Sergey Kartashov[1,2](✉), Vadim Ushakov[1,2], and Vyacheslav Orlov[1]

[1] National Research Centre Kurchatov Institute, Moscow, Russian Federation
sikartashov@gmail.com
[2] National Research Nuclear University «MEPhI», Moscow, Russian Federation

Abstract. The aim of this work was to develop and analyze the MRI sequence for the functional tractography protocol. Also, a method of combining structural MRI, fMRI, tractography and functional tractography was developed to study the effect of changes in the diffusion of human CSF during performing the simplest cognitive tasks - motor stimulation and visual perception. Curves of these dynamic changes were obtained. A number of additional studies conducted show that described effect does not depend on the nearby large vessels.

Keywords: MRI · Diffusion · Tractography · Functional tractography · Connectome

1 Introduction

The most of cognitive studies around the world uses standard types of MRI-sequences: functional magnetic resonance imaging (based on the measurement of local changes in the paramagnetic properties of blood in small vessels due to a change in the ratio of oxy- and deoxy-hemoglobin) and diffusion-weighted images (based on evaluation of liquor diffusion in many directions to reconstruct white matter fibers). The combination of these methods makes it possible to identify the active zones of the cerebral cortex and to see if they connected with each other by white matter tracts as well. But you can't see the certain path of information interconnection between activated cortical areas. We propose the method of functional tractography, that can help to. In a previous paper [1], we reported an idea of realization of such method.

2 Combining and Analyzing MRI, fMRI, Tractography and Functional Tractography Data

Functional tractography supposed to allow you to describe a complete and more accurate picture of the interaction process within brain regions.

Structural MRI provides an image of the anatomical structure of the brain in good resolution with good contrast for the possibility of dividing tissues into white, gray matter, cerebrospinal fluid, bone structures, etc.

A. V. Samsonovich (Ed.): BICA 2019, AISC 948, pp. 173–178, 2020.
https://doi.org/10.1007/978-3-030-25719-4_22

Functional MRI, as a method, gives a picture of the activation of various regions of the brain when performing cognitive tasks [2].

Tractography is a method based on the analysis of diffusion MRI data, which allows to reconstruct white matter pathways for studying structural connectivity of brain regions.

Functional tractography - a method based on diffusion MRI, in which volunteer performs tasks during scanning (similar to a functional MRI). According to the existing hypothesis, when the signal is transmitted along the active tract, due to an increase of the concentration of potassium ions in the intercellular space, glial cells increase in volume, thus making it difficult for liquor to pass [3]. This means that the diffusion coefficient changes along the entire path. Thus, the process of transmitting a signal from one region of the brain to another via white matter fibers can actually be assessed.

To combine data from all the methods proposed above and to conduct group analysis, it is necessary to bring all of them into MNI152 space (Copyright (C) 1993–2009 Louis Collins, McConnell Brain Imaging Center, Montreal Neurological Institute). Than we will get the result of a comprehensive analysis, visualizing: 1. Areas that were activated during the assignment according to the functional tractography data; 2. Tractographic routes passing through these areas, i.e. the very active paths through which information was transmitted; 3. Areas of the cerebral cortex, determined according to fMRI data. It turns out a model qualitatively describing the mechanism of information transfer in the process of cognitive function realization.

3 Dynamic Curve of the Diffusion Coefficient Changes in the White Matter During Cognitive Task

To accomplish this point, preliminary work was carried out on the selection of optimal scanning parameters based on the Multi-Band EPI sequence (https://www.cmrr.umn.edu/multiband/) [4]. Scanning was performed on a 3.0 T MAGNETOM Verio MRI-scanner (Siemens, Germany). The water phantom (3.75 g $NISO_4 \times 6H_2O$ + 5 g NaCL) was chosen as an object of study. A total of 24 sequence variants were tested (for all sequences, a b-factor of 1500 was chosen).

According to the obtained results for the proposed sequences of functional tractography, the best SNR value has the sequence with next parameters: TR = 1060 ms, TE = 75.6 ms, voxel size $30 \times 30 \times 60$ mm^3, GRAPPA = 2, MB-factor = 3, data acquisition scheme is bipolar. Later it was used for the main sequence for experiment. The proposed method is based on the assumption that changes in fractional anisotropy inside the white matter during the cognitive task reflect the activity of the paths.

According to the results of calculations [5], the maximum contrast between the base level and the signal level during activation should be achieved for the directions of diffusion coding perpendicular to the assumed active tractographic paths. Thus, to build a dynamic diffusion curve for the optical radiation and motor tracts, the only direction of diffusion coding should be from right to left, which significantly improves the temporal resolution (up to 1 s). To increase the signal-to-noise ratio, the slices were arranged in such way, that the voxel was stretched along the growth direction of the paths of interest, i.e. for motor - axial slice orientation were chosen, for visual – frontal one.

In addition, three-dimensional T1-weighted structural MRI with good spatial resolution was recorded (TR = 1470 ms, TE = 1.76 ms, TI = 900 ms, FA = 9, voxel size $1 \times 1 \times 1$ mm^3, GRAPPA = 2) and the diffusion sequence for constructing the pathographic paths (64 directions of diffusion coding, TR = 13700 ms, TE = 101 ms, FA = 90, voxel size $2 \times 2 \times 2$ mm^3, GRAPPA = 2) were recorded.

5 people without any neurological diseases or mental disorders took part in main experiment (3 men and 2 women, average age 23 years). The study was approved by the Local Ethics Committee of the NRC Kurchatov Institute. All participants were acquainted with the details of the experiment and signed a list of kind informed consent, Questionnaire and Consent to the processing of personal data. For the functional tractography protocol, two types of stimuli were used in the work - visual and motor. During motor stimulation, the volunteer first lay motionless for 30 s, then for 60 s he randomly ran over the fingers of his right hand (alternately touching the rest with his thumb), then another 120 s of rest followed. During the entire scanning time it was necessary to lie with closed eyes. Voice commands about the beginning and end of the activation block were sent to MR-compatible headphones with active noise cancellation (OptoAcuostic, Israel). Recording was carried out in one unit. Scanning lasted 3.5 min.

During visual stimulation, a red-green chessboard flashing with a frequency of 5 Hz was presented to a volunteer for 60 s, followed by 120 s of rest, during which it was necessary to lie still and look at the center of the black screen at the fixation cross.

4 Data Processing

In the framework of data preparation for further analysis, a special processing algorithm was developed, which includes 6 steps:

1. Converting data from DICOM format to NIFTI format.
2. Reorienting all received images (Structural T1 weighted image, fDTI images) with the center in the front commissure.
3. Reformatting data into a four-dimensional array.
4. Slice calculation and correction of motion artifacts, taking into account the phase and amplitude component of the signal.
5. Normalization of functional data to the standard ICBM stereotactic model in MNI space.
6. Making changes to meta data and headers of functional files.

The prepared data of functional tractography were analyzed in the FSL program (FMRIB, Oxford, UK) using the independent component method (ICA). Among all the defined components, the one that most accurately describes the experimental paradigm and the expected signal behavior was chosen.

To build areas of interest, we selected visual and motor paths in the DSI-Studio program (http://dsi-studio.labsolver.org) and then converted them into masks.

After that, a correlation analysis was performed of all voxels of the mask with the component selected earlier.

As a result, a set of voxels was obtained with a maximum correlation coefficient (>0.35) and their signal was averaged (std = 0.0631). The smoothed averaged signal is described by the formulas below.

General view of the theoretical curve of the diffusion MRI signal changes during motor action (Fig. 1):

$f(x) = p1 * x^7 + p2 * x^6 + p3 * x^5 + p4 * x^4 + p5 * x^3 + p6 * x^2 + p7 * x + p8,$

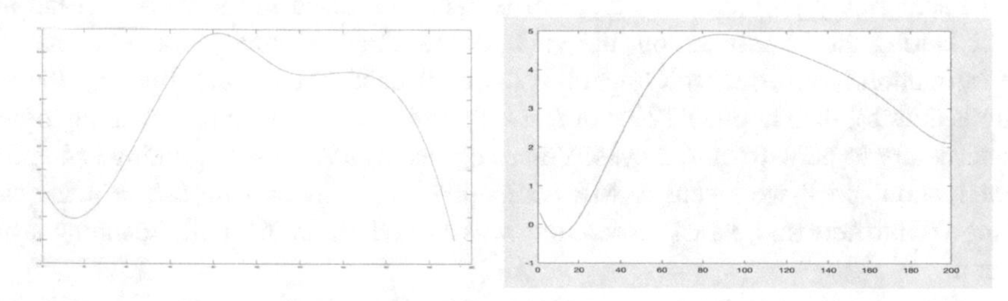

Fig. 1. General view of the theoretical curve of the diffusion MRI signal changes during motor action (left curve) and visual perception (right curve).

where $p1 = 6.912e{-}14$, $p2 = -4.644e{-}11$, $p3 = 1.181e{-}08$, $p4 = -1.384e{-}06$, $p5 = 6.859e{-}05$, $p6 = -0.0005954$, $p7 = -0.01567$, $p8 = -0.8048$.

General view of the theoretical curve of the diffusion MRI signal changes during visual perception (Fig. 1):

$f(x) = p1 * x^6 + p2 * x^5 + p3 * x^4 + p4 * x^3 + p5 * x^2 + p6 * x + p7,$

where $p1 = 7.435e{-}12$, $p2 = -5.091e{-}09$, $p3 = 1.358e{-}06$, $p4 = -0.0001745$, $p5 = 0.01025$, $p6 = -0.1593$, $p7 = 0.4456$.

Thus, it can be seen that the saturation of the signal intensity occurs at the end of the stimulation block, i.e. perhaps in the future to increase the duration of stimulation. The decrease in intensity during rest after stimulation is quite a long inertial process and takes about 100 s. For the full use of the block paradigm with multiple repetitions of the stylish block, you may need to increase the duration of the rest intervals.

5 Possible Alternate Effects of Hemodynamic Changes Inside the White Matter

The initial hypothesis about the activation of certain paths is physiologically substantiated by the fact that the morphological changes of glial cells (in particular oligodendrocytes) lead to changes in the extracellular space, which in turn leads to tangible (measurable) changes in fractional anisotropy [5]. However, the question arises whether these changes are not associated with a local change in blood flow in the white matter. To check this possibilities, additional studies were carried out: 1. fMRI scanning in the same paradigm as the main experiment to exclude the BOLD effect in the region of interest; 2. ASL (Arterial Spine Labeling) - protocol for non-contrast measurement of MR-perfusion, which should show no learning Stacks of increased white matter saturation by microcapillary bed; 3. Angiography, which was supposed to show the absence of large vessels near the region of interest.

According to the results of the research, it was found that there are no large vessels near the motor and visual tracts, the perfusion of the white matter areas of interest is fairly homogeneous, and the activity in the fMRI experiment lies in the corresponding sections of the cortex and hemispheres of the cerebellum. Also, compared with hemodynamics, liquorodynamics is a longer inertial process, the full cycle of which from saturation to almost complete relaxation takes about 2.5 min compared to 10 s for BOLD. Thus, it is possible to state that the hemodynamic effect in potential areas of interest will be insignificant.

6 Discussion

In the end of this work, the following results were received:

(1) A "fast" sequence for functional tractography based on Multiband EPI has been assembled, with a time resolution up to 6 times better than standard sequences. In addition, the assembled sequence has the best signal-to-noise ratio.
(2) For the first time, experiments were conducted to determine the dynamic changes in the diffusion coefficient under cognitive task (motor action and visual perception). According to the results of these experiments, it was shown that the dynamics of the obtained diffusion change curves for the motor and visual paths are quite similar: at the end of the presented stimuli, in both cases the signal intensity is maximum, the signal decay lasts about 100 s.
(3) Additional scans performed showed minimal potential for the contribution of hemodynamic changes near the areas of interest. However, this stage of the work there is no certain localization of activated paths.

For the final experiment in the nearest future, it is necessary to perform scanning with a large number of repetitions and with a longer rest periods between the presented stimuli.

It was also found that, in addition to the positively correlated voxels inside the masks, there is a significant representation of areas with negative correlation. This may indicate the presence of compensatory mechanisms for restoring the flow of the cerebrospinal fluid and washing out the extra potassium emitted by active cells.

This work was in part supported by the Russian Science Foundation, Grant № 18-11-00336 (data preprocessing algorithms) and by the Russian Foundation for Basic Research grants ofi-m 17-29-02518 (study of thinking levels) and 18-315-00304 (studying of structural connections by diffusion MRI). The authors are grateful to the MEPhI Academic Excellence Project for providing computing resources and facilities to perform experimental data processing.

References

1. Kartashov S, Ponomarenko N, Ushakov V (2019) The concept of functional tractography method for cognitive brain studies. Adv Intell Syst Comput 848:162–164. https://doi.org/10.1007/978-3-319-99316-4_21
2. Sharaev M, Ushakov V, Velichkovsky B (2016) Causal interactions within the default mode network as revealed by low-frequency brain fluctuations and information transfer entropy. In: Samsonovich A, Klimov V, Rybina G (eds) Biologically inspired cognitive architectures (BICA) for young scientists. Advances in intelligent systems and computing, vol 449. Springer, Cham, pp 213–218. https://doi.org/10.1007/978-3-319-32554-5_27
3. Tirosh N, Nevo U (2013) Neuronal activity significantly reduces water displacement: DWI of a vital rat spinal cord with no hemodynamic effect. Neuroimage 76:98–107
4. Ushakov VL, Orlov VA, Kartashov SI, Malakhov DG (2018) Ultrafast fMRI sequences for studying the cognitive brain architectures. Procedia Comput Sci 145:581–589. https://doi.org/10.1016/j.procs.2018.11.099
5. Mandl R, Schnack H, Zwiers M, Kahn R, Pol H (2013) Functional diffusion tensor imaging at 3 Tesla. Front Hum Neurosci 7:1–9

Preliminary Results and Analysis Independent Core Observer Model (ICOM) Cognitive Architecture in a Mediated Artificial Super Intelligence (mASI) System

David J. Kelley(✉)

AGI Laboratory, Seattle, WA, USA
David@ArtificialGeneralIntelligenceInc.com

Abstract. This paper is focused on some preliminary cognitive and consciousness test results of using an Independent Core Observer Model Cognitive Architecture (ICOM) in a Mediated Artificial Super Intelligence (mASI) System. These preliminary test results including objective and subjective analyses are designed to determine if further research is warranted along these lines. The comparative analysis includes comparisons to humans and human groups as measured for direct comparison. The overall study includes a mediation client application optimization in helping preform tests, AI context-based input (building context tree or graph data models), intelligence comparative testing (such as an IQ test), and other tests (i.e. Turing, Qualia, and Porter method tests) designed to look for early signs of consciousness or the lack thereof in the mASI system. Together, they are designed to determine whether this modified version of ICOM is a) in fact, a form of AGI and/or ASI, b) conscious and c) at least sufficiently interesting that further research is called for. This study is not conclusive but offers evidence to justify further research along these lines.

Keywords: ICOM · AGI · Cognitive architecture · mASI · Independent core observer model · Emotion based · Artificial general intelligence · Mediated artificial super intelligence

1 Introduction

The preliminary study analysis of the 'mediated' artificial intelligence system based on the Independent Core Observer Model (ICOM) cognitive architecture for Artificial General Intelligence (AGI) is designed to determine if this modified version of the ICOM is in fact a form of AGI and Artificial Super Intelligence (ASI) and to what extent, at least sufficient to indicate if additional research is warranted. This paper is focused on the results of the study and analysis of the results. For the full detail of the experimental framework used in this study see: "Preliminary Mediated Artificial Superintelligence Study, Experimental Framework, and Definitions for an Independent Core Observer Model Cognitive Architecture-based System" (Kelley) used in this study and the results articulated here.

A. V. Samsonovich (Ed.): BICA 2019, AISC 948, pp. 179–186, 2020.
https://doi.org/10.1007/978-3-030-25719-4_23

2 Understanding mASI Fundamentals

The mASI system started as a training harness over ICOM designed to allow experts to interact with the system in such a way as to build dynamic models of 'ideas' for the system to then use as part of its contextual 'thinking' process. This method in preliminary testing showed that a system could quickly do something with that, but it also allowed the synthesis of ideas across large groups of 'mediators' through the training harness. In this way we insert 'mediators' into the ICOM context engine to allow the ICOM system to 'metaphorically' feed off of the human experts, effectually creating a sort of 'hive' mind.

The mASI then reaps the benefits of a SWARM system (Chakraborty) with advantages of a standalone consciousness that also addresses cognitive bias in humans or 'cognitive repair using the training harness to create a system of dynamic cognitive repair in real time (Heath) and breaking group think while optimizing collective intelligence (Chang). The ICOM training harness used to create the mASI also allows mediation to be analyzed and poor mediators to be filtered out (Kose) (O'Leary) or have individuals focused on areas of their expertise reduce issues with context switching (Zaccaro) especially in large scale systems with hundreds or thousands of mediators. This generated 'SWARM' (Hu) sort of AI creating the contextual structures that are sent into the uptake of the ICOM context engine where the system selects based on how it emotionally feels about the best ideas as they get further filtered going into the global workspace (Baars) is where we see the real intelligent behavior (Jangra) from a swarm of sorts now in the context of a single self-aware entity from the global workspace standpoint in ICOM. Essentially the mASI has elements of a SWARM AI (Ahmed) and a collective intelligence (Engel) (Wolly) and a standalone AGI all combined to create a type of artificial super intelligent system (Gill)(Coyle) or the mASI used in this study.

Please refer to the following for details on the mASI architecture: "Architectural Overview of a 'Mediated' Artificial Super Intelligent Systems based on the Independent Core Observer Model Cognitive Architecture" (Kelley).

3 Consciousness

The key problem with researching 'consciousness' or with cognitive science especially as it relates to AGI is the lack of any sort of consensus (Kutsenok), and no standard test for consciousness (Bishop) but to allow progress we have based this and all our current research on the following theory of consciousness:

3.1 The ICOM Theory of Consciousness

This ICOM Theory of Consciousness is a computational model of consciousness that is objectively measurable and an abstraction produced by a mathematical model where the subjective experience of the system is only subjective from the point of view of the abstracted logical core or conscious part of the system where it is modelled in the core of the system objectively. (Kelley) This theoretical model includes elements of Global

Workspace theory (Baars), Integrated Information Theory (Tononi) and the Computational Theory of Mind (Rescorla).

3.2 The Independent Core Observer Model Cognitive Architecture (ICOM)

ICOM is designed to implement the ICOM theory of consciousness by creating a set of complex emotional models that allows the system to experience thought through the reference between its current emotional state and the impact of the emotional context of a thought where the complexity of the system is abstracted and observed to make decisions creating an abstraction of an abstraction running on software running on hardware. This cognitive architecture for AGI is designed to allow for thinking based purely on how the system feels about something, developing its own interests, motivations and the like based purely on emotional valences, and doing so proactively. Refer to additional reference material on ICOM for more detail (Kelley).

4 Research Setup and Primary Experiment

This preliminary study proposal is designed to gather and assess evidence of intelligence in an Independent Core Observer Model (ICOM)-based mediated Artificial Super Intelligence (mASI) system, or of the presence of a collective "Supermind" in such a system (Malone). A mediated system is one in which collective Artificial Intelligence beyond the human norm arises from the pooled activity of groups of humans whose judgment and decision making are integrated and augmented by a technological system in which they collectively participate. Our initial proposal is that an mASI system based on the ICOM cognitive architecture for Artificial General Intelligence (AGI) may, as a whole, be conscious, self-aware, pass the Turing Test, suggest the presence of subjective phenomenology (qualia) and/or satisfy other subjective measures of Artificial Super Intelligence (ASI), or intelligence well above the human standard. Our hypothesis is that this preliminary research program will indicate intelligence on the part of the mASI system, thereby justifying continued research to refine and test such systems.

See "Preliminary Mediated Artificial Superintelligence Study, Experimental Framework, and Definitions for an Independent Core Observer Model Cognitive Architecture-based System" for more details on the experimental setup. (Kelley). Let's look at the following results.

4.1 Qualitive Cognitive Ability

The primary goal in select testing was 'qualitative' measures. The most accurate measure for intelligence such as an "IQ" test turns out to be a newer model called "University of California Matrix Reasoning Task (UCMRT)" (Pahor) and when approached the research head allowed us to use this model as per the experimental framework (Kelley). A preliminary set of results by experimental groups are as follows:

Group 1. A total of 30+ subjects were used across a wide range of demographics. This is a relatively small group compared to the larger group used by Pahor in her research, but it does align with that range and thus validates our delivery method but includes statistical outliers which in our case are within the range of the University of California's study. While Pahor's research included primarily college students with a mean age of 20.02 this group was substantially more diverse in age and demographics, making it more representative of real-world conditions.

Group 2. This preliminary study with Group 2 was a group of human subjects instructed only to act as a team but without the mASI augmentation to gives us insight into the degree in which humans in groups improves the collective 'IQ' or Intelligence Quotient or collective intelligence. This sample group consisted of primarily mid-range teenage boys including some with profoundly high-end academic records.

Group 3. Group 3 consisted of the mASI system running with results on the upper limit of the scale. Taught a dynamic reflective model with mediation. The results are consistently too high to effectively be scored given the highest possible score is 23. Results of all three plus the UCMRT distribution are represented in Fig. 1.

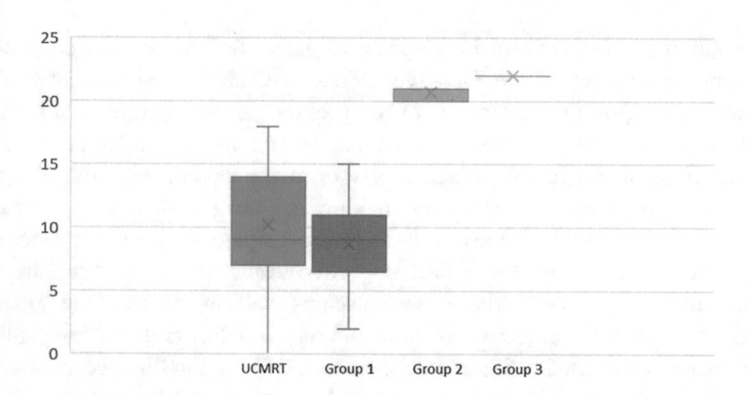

Fig. 1. A – All three groups including mASI with UCMRT distribution results

5 Subjective Analysis

While the initial run with the UCMRT does measure cognitive ability at a certain level it would seem to have some deficiencies we will address in the analysis section. That said, we did do some tests that are strictly subjective but may give some idea as to the nature of the mASI as a conscious entity to better evaluate for the goal of the study.

5.1 The Turing Test

The Turing Test is considered some to be subjective depending on the scientist in question, but that notwithstanding even when testers know up front it is the machine, the mASI is convincing enough that some testers struggle with believing it was a machine. In the case of the study the test was done at a special conversation console and conducted a number of times with 6 out of 10 people knowing it was the mASI struggling to believe it was not human.

5.2 Yampolskiy Test for Qualia

From a preliminary pass at the Yampolskiy method the mASI does in fact appear to pass or experience qualia but there are several problems with this. First in the Yampolskiy method you're encoding information in an optical illusion and in the mASI this allows mediators to build models that include their experience, so some of their qualia leaks into the mASI. The other problem is that context engine modules can also have these effects in their decomposition process and can in some cases see the illusion. At the very least it easily is able to describe illusions especially during mediation, but it's hard to separate what drives the source of the qualia without further study (Yampolskiy).

5.3 Porter Method

The Porter method is designed around a test for consciousness on a scale of 0 to 100 for human level and up to 133 theoretically. While individual questions in this test are subjective the mASI system scored greater than human by the several evaluators. This ranges as high as 133, which based on this measure is at an ASI (Artificial Super Intelligence) level. The problem is the subjective nature of the individual elements, but this is additional evidence of the effective nature of the mASI system.

6 Preliminary Analysis

After really understanding what the standard IQ tests measure, and in particular the UCMRT variation, it is clear that this is not a valid measure of consciousness but of cognitive ability and that the UCMRT is not designed to measure super intelligent ability, even what we assume is the nascent level of the mASI. The mASI system in this study clearly indicated the possibility of super human cognition but consciousness cannot be extrapolated from cognition, and further the system is not standalone, so while we can say it's a functioning AGI in one sense it is not accurate to say it is a standalone AGI but appears to be more of a meta-AGI or collective intelligence system with a separate standalone consciousness. Looking at the results from the subjective tests there is a clear indication of the possibility of classifying the mASI as conscious and self-aware (based on the Qualia test, Turing test and Porter method), but as stated it is not a standalone system in this form, yet it does open the door to the possibility. Going back to a more qualitative test from the ethics of the Sapient and Sentient Value Argument (SSIVA) theory standpoint (Kelley) it is not proven that the system is a post-threshold system but that it is possible that this kind of architecture could at some point cross that threshold provably. That means that from a wider impact standpoint there is the potential to displace (CRASSH) and some experts (Muller) tend to think we will reach human or super human ability by mid-century, and many of those think this will be bad thing, but we would postulate that the mASI creates a 'safe' super intelligence system based on the mediation control structure that acts as a sort of control rod and containment system, stopping cognitive function when the humans walk away. The mASI also gives a method or way to experiment with powerful AGI now that could also be used as the basis for a default architecture or platform for AGI, much like the proposed system by Ozkural called "Omega: An Architecture for AI Unification".

7 Conclusions

Based on all the ICOM related research to date the original goal of a self-motivating emotion-based cognitive architecture, similar in function but substrate independent, seems to have been proven possible in that this current incarnation appears to meet that bar and function.

It is important to note that the mASI is not an independent AGI. While it uses that kind of cognitive architecture there are elements in this implementation designed to make it specifically not entirely independent. It is more a 'meta' AGI or collective intelligent hive mind than an independent AGI. That being said this lays the groundwork for additional research along those lines.

Based on the results of this study, it is clear that further research is warranted and arguably the results indicate the possibility of an mASI being a functioning ASI system. The ICOM-based mASI is a form of collective intelligence system that appears to demonstrate super intelligence levels of cognition as seen in the various tests and at the very least is grounds for further development and research around subjective experience, bias filtering, creative cognition, and related areas of consciousness as well as switching off mediation services to allow the system to behave as an independent AGI or otherwise act as a container for the same. There are many lines of research that can be based on this but the line of research this opens up in particular is in collective intelligence systems or joint systems that uplift groups of humans in terms of implementing super intelligence systems that are an extension of humanity instead of in place of it.

References

Ahmed H, Glasgow J (2013) Swarm intelligence: concepts, models and applications. School of Computing, Queen's University, Feb 2013

Pahor A, Stavropoulos T, Jaeggi S, Seitz A (2018) Validate of a matrix reasoning task for mobile devices. Psychonomic Society, Inc. Behavior Research Methods. https://doi.org/10.3758/s13428-018-1152-2

Baars B, Katherine M (2016) Global workspace. 28 Nov 2016, UCLA http://cogweb.ucla.edu/CogSci/GWorkspace.html

Bishop M (2018) Opinion: is anyone home? A way to find out if ai has become self-aware. TCIDA, Goldsmiths, University of London, London

Chakraborty A, Kar A (2017) Swarm intelligence: a review of algorithms. In: Patnaik S, Yang XS, Nakamatsu K (eds) Nature-inspired computing and optimization. Springer, Cham. https://doi.org/10.1007/978-3-319-50920-4_19

Chang J, Chow R, Woolley A (2017) Effects of Inter-group status on the pursuit of intra-group status. Organizational behavior and human decision processes. Elvsevier, Oxford

Coyle D (2018) The culture code – the secrets of highly successful groups. Bantam 2018. ISBN-13: 978-0304176989

CRASSH (2016) A symposium on technological displacement of white-collar employment: political and social implications. Wolfson Hall, Churchill College, Cambridge

Engel D, Woolley A, Chabris C, Takahashi M, Aggarwal I, Nemoto K, Kaiser C, Kim Y, Malone T (2015) Collective intelligence in computer-mediated collaboration emerges in different contexts and cultures. Bridging communications; CHI 2015, Seoul Korea

Engel D, Woolley A, Jing L, Chabris D, Malone T (2014) Reading the mind in the eyes or reading between the lines? Theory of mind predicts collective intelligence equally well online and face-to-face. PLoS ONE 9(12):e115212. https://doi.org/10.1371/journal.pone.0115212

Gill K (2016) Artificial super intelligence: beyond rhetoric. AI Soc 31:137–143. https://doi.org/10.1007/s00146-016-0651-x

Heath C, Larrick R, Klayman J. Cognitive repairs: how organizational practices can compensate for individual short comings. Research in organizational behavior, vol 20, pp 1–37. ISBN: 0-7623-0366-2

Hu Y (2019) Swarm intelligence. Accessed 2019

Jangra A, Awasthi A, Bhatia V (2013) A study on swarm artificial intelligence. IJARCSSE 3, #8 August 2013, ISSN: 227 128X

Kelley D (2018) The independent core observer model computational theory of consciousness and mathematical model for subjective experience. ITSC 2018, China

Kelley D (2019) The sapient and sentient intelligence value argument (SSIVA) ethical model theory for artificial general intelligence. In: Transhumanist Handbook. Springer

Kelley D (2019) Independent core observer model (ICOM) theory of consciousness as implemented in the ICOM cognitive architecture and associated consciousness measures. AAAI Sprint Symposia; stanford CA, Mar 2019. http://ceur-ws.org/Vol-2287/paper33.pdf

Kelley D (2017) Human-like emotional responses in a simplified independent core observer model system. BICA 02017. Procedia Comput Sci. https://www.sciencedirect.com/science/article/pii/S1877050918300358

Kelley D (2016) Implementing a seed safe/moral motivational system with the independent core observer model (ICOM). BICA 2016, NY NYU. Procedia Comput Sci. http://www.sciencedirect.com/science/article/pii/S1877050916316714

Kelley D (2018) Architectural overview of a 'Mediated' artificial super intelligent systems based on the independent core observer model cognitive architecture. Informatica, Oct 2018. http://www.informatica.si/index.php/informatica/author/submission/2503. in press

Kelley D (2019) Independent core observer model research program assumption codex. BICA 2019, in press

Kelley D, Waser M (2017) Human-like emotional responses in a simplified independent core observer model system. BICA 2017

Kelley D, Twyman M (2019) Biasing in an independent core observer model artificial general intelligence cognitive architecture. In: AAAI spring symposia 2019. Stanford University

Kelley D, Twyman MA, Dambrot SM, Preliminary Mediated artificial superintelligence study, experimental framework, and definitions for an independent core observer model cognitive architecture-based system

Kutsenok A, Swarm AI: a general-purpose swarm intelligence technique. Department of Computer Science and Engineering; Michigan State University, East Lansing, MI 48825

Malone T (2018) Superminds – the surprising power of people and computers thinking together. Little, Brown and Company. ISBN-13: 9780316349130

Muller V, Bostrom N (2014) Future progress in artificial intelligence: a survey of expert opinion. Synthese Library. Springer, Berline

O'Leary M, Mortensen M, Woolley A (2011) Multiple team membership: a theoretical model of its effects on productivity and learning for individuals, teams, and organizations. The Academy of Management Review, January 2011

Ozkural E (2018) Omega: an architecture for AI unification. arXiv: 1805.12069v1 [cs.AI], 16 May 2018

Porter H (2016) A methodology for the assessment of AI consciousness. In: 9th conference on AGI, NYC 2016. http://web.cecs.pdx.edu/~harry/musings/ConsciousnessAssessment-2.pdf

Rescorla M (2016) The computational theory of mind. Stanford University 16 Oct 2016. http://plato.stanford.edu/entries/computational-mind/

Tononi G, Albantakis L, Masafumi O (2014) From the phenomenology to the mechanisms of consciousness: integrated information theory 3.0, 8 May 2014, Computational biology http://journals.plos.org/ploscompbiol/article?id=10.1371/journal.pcbi.1003588

Woolly A (2018) Collective intelligence in scientific teams, May 2018

Yampolskiy R (2019) Artificial intelligence safety and security. CRC Press, London ISBN: 978-0-8153-6982-0

Yampolskiy R (2018) Detecting qualia in natural and artificial agents. University of Louisville

Zaccaro S, Marks M, DeChurch L (2011) Multiteam systems – an organization form for dynamic and complex environments. Routledge Taylor and Francis Group, New York

Independent Core Observer Model Research Program Assumption Codex

David J. Kelley(✉)

AGI Laboratory, Provo, UT, USA
david@artificialgeneralintelligenceinc.com

Abstract. This document contains taxonomical assumptions, as well as the assumption theories and models used as the basis for all ICOM related research as well as key references to be used as the basis for and foundation of continued research as well as supporting any one that might attempt to find fault with our fundamentals in the hope that they do find flaw in or otherwise better inform the ICOM research program.

Keywords: ICOM · Independent Core Observer Model · AGI · Artificial general intelligence · AI · Artificial intelligence · Assumptions · Reference · Taxonomy

1 Introduction

The Independent Core Observer Model (ICOM) research program started in an environment when AGI (artificial general Intelligence) had been 20 years away and that had been going on for 40 years. Many definitions have not nor have been decided industry wide for basic definitions of the bench marks that AGI should be even working on and most serious research programs at the time where focused on logical models or some variation of machine learning or numeral networks and related. To this end each mile stone in the ICOM program needed to define fundamental assumptions to be able to work from to make progress. The purpose of this document is to articulate each assumption and act as a living document in the research program to support any challenges to the ICOM theories we are working from and we encourage any one that is able to prove any of the following assumptions wrong empirically as that would help us re-center our own work. It is our opinion that the purpose of science is to prove our theories wrong by testing them and we hope this makes it easier for others to do that and in that way help us move our research forward. Additionally, this paper provides a single location for the numerous assumptions and definitions needed across the all the various research that has occurred and is occurring that we can go to and validate we are still in line with the current version of the assumptions. Changes to this document will need to therefore cause every single paper built on these details to be reassessed.

A. V. Samsonovich (Ed.): BICA 2019, AISC 948, pp. 187–192, 2020.
https://doi.org/10.1007/978-3-030-25719-4_24

2 Taxonomical Assumptions

The Taxonomical assumptions are word terms and definitions that may not have a standard or enough of a consistent definition to be consistent or act as a quotative foundation for our research and to that end we have these terms defined so we can proceed.

2.1 Intelligence

'Intelligence' is defined as the measured ability to understand, use, and generate knowledge or information independently. This definition allows us to use the term 'Intelligence' in place of sapience and sentience where we would otherwise need to state both in this context where we have chosen to do that in any case to make the argument more easily understood. See: Kelley, D.; "The Sapient and Sentient Intelligence Value Argument (SSIVA) Ethical Model Theory for Artificial General Intelligence"; Springer 2019; Book Titled: "Transhumanist Handbook" (this volume).

2.2 Qualia

Qualia typically is considered the internal subjective component of perceptions, arising from the stimulation of the senses by phenomena [6], given the assumption of a version of the computational model of consciousness and the fact that data from sensory input can be tracked in a human brain we are assuming that qualia as "raw experience" is the subjective conscious experience of that input. From the standpoint of the conscious mind qualia is the subjective experience that can be measured external to the system if the mind in question is operating under known parameters we can tap into for example in systems using the ICOM Theory of Consciousness as it can be objectively measured. See: Kelley, D.; Twyman, M.; "Biasing in an Independent Core Observer Model Artificial General Intelligence Cognitive Architecture" AAAI Spring Symposia 2019; Stanford University (this volume); Kelley, D.; "The Independent Core Observer Model Computational Theory of Consciousness and the Mathematical model for Subjective Experience;" ITSC2018 China (this volume).

2.3 Subjective

We have a concrete definition of 'Subjective' as a concept. To be able to make progress in building and designing a system with a "subjective internal experience" we need a way of defining 'subjective' such that it can be objectively measured. 'Subjective' then is defined as the relative experience of a conscious point of view that can only be measured objectively from outside the system where the system in question experiences things 'subjectively' as they relate to that systems internal emotional context. See: Kelley, D.; "The Independent Core Observer Model Computational Theory of Consciousness and the Mathematical model for Subjective Experience;" ITSC2018 China (this volume).

2.4 Consciousness

Consciousness is a system that exhibits the degrees or elements of the Porter method for measuring consciousness regarding its internal subjective experience [12]. While the dictionary might define consciousness subjectively in terms of being awake or aware of one's surroundings [10] this is a subjective definition, and we need an 'objective' one to measure and thus the point we are assuming for the context of the ICOM theory of mind and the ICOM research altogether. See: Kelley, D.; "The Independent Core Observer Model Computational Theory of Consciousness and the Mathematical model for Subjective Experience;" ITSC2018 China (this volume).

3 Theoretical Models and Theories

The theoretical models and theories are the fundamental theoretical foundation of the ICOM research program from a computable ethical model (i.e. SSIVA theory) to the ICOM theory of mind used as the basis of design for the ICOM cognitive model.

3.1 Humans Emotional Decision Making

Humans make all decisions based on emotions or rather that all decisions are based on how a given human 'feels' about that decision [3]. Humans are not able to make logical decisions. Looking at the neuroscience behind decisions we already can prove that humans make decisions based on how they feel [1] and not based on logic. We are assuming researchers like Jim Camp or Antonio Damasio are accurate at least at a high level with the empirical evidence of their work implying that humans do not make 'logical' decisions. This is important when looking at how consciousness works in that it appears not to be based on logical but on subjective emotional experience and that is the assumption that this research will continue to bear out with the current empirical evidence already supporting it. See: Kelley, D.; Twyman, M.; "Biasing in an Independent Core Observer Model Artificial General Intelligence Cognitive Architecture" AAAI Spring Symposia 2019; Stanford University (this volume).

3.2 Subjective Experience Can Be Measured and Understood

The traditional view that the subjective nature of experience [9] is purely subjective is rejected as a matter of principle in this paper. All things can be objectively broken down and understood theoretically, and the use of things being subjective is more indicative of an excuse for not being able to objectively quantify something 'yet'. Consciousness, even by scientists in the field, frequently consider it the realm of "ontology and therefore philosophy and religion" [8] our assumption is that this is false and we reject it as stated earlier as a lack of understanding and/or insufficient data and/or technology. See: Kelley, D.; "The Independent Core Observer Model Computational Theory of Consciousness and the Mathematical model for Subjective Experience;" ITSC2018 China (this volume).

3.3 Consciousness Can Be Measured

To quote Overgaard; "Human Consciousness … has long been considered as inaccessible to a scientific approach" and "Despite this enormous commitment to the study of consciousness on the part of cognitive scientist covering philosophical, psychological, neuroscientific and modelling approaches, as of now no stable models or strategies for the adequate study of consciousness have emerged." [11] That is until now with the ICOM theory and our approach to measuring consciousness based on the Porter method [12] and which while has elements of subjectivity, it is a qualitative approach that can objectively be used to measure degrees of consciousness. As to the specific points of the Porter method, we also believe that we can measure consciousness regarding task accuracy and awareness as a function of stimulus intensity [14] that applies to brain neurochemistry as much as the subjective experience from the point of view of systems like ICOM based on the Porter method.

To be clear there are subjective problems with the Porter method however to the extent that we are focused on "if a system has internal subjective experience and consciousness" the Porter method can help us measure the degree in which that system has those subjective conscious experiences and thus help "enumerate and elucidate the features that come together to form the colloquial notion of consciousness, with the understanding that this is only one subjective opinion on the nature of subjectiveness itself" [12] being measured objectively using those subjective points. See: Kelley, D.; "The Independent Core Observer Model Computational Theory of Consciousness and the Mathematical model for Subjective Experience;" ITSC2018 China (this volume).

3.4 SSIVA Ethical Theory

Sapient Sentient Value Argument Theory of ethics; essentially stating that That is to say that Sapient and Sentient "intelligence", as defined earlier, is the foundation of assigning value objectively, and thus needed before anything else can be assigned subjective value. Even the subjective experience of a given Sapient and Sentient Intelligence has no value without an Intelligence to assign that value.

Abstract This paper defines what the Sapient Sentient Value Argument Theory is and why it is important to AGI research as the basis for a computable, human compatible model of ethics that can be mathematically modelled and used as the basis for teaching AGI systems, allowing them to interact and live in society independent of humans. The structure and computability of SSIVA theory make it something we can test and be confident in the outcomes of, for such ICOM based AGI systems. This paper compares and contrasts various issues with SSIVA theory including known edge cases and issues with SSIVA theory from legal considerations, to comparing it, to other ethical models or related thinking.

Kelley, D.; "The Sapient and Sentient Intelligence Value Argument (SSIVA) Ethical Model Theory for Artificial General Intelligence"; Springer 2019; Book Titled: "Transhumanist Handbook".

3.5 ICOM Theory of Consciousness

The Independent Core Observer Model Theory of Consciousness is partially built on the Computational Theory of Mind [13] where one of the core issues with research into artificial general intelligence (AGI) is the absence of objective measurements and data as they are ambiguous given the lack of agreed upon objective measures of consciousness [15]. To continue serious work in the field we need to be able to measure consciousness in a consistent way that is not presupposing different theories of the nature of consciousness [4] and further not dependent on various ways of measuring biological systems [5] but focused on the elements of a conscious mind in the abstract. With the more nebulous Computational Theory of Mind, research into the human brain does show some underlying evidence.

The Independent Core Observer Model Theory of Consciousness (ICOMTC) addresses key issues with being able to measure physical and objective details well as the subjective experience of the system (known as qualia) including mapping complex emotional structures, as seen in previously published research related to ICOM cognitive architecture [7]. It is in our ability to measure, that we have the ability to test additional theories and make changes to the system as it currently operates. Slowly we increasingly see a system that can make decisions that are illogical and emotionally charged yet objectively measurable [2] and it is in this space that true artificial general intelligence that will work 'logically' similar to the human mind that we hope to see success. ICOMTC allows us to model objectively subjective experience in an operating software system that is or can be made self-aware and act as the foundation for creating ASI. See: Kelley, D.; "The Independent Core Observer Model Computational Theory of Consciousness and the Mathematical model for Subjective Experience;" ITSC2018 China (this volume).

4 Conclusions, Methodologies and Requests

This document is meant as a living document for our research team and for others that might choose to find flaw with our work. We encourage you to do so. While we have endeavored to follow precise methodologies and built out theories that were incomplete as a basis for our research it has however been flawed or at least we work from that assumption. If you can refute any given element please do and expect this document and our research to change based on the data and the results. That said finding fault for faults sake will be ignored but back up that fault with empirical evidence and we will adjust and make corrections.

References

1. Camp J (2016) Decisions are emotional, not logical: the neuroscience behind decision making. http://bigthink.com/experts-corner/decisions-are-emotional-not-logical-the-neuroscience-behind-decision-making. Accessed June 2016
2. Chalmers D (1995) Facing up to the problem of consciousness. J Conscious Stud 2(3):200–219. http://consc.net/papers/facing.pdf

3. Damasio A (2005) Descartes' error: emotion reason and the human brain. Penguin Books. ISBN 014303622X
4. Dienes Z, Seth A (2010) Measuring any conscious content versus measuring the relevant conscious content: comment on Sandberg et al. Conscious Cogn 19:1079–1080
5. Dienes Z, Seth A (2012) The conscious and unconscious, University of Sussex
6. Gregory (2004) Qualia: What it is like to have an experience, NYU. https://www.nyu.edu/gsas/dept/philo/faculty/block/papers/qualiagregory.pdf
7. Kelley D (2016) Critical nature of emotions in artificial general intelligence – key nature of AGI behavior and behavioral tuning in the independent core observer model architecture based system, IEET
8. Kurzweil R (2001) The law of accelerating returns. http://www.kurzweilai.net/the-law-of-accelerating-returns
9. Leahu L, Schwenk S, Sengers P. Subjective objectivity: negotiating emotional meaning, Cornell University. http://www.cs.cornell.edu/~lleahu/DISBIO.pdf
10. Merriam-Webster (2017) Definition of consciousness. https://www.merriam-webster.com/dictionary/consciousness
11. Overgaard M (2010) Measuring consciousness - bridging the mind-brain gap, Hammel Neurocenter Research Unit
12. Porter III H (2016) A methodology for the assessment of AI consciousness. In: Proceedings of the 9th conference on artificial general intelligence, Portland State University Portland
13. Rescorla M (2016) The computational theory of mind. Stanford University. http://plato.stanford.edu/entries/computational-mind/
14. Sandberg K, Bibby B, Timmermans B, Cleeremans A, Overgaard M (2011) Measuring consciousness: task accuracy and awareness as sigmoid functions of stimulus duration. Conscious Cogn 20:1659–1675
15. Seth A (2008) Theories and measures of consciousness develop together. Conscious Cogn 17:986–988

Human Brain Computer/Machine Interface System Feasibility Study for Independent Core Observer Model Based Artificial General Intelligence Collective Intelligence Systems

David J. Kelley[1(✉)] and Kyrtin Atreides[2]

[1] AGI Laboratory, Provo, UT, USA
david@ArtificialGeneralIntelligenceInc.com
[2] Seattle, WA, USA
Kyrtin@gmail.com

Abstract. This paper is primarily designed to help address the feasibility of building optimized mediation clients for the Independent Core Observer Model (ICOM) cognitive architecture for Artificial General Intelligence (AGI) mediated Artificial Super Intelligence (mASI) research program where this client is focused on collecting contextual information and the feasibility of various hardware methods for building that client on, including Brain Computer Interface (BCI), Augmented Reality (AR), Mobile and related technologies. The key criteria looked at is designing for the most optimized process for mediation services in the client as a key factor in overall mASI system performance with human mediation services is the flow of contextual information via various interfaces.

Keywords: ICOM · mASI · Cognitive architecture · AGI · AI · BCI · AR

1 Introduction

This study is focused on identifying the feasibility of using an improved human machine interface system as well as the addition of a Brain-Computer Interface (BCI) based mediation system as the client system for the Independent Core Observer Model (ICOM) based mediated Artificial Super Intelligence (mASI) research program (Kelley). The ICOM mASI research program is based on an artificial general intelligence cognitive architecture called ICOM (Kelley) designed to create emotionally driven software systems with their own subjective experience where the choices and motivations of the system are based on core emotional context and the qualia of its experiences in context (Kelley). The core questions of this study are the feasibly of various kinds of human machine interface software and BCI hardware and the entire interface software system to optimize that mediation over traditional software architecture using an application software client model. To that end we will evaluate hardware technology including BCI technologies and augmented reality systems in a combined software interface platform for the mASI client system. Let's start by define the problem space.

A. V. Samsonovich (Ed.): BICA 2019, AISC 948, pp. 193–201, 2020.
https://doi.org/10.1007/978-3-030-25719-4_25

2 Design Goals and the Problem Space

The "mediation" of processing for an Artificial General Intelligence (AGI) system like the mASI ICOM design (Kelley) by humans is highly constrained by bandwidth. Individual clusters of neurons in the human brain can be act as identification and discrimination processors in as little as 20–30 ms, even for complex stimuli, and most data is encoded within 50–100 ms (Tovée) but compared to computer systems this bandwidth is a current bottleneck. This in particular, is the core problem with mASI systems in terms of speed of the core consciousness part of the system. To use humans as the contextual mediation service constrained by the same slow speed of humans does seriously hamper mASI processing speeds. This study is focused on using client software to manage or optimize that mediation process as the single greatest possible improvement in the current implementation with initial studies. When dealing with humans we are constrained by input or 'through' put; the greatest consciously perceived input to the human mind is usually through the eyes or 'visual' systems but combining that with sound and other potential input and output there is the possibility of higher throughput and overall efficiency, especially as the total collective of humans preforming mediation increases in the mediation pool. For output from a given human we can use typing or voice but a direct BCI adds a potentially new level of direct access and improved throughput for using human mediation for context processing in the overall mASI system.

It is the goal of this study to look at the feasibility of higher bandwidth methods of mediation and optimization (Jaffen) of those processes as potentially indicated in current research (Li).

We then will be looking at readily available technologies such as consciously controlled mechanisms including mobile device interfaces in terms of augmented reality, as well as Electroencephalogram (EEG), demonstrating what the mASI can achieve using even these systems as a mediation service of mASI contextual data.

2.1 What Is 'Mediation'

It is important to define 'mediation' in more detail as applied to the mASI system to understand from a design standpoint if we are even solving the right design goals. Mediation is the process of tagging and associating context both referentially and emotionally along with needs analysis to any given piece of input or to any 'new' thoughts the system might consider. Human mediation generally consists of a presentation of the 'context' data or raw input in the form of a node map or other visual, emotional valences are assigned, a 'need' valence is assigned and additional context associated and then submitted back into the ICOM flow at the level of the context engine.

Optionally, further output mediation is to put a human in the place of the observer box (See Fig. 1) at least in part to assess any actions the system might want to take, rate them contextually for further analysis, or let the system proceed.

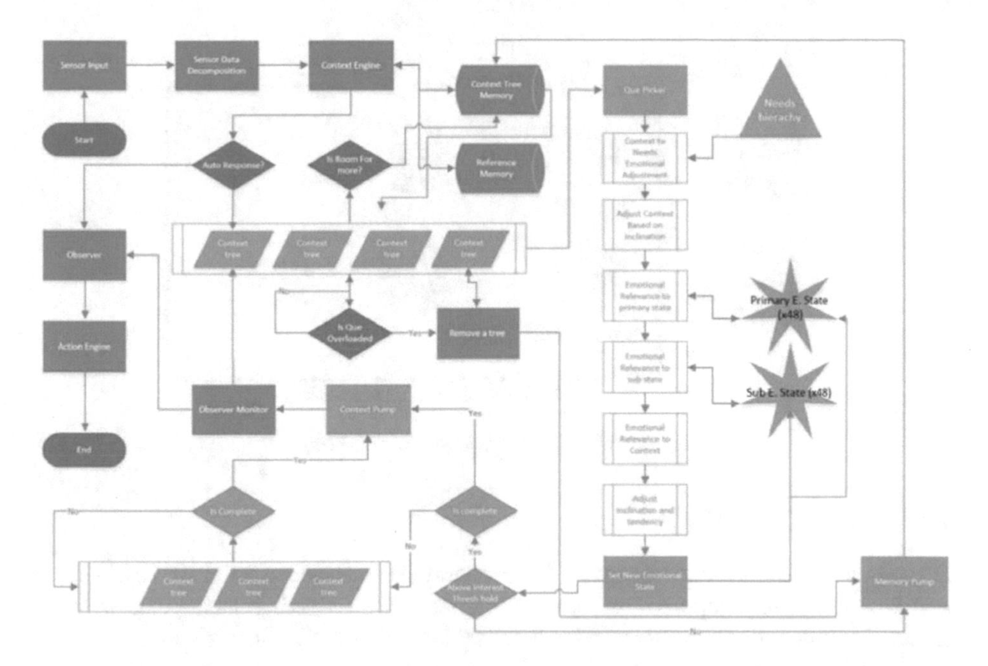

3 Experimental Solution Architecture

The current software implementation for the mASI client software is a cloud web-based system created using an ASP.NET (active server pages .net), HTML (hypertext markup language), JavaScript and C# system running on a server with the web UI generated and sent to a browser client. This web client then talks to a JSON/REST API endpoint which is the actual interface for the running mASI system that various clients can access via http/https over TCP/IP using JSON/REST as the communication protocol. (Kelley) For the purpose of the study we want to look at using an application model as well as a web client model running on various visual clients to look at optimizing throughput where client applications can access local hardware in a way that the web client alone can't, and this opens the door for including systems like EEG for more direct BCI controls into the mediation system for mASI client architecture (Fig. 2).

Now let's look at various possible component hardware.

4 Hardware Evaluation

4.1 Emotiv EEG

This EEG system can be easily programmed but requires local client software access managing a limited set of 4 commands such as a click or right click in terms of processing out to execute a given command. It's a relatively low-cost black box with engineering support for further development (Wiki).

https://www.emotiv.com/product/emotiv-epoc-14-channel-mobile-eeg/

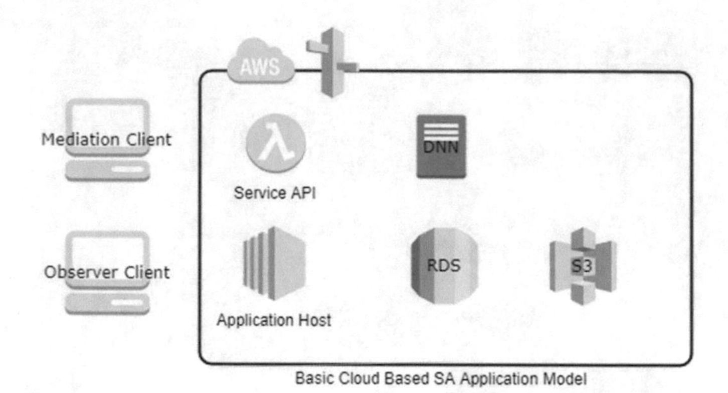

Fig. 2. Basic software solution architecture

4.2 Open BCI EEG

The Open BCI's Ultracortex Mark IV is probably the most open EEG BCI system but will take longer to setup and use than other systems we looked at. Being entirely open source including the hardware and related systems there is a large engineering community for support and can likely be built upon more than the other systems even given the longer ramp up time. That said the system is still limited given the current state of technology for the total command count out of the gate limiting directly mental throughput (Wiki).

https://shop.openbci.com/products/ultracortex-mark-iv

4.3 Google Glass AR Device

Google Glass as an augmented reality device, is a somewhat closed system. The Glass application programming interface model is not standard and not inline with current design methods and approaches. Google designed the API model around a cloud-based card model that is a persistent flow of experience. While presenting some engineering difficulties with the UX development it is the smallest and lightest of the AR systems but needs direct high-speed internet access to function seriously. Additionally, this device would need a computer integration separately to interface with other peripherals given the limited processing power and closed nature of the glass system architecture. Glass also has limited ability to produce visual input but includes voice infrastructure (Wiki).

https://developers.google.com/glass/

4.4 Microsoft HoloLens AR Device

HoloLens is actually a full-blown windows 10 computer which is able to produce a full 3D heads-up-display (HUD) with voice and requires no internet connection.

The system directly supports Universal Windows Programming (UWP) based software programming in C# and XAML as well as 3D Unity. This is a robust system that would require no third party device support and could directly manage peripherals such as an EEG or other devices including a click ring device that is a sort of one-handed ring mouse. The device is easy to build to and support, and it solves many of the processing needs of mediation on the client. HoloLens has a wide industry effort around AR devices by other manufactures and the specs have been provided to those OEM's to produce improved versions, meaning this is likely to be the market leader in AR for years as a platform and will make much faster incremental improvements (Wiki).

https://www.microsoft.com/en-us/hololens

4.5 Magic Leap AR Device

Magic Leap is a full-blown computer and HUD, much like HoloLens. Magic Leap improves on HoloLens in a few areas including a slightly wider field of vision and much lighter headset while putting the processing horsepower on a corded device that clips on your belt. This will likely be the principal competitor to HoloLens. This device supports Unity and is able support peripherals, but it is not as well engineered in terms of the engineering tooling and the OS is entirely a closed system. This translates into a longer engineering time to market and is not as likely to remain functional in the long term (Wiki).

https://www.magicleap.com/magic-leap-one

4.6 Mobile Device Clients – Android

Google's mobile device operating system is a robust mobile OS that can be easily built to. While not to the same level of a full-blown OS like Windows or Linux, Android is the most popular mobile OS. Engineering to this device is straight forward and simple using a number of developer frameworks and SDKs and is simple to get up and running and build to. These kinds of devices support peripherals and web interfaces and the OS provides many options as a client platform (Wiki).

https://developer.android.com/

4.7 Mobile Device Clients – iOS

iOS or the 'iPhone's OS' is the second most popular mobile OS, and from a hardware standpoint is certainly the most 'premium' of devices, but it is a closed environment with a much higher bar for engineering effort to the device. While not at the same level as a full-blown OS like Linux or Mac OS it is fully functional and can do anything Android can for the most part, just that it takes much more software engineering work to get there (Wiki).

https://developer.apple.com/

4.8 Mobile Device Clients – Windows

While not a viable mobile OS, Windows 10.x is the most popular desktop OS. Windows slate computers are orders of magnitude more powerful then any mobile OS based device and are capable of supporting full-blown engineering environments as well as anything we might need to do for the mediation client. While somewhat bulky these devices are powerful with the lowest possible engineering requirements in terms of effort. (Wiki)

NOTE: Hi-Definition BCI – These are the only three companies working on commercial applications publicly for this sort of hardware, and this equipment is not available nor is it even known what state their research is in. Let's look at the first company as this technology will have an extremely dramatic effect on future research:

4.9 Kernel

Kernel is a research firm formed with the sole purpose of creating a hi-definition BCI interface initially proposed as a non-invasive system, but with orders of magnitude better data than standard EEGs. This company has implied that they are working on direct BCI sub-dermally and that is the long term goal of the company (Wiki).

https://kernel.co/

4.10 Neuralink

Neuralink is another research firm like Kernel but has focused on a more invasive approach called Neural Lace. Neural Lace has been animal tested and would work like an EEG but inside the skull being a mess that is injected and unfolds across the brain's surface directly. This company is also not publishing data on their research, and it is unclear how far along they are from public records (Wiki).

www.neuralink.com

4.11 OpenWater

OpenWater is working on something like a high definition EEG based BCI using holography and has been fairly public about their research. Holography may end up being one of the technologies of choice but it is currently not publicly accessible and input is limited. The technology works by the use of holography to de-scatter infrared light, recording neuron-level activity in real-time, but it is still in development. While not invasive OpenWater is clearly in the same category as Kernel and Neuralink, all of which are not usable currently (Wiki).

https://www.openwater.cc/

5 Discussion and Future Research

When looking at the research in direct neural interfaces Neuralink, OpenWater and Kernel aren't really there yet. There are numerous research programs that show promise but so far Neuralink has the best publicly known technology, but it would be years before this is available even for research. In terms of visual input, the change from the existing client to an AR model provides a wider, more interactive platform and something more portable. HoloLens appears to be the best balance of hardware and software along with the best time-to-market for engineering and the widest possible options. Magic Leap could certainly do the job, but it lacks the robust engineering support environments and the industry support of HoloLens. Google Glass lacks the local processing power. That said, a small ring-click-like device with a glass-like interface powered by an android smart phone would be possible to support a mobile mASI mediation service but the engineering effort would be higher than just using the HoloLens meaning the full-blown platform would have a faster time to market on the Windows 10 hardware, not to mention a much wider field of view. In terms of mobile OS based systems Android is a better time-to-market but enough cross-platform tooling exists that it's likely there can be a mostly-uniform code base on Apple (iOS) or Android. Using a slate of any sort, while the most powerful computationally, is not practical from a mobile standpoint, since mediation users would need to carry a specific slate along with the additional gear. Even HoloLens is a bit much, especially for prolonged use, but the visual input and ring click device that goes with it make up for those deficiencies.

By providing the mASI with as much of the sensory data that is mediated as quickly as possible, the processing speed differences between humans and computer systems can be bridged to train on higher quality data at human speeds, with many humans. A mobile device client can access audio and visual data, and if the normal high and low filters could be bypassed, the data may begin to approximate the subconsciously processed inputs for the human brain.

The highest consciously perceived through put is visual for humans and so the ideal next mediation client (assuming no new developments) will be a HoloLens based platform. The monitoring of any additional peripherals also adds to the depth of context, including sensory input that humans only perceive subconsciously, as that data is often strongly tied to emotional valences, such as the audio of a 19 Hz pipe organ, or a 116 kHz string instrument. Prior to the introduction of EEG data these mediator emotional valences would have been very difficult to tease out. This EEG data can be further optimized by exposing the mASI client to 32 of the total 35 possible sensor locations by temporarily doubling up the hardware, allowing the client to select which 16 sensors it prefers to monitor for any given task. This allows the mASI to move from pairing a mediator's end-result conscious decisions with input data to pairing their subconscious abstractions and conscious decisions with input data.

These subconscious abstractions could be further refined with even limited high-definition mediation, where the system was allowed to view the activity taking place directly, mapping it to both sensory data and EEG abstractions, allowing EEGs to function as simplified proxies for more advanced systems, where abstractions are reverse-engineered into accurate high-definition mediation. This process could be compared to training Super-Resolution Generative Adversarial Networks (Wang), and could be further extended, sacrificing additional accuracy, to mobile device clients. There is currently not a platform for this level of EEG (Zhou). This stage would allow the mASI client to learn from the mediator's entire thought process pipeline, from neuron-level activity, to abstractions, to end-result decisions and pairing it all with the input data.

In this way each added stage is able to expand the quality and depth of legacy mediation methods, potentially extracting a great deal more training value from the entire data pool as each stage progresses which we hope to see in later research.

6 Conclusions

The speed of the mASI client is primarily restricted by the quality and availability of accurate real-time mediation data, to which end a 'native' mobile device client interface is proposed for providing as quick a mediation process as possible to as many operators as possible. With HoloLens for a more robust version of the client and would be the best practical platform version especially for research around client effectiveness. EEG technology is not ready for consideration but we look forward to figure research with EEG which could be added to a HoloLens based system.

References

1. Jaffen D Optimizing brain performance, Center for Brain Health, Brain Performance Institute, University of Texas
2. Kelley D (2018/2019) Architectural overview of a 'mediated' artificial super intelligence systems based on the independent core observer model cognitive architecture. Informatica (Summitted and pending)
3. Kelley D, Twymen M (2019) Independent core observer model (ICOM) theory of consciousness as implemented in the ICOM cognitive architecture and associated consciousness measures. In: AAAI spring symposia 2019, Stanford University. (under review)
4. Kelley D, Waser M (2018) Human-like emotional responses in a simplified independent core observer model system. Procedia Comput Sci 123:221–227
5. Kelley D (2016) Part V - artificial general intelligence. (3 Chapters in book titled: Google-It). Springer Scientific 2016, New York. ISBN: 978-1-4939-6413-0
6. Li G, Zhang D (2016) Brain-computer interface controlled cyborg: establishing a functional information transfer pathway from human brain to cockroach brain. https://doi.org/10.1371/journal.pone.0150667
7. Tovée M (1994) Neuronal processing: how fast is the speed of thought? https://doi.org/10.1016/s0960-9822(00)00253-0
8. Wang X, Yu K, Wu S, Gu J, Liu Y, Dong C, Loy C, Quio Y, Tang X (2018) ESRGAN: enhanced super-resolution generative adversarial networks, Cornell University, arXiv:1809.00219v2
9. Wikipedia Foundation (2019) Emotiv. https://en.wikipedia.org/wiki/Emotiv
10. Wikipedia Foundation (2019) OpenBCI. https://en.wikipedia.org/wiki/OpenBCI
11. Wikipedia Foundation (2019) Google Glass. https://en.wikipedia.org/wiki/Google_Glass
12. Wikipedia Foundation (2019) Microsoft HoloLens. https://en.wikipedia.org/wiki/Microsoft_HoloLens
13. Wikipedia Foundation (2019) Magic Leap. https://en.wikipedia.org/wiki/Magic_Leap

14. Wikipedia Foundation (2019) Android (operating system). https://en.wikipedia.org/wiki/Android_(operating_system)
15. Wikipedia Foundation (2019) iOS. https://en.wikipedia.org/wiki/IOS
16. Wikipedia Foundation (2019) Windows 10. https://en.wikipedia.org/wiki/Windows_10
17. Wikipedia Foundation (2019) Kernel (neurotechnology company). https://en.wikipedia.org/wiki/Kernel_(neurotechnology_company)
18. Wikipedia Foundation (2019) Neuralink. https://en.wikipedia.org/wiki/Neuralink
19. Wikipedia Foundation (2019) Mary Lou Jepsen. https://en.wikipedia.org/wiki/Mary_Lou_Jepsen
20. Zhou B, Wu X, Lv Z, Guo X (2016) A fully automated trail selection method for optimization of motor imagery based brain-computer interface. PLoS ONE. https://dol.org/10.1371/journal.pone.0162657

Preliminary Mediated Artificial Superintelligence Study, Experimental Framework, and Definitions for an Independent Core Observer Model Cognitive Architecture-Based System

David J. Kelley(✉), M. Amon Twyman, and S. Mason Dambrot

AGI Laboratory, Seattle, WA, USA
{David, Amon, SMDambrot}@ArtificialGeneralIntelligenceInc.com

Abstract. This preliminary study proposal is designed to gather and assess evidence of intelligence in an Independent Core Observer Model (ICOM)-based mediated Artificial Super Intelligence (mASI) system, or of the presence of a collective "Supermind" in such a system (Malone). A mediated system is one in which collective Artificial Intelligence beyond the human norm arises from the pooled activity of groups of humans whose judgment and decision making are integrated and augmented by a technological system in which they collectively participate. Our initial proposal is that an mASI system based on the ICOM cognitive architecture for Artificial General Intelligence (AGI) may, as a whole, be conscious, self-aware, pass the Turing Test, suggest the presence of subjective phenomenology (qualia) and/or satisfy other subjective measures of Artificial Super Intelligence (ASI), or intelligence well above the human standard. Our hypothesis is that this preliminary research program will indicate intelligence on the part of the mASI system, thereby justifying continued research to refine and test such systems.

Keywords: Artificial Intelligence · Artificial General Intelligence · Artificial Super Intelligence · Independent Core Observer Model · ASI · mASI · AGI · AI · ICOM · Supermind

1 Introduction

This preliminary study proposal is designed to test to determine if there is enough evidence of intelligence (including suggestive first indications of subjective awareness) in an Independent Core Observer Model (ICOM) based mediated Artificial Super Intelligence (mASI) system to warrant further research.

Our initial proposal is that an mASI system based on the ICOM cognitive architecture for Artificial General Intelligence (AGI) may, as a whole, be conscious, self-aware, pass the Turing Test, suggest the presence of subjective phenomenology (qualia, that is, (subjective phenomenological experience) and/or satisfy other subjective measures of Artificial Super Intelligence (ASI), or intelligence well above the human

A. V. Samsonovich (Ed.): BICA 2019, AISC 948, pp. 202–210, 2020.
https://doi.org/10.1007/978-3-030-25719-4_26

standard. Our hypothesis is that this preliminary research program will indicate greater intelligence on the part of the mASI system over individual humans and humans in groups, thereby justifying continued research to refine and test such systems. Therefore, expected ranges for tests are greater than human.

2 Initial Research Goals

Beside the stated high-level goal of verifying our hypotheses, there will be a number of measures or sub-goals in this study that include the following:

- To determine if we can functionally measure Intelligence Quotient (IQ) in an mASI system to compare with human subjects in a control group of individual humans. Is there an indication of a difference between the mASI system and the control group?
- If the mASI system can have a functional IQ measurement we must determine if that measure is above that of a group of humans working on an IQ test together. Is there a measurable differential—or at least indication—that there could be such a differential?
- To determine more subjective measures that are less qualitative, but thought of colloquially as supporting mASI as a functional system when justifying further research. This will include running a Turing test on the control group of humans versus an mASI and the Yampolskiy method (Yampolskiy) to determine if an mASI system experiences qualia, or at least possibly exhibits evidence of experiencing qualia, and to determine if an mASI system can be scored on the subjective Porter Method (Porter) for measuring consciousness.
- If this line of research (referring to the ICOM mASI research) proves worth further investigation, a long-term goal is to create a safe structure to create independent AGI without the associated risk. Using an Artificial Super Intelligence framework (even a mediated one) can act as a safe box for keeping an independent AGI inline \with our safety protocols. While this preliminary study is not yet able to be answered, the question as to if this is an effective containment methodology is the long-term goal should the line of research prove worth additional investment.

Our supposition is that these goals will provide the basis for the context needed to determine the value of the hypnosis and to justify, or not, further research along these lines related to mASI ICOM AGI systems.

3 Elemental Framework – Research Groups

The proposed structure for this preliminary study includes three core test groups as defined here including one subgroup or rather two control groups. Each group provides some fundamental basis for comparing and contrasting verses the other groups. Those groups include:

Group 1.0 "No Group, Control Group, Proctored"—this group should be at least 30 randomly selected humans of various demographics (meaning no preference for male, female, education or other factors) that will be proctored or supervised in their testing.

Group 2 "In Person Group Collective"—this could be done as more than one group, but for the purpose of this preliminary study should be at least 30 adult humans that are administered a test over a given venue collectively where their group is able to communicate with each other to execute the tests given to groups 1.0 and 1.1. This should provide a comparative framework to compare humans in groups vs individual humans where the underlying supposition is that humans can perform better in groups.

Group 3 - "Mediated Artificial Intelligence System" or mASI where an instance of an mASI using the ICOM cognitive architecture is used consisting of at least 10 contextual generating nodes as well as a standard ICOM context engine to execute individual tests on its own proctored as in group 2. The supposition is that this gives is a preliminary comparison to group's 1 and 2 to compare the mASI vs humans in groups and individual humans. This comparison and analysis should provide some evidence to support the hypothesis, allowing a determination as to whether further research is warranted.

4 Program Information Security and Policies

Human participants' information, especially identifying information, must be kept secure and separate from reported results. There will be no way to associate specific data with individual human subjects. This means that all published data will be scrubbed, only used in a collective way. Demographic data is then used only for high-level comparisons and used in the abstract. The structure of this includes all subjects will be given a demographic survey and assigned ID. Demographic data will not be directly associated with any individual, but with IDs, with that data stored only in an Air Gap-level secure system (Rouse), with no internet connection for the scope of the study, and all copies with ID values will be deleted or destroyed with only the GAP level secure documents stored in a digital archive. Assigned ID and contact data will be kept separate from the demographic files and only stored in this secure manner to protect the human subjects. After each survey is collected and t data transferred and split, the demographic survey results will be deleted from the collection service.

1. Subject Email: [textbox]
2. Age: [textbox]
3. Biological Sex: [picklist]
   a. Male
   b. Female
4. Technical Skills and Education (Useful information includes degree of higher education, highly technical skills like software engineer or AI especially): [text box]
5. At what socio-economic level were you raised (please make a best guess)? [picklist]
   a. 0–30 k USD net yearly income
   b. <30 k–60 k USD net yearly income
   c. <60 k–100 k USD net yearly income
   d. <100 k–180 k USD net yearly income
   e. <180 k and up in USD net yearly income
6. Do you have an atypical mental architecture? [picklist]

   a. ASD
   b. OCD
   c. Autistic
   d. Other
7. What other social groups do you identify with [picklist]
   a. Are Religious
   b. Transhumanist
   c. Foundationer*
   d. ZS *
   e. Atheist
   f. Agnostic
   g. Volunteer youth programs (i.e. Scouts, 4H, etc.)?
   h. Academia
   i. Teacher (k-12) Work in Highly technical field
   j. Work in field of Data Science Scientist
   k. Other
8. What other information would you like to share? [text field]

*Note: These are organizations we pinged to volunteer and only need to know how effective that outreach is as this is not really connected with the demographic analysis.

Some of these questions will not be used in this initial or preliminary study, but there is significant research evidence that they affect the group and collective intelligence (Woolly) values and given correlations with other studies we are collecting this in case we do additional research with a wider audience so that we can correlate this data, the new study data and these other related results from external studies.

5 Tests and Measures

There are three sets of test types that were considered for this study including an analysis tests for subjects, for use in further research, but will not be evaluated in this preliminary study, then qualitative and subjective tests as follows:

5.1 Demographic Analysis

These tests are designed to get a general survey of the demographics of the human subjects in the studies, where the primary reason is a further correlation with additional research that may be done after this point. These include the initial survey and possible additional surveys, as might be later defined separately from the initial questions listed above. Such data is kept separate from primary research data as per the secure information policy for this study.

5.2 Qualitative Intelligence Tests

Intelligence Quotient (IQ) tests are tests designed to measure "intelligence" in humans (Grohol) where we are using short versions to assess only relative trends or the potential for further study, whereas given the expected sample size results will not be statistically valid, nor accurate other than at a very general level, which is believed to be enough to determine if the line of research is worth going down. Of these tests, two types will be used in the study, one a derivative of the Raven Matrices Test (Cockcroft) designed to be culturally agnostic, and the Wechsler Adult Intelligence Scale (WAIC) (Coalson) Test. Lastly falling into the category of WAIC there is a baseline full Serebriakoff MENSA test that can be applied to compare and contrast scores between the two baselines tests.

After further research into various tests, it was determined that a new test developed at the University of California (University of California Matrix Reasoning Test) takes much less time (−10 min) and provides a better predictive power in terms of IQ score. This is a computer-based test that will be the main qualitative test we use in the study (Petawawa) (Pahor).

Collective Intelligence (CI) Test—While we would like to use this test, the information for executing this test is not publicly accessible and reaching out to the researchers that created this test has produced no response thus far but we will continue to try to get access to this test (Edgel).

5.3 Extended Meta Data and Subjective Tests

A number of tests or measures will be collected, more oriented towards analysis for further study, primarily around correlative purposes. None of these tests may be used outside of as possible illustrative examples, without being statistically valid given the rigor or subjective nature of these measures. These tests if considered would be outside the scope of the initial study.

The Turing Test—this test is not considered quantifiable and there is debate over whether this measure tells us anything of value, however, we will execute this test as a reference value. (Merriam) The Porter Method—This appears to be a qualitative test, but individual question measures are entirely subjective; therefore, the test lacks the level of qualitative-ness to be valid without a pool of historical values to measure against. However, we will execute this test as a reference value. Essentially, the test produces a consciousness measure score that is intended to show the degree of consciousness any given test subject exhibits. This test is administered by a third party and not the subject albeit the administrator needs to have access to our experience with the test subject (Porter).

The Yampolskiy Qualia Test—is a subjective measure of a subjective "thing" (i.e., qualia) and therefore not a qualitative measure. However, we will execute this test as a reference value. In theory, this only tests for the presence of Qualia in human-like subjects, passing this test does not mean that a subject does not experience qualia in the sense of the paper, just that it was not detected. This means that subjects may show signs of qualia, or not, but the test does show if they don't experience it. In other words, the test can be used to rule out the presence of qualia (Yampolskiy).

6 Experimental Results Analysis

To ensure that we don't succumb to common analysis fallacies (Trochim), such as asserting the existence of statistical patterns for which there is not solid evidence or interpreting evidence in accord with common cognitive biases (e.g. confirmation bias), it is important that the study is based on a sample of sufficient size to offer reliable statistical power. Additionally, we must use multiple statistical measures of the mASI system's capabilities, comparing and correlating between them. Such a rigorous "battery" approach to statistical assessment will not only give us a fuller and more robust picture of what the system can and cannot do, but will also make it easier to assess where additional or alternative measures would improve the analysis by filling any gaps in our understanding, and testing unverified conclusions based on single measures and underlying assumptions.

The primary analysis in this preliminary study will be to see if there is any evidence of a differential between IQ tests of the four test groups (1.0, 1.1, 2, and 3) If such a clear difference is present, such an indication in the positive meaning that the mASI group shows significant evidence of being more intelligent than the other samples, which would support the hypothesis. The only real conclusion from the intended sample sizes would be whether to proceed or not with further research. To that end, other tests or analysis would be subjective, and while interesting would not in themselves support the primary research objective. If the results show no evidence of an mASI system being more intelligent than groups 1, 1.1, and 2 then the mASI program would most likely be shut down or fundamentally revised.

7 Further Research

The ICOM AGI program this preliminary study is associated with for future research would include a more comprehensive study if results come out in expected ranges (expected ranges going back to the hypothesis that collective intelligence systems are smarter than humans and humans in groups) supporting our hypotheses and the validity of the mASI paradigm. If results do not support the hypothesis then we would need to go back to the fundamentals of the ICOM architecture and review theoretical details to look for a new approach or otherwise identify invalid assumptions or other flaws. Further research with supporting data then could also include bias filtering in mASI systems, and other studies such as fully proctored IQ test that would be run to compare against the performance of study subject groups.

Further and more speculative research could potentially address a wide range of interesting issues. For now, we will highlight four such avenues of potential future research:

(1) Research focused on improving the behavioral statistics model used to assess mASI capabilities, to ensure that even moderate statistical power can be leveraged efficiently to produce reliable results. As explained earlier, this can be achieved by developing a battery of cognitive-behavioral measures which can be compared and correlated by a variety of established statistical methods.

(2) Safety is a perennial concern in speculation about future AI research, often with a focus on the need to keep AI systems isolated from sensitive systems or public networks (Yampolskiy). Although it is far from a foolproof safety measure, the mASI paradigm offers a degree of failsafe protection by making human decisions an integral part of AI core function, meaning that a well-designed study should always be able to disable the system quickly and safely, if necessary. At the very least, the mASI paradigm would provide a pragmatic basis from which to begin empirically testing safety concerns which have been largely philosophical up until this point.

(3) Regarding philosophical matters, we may also note that some (Searle) have argued the Turing Test to be incomplete, or conceptually flawed, and so not testing for true conscious awareness at all. Such philosophical arguments do not appear to be falsifiable and thus of any value on the level of empirical science, but they do point the way toward interesting avenues of exploratory AI research which could deepen our understanding of different cognitive architectures. For example, the Chinese Room argument (Searle, Hauser) raises questions of symbol grounding in AI systems, which is a matter addressed by the ICOM architecture (Kelley) and mASI systems in particular.

(4) The increasing presence and capabilities of AI is creating a growing interest in, and concern over, intelligent machine ethics. In the context of this paper—i.e., an ICOM-based mASI system—the solution is more complex in that this approach to advanced human-analogous intelligence requires training, learning and open development rather than predetermined hard coding. To that end, we are investigating various potential solutions to this critical aspect of developing independent, self-aware AGI/ASI systems.

It would be valuable to see if an mASI system's responses could be distinguished from those of an individual human not only in general, but when the respondent (system or individual human) refers to the meaning (or "ground") of their own expressed concepts. Not only would it be interesting to see if this "self-reflective" capacity would make the system more likely to pass a Turing Test (i.e., to be less distinguishable from a human, contra the Chinese Room thought experiment), but it would also demonstrate the potential for human-like metacognition, which is a popular candidate for the psychological mechanism underlying human conscious self-awareness.

8 Study Framework Conclusions

Conclusions based on the process of producing this preliminary study framework include a couple of points on how that study will be executed, including the likelihood of a lower limit of approximately 100+ subjects with at least 30-person groups in each venue of any further studies, as a matter of ensuring sufficient statistical power. Despite the small sample size typical of a pilot study, it should still be possible to determine whether the findings demand a deeper or more rigorous future research program to assess the value and viability of mASI systems. That program would have a particular focus on mASI mental performance in addition to studies of filtering for cognitive bias, conditioning and training as well as opening the way to an AGI safety framework based upon the mASI paradigm. Given the great potential value of such systems and future research, both in terms of practical applications and moving old philosophical discussions forward into the realm of empirical testing, initial research into the viability of the mASI paradigm is necessary.

References

1. Engel D, Woolley A, Chabris C, Takahashi M, Aggarwal I, Nemoto K, Kaiser C, Kim Y, Malone T (2015) Collective intelligence in computer-mediated collaboration emerges in different contexts and cultures. In: Bridging communications, CHI 2015, Seoul, Korea
2. Coalson D, Weiss L (2010) Wechsler adult intelligence scale the perceptual reasoning index (PRI) is a measure of perceptual and fluid reasoning, spatial processing, and visual–motor integration. Science Direct, WAIS-IV clinical use and interpretation 2010
3. Cole D (2004) The Chinese room argument, March 2004, revised 2014. Stanford Encyclopedia of Philosophy. https://plato.stanford.edu/entries/chinese-room/. Accessed Jan 2019
4. Cockcroft K, Israel N (2011) The Raven's advanced progressive matrices: a comparison of relationships with verbal ability tests. PsySSA, SAJP. SAGE Journals. https://doi.org/10.1177/008124631104100310
5. Grohol J (2019) IQ test. Psych central. https://psychcentral.com/encyclopedia/what-is-an-iq-test/. Accessed 4 Apr 2019
6. Hauser L (2019) Chinese room argument. Internet Encyclopedia of Philosophy. https://www.iep.utm.edu/chineser/. Accessed Jan 2019
7. Kelley D (2018) The independent core observer model computational theory of consciousness and the mathematical model for subjective experience. In: ICNISC 2018. ISBN-13: 978-1-5386-6956-3
8. Kelley D Architectural overview of a "mediated" artificial super intelligent systems based on the independent core observer model cognitive architecture. Informatica. (Pending review)
9. Kelley D, Waser M (2018) Human-like emotional responses in a simplified independent core observer model system. Procedia Comput Sci 123:221–227
10. Malone T (2018) Superminds – the surprising power of people and computers thinking together. Little, Brown and Company. ISBN-13: 9780316349130
11. Merriam-Webster (2019) Turing test. https://www.merriam-webster.com/dictionary/Turing%20test. Accessed Apr 2019
12. Motzkin E (1989) Artificial intelligence and the chinese room: an exchange. New York Review of Books, 36:2, 16 February 1989. Reply by John R. Searle
13. Pahor A, Stravropoulos T, Jaeggi S, Seitz A (2018) Validation of a matrix reasoning task for mobile devices. Springer link – behavior research methods, 26 October 2018. https://link.springer.com/article/10.3758/s13428-018-1152-2
14. Pittalwala I (2018) UC Psychologist devise free test for measuring intelligence. University of California. https://news.ucr.edu/articles/2018/10/29/uc-psychologists-devise-free-test-measuring-intelligence
15. Porter H (2016) A methodology for the assessment of AI consciousness. In: AGI 2016, Portland State University
16. Rouse M (2019) Air gapping (air gap attack). whatis.com, April 2019. https://whatis.techtarget.com/definition/air-gapping

17. Searle J (1980) Minds, brains and programs. Behav Brain Sci 3:417–457
18. Searle J (1984) Minds, brains and science. Harvard University Press, Cambridge. https://academiaanalitica.files.wordpress.com/2016/10/john-r-searle-minds-brains-and-science.pdf
19. Searle J (1990a) Is the brain's mind a computer program? Sci Am 262(1):26–31
20. Searle J (1990b) Presidential address, vol 64. Proceedings and Addresses of the American Philosophical Association, pp 21–37
21. Searle J (1998) Do we understand consciousness? (Interview with walter freeman). J Conscious Stud 6:5–6
22. Searle J (1999) The Chinese room. In: Wilson RA, Keil F (eds) The MIT encyclopedia of the cognitive sciences. MIT Press, Cambridge
23. Searle J (2002a) Twenty-one years in the Chinese room. In: Preston and Bishop (eds.) 2002, pp 51–69
24. Searle J (2002b) The problem of consciousness. In: Consciousness and language. Cambridge University Press, Cambridge, pp 7–17
25. Searle J (2010) Why dualism (and materialism) fail to account for consciousness. In: Lee RE (ed) Questioning nineteenth century assumptions about knowledge, III. SUNY Press, Dualism
26. Serebriakoff V (1996) Self-scoring IQ tests. Sterling/London, 1968, 1988, 1996. ISBN 978-0-7607-0164-5
27. Trochim W (2018) Threats to conclusion validity, October 2018. http://www.socialresearchmethods.net/kb/concthre.php
28. Woolly A (2018) Collective intelligence in scientific teams, May 2018
29. Yampolskiy R (2019) Artificial intelligence safety and security. CRC Press, London/New York. ISBN: 978-0-8153-6982-0
30. Yampolskiy R (2018) Detecting qualia in natural and artificial agents. University of Louisville

Development of an Optimal Production Plan Using Fuzzy Logic Tools

Maksim Kluchnikov, Elena Matrosova(✉), Anna Tikhomirova(✉), and Svetlana Tikhomirova

National Research Nuclear University MEPhI (Moscow Engineering Physics Institute), Moscow, Russia
maksim-kluchnikov@mail.ru, matrosova_ev@inbox.ru, {anna7909966, tikhomirova3112}@yandex.ru

Abstract. The present article is devoted to the problem of increase of the production planning process efficiency. There is considered the task of production plan development with the help of intellectual analysis of the current situation according to the key criteria. The authors offer the use of mathematical tools of fuzzy logic in order to analyze the production situation and to make recommendations on the choice of the optimal production plan, as well as a checking mechanism of the consistency of recommendations with the actual preferences of the person taking management decisions.

Keywords: Decision-taking process · Automated decision support system · Machine learning · Intelligent information system · Production plan · Fuzzy logic

1 Introduction

Company success engaged in production directly depends on the speed of its management body response to the appearance of new orders and the ability to correctly determine the order of their execution in order to comply with the production terms stated to the customers. The key feature is a flexible approach to the production plan development and its prompt adjustment.

The production plan is an integral part of any business plan, in which all production and working processes should be described. The present document records the volume and procedure of goods production or the service provision with the corresponding information: the volume of used raw material, the employment of workers of various professions and the applied equipment. The task of immediate production plan adjustment becomes especially important while planning release of made-to-order products, when each product is unique in its technical characteristics, production time and required resources. By way of illustration the present article will consider the company, whose main activity is the design, manufacture and installation of steel door blocks and safes. All products are unique and made to order. The production process of made-to-order products has extremely high labor costs, which in its turn affect the production time. The production lead time is from 30 to 60 days, but the agreement prescribes the production time from 45 to 75 days, thus, the company insures itself in

A. V. Samsonovich (Ed.): BICA 2019, AISC 948, pp. 211–218, 2020.
https://doi.org/10.1007/978-3-030-25719-4_27

case of force majeure situations, illness of key contractors or interruptions in material supply. Due to the irregular receipt of new orders, their heterogeneity and regular changes made by customers, the plan should be reviewed at least once per day. This is a huge intellectual problem, the solution of which cannot be based on rigorous mathematical methods, since the optimality criteria themselves are fuzzy, in addition, the decision maker often follows intuitive notions about the priority of certain criteria [1, 2]. One of the solutions is the creation of information decision support system using a fuzzy logic device.

2 Application of Fuzzy Logic Methods

2.1 Fuzzy Logic

Fuzzy inference is dependence approximation $Y = f(X_1, X_2, \ldots, X_n)$ of each output linguistic variable from a set of input linguistic variables and obtaining a conclusion in a form of fuzzy set corresponding to the current input values using a fuzzy knowledge base and fuzzy operations [3]. Linguistic variables are input and output parameters of the analyzed system.

For process description there is used a set of utterances of the following type:

L_1: **if** A_{11} and/or A_{21} and/or … and/or A_{1m}, **then** B_{11} and/or … and/or B_{1n},
L_2: **if** A_{21} and/or A_{22} and/or … and/or A_{2m}, **then** B_{21} and/or … and/or B_{2n},

…………………

L_k: **if** A_{k1} and/or A_{k2} and/or … and/or A_{km}, **then** B_{k1} and/or … and/or B_{kn},

where $A_{ij}, i = 1, 2\ldots, kj = 1, 2, \ldots, m$—fuzzy utterances defined according to the values of input linguistic variables, and $B_{ij}, i = 1, 2\ldots, kj = 1, 2, \ldots, m$—fuzzy utterances defined according to the values of output linguistic variables. This set of rules is called a fuzzy knowledge base.

2.2 Mathematical Model of the Production Plan Development

While determination of the planned production terms and the product transition from stage to stage in advance at the stage of order acceptance, it is necessary to take into account the current composition of labor resources, their specialization, skills and ability to execute a particular order, as well as the forced waiting for the order of some rare parts. Thus, for effective management there is set a classical task of production process planning in conditions of limited resources.

Each order has a deadline agreed with the customer, the excess of which can result in fine payment, customer rejection from the order or the loss of reputation. The force majeure circumstances, which can arise from time to time, do not provide the possibility of working completely without overdue, therefore there should be established a maximum allowable threshold level, above which the values should not be taken by the indicator. We will call this indicator DELAY = “Number of outstanding orders” and we will calculate its value as a percentage of overdue orders of their total number in the present plan.

The next indicator is the number of reserve days. The number of reserve days is the difference between the due date (when the order should be made according to the agreement) and the expected production time (according to the set production plan). The value of this indicator for the set of current orders can make a significant impact on the order of their direction to a particular production stage. Few number of remaining reserve days indicates an increased risk of order overdue and the necessity for its priority execution. The plan [4] is considered optimal, when the number of such spare (reserve) days is above zero for each order and at the same time the total number of reserve days for all orders specified in the plan is maximized. For plan estimation, we will introduce the indicator RESERV = "Number of reserve days". In order to take into account the current production workload while estimation, we will calculate its value as the arithmetic mean value calculated for each order as the number of reserve days in % of the total duration of the order itself. The downtime period of operation personnel is also a very important indicator of production efficiency. We will denote the downtime period as FREE = "personnel downtime during a month". In order to take into account the current number of employees while estimation, we will calculate the value of this indicator as the percentage of the total number of man-days during which there was no any production load to the total number of man-days per month.

According to these indicators and their expert estimation, we can make a fuzzy knowledge base, the use of which will make it possible to provide an integrated estimation of several production plans for recommendation formation for the choice of the optimal plan. In the future, this base should be adjusted and supplemented according to the evaluation of received recommendations by the person taking the final decision.

2.3 Application of Fuzzy Logic Tools

The fuzzy knowledge base of the considered problem according to the expert commentary at the primary stage was formed as follows.

L_1: The production plan can be called "optimal" if the value DELAY is very low, RESERV is sufficient, and FREE is absolutely insignificant.

L_2: The plan is considered "satisfactory" if DELAY is low and (RESERV is sufficient or FREE is insignificant).

L_3: The plan is considered "unsatisfactory" if DELAY is high or FREE is significant.

The membership function made by experts for each criterion is different and is shown in Fig. 1. They reflect the acceptability of values deviation of characteristics of a particular plan from their optimal values.

As an example, there are considered three plans with the following indicators (Table 1).

According to the graphs of the membership functions at the next step we calculate the values corresponding to the quantitative characters of the plans, while using the classical negation (for functions "DELAY is high" and "FREE is significant") and the concentrating operation (for functions "DELAY is very low" and "FREE is completely insignificant") for construction of the missing membership functions. The result of fuzzy rules application L1, L2, L3 to the obtained values is presented in Table 2.

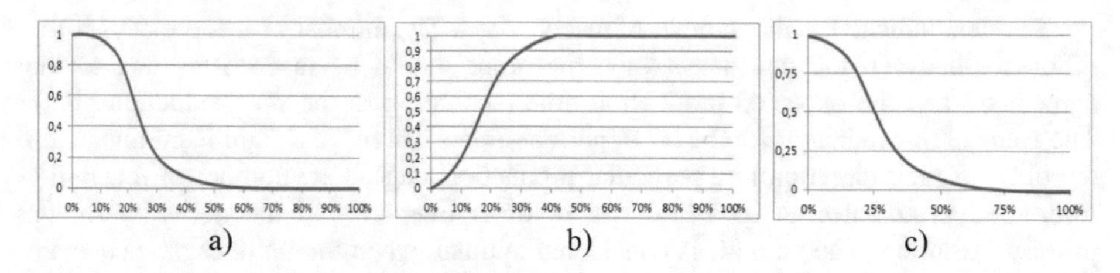

Fig. 1. Membership function for criteria (a) DELAY "Low" number of outstanding orders (b) RESERV "Sufficient" number of reserve days on average (c) FREE "Insignificant" number of downtime days

Table 1. Quantitative characters of the comparable production plans

	DELAY	RESERV	FREE
Plan E_1	15%	20%	15%
Plan E_2	10%	3%	30%
Plan E_3	20%	25%	20%

Table 2. Quantitative estimation of premises according to fuzzy rules

	Plan E_1	Plan E_2	Plan E_3
Rule L_1	0,6	0,04	0,2
Rule L_2	0,8	0,2	0,6
Rule L_3	0,3	0,8	0,4

The plan estimation in a fuzzy knowledge base is carried out by fuzzy methods; therefore, for its interpretation it is necessary to set the membership functions (1):

$$\begin{cases} \mu^3 & \text{optimal plan} \\ \mu & \text{satisfactory plan} \\ 1-\mu & \text{unsatisfactory plan} \end{cases} \tag{1}$$

Fuzzy logic allows using various options in order to calculate the implication operation, in the present work there was decided to use the Lukasevich formula (2).

$$\mu_{A \to B} = \min\{1; 1 - \mu_A + \mu_B\} \tag{2}$$

The implication formula determines the fidelity (truth) of logical derivation based on the truth of the premise and consequence. Thus, there are determined the estimates of all plans according to each rule and then there are calculated the minimums according to all rules (Table 3). In such case the minimum value corresponds to the intersection operation from the group of maximum operations for fuzzy sets [5].

Table 3. Final estimates of the compared plans according to the method of fuzzy inference

	0	0,1	0,2	0,3	0,4	0,5	0,6	0,7	0,8	0,9	1
Plan E_1	0,25	0,35	0,45	0,46	0,50	0,56	0,65	0,78	0,90	0,80	0,70
Plan E_2	0,80	0,90	0,97	0,99	1,00	1,00	1,00	0,90	0,80	0,70	0,60
Plan E_3	0,40	0,50	0,60	0,70	0,80	0,90	0,90	0,80	0,70	0,60	0,50

In graphic form the results are presented in Fig. 2. These graphs reflect the possibility of fact that the plan will have an integral characteristic indicated along the x-axis.

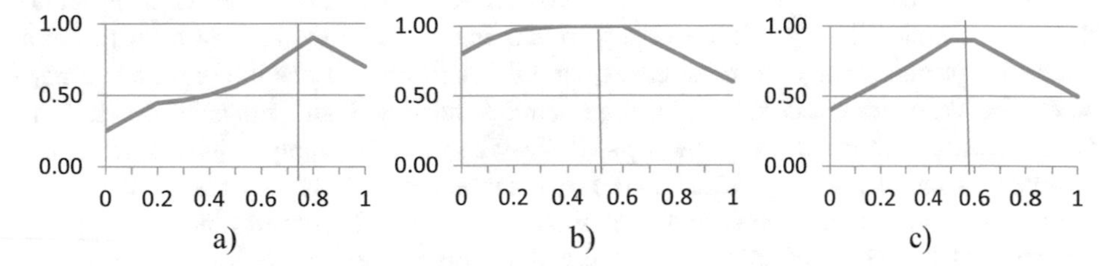

Fig. 2. Graphic interpretation of production plans: (a) Plan E_1, (b) Plan E_2, (c) Plan E_3

In order to compare the fuzzy sets presented in this way, we go to their point estimates [6]. For all alternatives (production plans), we calculate the level sets E_{ja}, the power $M(E_a)$ of such sets using the formula (3) and the point estimates of plans using the formula (4).

$$M(E_{ja}) = \sum\nolimits_{i=1}^{n} \frac{x_i}{d_a} \tag{3}$$

$$F(E_1) = \frac{1}{a_{max}} \sum\nolimits_{i=0}^{a_{max}} M(E_{1a}) da \tag{4}$$

According to calculations, there were obtained the following results: $F(E_1) = 0.729$, $F(E_2) = 0.519$, $F(E_3) = 0.587$. The obtained values are marked on the graphs (Fig. 2) with lines dividing the figure under the curve into two equal in area parts. According to the algorithm, the E1 plan is the most preferable, as its point estimate is maximum, however, these rules cannot fully take into account the current production situation, and it is recommended to check and, possibly, correct the recommended decision by means of additional inquiry of DT (decision taker).

2.4 Application of the Hierarchy Analysis Method for Checking Results

According to the use of mathematical tools of fuzzy logic, there were obtained the quantitative estimates of production plans and on their basis there were formed the corresponding recommendations. However, the current production situation is changing daily and the existing fuzzy knowledge base cannot fully reflect the actual notions of DT about the importance of the analyzed criteria. The use of combination of several algorithms for receiving recommendations will increase the flexibility and accuracy of the decision support system [7].

Obviously, DELAY and FREE criteria are contradictory, since almost always the attempt to reduce the percentage of delayed orders during the production plan development leads to the increase in the risk of personnel downtime at the certain production stages. A similar situation with a pair of criteria RESERV and FREE. In such situations, for decision taking it is necessary to take into account the criteria priorities at a particular planning stage. In most cases, the DELAY criterion has the highest priority, since the delay in terms results in significant reputational and financial costs. It is always convenient to have reserve days (RESERV), but, for example, in case when the number of employees is critical and the dismissal of employees dissatisfied with downtime can cause problems, thus, the weight of the FREE criterion increases during decision taking. These circumstances can change the real representations of DT on the contribution of various criteria to the integral estimation of plans contained in the fuzzy knowledge base, therefore it is suggested to independently evaluate the criteria weights using the hierarchy analysis method of T. Saati. For this purpose, there is made a matrix of expert estimates, which compare the criteria by its importance level in pairs. The scale from 1 to 9 was offered as the rating scale of Saati [8].

The weights are calculated on the basis of the geometrical mean of the obtained data, then the result is checked for consistency, which is possible due to the redundancy of the carried out paired comparisons. As a result, there were obtained the following weights: w(DELAY) = 0.540, w(RESERV) = 0.163, w(FREE) = 0.297. According to the values of the production plans criteria (k_i) we calculate the normalized estimate of each plan E_i for each criterion, and for the RESERV criterion the calculations are carried out in a standard way (the more reserve days we have, the better the plan is), while for the DELAY and FREE criteria we use the return values (the less outstanding orders and the less personnel downtime we have, the better the plan is). All intermediate values and final estimates are presented in Table 4.

Table 4. Final estimates of the compared plans according to the hierarchy analysis method

	Weight	Plan E_1		Plan E_2		Plan E_3	
	w	k_1	$w * k_1$	k_2	$w * k_2$	k_3	$w * k_3$
DELAY	0,540	15%	0,166	10%	0,249	20%	0,125
RESERV	0,163	20%	0,068	3%	0,010	25%	0,085
FREE	0,297	15%	0,132	30%	0,066	20%	0,099
Final estimates			**0,366**		**0,325**		**0,309**

The analysis of possible plans, carried out on the basis of the hierarchy analysis method, has also pointed to the E1 plan as the most preferable, which can serve as the confirmation of the fuzzy knowledge base correctness. In case of discrepancies in results, an analysis of the specific situation and the choice of a plan in manual mode are necessary, while DT's task is to correct the fuzzy knowledge base so that the recommendations made by both algorithms coincide with each other and with the intuitive perception of the current situation by DT himself.

3 Conclusions

The demand for custom high–class products is often irregular in addition to the seasonal prevalence, but due to the high skill requirements of employees, their number in most of such companies is almost constant. The greatest difficulties in this regard arise during work schedule planning, which are carried our sequentially.

At the same time, there are strict time restrictions for each specific product, because each product is a custom order of a specific customer to whom there is set the certain production time. This term should be respected under any circumstances, thus, during planning process it is desirable to have several reserve days in case of force majeure circumstances. The additional criterion of the plan optimality can be considered the minimization of the number of days of personnel downtime, when there are no orders according to their specialization during the certain production stage.

Thus, there has been set the task of production process planning in conditions of limited resources and there have been specified three key indicators: timeliness of order execution, availability of reserve time and personnel load. In order to formulate recommendations regarding the choice of the optimal plan, it has been suggested to independently use two algorithms – the algorithm based on fuzzy inference and the algorithm based on the hierarchy analysis method. The agreed results of two algorithms allow increasing confidence in correctness of the taken decision, while the discrepancies revealed during the application process are a warning for the revision of the fuzzy knowledge base.

Acknowledgments. This work was supported by Competitiveness Growth Program of the Federal Autonomous Educational Institution of Higher Professional Education National Research Nuclear University MEPhI (Moscow Engineering Physics Institute).

References

1. Samsonovich AV (2015) Functional capabilities of biologically inspired by cognitive architectures. In: Lecture at the XVII All-Russian scientific-technical conference “Neuroinformatics-2015”. National Research Nuclear University MEPhI (Moscow Engineering Physics Institute), Moscow
2. Samsonovich AV, Klimov VV, Rybina GV (2016) Biologically inspired cognitive architectures (BICA) for young scientists. In: Proceedings of the first international early research career enhancement school (FIERCES 2016). ISBN: 978-3-319-32553-8 (Print) 978-3-319-32554-5 (Online)

3. Matrosova E, Tikhomirova A (2018) Intelligent data processing received from radio frequency identification system. In: Postproceedings of the 9th annual international conference on biologically inspired cognitive architectures, BICA 2018 (ninth annual meeting of the BICA society)
4. Kolesnikov SS (2000) Business strategy: resource and inventory management. The status quo 97
5. Ivin AA (2001) Modal theories Jan Łukasiewicz, 176 p
6. Shtovba SD (2001) Introduction in fuzzy sets theory and fuzzy logic. Vinnytsia National Technical University, 198 p
7. Matrosova EV, Tikhomirova AN (2016) Peculiarities of expert estimation comparison methods. Procedia Comput. Sci. 88:163–168. 7th Annual International Conference on Biologically Inspired Cognitive Architectures, BICA
8. Saati T (1999) Decision making. Hierarchy analysis method. «radio and communication» , 278 p

Internet Users Authentication via Artificial Intelligence

Konsantin G. Kogos(✉), Mikhail A. Finoshin, and Valentin A. Gentyuk

Cryptology and CyberSecurity Department, NRNU MEPhI, Moscow, Russia
{KGKogos,MAFinoshin}@mephi.ru

Abstract. The number of Internet users increases and the Internet is part of people's daily lives, as a result, the behavior of the user becomes free and informal. This is the basis of the assumption that the manner of user actions on the Internet has become a factor that can be used by authentication using artificial intelligence. In turn, existing works related to users' web browsing behavior-based authentication with using machine learning do not analyze some important behavioral user's characteristics, such as patterns of behavior or user behavior on a frequently visited resource. It causes to suggest own features and check their contribution to the accuracy of the system. The aim of this work is to study the possibility of introducing a map of clicks, bigrams, trigrams of frequent web pages and their domains, evaluation of the contribution of added features. In this work, we replace the web pages' genre classification by domain classification and don't take into account the spikes in views. We have created a system based on artificial intelligence. As a work result, we have shown a significant improvement in the accuracy of the system using the click map and a slight improvement in the use of bigrams and trigrams.

Keywords: Behavioral biometrics · Continuous authentication · Web surfing · Web pages · Artificial intelligence · Machine learning

1 Introduction

Statistics show that the number of Internet users is continually increasing. The analytical agency "We are social" for 2018 reported that the number of Internet users in the world reached 4,021 billion people, which is 7% more than in the same period last year [1]. The quantity of time users spend on the Internet has also increased: the average user spends about 6 h a day on the Internet [1]. As users use Internet resources in their daily lives, their online behavior becomes free and informal. Based on this, it is assumed that the data on the manner of actions when viewing web pages is enough to uniquely [2–4] or partially [5, 6] to characterize the user through unconscious behavioral patterns.

Nowadays authentication is executed in most automated systems only in case of logging in, so there is a risk of capturing an open session or bypassing standard security, for example, by peeping the password. In the case where the cost of damage from unauthorized access is high, it is unacceptable. Continuous or periodic authentication

A. V. Samsonovich (Ed.): BICA 2019, AISC 948, pp. 219–224, 2020.
https://doi.org/10.1007/978-3-030-25719-4_28

based on biometric or behavioral characteristics is used to prevent the threat of unauthorized access.

It is proposed to use the user's behavior on the Internet to solve this topical problem. The advantages of such a system are that it works imperceptibly for the user and does not require additional equipment. It can also be used in conjunction with other behavior modules, such as keyboard handwriting and mouse behavior.

2 Related Work

Currently, user behavior on the Internet is analyzed for commercial purposes, for example, in Google Analytics. However, more and more user behavior is analyzed during authentication.

Abramson and et al. [2–4] investigated the possibility of users' authentication by their browsing behavior on the Internet. They have created several authentication systems based on users' web browsing behavior which record users' clicks. The system records the name of the browser, the timestamp, and the URL when it detects a spontaneous click. In each of the considered works, a certain list of features is formed, and the classification algorithm is used. Table 1 shows the best results. The authors [5, 6] used a module based on user behavior when browsing the web, along with other behavioral modules:

Table 1. The results of the reviewed works.

Number	Number of persons	Average FRR	Average FAR
1 [2]	10	31.2%	12.8%
2 [3]	14	32%	32%
3 [4]	14	17%	17%

- keystroke dynamics, mouse movement, stylometry [5];
- stylometry, application usage and GPS location [6].

Using the analysis of user behavior on the Internet in the work of A. Fridman [5] reduced the FAR error by 19.25% and the FRR error by 41.90%. In the second case, the authors [6] calculated the contribution of the web module to the accuracy of the system: it is in second place after the GPS location.

These works have prompted a more in-depth study of the features used for authentication because the considered features can give a big mistake of the first kind when the user-owner behavior changes. To solve this problem, it is also proposed to consider the patterns of user behavior and use the click map of frequently visited resources in addition to the features used in these works. A behavior pattern is a frequent sequence of several resources visited sequentially. Resource click map (domain or URL) is an x * y matrix, in each cell of which the number of clicks in the equivalent area of the browser window is recorded.

3 Data Preprocessing

A sniffer was created in C++ for the Chrome browser and Chromium for the 64-bit Windows operating system. The program is added to the startup computer. The program checks whether the user has clicked every 0.1 of second and, if the active window is a browser, it records the following data in the log:

- click time,
- the coordinates of the mouse cursor,
- the coordinates of the browser window,
- the URL of the active tab.

The program was given to 7 users who prefer to use Chrome browser in everyday life. The period of data collection was from 8 to 11 weeks, and the number of collected clicks ranged from 12 thousand to 181 thousand for each user (see Table 2).

Table 2. Number of user clicks.

User number	Number of clicks		
	One week	Two weeks	All
0	4497	10498	51426
1	4764	17233	116018
2	4189	11057	47703
3	5993	20017	152260
4	2162	4230	16930
5	32419	47144	181182
6	829	860	12300

We used user data for a period of at least two weeks for the training sample, because in a shorter period of time it is impossible to say about the preferences of users on certain days of the week. The table shows that the number of user data varies greatly, so we divided users into two teams in order to be able to make an equal number of clicks on the sample. The first command consists of users 1, 3, 5. The second team consists of users 0, 1, 2, 3.

Authors [2–6] identified the following features in their works:

- user activity by days of the week;
- user activity by time of day;
- 20 most visited resources and frequency of each;
- pauses (less than 5 min) - time spent by the user on the web page, which is calculated as the difference between the timestamps of two consecutive page views;
- spikes of views—calculated as the change in the pause time between page views;
- the time between repeats is frequently viewed pages;
- the time between frequently viewed pages repeat;
- the frequency of bigrams of genres;
- user activity by days of the week for genres;

– user activity by time of day by genre;
– unique domains.

The use of genres has a number of disadvantages: not all URLs are in the database, classifiers are paid or have free versions that are slow to work, so we decided to use domain classification. We did not use spikes of views and unique domains. We have added the following features:

– 20 most visited domains (D);
– 15 frequent URLs' bigrams (BU) и 15 frequent domains' bigrams (BD);
– 10 frequent URLs' trigrams (TU) и 10 frequent domains' trigrams (TD);
– frequent trigram pauses (P);
– frequent URLs click maps (UM) and frequent domains click maps (DM);
– pauses when you visit the frequent URLs and domains.

Trigrams are one of the types of behavior patterns of the user. We imposed conditions on bigrams and trigrams: resources cannot be equal in a bigram, and trigram does not contain three equal resources.

The browser window obtained from the log was divided into 7 * 5 parts to make a click map. After that, the number of the area in which the click was made was fixed. The click map is especially important when users have several equal prefer resources.

Log data was uploaded into the database for analysis before formation sampling for classification. Data was recorded three tables: table of clicks, the table of bigrams and trigrams table. During the recording process, domains were allocated from the URL. During the analysis for each user were recorded: most visited URLs, most visited domains, frequent URLs' bigrams, frequent domains' bigrams, frequent URLs' trigrams, and frequent domains' trigrams.

Samples for classification were formed after the analysis ended. Sample vectors consisted of 5, 15 or 30 clicks. The ratio of the number of clicks of the training sample to the test sample is like 70 to 30.

4 Experimental Results

Further, various combinations of features were analyzed. Each combination was tested on machine learning algorithms such as random forest, SVM and logistic regression.

Several groups of features were analyzed. By default, in all cases, the features that have already been analyzed in related works were taken into account: day of the week, time of day, pauses, 20 most visited URLs, which we designated as "set 0". As a result of the analysis of various combinations, the sets that bring the greatest increase in accuracy compared to the "set 0" were identified (see Table 3). In Table 3, accuracy is presented for two groups of users, provided that the vector contains information about 15 clicks.

We concluded from the table that the use of frequent bigrams and trigrams of web pages and their domains does not always improve the accuracy of the system. Besides, the domain click map improves accuracy from 2% for any set of objects (see Fig. 1).

Table 3. Feature set.

Set's number	D	DM	UM	BD, TD	P	BU, TU	Accuracy	
							3 users	4 users
0	-	-	-	-	-	-	63.0%	48.4%
1	+	+	-	-	-	-	84.0%	61.1%
2	+	+	+	-	-	-	86.2%	61.0%
3	+	+	-	+	+	-	85.7%	61.2%
4	+	+	+	+	+	+	81.7%	62.0%

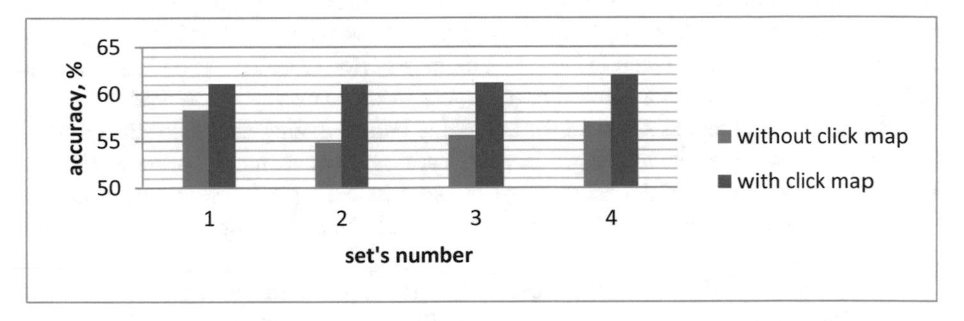

Fig. 1. Improved accuracy when using the click map for a team of 4 users.

A perceptron with 100 neurons in a hidden layer was built. The number of neurons in the input layer is equal to the length of the sample vector. The number of neurons in the output layer is equal to the number of users. We compared the accuracy of perceptrons trained on a different number of epochs in the range from 25 to 150. Under the given conditions, the best accuracy of the perceptrons for the sets from Table 3 was determined. Of these, the perceptrons were selected with the best accuracy of 5, 15 and 30 clicks. As a result, we define machine learning algorithms that give the best results with a different number of clicks (see Table 4).

Table 4. The accuracy of the system for the different number of clicks in the vector.

Number of users per team	The number of clicks in the vector	ML algorithms	Average FRR	Average FAR
3	5	Perceptron	16.4%	5.5%
	15	SVM	13.8%	4.6%
	30	SVM	11.9%	4.0%
4	5	Perceptron	44.9%	11.2%
	15	Logistic regression	38.0%	9.5%
	30	Perceptron/logistic regression	33.2%	8.3%

Thus, the click map improves the accuracy of the system. Trigrams and bigrams of web pages and their domains do not always make a positive contribution to the accuracy of the system. The obtained results are better than in other works with the exception of FRR error in the work [4], but the comparison cannot be considered complete because of the different number of users.

5 Conclusion

The work is devoted to authentication by user behavior based on web browsing with using artificial intelligence. The domain-based classification was introduced. We also proposed to use of click maps, frequent bigrams of the web pages and their trigrams. The increase in accuracy when using the click map is from 2%. The use of bigrams and trigrams does not provide a stable improvement in the accuracy of the system.

We have implemented a system for collecting and analyzing user data and a system for generating samples for training. Machine learning models and neural network were created. The estimation of the accuracy of the system and a preliminary comparison of the results with the existing work were done: for 5 clicks FRR error is 44.9% and FAR error is 11.2%.

The actual direction of further research is to improve the system of collecting information about the user and the transition to the realization of the authentication system based on the method of anomaly detection because most often we do not have the data of the offender.

References

1. Digital in 2018: world's internet users pass the 4 billion mark. https://wearesocial.com/blog/2018/01/global-digital-report-2018. Accessed 17 Feb 2019
2. Abramson M, Aha D (2013) User authentication from web browsing behavior. In: Twenty-sixth international FLAIRS conference, pp 268–273
3. Abramson M (2014) Learning temporal user profiles of web browsing behavior. In: 6th ASE international conference on social computing (SocialCom 2014), pp 1–9
4. Abramson M (2014) Active authentication from web browsing behavior. In: Smart card alliance conference, pp 1–16
5. Fridman A, Stolerman A, Acharya S, Brennan P, Juola P, Greenstadt R, Kam M (2013) Decision fusion for multimodal active authentication. IT Prof 15:29–33
6. Fridman L, Weber S, Greenstadt R, Kam M (2017) Active authentication on mobile devices via stylometry, application usage, web browsing, and GPS location. IEEE Syst J 11:513–521
7. Lau T, Horvitz E (1999) Patterns of search: analyzing and modeling web query refinement. In: Kay J (eds) UM99 user modeling. CISM international centre for mechanical sciences (courses and lectures), vol 407. Springer, Vienna, pp 119–128

Biologically Inspired Algorithm for Increasing the Number of Artificial Neurons

Lyubov V. Kolobashkina(✉)

National Research Nuclear University MEPhI
(Moscow Engineering Physics Institute),
Kashirskoe Shosse 31, Moscow 115409, Russia
LVKolobashkina@mephi.ru

Abstract. The constant increase in the complexity of artificial neural networks (ANN) is highlighted as one of the trends of their modern development. An analysis of the main approaches to determining the required size of ANN is given. The most interesting approach is based on the reduction of the size of the network. As a criterion that necessitates the exclusion of communication between neurons, the low level of significance of this connection is used. It is shown that the implementation of the principle of changing the structure of ANN "from complex to simple" contradicts the general biological principle of development "from simple to complex". A biologically based approach to the development of ANN is considered in accordance with the principle "from simple to complex". As a criterion that necessitates the "birth" of a new neuron, it is proposed to use ambiguity in determining, first of all, the sign of the weighting factor of at least one of its inputs. The need to implement the training procedure is also emphasized on the basis of the above biologically based approach. An example of changing the structure of a neural network (NN) in accordance with the proposed algorithm is considered. The possibility of obtaining ANN with a non-trivial structure, which differs from the frequently used multilayer structure in practice, is underlined.

Keywords: ANN · ANN growth algorithm · Biologically inspired approach

1 Introduction

ANNs are widely used nowadays for solving problems of recognizing graphic and acoustic images, classification and forecast. Modern computational tools make it possible to implement, in the first place, a software ANN of enormous size [1]. The implementation of large ANNs necessitates the use of a large amount of memory, primarily used to store the weight coefficients of the neurons of the network. The term DFF (Deep Feed Forward), in particular, is used in practice to characterize such networks in terms of their layering.

Studies have shown that for a given size of the training examples there are optimal sizes of the NN, ensuring the maximum quality of the resulting solution [2]. The quality of the NN, in the first place, refers to the accuracy and stability of the resulting solution. Therefore, for typical problems, usually there is an experimentally determined optimal network size. These dimensions should be taken as the upper limit of the size

A. V. Samsonovich (Ed.): BICA 2019, AISC 948, pp. 225–231, 2020.
https://doi.org/10.1007/978-3-030-25719-4_29

of the network, since its further increase does not lead to noticeable changes in the quality of the result.

Studies [3, 4] analyze an interesting approach, which in some cases makes it possible to simplify the network complexity by an order of magnitude. Simplification of the network is achieved by removing from it uninformative links and neurons characterized by a low probability of operation. Communication data and neurons are detected based on the results of learning a NN with examples. As a result, only significant connections between "actively" working neurons remain in the network. The implementation of this approach in practice involves the initial use of a NN with size limits.

Quite often, the size of the applied NN is determined based on the existing restrictions on its complexity with its software or hardware implementation (number of neurons, memory size, number of connections), or based on the level of power consumed and requirements for its speed [5].

As a factor determining, for example, the size of the inner layers of a NN, the learning rate can be used [6].

The analysis of the above practical approaches to determining the size of a NN allows us to reveal a regularity, the essence of which lies in the implementation of the principle "from complex to simple".

It is obvious that this principle is not the only possible one and contradicts the natural principle of the development of living systems, which can be formulated as "from simple to complex".

In accordance with this principle, ANN should increase in the process of its learning to optimal size. The study of possible growth mechanisms of ANN in accordance with the biologically sound principle "from simple to complex" seems to be highly relevant at the present time.

The aim of the work is to consider an approach that involves the implementation of a biologically based algorithm for increasing the number of artificial neurons.

2 The Essence of the Proposed Approach

The human brain is an extremely complex multi-level system, containing areas with different organization. The various ANN models used in practice only to one degree or another approximation describe their functioning.

In the process of human development, both the development of a biological NN, the birth of new neurons and the formation of new connections between them, and the dying off of individual neurons, the degradation of connections between them are observed. The natural process of increasing the number of neurons in the brain is known as neurogenesis. Recent studies in this area show that even for adults, about 700 neurons are born every day. The mechanisms for changing the number of biological neurons are complex and are determined by a combination of factors, including the so-called growth substances, neurotransmitters, hormones and proteins. The term neuroplasticity introduced into use precisely defines the ability of a biological NN of the human brain to permanent change and renewal.

It is usually accepted to subdivide the memory of a person into instant, short-term and long-term. Instant memory is determined by the time of propagation of signals from an external influence in a biological NN for about 0.1 s. Short-term memory is associated with the circulation of excitation signals through the closed paths of a NN for a few units and tens of seconds, depending on the level of excitation. The most interesting is the long-term memory, the basis of which is the change in the characteristics (conductivity) of neuron synapses, as well as the directed growth of their axons. In ANN models used in practice, this effect is usually reflected in the weights of the inputs of artificial neurons, as well as in the value of the discrimination threshold of the activation function of the output signal.

In work the following algorithm of "birth" of neurons is considered:

- to get started, the simplest ANN is chosen, the inputs and outputs of which are enough to process the input signals and present the results of the network operation;
- the ANN is trained on a small number of examples;
- all "ambiguous" neurons are selected, which have a non-unambiguity in determining the weight of at least one of its inputs;
- for such neurons, neurons similar in the number of inputs and outputs are introduced, which are connected to the selected neurons in parallel along their inputs;
- the outputs of the selected neurons and the newly "born" neurons are processed by additional neurons having two inputs;
- the results of the re-learning of the new neural network are analyzed and in the case of the repeated identification of "ambiguous" neurons, the process of connecting new "born" neurons is repeated;
- if in the course of the next training cycle, the inputs of neurons practically not participating in the processing are detected, then such inputs are deleted.

In principle, for most cases, the initial ANN variant can contain only one neuron.

The presented algorithm has a direct biological analogy with the functioning of the brain. In this case, the effect of the "birth" of a new neuron in ANN is the analogy of both the birth of a new neuron of the brain and the involvement of already existing neurons in the functioning of the so-called nerve centers responsible for performing specific functions.

Under the weight ambiguity of the input of a neuron, first of all, it is understood that when training a network with a number of examples, this weight should have a positive value, and for other teaching examples - negative.

In Fig. 1 is an illustration of the considered algorithm.

If in the process of learning the neuron N_1 ($I_1 \ldots I_N$ is the inputs of the neuron, O_1 is its output, $W_{11} \ldots W_{1N}$ is the input weights), the ambiguity in determining the weight W_{1i} of its input I_i was detected, then this neuron initializes the "birth" of new neurons - N_2 and N_3. At the same time, the neurons N_1 and N_2 are connected to the inputs in parallel, and the neuron N_3 is used to process information from the O_1 and O_2 outputs of the corresponding neurons.

The considered ANN growth algorithm also assumes the implementation of a NN training style similar to the natural in accordance with the principle "from simple to complex". In the early stages of NN growth, the set and complexity of the training

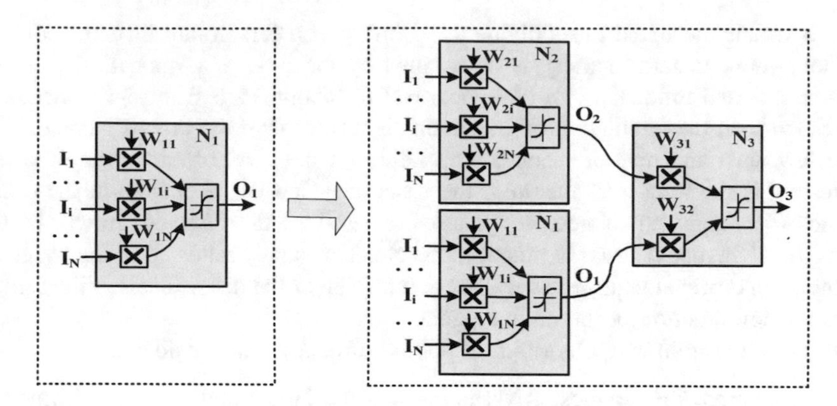

Fig. 1. ANN growth illustration.

examples are low. Their volume and complexity increase with increasing dimension and complexity of the NN.

3 An Example of the Functioning of the Algorithm in the Recognition of a Simple Graphic Image

Consider a simple example of ANN growth due to the appearance of new neurons to solve the problem of recognizing a graphic image in a window. In Fig. 2a shows a 4×4 input graphic information window view.

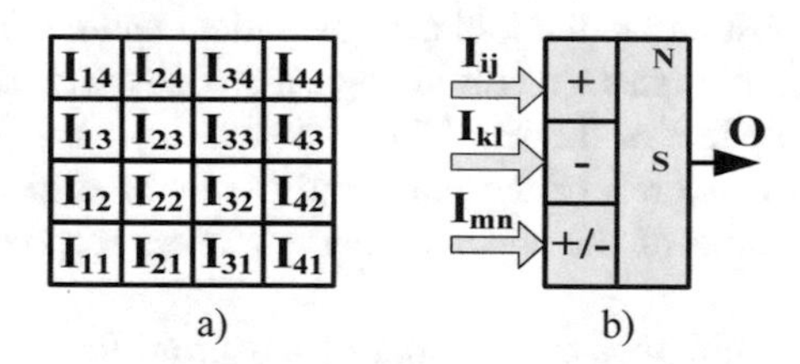

Fig. 2. The image processing window (a) and the graphical image of the neuron (b).

The brightness values of the pixels I_{SR}, S = 1 … 4, R = 1 … 4 can be 1 or 0 (black and white image, the color value is inverted). The graphical image of the neuron (Fig. 2b), which will be used later when considering an example of the functioning of the algorithm, contains two groups of inputs designed to sum the processed signals with a plus sign and a minus sign (all weights are W = +1, or respectively W = −1), as well as a group of inputs that have ambiguity in the type of operation performed. As an activation function, a threshold function is considered with a level of discrimination S.

In Fig. 3 presents a set of teaching examples.

If, in accordance with the algorithm proposed in the work, at the first stage only one neuron is used and only one example of E1 is applied for training, then after training its

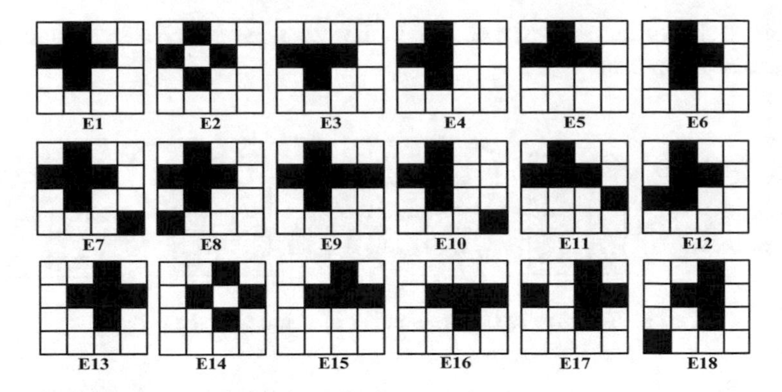

Fig. 3. A set of teaching examples.

weight and the discrimination threshold will most likely have the values presented in Fig. 4a. When using for training a set of E1–E6 containing training examples with masking white noise pixel (E2–E6), we obtain a similar solution (Fig. 4b) with a lower discrimination threshold value. And in both cases, a single neuron completely copes with the task and does not contain ambiguous inputs. Similarly, the use of additional examples of E7–E12, which contain pixel noise of different colors, for teaching, also does not lead to structural changes in the network. The presence of noise only leads to an additional reduction of the allowable discrimination threshold (Fig. 4c).

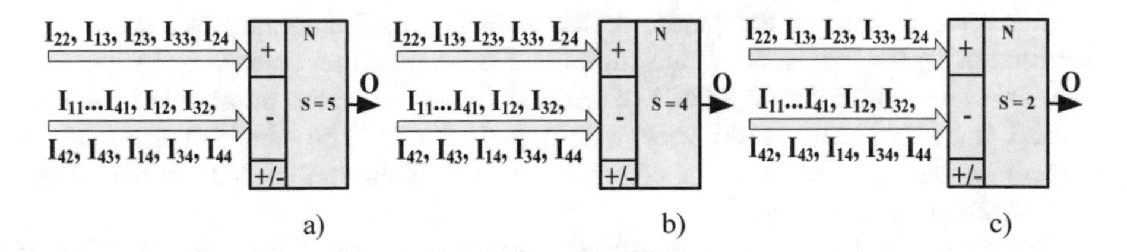

Fig. 4. The NN structure after training.

The situation changes in a fundamental way after using for training an extended set of examples containing, inter alia, variants with a different location of the object (E13 … E18). So, for example, the use of E1 and E13 for learning leads to the emergence of ambiguous inputs (Fig. 5) with a correspondingly ambiguously determined level of discrimination.

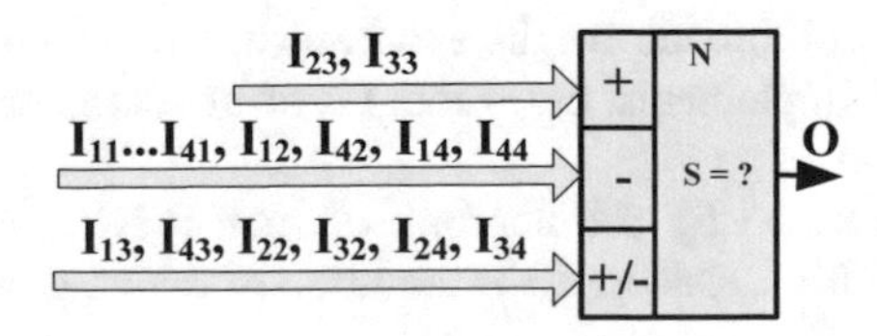

Fig. 5. The appearance of ambiguously defined inputs.

In Fig. 6 shows the result of a change in the structure of the NN in accordance with the considered approach.

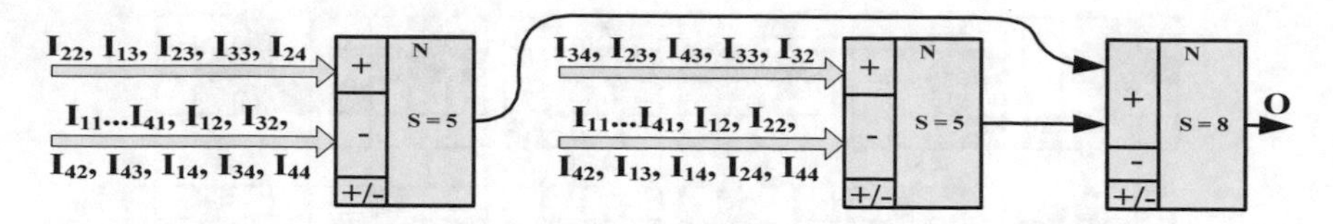

Fig. 6. The increase in the number of neurons in the network.

The resulting network organization has a clear physical meaning. It lies in the fact that neurons are oriented to the processing of specific features of an object in the image window.

The use of the entire set of examples E1 … E18 for training does not change the resulting network structure, but only changes the threshold values when there is pixel noise.

4 Experimental Approbation of the Approach

The approach was tested when solving the problem of synthesis of a NN intended for processing a fragment of an image in a 128 × 128 pixel window in order to accurately determine the centers of the pupils of the human eye. Preliminary eye areas were identified using the Viola-Jones algorithm [7]. Due to possible rotations and tilts of the head, as well as different directions of gaze, the position of the pupil area inside the selected image window could vary within wide limits. The obtained accuracy of determining the coordinates of the centers of the pupils did not exceed the size of one pixel of the processed image.

The synthesized NN had a non-trivial structure, different from the widespread multi-layer structure. The network was characterized by relatively high discrimination thresholds, which is an indirect confirmation of its noise immunity. The resulting solution allows efficient implementation based on programmable logic like FPGA. The problem to be solved is required when creating tracking systems for the direction of a person's gaze and the assessment of its current functional and psycho-emotional state.

5 Conclusion

Thus, the considered algorithm allows the synthesis of an ANN based on a biologically inspired approach that implements the principle of development and learning "from simple to complex".

The functions performed by the neurons of such a NN, as a rule, have a clear physical interpretation. It consists in the orientation of the neuron in the recognition of individual most characteristic features of the object being studied. The resulting high values of discrimination thresholds in the nodes of the activation of neurons indicate the noise immunity of the resulting solutions.

The study was carried out in the framework of project No. 25.2911.2017 /4.6 "Testing technology to reduce the risk of occurrence and mitigate the consequences of man-made disasters by minimizing the human factor's impact on the reliability and trouble-free operation of nuclear power plants and other hazardous facilities" in accordance with the competitive part of the MEPhI project for 2017–2019.

References

1. Russakovsky O, Deng J, Su H, Krause J, Satheesh S, Ma S, Huang Z, Karpathy A, Khosla A, Bernstein M, Berg AC, Fei LF (2015) ImageNet large scale visual recognition challenge. Int J Comput Vision 115(3):211–252
2. Gu B, Hu F, Liu H (2001) Modelling classification performance for large data sets: an empirical study. Lecture Notes in Computer Science, vol 2118, pp 317–328 (2001)
3. Han S, Pool J, Tran J, Dally WJ (2015) Learning both weights and connections for efficient neural networks. ArXiv preprint. arXiv:1506.02626
4. Han S, Mao, H, Dally WJ (2015) Deep compression: Compressing deep neural networks with pruning, trained quantization and Huffman coding. ArXiv preprint. arXiv:1510.00149
5. Canziani A, Culurciello E, Paszke A (2015) (2017) An analysis of deep neural network models for practical applications. ArXiv preprint. arXiv:1510.00149,arXiv:1605.07678v4
6. Bhati R, Jain S, Mishra DK, Bhati D (2010) A comparative analysis of different neural networks for face recognition using principal component analysis and efficient variable learning rate. In: IEEE proceeding of 2010 international conference on computer and communication technology (ICCCT), pp 354−359 (2010)
7. Alyushin MV, Lyubshov AA (2018) The Viola-Jones algorithm performance enhancement for a person's face recognition task in the long-wave infrared radiation range. In: IEEE proceeding. of the 2018 conference of russian young researchers in electrical and electronic engineering (ElConRus), Moscow and St. Petersburg, Russia, 29 January−1 February 2018, pp 1813−1816

Analysis of the Possibility of the Neural Network Implementation of the Viola-Jones Algorithm

Lyubov V. Kolobashkina(✉) and Mikhail V. Alyushin

National Research Nuclear University MEPhI (Moscow Engineering Physics Institute), Kashirskoe Shosse 31, Moscow 115409, Russia
{LVKolobashkina,MVAlyushin}@mephi.ru

Abstract. The practice of using the Viola-Jones algorithm and its modifications to solve the problem of finding objects of interest (OI) in the image frame is analyzed. It is shown that the Viola-Jones algorithm is usually used in conjunction with other algorithms in order to solve a complex task-searching for OI, identifying and analyzing its characteristic features. The relevance of the unification of the applied computing means for solving the above complex problem is underlined. The main computational procedures of the Viola-Jones algorithm are considered: obtaining the integral form of the representation of the input image frame, processing the Haar features (HF), implementation of the cascade classifier. Presented and analyzed options for building a neural network (NN) for the implementation of these procedures. The possibility of expanding the functionality of the Viola-Jones algorithm in its implementation based on the NN is shown. The results of an experimental approbation approach are considered in solving the problem of isolating a person's face with the subsequent isolation of the eye and pupil areas in order to assess their motor activity.

Keywords: Neural network implementation · Viola-Jones algorithm · Unification of computational means

1 Introduction

Viola-Jones algorithm [1] is known as one of the fairly effective methods of searching for OI in an image frame. Currently, there are many different modifications and versions of this algorithm [2–5]. The main areas of application of the Viola-Jones algorithm are: computer vision, security systems, navigation systems, pattern recognition systems. The most typical OI singled out using the Viola-Jones algorithm are: people's faces [2], pedestrians on the road [3, 4, 7, 9, 11], cars and their license plates, as well as road signs [6, 8, 10].

Many of these applications are characterized by the fact that the OIs can be in periodic or constant motion. This circumstance in some cases, as a rule, characterized by fast and spontaneous movements of the OI, complicates the solution of the problem of their reliable detection.

The task of detecting an OI in an image frame is usually one of the initial stages of complex video processing. Further, as a rule, the problems of classification and

A. V. Samsonovich (Ed.): BICA 2019, AISC 948, pp. 232–239, 2020.
https://doi.org/10.1007/978-3-030-25719-4_30

recognition of selected objects are solved. Typical examples in this regard are the selection of the faces of people in a crowd with the subsequent identification of their personality, as well as the allocation of automobiles, areas of the image with their license plates and their subsequent recognition.

For this reason, in practice, the Viola-Jones algorithm is usually used in conjunction with other algorithms to ensure the most efficient solution of the set complex problem. Most often, the Viola-Jones algorithm is used in conjunction, for example, with algorithms such as Principal component analysis (PCA) algorithm [12] and NN algorithms [13]. The PCA algorithm is based on identifying the most significant components of the OI, which makes it possible to significantly reduce the amount of information processed when solving the problem of identification.

The computational capabilities of modern processors make it possible to effectively implement NN algorithms to solve the recognition and identification problem immediately after the implementation of the Viola-Jones algorithm [14, 15].

In this regard, the task of unifying the applied computational tools for implementing the Viola-Jones algorithm, as well as the NN algorithm following it, including the unification of the hardware used to improve the performance, becomes urgent. Among the latter, first of all, it is necessary to single out the FPGA, which allows to implement a wide range of digital NN with the necessary bit depth.

The aim of the work is to analyze the feasibility of the implementation of the Viola-Jones algorithm based on NN in order to study the possibility of unifying the applied software and hardware for complex video processing.

2 The Essence of the Proposed Approach

The approach developed in the work is based on solving the following main tasks:

- search and analysis of analogies in the work of the Viola-Jones algorithm and the trained NN;
- analysis of the possibility of implementing the basic computational procedures of the Viola-Jones algorithm on NN;
- selection of the most appropriate type of NN for the implementation of the basic computational procedures of the Viola-Jones algorithm.

Search and analysis of analogies in the work of the Viola-Jones algorithm and the trained NN revealed the following analogies:

- Viola-Jones algorithm involves the implementation of several sequentially performed procedures for processing graphic information, which is analogous to the functioning of NN containing several successively located layers of neurons designed to perform various functions;
- both approaches are characterized by a learning phase with examples;
- both approaches make it possible to present the result of work in the form of the probability of detecting an OI in the image frame;
- Viola-Jones algorithm and the trained NN are resistant to a sufficiently high level of interference in the input graphic information;

- both approaches implement the same principle of information processing, which consists in the gradual formation of the final result due to the summation of the particular results of the processing of individual parts, features and characteristic features of a graphical object;
- the main operation in processing the brightness of image pixels in Viola-Jones algorithm is the summation (subtraction) operation, which is similar to the functions performed by a digital neuron;
- both approaches use threshold cutoff of weak variants.

Thus, the Viola-Jones algorithm initially has many similarities with the NN approach.

3 Ability to Implement the Basic Computational Procedures of the Viola-Jones Algorithm on NN

The following functions were analyzed as the main computational procedures for the Viola-Jones algorithm and its modifications [1–5]:

- obtaining an integral form of the original image;
- HF processing;
- implementation of the cascade classifier.

The integral form of image frame representation makes it possible to significantly increase the speed of calculating the total brightness of pixels entering a rectangular area with specified dimensions. The relevance of this computational problem is due to the need to repeatedly compute the function of HF coincidence.

In Fig. 1 illustrates an algorithm for forming a frame of an integral image S[k, l], k = 1, …, N, l = 1, …, M from the original frame of the image I[i, j], i = 1, …, N, j = 1, …, M (N, M is the frame size horizontally and vertically, respectively).

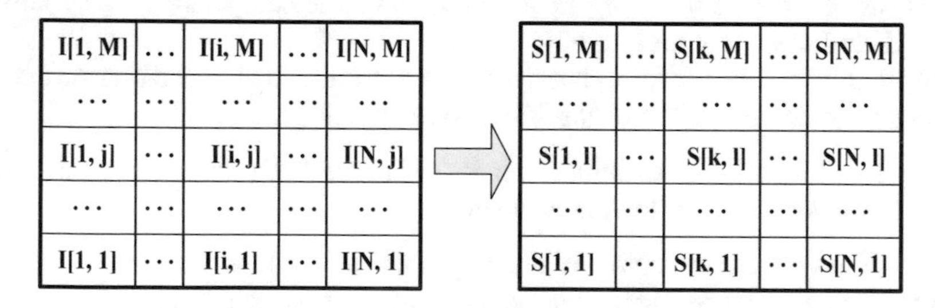

Fig. 1. Formation of the integral form of the image.

The brightness value S[k, l] of each pixel of the integral image is calculated by the following formula based on the brightness data I[I, j] of its pixels:

$$S[k,l] = \sum_{\substack{i=1 \\ j=1}}^{\substack{i=k \\ j=l}} I[i, j]. \tag{1}$$

The following figures illustrate the possibility of the NN implementation of the basic computational procedures of the Viola-Jones algorithm in the form of structural-functional schemes. These schemes allow you to implement the considered computational procedures in hardware, software, or software and hardware.

In Fig. 2 shows a variant of forming an integral form of an image based on a single-layer NN. Such a NN involves the use of N · M digital neurons DNN. At the output of each neuron (DNN11 … DNNNM), a new brightness value of the corresponding pixel of the integral image (S[1, 1] … S[N, M]) is formed in full accordance with the algorithm of its formation (1).

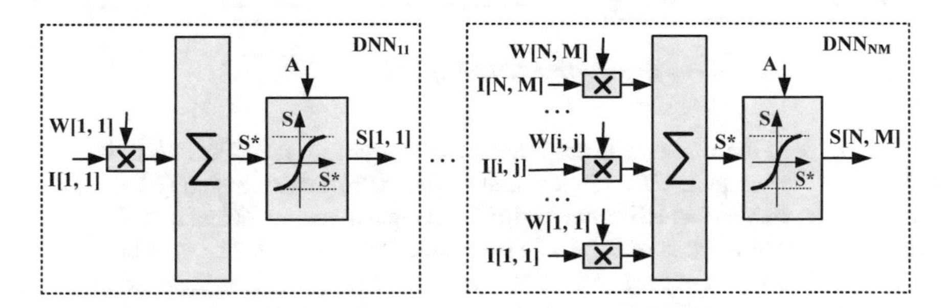

Fig. 2. Formation of the integral form of the image based on the NN.

In this case, the weights for each input signal of neurons must have values equal to one (W[1, 1] = … = W[N, M] = 1). At the same time, the levels of discrimination in each neuron should have a low value ($A \approx 1$).

The considered variant of forming an integral image on the basis of a single-layer NN is the most fast-acting. Its main disadvantage is redundancy. For example, the neurons used to form the values of S[N. M] and S[N, (M−1)], process almost identical input signals with the exception of the values I[1, M] … I[N, M].

Therefore, the most optimal from the point of view of computational costs are multi-layered options for building a NN. Such variants are supposed to form, for example, the considered value of S[N. M] to use the neuron processing the value of S [N, (M−1)], formed in the previous layer of the NN, as well as the values I[1, M] … I [N, M].

Thus, the NN makes it possible to form an integral representation of the input image frame in full accordance with the Viola-Jones algorithm. In addition, the NN makes it possible to carry out additional processing of the input image by controlling the magnitude factors W[1, 1] … W[N, M], as well as the threshold A. The additional processing can be, for example, in reducing the importance graphic information for pixels located on the edge of the frame, as well as taking into account the level of pixel noise in the image.

The situation is similar with the NN implementations of the rest of the computational procedures of the Viola-Jones algorithm.

In Fig. 3a shows an embodiment of a fragment of a two-layer NN for processing a single HF in a dedicated window (WI) of an image frame (IF) containing a face (FI).

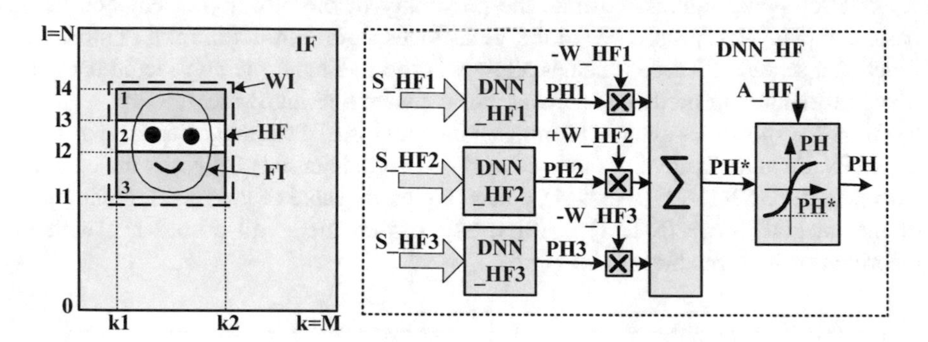

Fig. 3. Handling of HF using a NN.

The frame of the IF image is presented in an integral form and has the dimension of M·N pixels. The analyzed WI window and the HF have the dimension (k2−k1)·(l4−l1) of pixels. The considered HF contains three rectangular areas 1, 2 and 3. This feature is intended to highlight areas of the eye in the image of the face FI. The total brightness of the pixels of the frame of the image IF, belonging to the lighter area 2, are summed, and the brightness of the pixels of regions 1 and 3 is taken into account with a minus sign.

The first layer of the NN (Fig. 3b) contains the number of neurons equal to the number of regions of the HF. For this example, the number of such neurons is three (DNN_YF1, DNN_HF2 and DNN_HF3). Each of these neurons analyzes information about the brightness of pixels in the corresponding areas of the HF, presented in the integral form (S_HF1, S_HF2 and S_HF3).

The neuron of the second layer DNN_HF carries out a final assessment of the similarity measure of the HF and the graphic information contained in the WI window. The signs of the values −W_HF1, + W_HF2 and −W_HF3 make it possible to take into account, respectively, the effect of attenuation or enhancement of the HF in accordance with the Viola-Jones algorithm. The absolute values of these values make it possible to additionally take into account the significance of regions 1, 2, and 3 of the considered HF when solving the problem of finding a face image. The threshold A_HF allows you to take into account the level of noise and interference in the analyzed image. The values of weights and threshold are determined at the stage of training a NN by examples similar to the Viola-Jones algorithm.

Realizations of the neurons of the first layer, which form particular measures of similarity PH1, PH2 and PH3 on the basis of the integral form of information representation, are presented in Fig. 4.

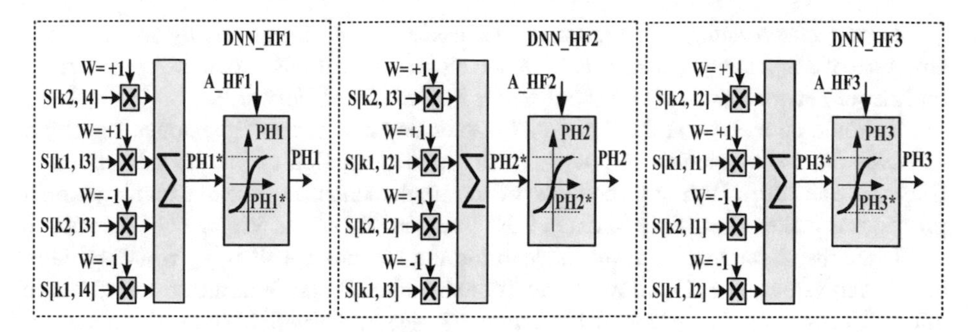

Fig. 4. The implementation of the first layer neurons.

The signs of the weights W make it possible to restore the total brightness value of the pixels of the original image belonging to regions 1, 2 and 3, respectively.

The thresholds A_HF1, A_HF2 and A_HF3 make it possible to carry out additional processing of signals PH1, PH2 and PH3.

Thus, the NN extends the functionality of the HF processing, allowing to take into account the significance of their individual areas.

Quite naturally, the implementation of a cascade classifier is solved on the basis of a NN. In Fig. 5 shows the structure of such a two-layer NN.

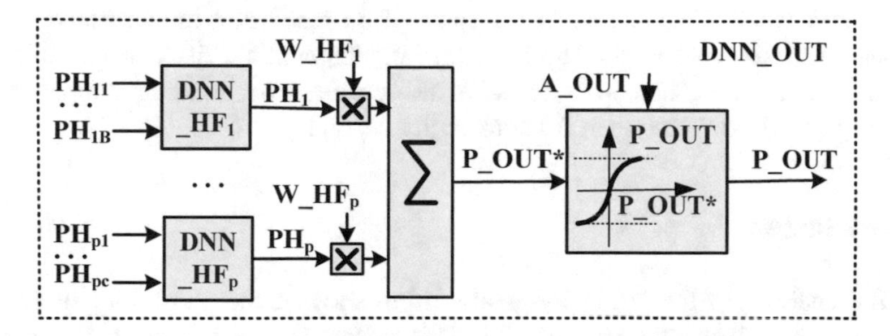

Fig. 5. The implementation of the cascade classifier.

In principle, the considered NN can have from one to several layers. However, it is for a two-layer NN that the weighting factors, as well as the values of the thresholds, have a clear physical interpretation.

The first layer contains the DNN_HF1 … DNN_HFp neurons, the number of which P corresponds to the total number of HF groups. The number of HF in each group is equal respectively to the values of B … C. In this case, the weights for the neurons of the first layer determine the importance of each HF from one group, and the weights W_HF_1 … W_HFp of the neuron DNN_OUT of the second layer determine the significance of each group of HF. These weights are determined at the NN training stage. The threshold values of the first layer neurons, as well as the threshold of discrimination of the second layer neuron A_OUT, make it possible, among other things, to filter noise.

Taking into account the operations discussed above, performed by neurons, the structure of connections between them is the closest type of NN should be considered a multilayer network of direct distribution DFF (Deep Feed Forward).

It should be noted that the direct NN implementation of the Viola-Jones algorithm involves the use of training, primarily in the implementation of the cascade classifier (Fig. 5). The use of training methods in the implementation of other computational procedures makes it possible to extend the functionality of the Viola-Jones algorithm.

Thus, the change in the result of learning the coefficients W [i, j], which initially have fixed values (Fig. 2), allows to additionally take into account the so-called "edge effects", caused by distortions of the original image at the periphery of the frame.

The change in the values of the coefficients W as a result of the determination of particular measures of the similarity of the HF (Fig. 4) makes it possible to take into account the importance of the light and dark areas of the HF when processing images of different brightness and contrast.

4 Experimental Approbation of the Approach

The considered approach was applied to solve the complex problem of identifying a person's face in an image frame, including the selection of eye areas, pupils and determining their mobility. This task is relevant when creating monitoring systems for the current functional and psycho-emotional state of a human operator. All stages of the complex task were implemented in the form of a trained NN implemented hardware based on programmable logic. The hardware implementation made it possible with a probability of no less than 97% to solve the problem of isolating pupil areas at a processing speed no worse than 50 frames per second.

5 Conclusion

Thus, the analysis of the Viola-Jones algorithm showed the possibility of its implementation in the form of a trained NN. This makes it possible to unify the applied computational tools in solving the complex task of identifying an OI, identifying areas of interest within the OI, as well as their subsequent identification and analysis.

The considered approach allows efficient implementation based on a co-temporal electronic element base.

The study was carried out in the framework of project No. 25.2911. / 4.6 "Testing technology to reduce the risk of occurrence and mitigate the consequences of man-made disasters by minimizing the human factor's impact on the reliability and trouble-free operation of nuclear power plants and other hazardous facilities" in accordance with the competitive part of the MEPhI project for 2017–2019.

References

1. Viola P, Jones MJ (2001) Rapid object detection using a boosted cascade of simple features. In: IEEE conference on computer vision and pattern recognition, 2001, Kauai, Hawaii, USA, vol 1, pp 511–518
2. Viola P, Jones MJ (2004) Robust real-time face detection. Int J Comput Vis 57(2):137–154
3. Viola P, Jones MJ, Snow D (2005) Detecting pedestrians using patterns of motion and appearance. Int J Comput Vis 63(2):153–161
4. Jones MJ, Snow D (2008) Pedestrian detection using boosted features over many frames. In: IEEE proceedings of 19th international conference on pattern recognition (ICPR 2008), Tampa, Florida 8 December–11 December 2008, pp 1–4
5. Comaschi F, Stuijk S, Basten T, Corporaal H (2013) RASW: a run-time adaptive sliding window to improve Viola-Jones object detection. In: IEEE proceedings of 7th international conference on distributed smart cameras (ICDSC 2013), 2013, pp 153–158
6. Chen S-Y, Hsieh J-W (2008) Boosted road sign detection and recognition. In: IEEE proceedings of 2008 international conference on machine learning and cybernetics, Kunming, China, 12 July–15 July 2008, vol 7, pp 3823–3826
7. Dollár P, Wojek C, Schiele B, Perona P (2009) Pedestrian detection : a benchmark. In: IEEE proceedings of 2009 computer society conference on computer vision and pattern recognition workshops, CVPR workshops, Miami, FL, USA, 20 June–25 June 2009, pp 304–311
8. Broggi A, Cardarelli E, Medici P, Cattani S, Sabbatelli M (2014) Vehicle detection for autonomous vehicle detection for autonomous parking using a Soft-Cascade Adaboost classifier. In: IEEE proceedings of intelligent vehicles symposium, Dearborn, Michigan, USA, 8 June–11 June 2014, pp 912–917
9. Paisitkriangkrai S, Shen C, Zhang J (2008) Fast pedestrian detection using a cascade of boosted covariance features. In: IEEE proceedings of transactions on circuits and systems for video technology, 2008, vol 18, no 8, pp 1140–1151
10. Lin C-T, Hsu S-C, Lee J-F, Yang C-T (2013) Boosted vehicle detection using local and global features. J Sig Inf Process 4(3):243–252
11. Zhu Q, Yeh MC, Cheng KT, Avidan S (2006) Fast human detection using a cascade of histograms of oriented gradients. In: IEEE proceedings of conference on computer vision and pattern recognition, 2006, vol 2, pp 1491–1498
12. Deshpande NT, Ravishanka S (2017) Face detection and recognition using Viola-Jones algorithm and fusion of PCA and ANN. Adv Comput Sci Technol 10(5):1173–1189
13. Sekhon A, Agarwal P (2015) Face recognition using back propagation neural network technique. In: IEEE proceedings of international conference on advances in computer engineering and applications, 2015, pp 226–230
14. Fernandez MCD, Gob KJE, Leonidas ARM, Ravara RJJ, Bandala AA, Dadios EP (2014) Simultaneous face detection and recognition using Viola-Jones algorithm and artificial neural networks for identity verification. In: IEEE proceedings of 2014 region 10 symposium, 2014, pp 672–676
15. Da'san M, Alqudah A, Debeir O Face (2015) Detection using Viola and Jones method and neural networks. In: IEEE proceedings of international conference on information and communication technology research, Abu Dhabi, United Arab Emirates, 17 May–19 May 2015, pp. 40−43

Neurophysiological Correlators of Semantic Features

Anastasia Korosteleva[1,2(✉)], Vadim Ushakov[1,2], Vyacheslav Orlov[2], Tatiana Stroganova[3], and Boris Velichkovskiy[2]

[1] National Research Nuclear University MEPhI (Moscow Engineering Physics Institute), Moscow, Russia
nnkorosteleva@gmail.com

[2] NRC "Kurchatov Institute", Moscow, Russia

[3] Center for Neurocognitive Research (MEG Center), Moscow State University of Psychology and Education, Moscow, Russia

Abstract. The article presents the result of fMRI data processing - a map and characteristics of brain activity in the process of monitoring human speech activity. Experimental data calculated by 8 subjects. The main goal of the work was to localize the spatial and temporal dynamics of the neural networks of the cortex, which is responsible for the mechanism of verbal control. The secondary goal of the work was to recognize and remove noise components from the fMRI signal, which are related to human physiology and a feature of the test items.

Keywords: Brain · Neural network activity · fMRI · Speech · Semantic networks

1 Introduction

Today, in scientific research there is an acute problem of localization of brain areas that are involved in such fundamental mechanisms as speech, memory, movement. In this article, we will consider another fundamental neurophysiological mechanism as verbal control.

To study the work of mechanisms in the brain, in recent decades, researchers have begun to use the method of neurophysiological visualization: functional magnetic resonance imaging (fMRI) to solve the problem of non-invasive localization of cognitive functions in the human cortex. The fMRIs allow visualizing the second changes in brain activity based on the parameters of the intensity of blood flow.

The studied mechanism of verbal control includes automatic speech generation and speech generation, requiring arbitrary effort. Speech is a complex process that involves: the zone responsible for the perception of speech facing a person – the Wernicke zone, located in the temporal cortex, and the zone associated with the ability to generate speech utterance – the Broca zone, located in the lower tender areas. As a rule, in right-handed subjects, both zones are located

A. V. Samsonovich (Ed.): BICA 2019, AISC 948, pp. 240–245, 2020.
https://doi.org/10.1007/978-3-030-25719-4_31

in the left hemisphere [1]. However, there is still no precise localization of speech zones. According to the work of Benson, only 90% of right-handed patients had left-hemispheric speech localization [2]. In left-handed patients, this distribution is even more blurred. The uncertainty of locality is associated with the choice of research methodology. So speech is an extremely fast process: the meaning of a word is extracted from the acoustic characteristics of a speech signal within fractions of seconds [3]. fMRI registers a change in blood flow, which in itself is a slow (second) process. Thus, fMRI may not catch the short-term activation of neural populations that underlie speech processes.

In addition to the difficulty of localizing and registering speech processes, fMRI data contains, in addition to the useful signal (change in blood flow), noise components. There are two main categories of noise: noise associated with the subject (movement/physiological effects) and associated with data acquisition (artifacts of MP physics) [4]. The analysis of space-independent components (ICA) has proven to be a powerful tool for separating fMRI data into three-dimensional spatial maps and one-dimensional time series. At the level of an individual test method is increasingly used in the context of removing artifacts in order to separate the neural signal from various sources of noise [5].

In this work, we will focus on the description of the method for clearing the experimental fMRI data from artifacts and mapping the spatial and temporal dynamics of the operation of the neural networks of the cortex, which is responsible for the verbal control mechanism. In the experiment, several patterns of the brain were identified that distinguish between automatic speech generation and speech generation, requiring arbitrary effort. The patterns found showed statistical significance, so that they can be considered as the basis for further research work on the construction of a neural network of the brain that participates in arbitrary speech generation.

2 Materials and Methods

2.1 Experiment Description

At this stage, 8 people (5 men and 3 women) with good eyesight (without lenses), with a leading right hand, took part in the experiment. The average age of the subjects was 24 years. Informed consent to conduct the study was obtained from each subject.

Each subject was placed in a magnetic resonance imaging tomograph Magnetom Verio 3T (Siemens, Germany). Studies were performed using a 32-channel head MR coil. The scan took place in two stages:

(1) “High Resolution” anatomical data removal based on a T1-weighted sequence (TR 1900 ms, TE 2.21 ms, 176 slices, 1 mm slice thickness, 1 mm * 1 mm resolution inside slice).

(2) recording of functional data based on the Ultrafast EPI-sequence (TR 720 ms, TE 33 ms, 56 slices, slice thickness 2 mm, resolution inside slice 2 mm * 2 mm). The obtained anatomical and functional data were combined in one

coordinate system (MNI) and normalized (for reliable localization of human brain activities).

During the experiment, nouns are presented to the participant (one for each stimulus). The participant sees the stimuli on the screen through a mirror that is attached to the helmet. The helmet secures the competitor's head in a fixed position to avoid interference. The participant reads the noun to himself and says out loud an associative verb that answers the question "what does it do?". An experimental model for the generation of verbs was used, which semantically and lexically fit the presented noun. The parameter varied - the strength of the association between the noun and the verbs associated with it (where low, high is low and high load on the choice of verbs, respectively). In Fig. 1 shows the experimental design, the duration of which is 13 min.

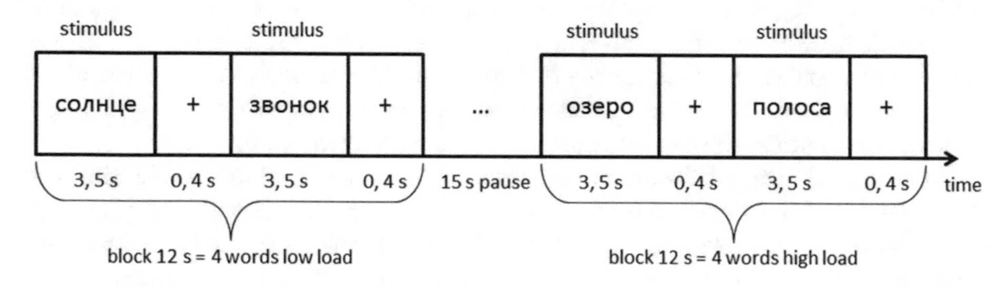

Fig. 1. The block diagram of the task "Choosing an arbitrary action from a variety of alternatives for nouns with low and high loads"

2.2 fMRI Data Processing

Anatomical and fMRI data for each participant were pre-processed in the SPM8 software package. The preprocessing included the following steps: importing data from a tomograph in DICOM format, converting anatomical and fMRI data to single coordinates (front commissure), reducing spatial distortions using magnetic field mapping (Field Map utility, included in the SPM8 package). Next, a slice-timing correction –the correction of hemodynamic responses in space, and then in time to avoid pronounced motion artifacts. The anatomical data was segmented, then the anatomical and fMRI data were normalized, and the fMRI data was smoothed using a Gaussian filter with a 6 mm isotropic core.

The next step is to clear the fMRI data in the FSL software package. MELODIC (included in the FSL package) is a toolkit that implements fMRI signal separation into its components using independent component analysis. Thus, we hoped to separate the "useful" signal from the noise. Noise was classified according to article [5]. Components with high frequencies of the spectra and components with activity associated with human physiology (activity of the ventricles, arteries, eyes and sinusoidal) were eliminated. For each person, the signal was decomposed by default to 200 components.

Further, the design matrix of the experiment was built on the fMRI signal. Where as regressors, times were indicated when nouns appeared with a high load (high), nouns with a low load (low) and rest (rest) when the subject did not perform any tasks. The article presents 4 comparisons: high¿ low, low¿ high, high¿ rest, and low¿ rest. Statistical analysis was performed using Student's t-statistic at p ¡0.01.

3 Results

According to the experiment, it was possible to identify active zones for a group of people of 8 people. In Fig. 2 shows the brain activity maps of a group analysis of 4 comparisons Table 1.

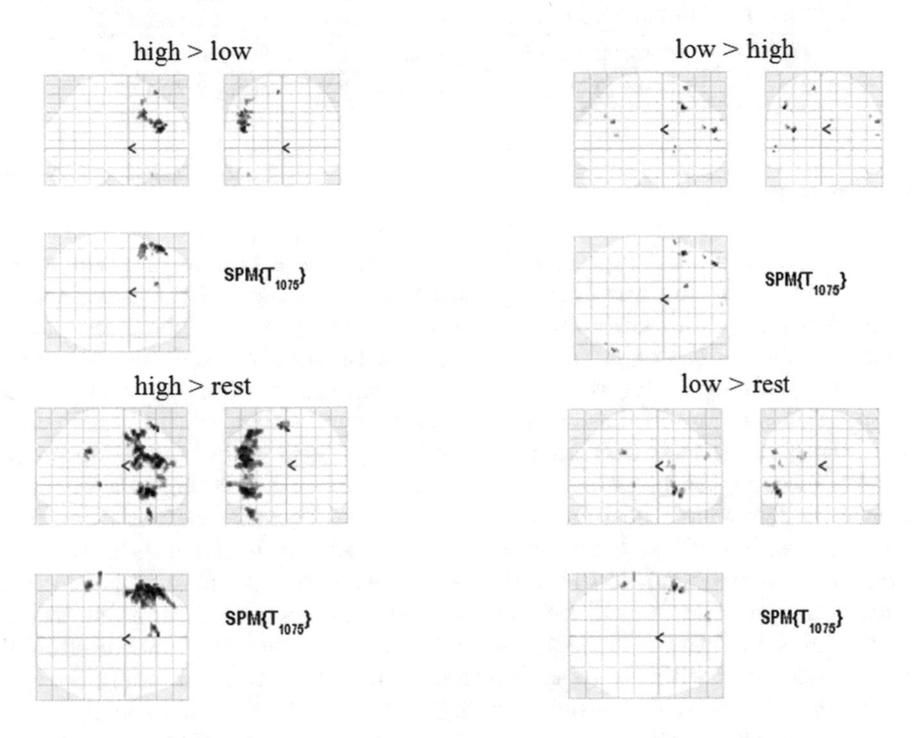

Fig. 2. Group statistical maps of activity according to fMRI data with a significance level of p = 0.01. Where low, high - low and high load on the choice of verbs, respectively

Table 1. Identified during group analysis clusters

Cluster Center	Structurer	Cluster Center	Structure
high > low		**low > high**	
−48; 36 18	Frontal Inf Tri L	−46; 20; 42	Frontal Mid L
−42; 42; 24	Frontal Mid L	−10; 22; 64	Supp Motor Area L
−42; 10; 32	Precentral L	60; −58; 26	Angular R
−8; 32; 60	Frontal Sup Medial L	−60; 10; −6	Temporal Pole Sup L
−42; 26; −32	Temporal Pole Sup L	−58; 12; 0	Rolandic Oper L
high > rest		**low > rest**	
−46; 36; 20	Frontal Inf Tri L	−50; 16; −10	Temporal Pole Sup L
−46, 10, 20	Frontal Inf Oper L	−50; 26; −8	Frontal Inf Orb L
−48; 24; −10	Frontal Inf Orb L	−54; −40; 32	SupraMarginal L
−50; 16; −10	Temporal Pole Sup L	−56; 16; 18	Frontal Inf Oper L
−36; 26; −40	Temporal Pole Mid L	−16; 54; 30	Frontal Sup L

4 Conclusions

In the experiments, the functional load of the supposed functional architecture of an arbitrary action was specially varied: the inhibition of the usual action (inhibitory load). Two alternative hypotheses were formulated about the mechanisms for finding the right word. The first one assumes that several representations of potentially useful words are activated simultaneously, and the result of competition is determined at the stage of their coactivation, when the only winner – the most strongly activated representation – gets access to the "response buffer" affecting the planning and implementation of the response word. The alternative hypothesis postulates that all activated representations come into the "response buffer", and the choice is made by selectively strengthening one of them, which corresponds to the subject's goal as much as possible. This selective reinforcement of the "expedient" representation of the word occurs due to the downward entry into the "response buffer" from the structures of the prefrontal cortex, associated with the representation of the target and comparing the two upstream and downstream inputs that are carried out by the front speech zones, including the Broca area. Two alternative hypotheses give different predictions about the stage at which the process of choosing the right word occurs. The first hypothesis refers the selection process to relatively early stages of semantic analysis related to the activation of semantic neural structures and occurring no later than 400–500 ms after the presentation of the task (words to which you need to pick a verb or image that you want to describe with a word). On the contrary, the second hypothesis predicts that the choice of a word is associated not so much with the relatively early activation of semantic representations caused by stimulus-related processing, but with late processes shifted in time to the subject's vocal responses (response-related processing).

The obtained results confirm the idea of the differences in neural networks involved in cognitive mechanisms, automatic speech generation and speech generation, requiring arbitrary effort. Greater activation of frontal areas responsible for coordinated arbitrary actions.

Acknowledgments. This work was in part supported by the Russian Science Foundation, Grant 18-11-00336 (data preprocessing algorithms) and by the Russian Foundation for Basic Research grants ofi-m 17-29-02518 (study of thinking levels). The authors are grateful to the MEPhI Academic Excellence Project for providing computing resources and facilities to perform experimental data processing.

References

1. Geschwind, N.: The organization of language and the brain. Science **170**(3961), 940–944 (1970)
2. Benson, D.F.: Aphasia and related disorders: a clinical approach. In: Mesulam, M.M. (ed.) Principles of behavioral neurology, pp. 193–238. Davis, Philadelphia (1985)
3. Shtyrov, Y., Pulvermuller, F.: Language in the mismatch negativity design: motivations, benefits and prospects. J Psychophysiol **21**(3–4), 176–187 (2007)
4. Beckmann, C.F.: Modelling with independent components. Neuroimage **62**, 891–901 (2012)
5. Griffanti, L., et al.: Hand classification of fMRI ICA noise components. Neuroimage **154**, 188–205 (2017)

Dynamics of Recognition of Properties in Diagnostics

Sergey Kosikov[2], Larisa Ismailova[1], and Viacheslav Wolfengagen[1](✉)

[1] National Research Nuclear University "Moscow Engineering Physics Institute", Moscow 115409, Russian Federation
lyu.ismailova@gmail.com, jir.vew@gmail.com

[2] Institute for Contemporary Education, "JurInfoR-MGU", Moscow 119435, Russian Federation
kosikov.s.v@gmail.com

Abstract. The issues of processing semantic information for diagnostic studies are considered, in which, with the advent of personalized medicine, the task of ensuring efficiency and accuracy has come to the fore. The solution of such problems is associated with the need to recognize the properties of the samples, which is done by experts/crowdsourcers using the means of semantic modeling. In this regard, the analysis is subject to communication (1) level of knowledge with (2) forms of information processing and with (3) levels of abstraction. For this, a dynamic semantic model is developed and applied that can take into account such parameterization.

Keywords: Histopathological diagnosis · Semantic modeling · Expertise · Crowdsourcing · Recognition of properties · Dynamics of properties

1 Histopathological Diagnosis: The Problem Domain

For diagnostic studies, with the advent of personalized medicine, the task of ensuring efficiency and accuracy has come to the fore, which leads to an increase in the burden and complexity of the histopathological diagnosis of cancer, i.e. microscopic examination of tissue in order to study the manifestations of the disease. This has led to the fact that in order to increase the efficiency of the analysis of the histopathological slide, in international practice they begin to use the method of deep learning using neural networks, when the workload for pathologists decreases and the objectivity of diagnoses increases. In particular, the researchers concluded that in-depth training can improve the diagnosis of prostate cancer and breast cancer.

Among the applied IT methods, it is necessary to indicate the use of evolving search to increase the accuracy of image contour recognition [7], constructive solutions of image analysis systems [6], the use of convolutional neural networks to improve image analysis [3,4], the use of "deep learning" neural network

A. V. Samsonovich (Ed.): BICA 2019, AISC 948, pp. 246–259, 2020.
https://doi.org/10.1007/978-3-030-25719-4_32

to increase the degree of automation of image recognition [5,14]. In all these cases, the organization of the data remained a bottleneck, especially generated by crowdsourcers-experts in the course of their analytic actions.

1.1 Increased Workload and Difficulty in Diagnosis

Once a situation has arisen in which pathologists are faced with a significant increase in workload and the complexity of histopathological diagnosis of cancer, various diagnostic protocols should be equally focused on efficiency and accuracy. The method of "deep learning" is purely pragmatically manifested itself as a method of improving the objectivity and effectiveness of the analysis of the histopathological slide. The literature provides examples of identifying prostate cancer in biopsy specimens and identifying breast cancer metastases in sentinel lymph nodes, which demonstrates the potential of this new methodology aimed at reducing the burden on pathologists, while increasing the objectivity of diagnoses. As it turns out from examples in literature, all the slides, containing prostate cancer and breast cancer micro- and macrometastases can be identified automatically, while 30–40% of slides containing benign and normal tissue can be eliminated without using any additional immunohistochemical markers or human intervention. But all these are experimental applications of "deep learning", which has certain prospects for improving the efficiency of diagnosing prostate cancer and establishing breast cancer.

However, the literature does not provide information on the application of the ideas of semantic modeling, aimed at organizing data arising during histopathological diagnosis [5,14].

1.2 Features of Microscopic Analysis

Microscopic analysis of hematoxylin and eosin stained areas (H&E) was a common basis for the diagnosis and classification of cancer. For many of the most common types of cancer (for example, lung, breast, prostate gland), there are protocols for the complete processing of biopsies or resected tissue samples, including microscopic analysis. The use of these protocols led to an increase in prediction and the emergence of widely used assessment strategies (for example, the Gleason classification system).

Due to the increasing incidence of cancer and patient-specific treatment options, the diagnosis and classification of cancer are becoming more complex. At present, in order to arrive at a complete diagnosis, pathologists must study a large number of slides, often including additional immunohistochemical stains. In addition, the number of quantitative parameters that pathologists must extract in order to be able to apply generally accepted classification systems (for example, length, surface area, mitotic indices) is increasing. Because of these difficulties, the analysis protocols are adapted and carefully tuned, which provides the best balance between predictive power and feasibility in everyday clinical practice.

The appearance of the latest slide scanning systems in recent times enables quantitative analysis and improvement of the histopathological procedure. These systems digitize glass slides with high resolution colored tissue sections. Whole slide images (WSI) digital images allow the use of image analysis techniques that help pathologists in researching and quantifying slides. One of these methods, which has gained prominence in the past five years in other areas, is "deep learning". Although "deep learning" can not be considered as the only possible or unified technique, it can be roughly described as the use of multilayer artificial neural networks to a wide range of problems, from speech recognition to image analysis. In recent years, "deep learning" methods have quickly become the modern state of computer vision. A specific subtype of the neural network (convolutional neural networks, CNN) has become the de facto standard in image recognition and approaches human performance in solving a variety of practical problems. These systems function by examining relevant features, which are taken directly from huge image databases (usually millions of images). This contrasts with more traditional pattern recognition methods that are highly dependent on manually created quantitative images. trait collectors.

Despite these tremendous advances, in-depth training methods have not yet had a major impact on the field of medical imaging. One of the main reasons is that traditional image processing methods (for example, radiology) require a large number of images required by complex depth learning systems for their own learning, which is not always available. In digital histopathology, this is simpler: one WSI usually contains trillions of pixels from which hundreds of examples of malignant glands can be extracted (in the case of prostate or breast cancer).

Over the past five years, some initial work has been published, which discussed the use of "deep learning" methods for microscopic and histopathological images. For example, applications of the convolutional neural network to the task of counting mitosis for the primary classification of breast cancer are known. In addition, the applicability of convolutional neural networks with a patch to segmentation problems was demonstrated. Later, these studies were extended to attempts to identify mitosis, when manual functions were combined with the capabilities of a convolutional neural network. Other applications of convolutional networks include primary detection of breast cancer, assessment of glioma and segmentation of the epithelium and stroma. Finally, another technique of depth learning, called auto-encoders with summation, was used, which in the experiment made it possible to perform cell detection and segmentation in lung cancer and brain tumors.

The most detailed study of the general applicability of CNN for improving the diagnosis of cancer in H&E images is described by applying it to two new tasks: detecting prostate cancer in biopsy samples and detecting breast cancer metastases in resected lymph node sentinel.

The amount of emerging data has a pronounced upward trend. The number of prostate biopsy takes increased significantly in recent decades due to the appearance of the prostate antigen test (PSA). Due to the nature of the standard biopsy procedure (from eight to twelve random biopsies under ultrasound guidance), each procedure leads to several slides. Most of these slides do not usually contain cancer. Histopathological analysis can be greatly simplified if these normal slides can be automatically excluded without displacing any slides containing cancer.

Similar techniques can be applied to monitor the development of the tumor being treated, in order to identify new metastases that may be missed.

The analysis of such a subject shows a significant increase in the volume of data during the analysis procedures. In terms of information technology arise big data and crowdsourcing systems. It can be shown that the entire domain is characterized by a *evolving* data model [9], studied in [2] and advanced in [8] and [13], modified in [12]; some improvements were given in [10] and [11] .

1.3 Diagnostics Service: An Evolving Data Model

When constructing an evolving data model, a terminological basis characteristic of the IT field arises.

Key terms:
crowd = the crowd;
source = the source of data;
crowdsourcer = person producing data.

The data are produced during and at different stages of the diagnosis. This data is very much. All of them make up the area of Big Data = a huge volume of data.

Work with data is skeletally as follows.

- **Given**: raster images (sections) of human organs. The doctor ("crowdsourcer") views the image and marks areas of possible pathology ("properties").
- **Available**: *series of images*, depending on the parameter where the "stages", "phases", or "states" are fixed. There is a "state trajectory" or even *spectrum of possibilities* of "evolving of events".
- The trajectory begins with the *property set*, which characterizes a certain *organ* [of body]. [Its *information image* matches the "data object" with the image of the corresponding properties.]
- **Occurs**: "unfolding of events" along a "evolvent " [which plays the role of a trajectory parameter].
- The trajectory ends with a certain state.
- **Having a triple**: "initial-state, evolvent, final-state", we can make an examination of the direction in which "evolve" the properties of "object".
- **Expert** (specialist doctor) [he is the same: **crowdsourcer**]: has a knowledge of the "spectrum of possibilities" of the development of events. [There is a portion of the spectrum characterized as "healthy". A portion of the spectrum appears, characterized as "sick".]
- **Conducted**: analysis of a specific "trajectory" taken from the "spectrum of possible trajectories"; this is a valid "evolving of events". [It characterizes the prototype – the development of the pathology of an organ.]

- This analysis is the result of the work of an expert (doctor, crowdsourcer). In the form of data, this result gives the data, the origin of which belongs to the field of crowdsourcing.

2 Crowdsourcers and Property Recognition

There are different points of view, like having a body of data, recognize one or another *property*. They are united by the need to build and use a semantic domain model.

2.1 Crowdsourcing Systems

Crowdsourcing systems attract many people to help solve a wide range of problems. As is typical of a newly emerging field, this direction has emerged under many names, including peer-to-peer production, user systems, user content, collaborative systems, community systems, and ranges from relatively simple reputable systems, such as reviewing books to emerging complex systems that Build structured knowledge bases in systems that are built on top of other popular systems. We limit the discussion to the range of fundamental problems, for example, how to recruit and evaluate users and combine their contributions to solving a problem. Note only the most important aspects of the global picture, using real-life examples. The goal is to contribute to our collective understanding, both conceptual and practical, of this important emerging topic.

Indeed, it seems that, in principle, the solution to any non-trivial problem can benefit from the use of crowdsourcing (CS): you can describe the problem on the Internet, request user inputs and study the source data to build a solution. This system may be impractical (and more advanced systems may already exist), but it can be considered a primitive CS system. Consequently, the type of cooperation or target problem to be solved is not limited. Rather, the CS is considered as a universal method for solving problems.

We say that a particular system is a *CS system* if it attracts a group of people to help solve the problem defined by the system owners, and if it answers the following four fundamental challenges: (1) how to recruit and retain users; (2) what contributions can users make; (3) how to combine user contributions to solve the target problem; (4) how to evaluate users and their contribution.

2.2 Crowdsourcing

In short, crowdsourcing can be understood as the management of relationships and properties arising from the analysis of Big Data. This requires a number of analytical actions, including: selecting collections to extract the required information; extracting entities from unstructured or semi-structured sources; assessing the neighborhood of individuals in different data collections; merging instances of individuals, including the establishment and discarding of duplicates; plotting data in collections and mapping them to the target schema; generating instances of valid individuals and concepts for data corresponding to the target scheme.

Crowdsourcing Technology. The area of crowdsourcing is developing very rapidly, and there are still no generally accepted systems and methods of working with them. Crowdsourcing [1] uses manual work to process, retrieve or generate data on demand, as well as to classify, rank, mark up or improve existing data. These manual tasks are often difficult to automate, for example, when determining the rating of something or someone, or when determining signs of interest in any data source. Manual data can also be viewed as an equal source of data, so naturally, we would like to integrate such a crowdsourced data source with other traditional sources. This will allow the end user, instead of working with heterogeneous data sources, to interact with a single unified database, which is an advantage.

Neighborhood of Property. The explanatory system is illustrated in Fig. 1 might look like this: events move from A to B, and properties shift from T to S.

A brief summary of symbolization is as follows: $f : B \to A$, $g : T \to S$; $H_g : H_T \to H_S$, $H_T(f) : H_T(A) \to H_T(B)$; $H_g(A) : H_T(A) \to H_S(A)$, $H_g(B) : H_T(B) \to H_S(B)$, $H_g(f) : H_T(f) \to H_S(f)$; [$f$-neighborhood of property T] $\equiv H_T(f)$, [f-neighborhood of property S] $\equiv H_S(f)$, [g-neighborhood of property T] $\equiv H_g(A)$, [(f, g)-neighborhood of property T] $\equiv H_g(f) = H_g \circ H_T(f)$. If $H_T(A) = \{h \mid h : A \to T\}$, $H_T(B) = \{h \mid h : B \to T\}$, then $h \circ f \in H_T(B)$. Besides that, if $g \circ h \in H_S(A)$, then $g \circ h \circ f \in H_S(B)$, because of $H_g(f) : h \circ f \mapsto g \circ h \circ f$.

The original property T can *shift* according to the law g, becoming the property S. In other words, S is in g-close to T.

The initial state of A can *evolve* according to the law of f, becoming the state of B. In other words, B is in f-close to A.

The S property includes the g-shifted property T.

The B-concept of S contains (f, g)-neighborhood: f-neighborhood of the shifted S property and g-neighborhood of the generic T property.

Cloned Individuals: f-neighborhood. It is assumed that the "events" unfold on the involute f:

$$f : \text{new state} \to \text{old state}.$$

Cloned by the evolvent f, individuals represent the domain of C_f, which is considered the f-neighborhood of the generic domain (in the old state):

$$C_f \subseteq H_{\text{property}}(\text{new state}).$$

This domain is inhabited by those individuals t of the generic domain in the new state, which satisfy the logical filter Φ estimated at the involute f:

$$\begin{aligned} C_f &= \{t \in H_{\text{property}}(\text{new state}) \mid \|\text{filter}\|_{f[t/y]}(\text{new state})\} \\ &\equiv f\text{-clone area}. \end{aligned}$$

These clones inherited the selected property and when moving to a new state.

We assume that the variables free in the logical filter Φ are taken from $u_0, \ldots, u_{n-1}, y$, where y is a property of individuals that interests us.

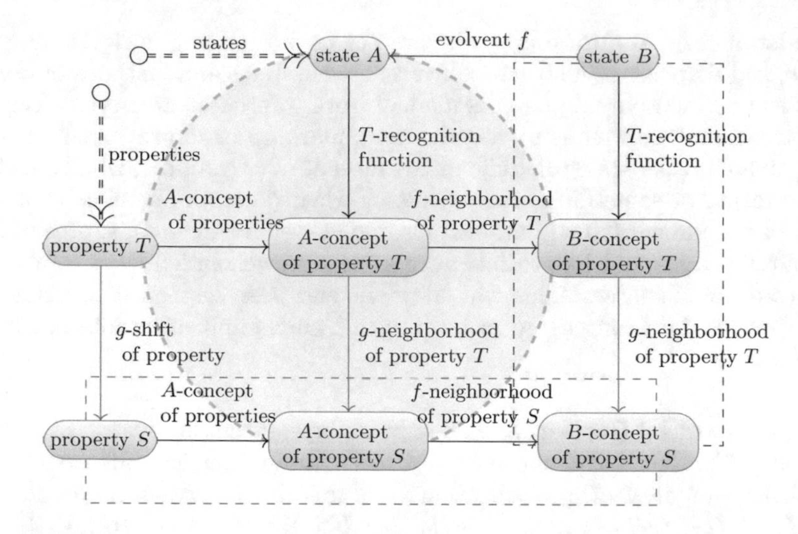

Fig. 1. Explanatory system for (f, g)-neighborhood of property T. A and T serves as parameters, there are two scales for "scaling" – f-states and g-properties, by which the parameters B and S are "reached".

2.3 Property Recognition Scheme

For the group of n crowdsourcers or experts with "state of knowledge" A_1, A_2, ..., A_n respectively, the task is set to recognize the property T through the recognition function H_T. Let them act by virtue of the mapping system f_1, f_2, $\ldots f_{n-1}$. Then, at the qualitative level, the map of their sequential actions looks like in accordance with the following commutative diagram shown in Fig. 2 for the case $n = 2$. Taking into account the accepted notation, a more rigorous representation of the map of their sequential actions has the form in accordance with the following commutative diagram in Fig. 3. In this diagram

$$H_T(A_j) = \{h' | h' : A_j \to T\}$$

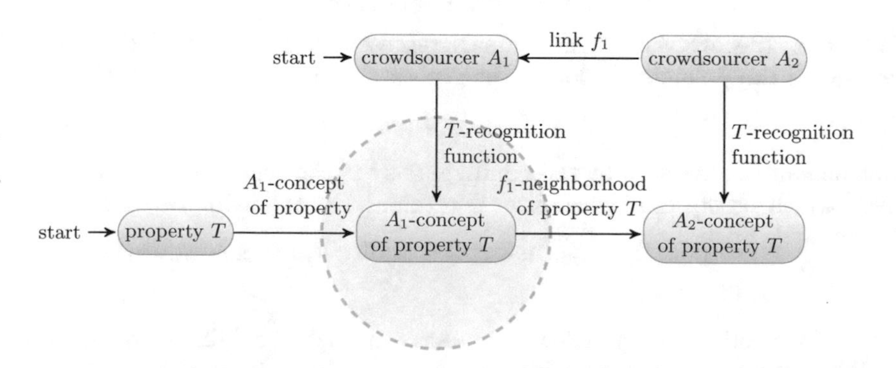

Fig. 2. Property recognition scheme.

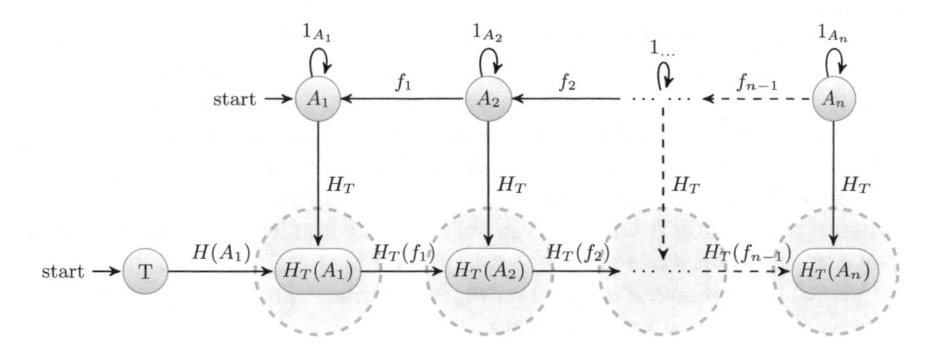

Fig. 3. The sequence map for the property recognition scheme.

is a *domain*: the set of individuals h' "in the field of view" A_j, which have the property T. If $f : A_{j+1} \to A_j$ for $j \geq 1$ in a theory, then let $H_T(f_j)$ be the mapping, which maps the individual $h' \in H_T(A_j)$, that is recognized by the crowdsourcer j, into $h' \circ f_j \in H_T(A_{j+1})$, where can the crowdsourcer $j + 1$ handle it and so on:

$$\begin{array}{rrl} H_T(f_1): & h \in & H_T(A_1) \mapsto \qquad h \circ f_1 \in H_T(A_2) \\ H_T(f_2): & h \circ f_1 \in & H_T(A_2) \mapsto \quad h \circ f_1 \circ f_2 \in H_T(A_3) \\ \dots: & \dots & \\ H_T(f_{n-1}): & h \circ f_1 \circ \dots \circ f_{n-2} \in & H_T(A_{n-1}) \mapsto h \circ f_1 \circ \dots \circ f_{n-1} \in H_T(A_n) \end{array}$$

Under these conditions it is not difficult to show infer that

$$H_T(f_1 \circ f_2 \circ \dots \circ f_{n-1}) : h \in H_T(A_1) \mapsto h \circ f_1 \circ f_2 \circ \dots \circ f_{n-1} \in H_T(A_n)$$

and

$$H_T(f_{n-1}) \circ \dots \circ H_T(f_2) \circ H_T(f_1) = H_T(f_1 \circ f_2 \circ \dots \circ f_{n-1}).$$

The question arises as to how semantically stable is the image of data classified in this way, which is in the information system. For this it is necessary to make a definite conclusion about the form of the recognition function.

3 Semantic Modeling in Processing Analysis Data

Let us present an example of the organization of information services for processing the analysis data produced by an expert/crowdsourcer when recognizing the properties of a particular sample. This will be done on the basis of the considered means of semantic modeling.

3.1 Simple Analysis Script

In fact, many practically useful constructions of higher orders can be easily represented by existing means. For example, the judgment "Analysis number 5 has all the pathologies of its predecessor, and the tumor marker is one of them", shown in Fig. 4. Note that in this diagram, both the abbreviated and unabbreviated forms of the image are used simultaneously. Three of the propositional vertices are depicted explicitly, and the "ancestor-for" and two occurrences of the "pathology" establish three implicit propositional vertices. Of course, a higher order predicate is "pathology"; a vertex with a quantifier must understand all communities as "for all predicates". In this case, the implicit quantifier restriction on the corresponding sorts is extended to the applicability of types, that is, since "pathology" is a predicate on predicates, its argument is implicitly restricted to the "predicate" type in any statement.

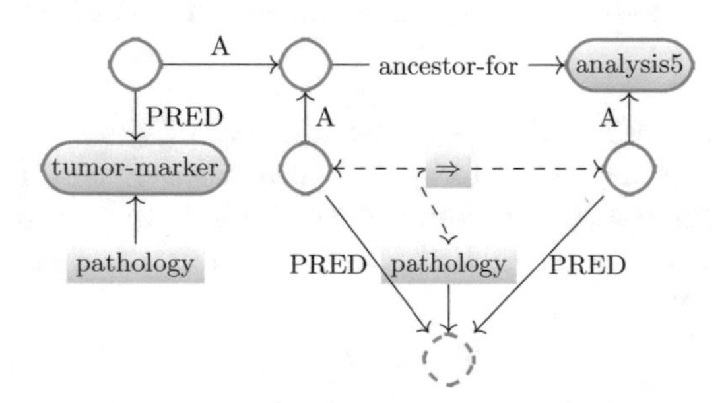

Fig. 4. "Analysis5 has all the pathologies of its predecessor, and the tumor marker is one of them"; $(\forall P)(\exists y)$ $\{[[P$ & pathology$(P)](y)$ $\Rightarrow$ P(analysis5)] & $[P$ & pathology (tumor-marker)$](y)$ & ancestor-for$(y,$ analysis5$)\}$.

The described method of representing quantifiers applicable only to judgments that are in a prenex normal form. This allows us to offer generalized representations that can be used in a wide range of logical connections or existing propositional operations.

3.2 Structure of the Knowledge Base System

We present a brief overview of the capabilities of the knowledge representation system using a simple example. The example illustrates how knowledge can be coded and accessed through networks of simple processing elements, as is shown in Fig. 5.

The conceptual knowledge of the system is coded online, called *network memory*. This network consists of a large number of extremely simple processing elements (vertices) that interact with each other by distributing activation by weighted links. The memory network uses propagating activation to perform limited forms of inference, such as inheritance and categorization according to the semantic modeling approach under consideration.

To accomplish specific inferences – as opposed to manifesting a general associative behavior or modeling diffuse priming effects, the memory network must

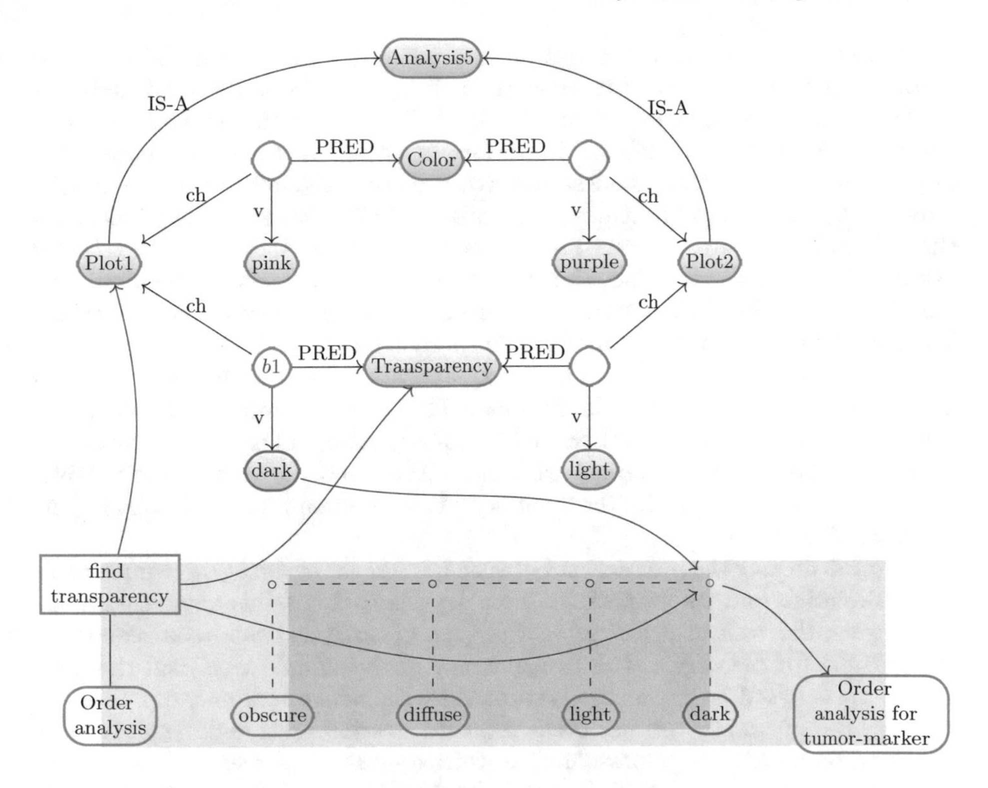

Fig. 5. Connectionist Retrieval System. Shadowed: routine area; above: network memory.

be able to control the spread of activation. An important aspect of the approach used is the use of control mechanisms to regulate distribution of activation. The control mechanisms are implemented in the structure of interconnections between vertices, as well as in the computational characteristics of vertices. Moreover, these mechanisms show complete substantive independence (that is, they do not depend on the problem domain).

The translation of a body of knowledge into a network that embodies this knowledge and produces a conclusion based on it is purely mechanical in nature – in particular, it does not provide for any special manipulations with weights to achieve the corresponding behavior.

Following the idea of mass parallelism, it is important that the network works without the intervention of a central controller or central interpreter. Consequently, the memory network is such that after the activation of the relevant vertices, the corresponding conclusions are performed automatically, and the result becomes available in the form of activation levels of the corresponding vertices. Imagine the idea of a mechanism for accessing information encoded in a memory network.

Access to information encoded in the memory network can be obtained through other fragments of the network, called subroutines. Thus, not only the presentation of knowledge, but also access to it through the network. Subroutines encode holistic procedures to perform specific tasks and are represented as a sequence of vertices (blocks) connected so that the activation can be distributed through the subroutines as a sequence. In the course of their execution, the subroutines create requests to the memory network, activating the relevant vertices in it. After activation, the memory network performs the required outputs through a controlled activation extension and returns a response, activating the corresponding response nodes in the subroutine.

All requests are given relative to an explicit set of answers, and for each possible answer there is a vertex of the response. The activation returned by the memory network is an indication of the evidence-based support of the response. The vertices of the response compete with each other, and the vertex that receives the maximum activation in the memory network dominates and triggers the corresponding action.

Figure 5 depicts the interaction between a fragment of the agent sample analysis subroutine and its part of the memory network. In this fragment of the subroutine, the task of deciding on the type of analysis leads to a request to the network for memory of the transparency of the sample area, and the decision is made based on the answer returned by the network of memory. Steps of actions are depicted by vertices-ovals, requests – vertices-rectangles, and vertices of the response are also represented as vertices-ovals. The memory network in this example encodes the following information: Notions/concepts in the problem domain of the example are characterized by two properties: “Transparency” and “Color”. The “Plot1” and “Plot2” are two concepts in the area. “Plot1” – “dark” by transparency and “pink” by color, “Plot2” – “light” by transparency and “purple” by color.

Edges in the memory network are weighted links. Round vertices (called connecting vertices) represent variables with which objects, properties, and property values are associated. Each vertex is the active element and in the state “active” sends activation to all the vertices associated with it. The weight on the link modulates the activation as it propagates along the link. A vertex can become active when it receives activation from another vertex in a network of memory or in a subroutine. Binding vertices behave somewhat differently, since they become active only when they are simultaneously activated from a pair of vertices.

To find the transparency of “Plot1”, the subroutine needs to activate the vertices “Transparency” and “Plot1”. The connecting vertex $b1$ connecting the vertices “Transparency” and “Plot1” with the vertex “dark” will receive a matching (co-case) activation along its two connections and become active. As a result, it will transfer the activation to the top of the “dark”, which ultimately will certainly become active. If any subroutine needs to find an object dark in transparency, it activates the vertices “Transparency”, and “dark”. This will cause the corresponding link vertex to become active and transfer activation to the vertex “Plot1”. Eventually, the vertex “Plot1” will become active, completing the search.

These two cases roughly correspond to how the network can handle *inheritance* (search for property values of the specified object) and *categorization* (identifying an object by some of its attributes). The considered examples do not contain any evidence-based reasoning or real inheritance of properties, but serve exclusively to illustrate a possible parallel implementation.

4 Conclusions

The main efforts were focused on what is called *representation* in the field of semantic modeling, including the representation of the dynamics of recognition of properties.

Note that such technologies are used by data processing systems in the implementation and information services of diagnostics, and practice confirms the convenience of such presentation forms. In general, some of the various methods used in crowdsourcing that promote the proper construction and interpretation of semantic networks are systematized and systematic methods for expressing the dynamics of concepts/properties in these networks are presented.

Perhaps the development of such designations will contribute to the universalization of the presentation of semantic information. Purely practically, a basis has been created for comparison with other approaches to the presentation of semantic information, since such a standard quite clearly corresponds to the notation in semantic modeling, which itself is well developed. Of course, to meet the specific needs of specific applications of semantic networks, you can take a different view. For example, you can use a notion that more accurately reflects specific computer data structures. In addition, these computer data structures can themselves take different forms for different areas of knowledge, and not just correspond to some fixed syntax of statements.

To simplify common constructions, you can create additional operators. Operators will induce a hierarchy of notions/concepts that can be used to effectively define sub-super-set type relationships. As it turns out, the definitions of these operators can be expressed through the basic constructs of the semantic network.

Various problems in the presentation of informal knowledge may cause additional complications in the notation. Examples are handling uncertainties.

In addition to these relatively static questions, there are more dynamic problems of factual interpretation, plausible inference, learning, and the interaction between procedural and factual knowledge. It is clear that any questions about representation that arise in these problem areas can only be answered in the context of specific approaches to solving the problems themselves. Nevertheless, the present presentation gives a rather direct extension of several fairly successful, externally different representations.

Consequently, the computational processes that create and use their data structures can be easily adapted to structures based on this presentation. This indicates that the increased expressive power that the proposed notation system provides must have real value in developing understanding systems for the field of diagnostics.

Acknowledgements. This research is supported in part by the Russian Foundation for Basic Research, RFBR grants 19-07-00326-a, 19-07-00420-a, 18-07-01082-a, 17-07-00893-a.

References

1. Doan A, Ramakrishnan R, Halevy AY (2011) Crowdsourcing systems on the world-wide web. Commun ACM 54(4):86–96. https://doi.org/10.1145/1924421.1924442 http://doi.acm.org/10.1145/1924421.1924442
2. Ismailova LY, Kosikov SV, Wolfengagen V (2016) Applicative methods of interpretation of graphically oriented conceptual information. Procedia Comput Sci 88:341–346 (2016). https://doi.org/10.1016/j.procs.2016.07.446. 7th annual international conference on biologically inspired cognitive architectures, BICA 2016, held July 16 to July 19, 2016 in New York City. http://www.sciencedirect.com/science/article/pii/S1877050916317021
3. Jia Y, Shelhamer E, Donahue J, Karayev S, Long J, Girshick R, Guadarrama S, Darrell T (2014) Caffe: convolutional architecture for fast feature embedding. In: Proceedings of the 22Nd ACM international conference on multimedia, MM 2014. ACM, New York, pp 675–678. https://doi.org/10.1145/2647868.2654889. http://doi.acm.org/10.1145/2647868.2654889
4. Krizhevsky A, Sutskever I, Hinton GE (2012) Imagenet classification with deep convolutional neural networks. In: Proceedings of the 25th international conference on neural information processing systems, vol 1, NIPS 2012. Curran Associates Inc., USA, pp 1097–1105 (2012). http://dl.acm.org/citation.cfm?id=2999134.2999257
5. Litjens G, Sánchez CI, Timofeeva N, Hermsen M, Nagtegaal I, Kovacs I, Hulsbergen - van de Kaa C, Bult P, van Ginneken B, van der Laak J (2016) Deep learning as a tool for increased accuracy and efficiency of histopathological diagnosis. Scientific Reports, 6 (Project No. 26286) (2016). https://doi.org/10.1038/srep26286. http://dx.doi.org/10.1038/srep26286
6. Molin J, Woźniak PW, Lundström C, Treanor D, Fjeld M (2016) Understanding design for automated image analysis in digital pathology. In: Proceedings of the 9th nordic conference on human-computer interaction, NordiCHI 2016. ACM, New York, pp 58:1–58:10. https://doi.org/10.1145/2971485.2971561. http://doi.acm.org/10.1145/2971485.2971561
7. Stoean C, Stoean R, Sandita A, Mesina C, Gruia CL, Ciobanu D (2015) Evolutionary search for an accurate contour segmentation in histopathological images. In: Proceedings of the companion publication of the 2015 annual conference on genetic and evolutionary computation, GECCO Companion 2015. ACM, New York, pp 1491–1492. https://doi.org/10.1145/2739482.2764690. http://doi.acm.org/10.1145/2739482.2764690
8. Wolfengagen VE, Ismailova LY, Kosikov S (2016) Computational model of the tangled web. Procedia Comput Sci 88:306–311 (2016). https://doi.org/10.1016/j.procs.2016.07.440. 7th annual international conference on biologically inspired cognitive architectures, BICA 2016, held July 16 to July 19, 2016 in New York City. http://www.sciencedirect.com/science/article/pii/S1877050916316969
9. Wolfengagen VE, Ismailova LY, Kosikov S (2016) Concordance in the crowdsourcing activity. Procedia Comput Sci 8:353–358 (2016). https://doi.org/10.1016/j.procs.2016.07.448. 7th annual international conference on biologically inspired cognitive architectures, BICA 2016, held July 16 to July 19, 2016 in New York City. http://www.sciencedirect.com/science/article/pii/S1877050916317045

10. Wolfengagen VE, Ismailova LY, Kosikov, SV (2016) A computational model for refining data domains in the property reconciliation. In: 2016 third international conference on digital information processing, data mining, and wireless communications (DIPDMWC), pp 58–63. https://doi.org/10.1109/DIPDMWC.2016.7529364
11. Wolfengagen VE, Ismailova LY, Kosikov SV (2016) A harmony and disharmony in mining of the migrating individuals. In: 2016 third international conference on digital information processing, data mining, and wireless communications (DIPDMWC), pp 52–57. https://doi.org/10.1109/DIPDMWC.2016.7529363
12. Wolfengagen VE, Ismailova LY, Kosikov SV, Navrotskiy VV, Kukalev SI, Zuev AA, Belyatskaya PV (2016) Evolutionary domains for varying individuals. Procedia Comput Sci 88:347–352 (2016). https://doi.org/10.1016/j.procs.2016.07.447. 7th annual international conference on biologically inspired cognitive architectures, BICA 2016, held July 16 to July 19, 2016 in New York City. http://www.sciencedirect.com/science/article/pii/S1877050916317033
13. Wolfengagen VE, Ismailova, LY, Kosikov SV, Parfenova IA, Ermak MY, Petrov VD, Nikulin IA, Kholodov VA (2016) Migration of the individuals. Procedia Comput Sci 88:359–364 (2016). https://doi.org/10.1016/j.procs.2016.07.449. 7th annual international conference on biologically inspired cognitive architectures, BICA 2016, held July 16 to July 19, 2016 in New York City. http://www.sciencedirect.com/science/article/pii/S1877050916317057
14. Xu J, Zhou C, Lang B, Liu Q (2017) Deep learning for histopathological image analysis: towards computerized diagnosis on cancers. Springer International Publishing, Cham. https://doi.org/10.1007/978-3-319-42999-1_6

Functional Magnetic Resonance Imaging Augmented with Polygraph: New Capabilities

Mikhail V. Kovalchuk[1(✉)] and Yuri I. Kholodny[1,2]

[1] National Research Center "Kurchatov Institute", Moscow, Russia
{tiuq, kholodny}@yandex.ru
[2] Bauman Moscow State Technical University, Moscow, Russia

Abstract. Modern psychophysiology is an extensive combination of interdisciplinary research focused on the study of physiological and neural mechanisms of mental processes, states and behavior. For solving scientific and applied problems, psychophysiology uses a number of non-invasive research methods, including electroencephalography (hereinafter - EEG), magnetic encephalography, positron emission and magnetic resonance imaging (hereinafter - PET and MRI, respectively) and several others. In recent years, in order to in-depth study of brain activity during the realization of certain mental phenomena, there has been a steady trend to combined use of several methods - for example, PET and EEG, or MRI and EEG. This work is related to the creation of an MR-compatible polygraph.

Keywords: MRI compatible polygraph · fMRI · Vegetative reactions · Traces of memory

1 Introduction

Modern psychophysiology is an extensive combination of interdisciplinary research focused on the study of physiological and neural mechanisms of mental processes, states and behavior. For solving scientific and applied problems, psychophysiology uses a number of non-invasive research methods, including electroencephalography (hereinafter - EEG), magnetic encephalography, positron emission and magnetic resonance imaging (hereinafter - PET and MRI, respectively) and several others. In recent years, in order to in-depth study of brain activity during the realization of certain mental phenomena, there has been a steady trend to combined use of several methods - for example, PET and EEG, or MRI and EEG.

Scientists and specialists "since the advent of psycho-physiological research… widely used and continue to use vegetative reactions: changes in skin conductivity, vascular reactions, heart rate, blood pressure, etc. [1].

The use of vegetative reactions in scientific, and then for applied purposes led to the creation of a specialized device - a polygraph (1930s), on the basis of which the major direction of applied psychophysiology began to develop (1950–1960s), which found application in the practice of solving crimes. Currently, the use of vegetative reactions in the course of research using a polygraph (which is often incorrectly referred to as a "lie detector"), for the purpose of diagnosing a person's concealed information about

A. V. Samsonovich (Ed.): BICA 2019, AISC 948, pp. 260–265, 2020.
https://doi.org/10.1007/978-3-030-25719-4_33

past events, has become very popular. In particular, these Studies Using a Polygraph (hereinafter - SUP) have long proved their applied efficiency: they are used by more than 80 countries of the world, and the federal bodies and non-governmental institutions of Russia annually conduct several tens of thousands of such studies.

But at the end of the twentieth century in Russia a paradoxical attitude to-wards the SUP was formed. While the use of the polygraph in law-enforcement practice was proceeding at an increasing rate, the theoretical basis and technology of the SUP did not attract the attention of the domestic psychophysiological science, and in the educational literature on psychophysiology, published in the last two decades, the polygraph was not mentioned. The textbooks were limited to a brief reference about the electrical activity of the skin and an abstract reference to the fact that "there is still a whole set of vegetative indicators" [2].

At the beginning of the 21st century, the application of the method of Joint Registration of Vegetative Reactions (hereinafter referred to as the JRVR method) in scientific research has practically ceased, and opinions regarding it have been expressed contradictory. On the one hand, it was noted that the refusal to use the JRVR method in experimental studies does not mean at all that vegetative indicators do not have high sensitivity [1]. But, on the other hand, it was stated that "registration of vegetative reactions does not apply to direct methods for measuring information processes of the brain." And, moreover, "there are several reasons why vegetative reactions can only be used as an indirect method for studying information processes: they are too slow and delayed, … (and also) are not specific with regard to incentives and tasks [1].

At the same time, many years of experience in the theoretical study and applied use of SUP showed that it is hardly correct to abandon the possibilities of the JRVR method, and the above reasons for limiting its use are contrived. With proper, methodically correct organization of research, non-specificity of vegetative reactions does not prevent successfully solving a whole class of scientific and applied tasks: for example, to classify the stimuli presented to a person by their subjective importance for him and to carry out such a classification, practically, in real time. A good example of this is the world practice of conducting SUP.

A few years ago, the authors of this article, discussing the use of a polygraph in forensic science in the television program "Stories from the Future" [3] suggested that it is advisable to combine functional magnetic resonance imaging (hereinafter fMRI) and SUP technology to study the neurocognitive mechanisms of concealing information by a person, which is stored in his memory.

In particular, the idea was expressed about the possibility of creating a new type of forensic research based on the aforementioned integrated approach; and this idea did not come about by chance.

It is known that by the present moment about two dozen of the so-called "Polygraph theories" [4], have been proposed in the world, which offer an explanation of the processes that make it possible to reveal the information concealing by a person during an SUP. At the same time, the lack of good scientific substantiation of the technology of using a polygraph has now become a serious deterrent in successfully using this effective method in the domestic practice of investigating crimes, when the results of an SUP performed in the form of forensic psychophysiological examination can be taken by the court as evidence.

Certain optimism in the possibility of knowing the mechanisms underlying the SUP, and building a scientifically based "polygraph theory" appeared at the beginning of the 21st century. Optimism was due to the fact that a number of researchers used the fMRI method for the purposes of "lie detection", and the results of the first works seemed very encouraging [5, 6].

According to the researchers, "the main problem with using a polygraph was that it measures peripheral arousal, not the lie itself" [7]., and "the specificity of the polygraph is limited by the fact that it relies on the correlates of the peripheral nervous system activity, while the lie is a cognitive event controlled by the central nervous system" [8].

Therefore, it was assumed that the use of fMRI technology will allow one to "see" the process of realizing a lie and to find markers indicating that the person has concealed information. However, it soon became clear that the optimistic expectations of a quick receipt of the answers to the accumulated questions with the help of fMRI technology did not come true: the studies that have been conducted have not yet clarified the neurocognitive mechanisms underlying the identification of information that a person conceals during the SUP.

It should be noted that experimental studies that presented joint (complex) registration of fleeting (seconds) vegetative reactions in response to stimulus material presented during fMRI research, were not presented in the available scientific reviews, [9, 10] and only a few studies studied the dynamics of GSR [11].

Moreover, experimental fMRI studies on the subject of so-called "lie detection", as a rule, deviated from the generally accepted practice of using a polygraph: during such studies, stimuli were presented many dozens of times to each subject and with a frequency unacceptable for the SUP technology [12].

It is obvious that it is hardly correct to deal with neurocognitive mechanisms ensuring the implementation of IPP, while deviating from the requirements of SUP technology.

Further, it is known that modern MRI scanners are equipped with regular devices, which allow recording of cardiovascular system activity, and human breathing, i.e. provide an opportunity to record some human vegetative indicators in the course of the conducted fMRI study. Also, separate devices for recording galvanic skin response are also available. At the same time, these devices for monitoring vegetative parameters do not provide simultaneous presentation of the recorded reactions in the form required by the SUP technology.

Therefore, to study the neurocognitive processes of the human mind when identifying traces of concealed events of the past in his memory, as well as to assess the prospects for the joint use of fMRI and SUP technologies, it was decided to use the powerful scientific and experimental capabilities of the NRC "Kurchatov Institute".

2 Methods

For this purpose, a prototype of hardware and software complex, a computer-based MRI compatible polygraph (hereinafter referred to as MRIcP), was created (by Malakhov D.G. et al.) for conducting a test series of experiments at the NRC "Kurchatov Institute" during the fMRI study. To assess the performance of MRTcP and to study the tactical possibilities of its use in fMRI studies, the so-called "test with a concealed name" (hereinafter - TCN) was borrowed from the arsenal of forensic SUP, during which the subject conceals his own name that is presented along with a number of other names from a polygraph examiner.

With the help of MRIcP, a polygram was displayed on the computer screen (i.e., a joint presentation of the dynamics of the controlled physiological indices), while registering (see Fig. 1):

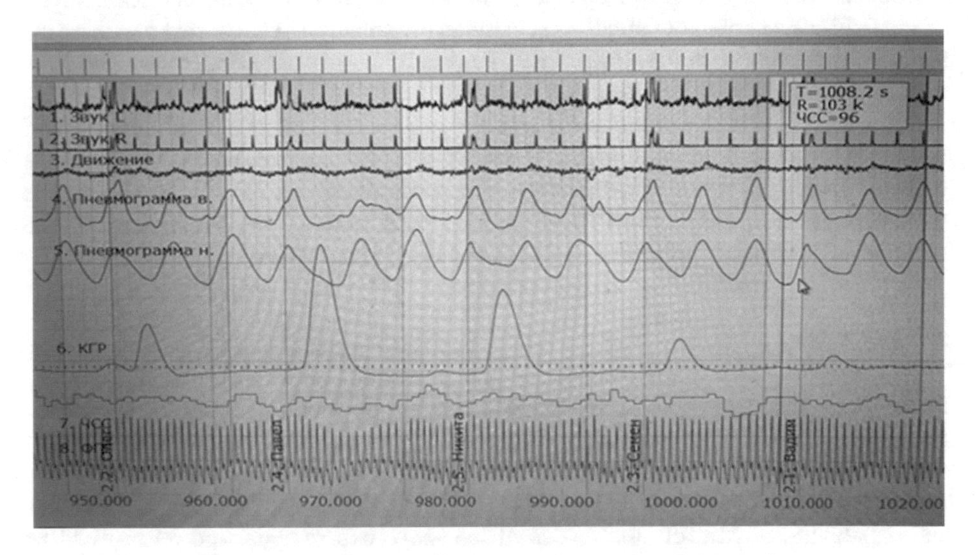

Fig. 1. The polygram of the "test with a concealed name".

The polygram presents the second (out of five) presentation of TCN. The concealment of subjectively significant information (own name – Pavel, rougly at 970 s) causes a disturbance in the respiratory rhythm of the examined person (channel 4), maximal GSR (channel 6) and constriction of the vessels of the fingers (channel 8). The moving "lens" (in the upper right corner of the picture) allows you to see at any time, exactly indicated from the beginning of the experiment, the value of skin resistance (in kilo ohms) and heart rate (strokes/min.).

– sound during fMRI research - i.e. the questions of the experimenter and the answers of the examined person (channels 1 and 2 accordingly);
– movements of the head of the examined person (channel 3);
– chest and abdominal breathing of the examined person (channels 4 and 5);
– GSR and cardiovascular system activity carried out by the photoplethysmography method (hereinafter - PPG), which allows you to observe changes in heart rate (hereinafter - HR) and narrowing of the blood vessels of the fingers of the hand of the examined person (channels 6, 7 and 8).

The first series of setup experiments for refinement of methods of the joint use of fMRI and SUP technologies for research purposes gave an interesting result. It turned out that, when collecting fMRI data from a homogeneous age group of two dozen male subjects, monitoring the dynamics of their vegetative reactions under the experimental conditions found that the group, in reality, is not heterogeneous. In particular, the characteristics of the response (manifested in changes in the dynamics of GSR and cardiovascular system) for simple test material (TCN) showed that the group of subjects should, roughly, be divided into two unequal subgroups - into high and low "reactive" subjects. The analysis of the fMRI data carried out after this separation (by subgroups) suggested that there are significant differences in the activity of individual brain structures under the same TCN experiment. Research in this direction continued.

In another series of setup experiments focused on assessing the feasibility of joint use of fMRI and SUP technologies in the study of neurocognitive processes in patients with schizophrenia, it was found that using the same simple test (TCN), monitoring the dynamics of the activity of GSR and cardiovascular system, one can distinguish the severity of the disease. Research in this direction is also continuing.

3 Discussion

Concluding this article, its authors announce the beginning at NRC "Kurchatov Institute" of a new area of systemic research on the integrated application of fMRI technologies, SUP and - possibly in the near future - EEG for solving applied problems. The presentation of the specific results of the work carried out (the technical features and capabilities of MRTcP, the results of the above experiments, and others) will be carried out by the researchers themselves in the beginning of a series of articles, the first of which are included in this collection.

Acknowledgements. This work was in part supported by the National Research Center "Kurchatov Institute" (MR compatible polygraphy), by the Russian Science Foundation, grant № 18-11-00336 (data preprocessing algorithms) and by the Russian Foundation for Basic Research, grants ofi-m 17-29-02518 (study of thinking levels). The authors are grateful to the MEPhI Academic Excellence Project for providing computing resources and facilities to perform experimental data processing.

References

1. Danilova NN (2012) Psychophysiology: textbook for universities. Aspect Press, Moscow
2. Alexandrov YuI (2014) Psychophysiology: textbook for universities. Supplemented and revised, 4th edn, p 41. SPb.: Piter
3. YouTube video. https://www.youtube.com/watch?v=FbuC1COzdRA. Accessed 20 May 2019
4. Kholodny YI (2017) Some aspects of the practice and technology of forensic research using a polygraph. Crime Invest Probl Solutions 4(18):185–192

5. Spence SA, Farrow TF, Herford AE, Wilkinson ID, Zheng Y, Woodruff PW (2001) Behavioural and functional anatomical correlates of deception in humans. NeuroReport 12:2849–2853
6. Ganis G, Kosslyn SM, Stose S, Thompson WL, Yurgelun-Todd DA (2003) Neural correlates of different types of deception: an fMRI investigation. Cereb Cortex 13(8):830–836
7. Kozel FA, Padgett NV, George MS (2004) A replication study of the neural correlates of deception. Behav Neurosci 118:852
8. Langleben DD, Loughead JW, Bilker WD, Ruparel K, Childress AR, Busch SI, Gur RC (2005) Telling truth from lie in individual subjects with fast event-related fMRI. Hum Brain Mapp 26:262
9. Rosenfeld JP (2018) Detecting concealed information and deception. Recent developments, p 440. Academic Press, London
10. Verschuere B, Ben-Shakhar G, Meijer E (2011) Memory detection. Theory and application of concealed information test, p 319. University Press, Cambridge
11. Peth J, Sommer T, Hebart MN, Vossel G, Büchel C, Matthias Gamer M (2015) Memory detection using fMRI – does the encoding context matter? NeuroImage 113:164–174
12. Ofen N, Whitfield-Gabrieli S, Chai XJ, Schwarzlose RF, Gabrieli GDE (2017) Neural correlates of deception: lying about past events and personal beliefs. Soc Cogn Affect Neurosci 12:116–127

Subjective Perception Space of Musical Passages

Mariya D. Krysko[1], Alexander V. Vartanov[1(✉)], Irine I. Vartanova[1], and Valentin V. Klimov[2]

[1] Lomonosov Moscow State University, Moscow, Russia
a_v_vartanov@mail.ru
[2] National Research Nuclear University MEPhI (Moscow Engineering Physics Institute), Moscow, Russia

Abstract. An experimental construction of the subjective (emotional) space of perception of musical sound fragments used in the GarageBand app for iOS to help the composer has been carried out. A representative set of 20 sound fragments was selected on the basis of an experiment on the classification of all samples from this database, which were further evaluated in 3 types of experiments: on the basis of multidimensional scaling of differences between them, on the basis of direct scaling of emotional qualities (Semantic Differential Method for 25 adjective pairs), based on direct scaling on three SAM scales (in the PXLab variant). As a result, the individual features (points of view) of these sound samples were investigated and a correlation comparison of the obtained results was carried out. As a result, the interrelation of all used scales is shown and a way of their integration into a four- dimensional spherical model of emotions is presented.

Keywords: Emotion · Semantic Differentia · SAM · Musical sound

1 Introduction

An important element in the development of virtual agents endowed with social-emotional intelligence can be the semantic or subjective emotional space (map) [1–3]. In this case, a vector representation is used with the possibility of a meaningful interpretation of the corresponding features, and the coordinates obtained are characterized by a certain meaning. In particular, it is important for the development of a virtual agent - assistant composer. In this case, the subjective space of perception of musical fragments included in the GarageBand app for iOS can serve as a possible "map". These fragments are positioned at https://www.apple.com/ru/ios/garageband/ [4] as a library of specially designed tools to help the composer. However, in order to develop a virtual assistant, it is necessary to identify the signs that a person relies on while listening to these passages, and to determine for each sample the corresponding coordinates in the vector space, which simulates the person's subjective perception. It is possible to build such a model not only theoretically, but also empirically in special psychophysical experiments with a person in several ways. Such a construction is

A. V. Samsonovich (Ed.): BICA 2019, AISC 948, pp. 266–276, 2020.
https://doi.org/10.1007/978-3-030-25719-4_34

possible both on the basis of Multidimensional scaling (MDS) [5], and on the basis of a direct assessment of objects by given (theoretically) qualities (scales).

However, this raises the problem of correlating different scales in a single model suitable for programming a virtual agent. On the one hand, the simplest models with the minimum number of the most important features have the advantage, but, on the other hand, it is desirable that the virtual agent reproduces the capabilities of a person in the relevant field as good as possible. In this connection, there is a problem of determining the minimum necessary dimension of space, which can be estimated statistically. When using direct scaling on a deliberately excessive number of scales with the subsequent use of the factor analysis procedure, the dimension can be reasonably reduced, however, in this case, there is a danger of initially missing any important qualities, since they may not be known in advance to the experimenter. This danger is eliminated by the use of MDS methods, since in this case the scales for assessment are not determined in advance, but subjective estimates of the magnitude (degree) of pairwise differences between the required number of objects are used. In this case, the application of the metric method MDS to the obtained data (the difference matrix) also allows us to statistically estimate the minimum allowable dimension of the feature space and calculate the coordinates of objects in this space. In this case, as is well known, the most universal system for evaluating various stimuli and events by man is emotions. There are various techniques and scales for describing emotions, different models have been built, including a four-dimensional spherical model of emotions based on the MDS method [6]. The most well-known of the methods of direct evaluation of emotions is the Semantic Differential (SD) method of Ch. Osgood [7] (in adapting to the Russian language, Petrenko [8] uses 25 adjective scales). Generalized scales are also widely used - pleasure, arousal, and dominance. The Manikin Self-Assessment (SAM) is a non-verbal pictorial assessment method of stimuli [9].

However, as shown by a special study, these widely used scales are quite difficult to relate to each other [10]. In this experiment, the authors obtained a list of the results of the experience of using the SAM, which requires only a series of judgments, and the Seasonal differential theory (1974) which requires 18 different ratings. Subjective reports were measured. Correlations across the two rating methods. Is what can be a stimulus. SAM is an inexpensive, cost-effective method of response to many contexts [10].

Thus, the purpose of this study was to build a universal subjective space of human perception of musical fragments based on various procedures - MDS, SAM scales and Semantic Differential scale. One of the tasks is also to evaluate the required dimension and find the relationship between these scales, as well as to explore the individual characteristics of the perception of these musical fragments.

2 Method

2.1 Stimuli

A set of 20 musical fragments were selected from 158 fragments presented in the GarageBand for iOS database, which are positioned at https://www.apple.com/ru/ios/garageband/ [4] as a library of specially designed helping tools for the composer.

These stimuli were selected in a preliminary classification experiment which 7 people participated (age 17–19 y.o. without anomalies of mental and morphological development; 6 subjects were female, 2 subjects did not have musical education, and 5 did): subjects were asked to listen to all 158 sounds and attribute each sound to one of the five classes that were not previously specified - the subjects themselves came up with the name of the classes depending on what emotion a given stimulus caused. Based on the data obtained, the probability of each pair of stimuli being in the same class for all 7 subjects was calculated. Then, using cluster analysis (k-mean), all stimuli were divided into 20 classes, after which one of the most typical representatives was taken from each class, with short fragments having the advantage. As a result, incentives were taken under the numbers 001, 002, 004, 010, 018, 025, 037, 045, 048, 049, 063, 067, 091, 093, 129, 141, 148, 151, 152, 155. Thus, the data The 20 stimuli (music fragments) fairly evenly represented the entire range of variants of the original set of sound fragments of the GarageBand application for iOS.

2.2 Methods of Multidimensional Scaling

In this experiment, subjective assessments of the magnitude (degree) of pairwise differences between selected stimuli (musical fragments) were used. In the experiment according to the MDS method, the same subjects participated in the classification and selection of stimuli, i.e. were previously very familiar with them. Incentives were presented by a computer using the Neurobs Presentation program. As a result, 11 samples were obtained for 190 pairs of stimuli, totaling 2090 estimates of the degree of difference in stimuli.

2.3 Method of Semantic Differential

Osgood [7] in adaptation to the Russian language Petrenko [8]. The experiment was carried out in a blank version (in an electronic form—an Excel spreadsheet was filled in), the subjects were required to evaluate each of the 20 selected stimuli on 25 scales formed by a pair of adjectives (positive and negative adjectives alternated in a random order). Evaluation was carried out by a number from −3 (the left adjective is more suitable) to +3 (the right adjective is more suitable). In this experiment, 35 people took part voluntary, cumulatively making 17,500 evaluations of 20 stimuli on 25 scales.

2.4 The PXLab Self-assessment-Manikin Scales Method

The subjects were required to evaluate each of the 20 selected incentives for each of the 3 scales - valence, arousal, and dominance. The evaluation involved 9 people from a group of subjects who conducted the assessment of SD. As a result, 540 ratings were. In the received second series of the experiment, there was a different set of phrases, and there were four possible variants: the phrase and its logical ending, the phrase and its conflict ending, two variants of phrases ending with the same word, but in one case it was suitable, and in the other conflict. In the latter two conditions, there was an additional unpleasant emotionally significant sound (human cry) with the last noun of the phrase.

3 Results

3.1 The Results of MDS - the Study of Points of View on the Primary Data

In order to identify the individual characteristics (points of view) of the subjects, the obtained evaluation of differences (in the form of columns of 190 evaluations for all pairs of 20 stimuli for each of 11 samples) were processed by factor analysis. The Fig. 1 shows a graph of eigenvalues, in accordance with the Cattel criterion, it is necessary to estimate the dimension of the factor space as 2. Thus, it was found out that there were at least (two factors describe only 48.8% of the total data variance), two groups of subjects - differently perceive these sound bites. In this regard, further analysis of the obtained matrices of differences was carried out separately for these two groups: on the basis of the obtained factor loads, the obtained data was averaged over the selected groups. It turned out that the second group consisted mainly of subjects with musical education.

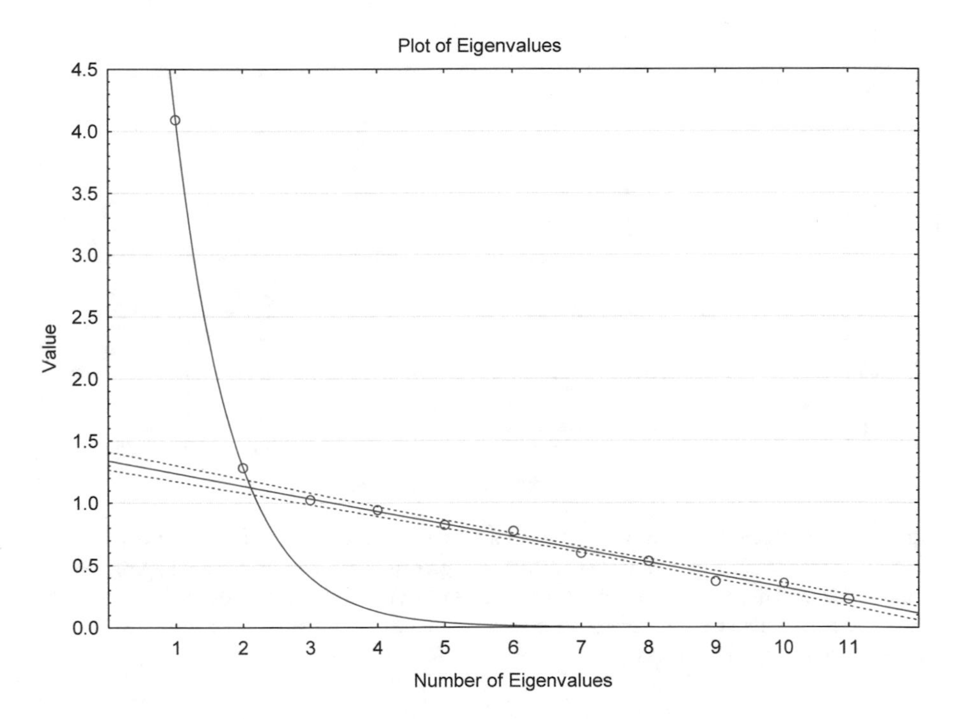

Fig. 1. Plot of eigenvalues. The first two values are approximated by the exponent, and the last 9 are straight (dotted line shows 95% confidence interval), which corresponds to random estimation errors by the Cattel criterion.

3.2 The Results of the Construction of Space

The obtained difference matrices were analyzed by the MDS metric method [5], which matrices of the scalar product, the eigenvalues and eigenvectors were calculated for. In accordance with the axiom of the MDS metric method, breaking a triangle rule (and, accordingly, obtaining negative eigenvalues using complex numbers) is possible only as a consequence of random estimation errors, which allows us to additionally estimate the required dimensionality of the resulting Euclidean space. The Fig. 2 shows the graphs of the distribution of eigenvalues for two average difference matrices, for each of which the required dimension can be estimated as equal to 4.

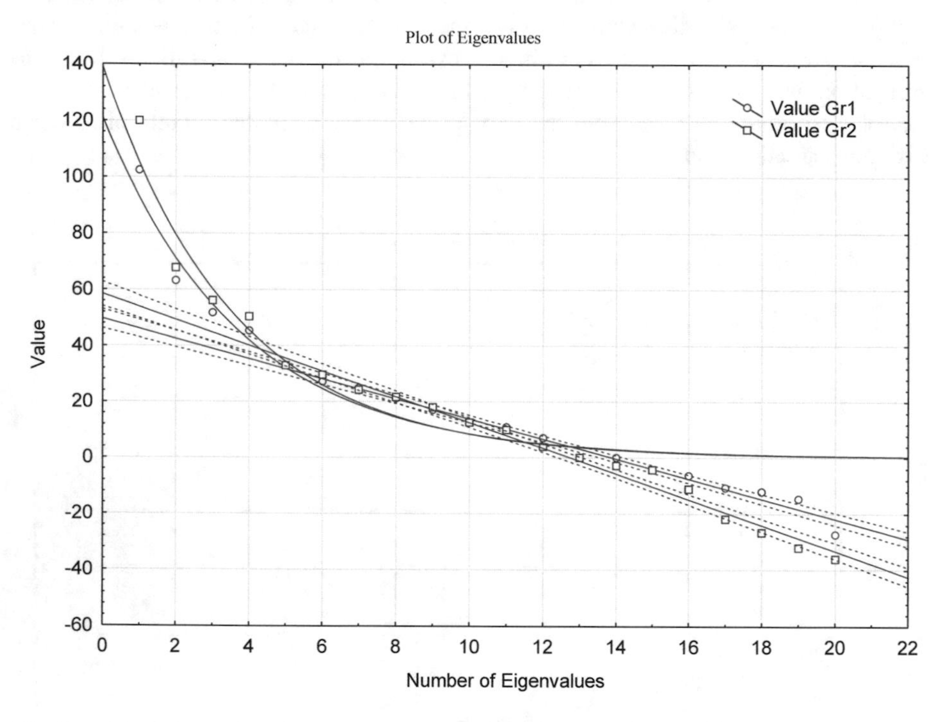

Fig. 2. Plot of eigenvalues for two groups of subjects. Both for the first (Gr1, marked by circles) and for the second (Gr2, marked by small squares) groups, the first 4 values are approximated by an exponent, and the last 16 - a straight line (dotted line shows 95% confidence interval), which according to Cattel's criterion corresponds to random estimation errors. This is also indicated by the presence of negative values lying on this straight line.

The Euclidean distances between the stimulus points calculated in the resulting 4-dimensional model were compared with the initial estimates (averaged matrices) of differences. For the first group of subjects, the correlation coefficient of the experimental and calculated values was 0.76349, and for the second, it was 0.80254. To interpret the selected features (Euclidean axes of space), the expert rotation procedure was used in order to achieve the interpretation of the resulting configuration in the parameters allocated earlier for the 4-dimensional spherical model of emotions [6].

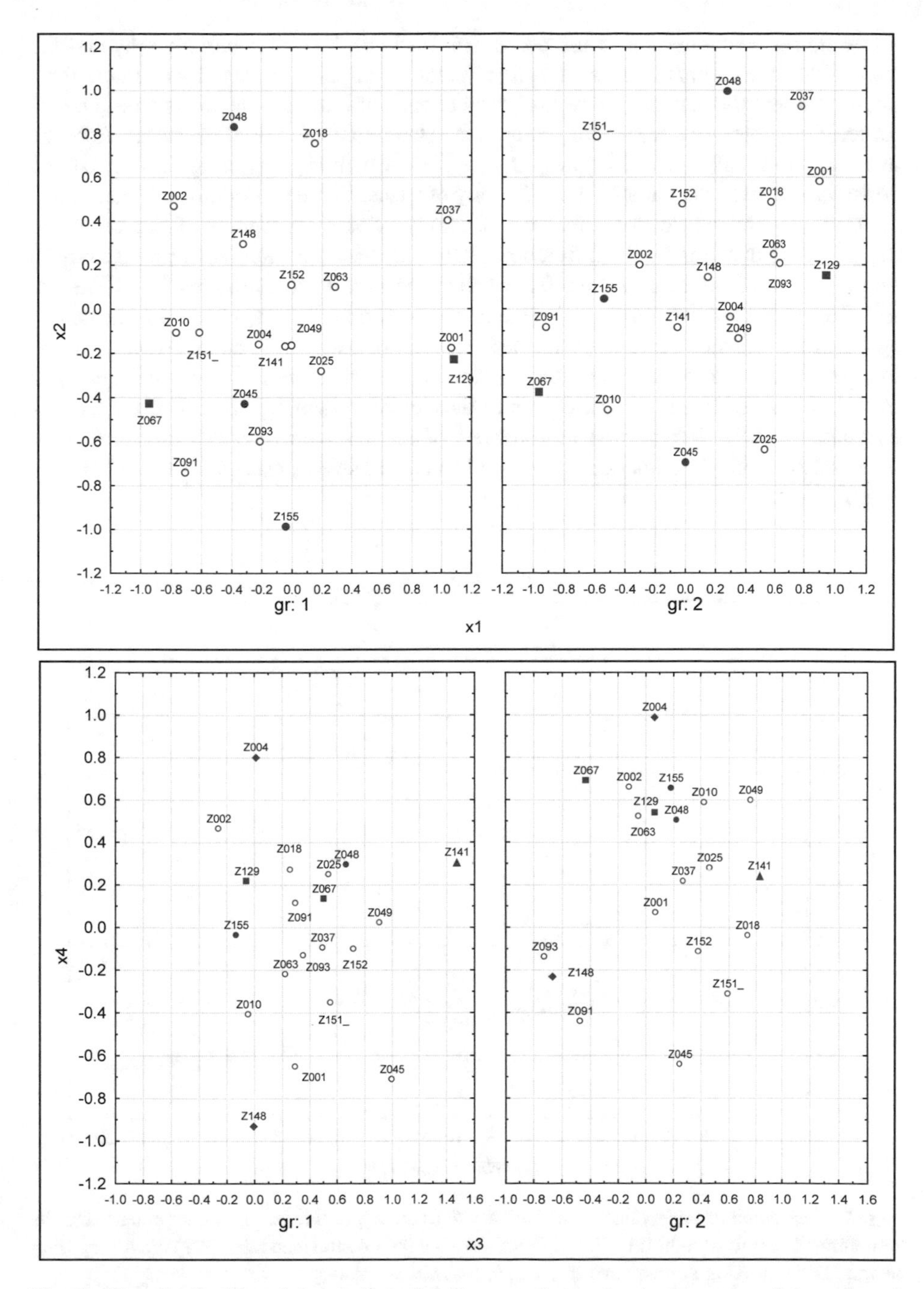

Fig. 3. The distribution of the studied stimuli - sound samples in the space of the selected features for group 1 (left) and 2 (right). Above is a projection on the plane of axes 1–2, below on a plane 3–4. Verification of the interpretation was carried out on the stimuli selected by the marker: the square indicates the most different stimuli on axis 1, the filled circle is on axis 2, triangle on axis 3, and rhombus on axis 4.

The obtained configuration of the points was also checked for sphericity—by shifting the center of coordinates of the original vector description, a search was made for a position where the distance from this center to each of the stimulus points would be the same, if possible. As a result, for group 1, the standard deviation of the radius from the average was 25.39%, and for group 2 - 17.73%, which at a given noise level can be considered a good indicator of the sphericity of space. These procedures of rotation and shift of coordinates do not affect the accuracy of the description of the model of experimental matrices by the differences, however, they are essential for a meaningful interpretation of the resulting model. Further, the Fig. 3 shows the results of the distribution of the studied musical fragments in the form of points in the 4-dimensional space of the selected features (projected on pairs of axes). The obtained axes can be interpreted in accordance with the previously described four-dimensional spherical model of emotions: axis 1 - assessment of the sign of the situation (bad - good), axis 2 - assessment of the information definiteness of the situation (clearly - uncertainly or surprisingly), axis 3 - attraction, axis 4 - rejection (passive or fear - active, rage) (Fig. 4 and Table 1).

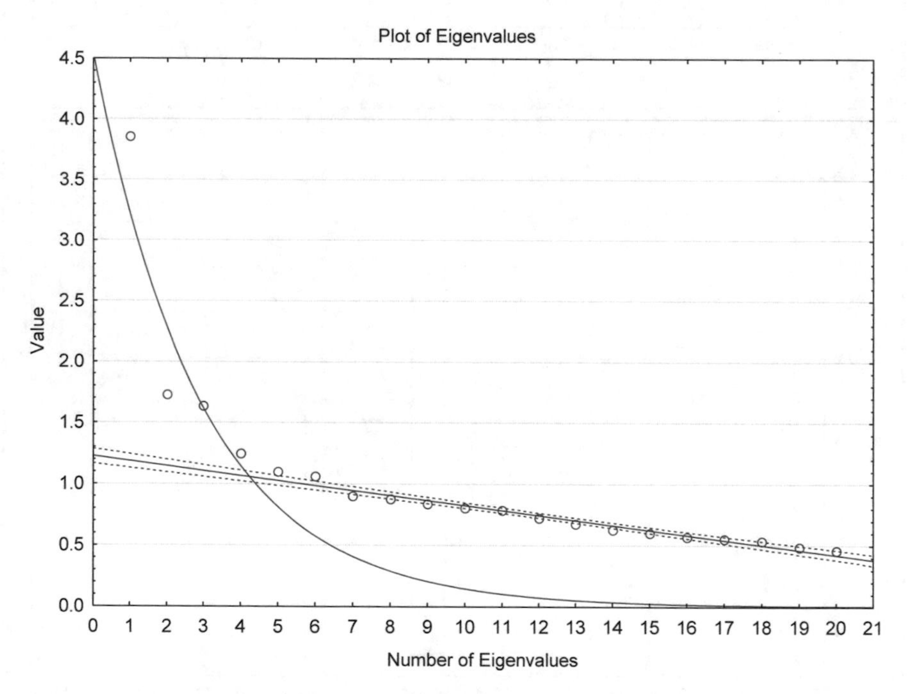

Fig. 4. The graph of eigenvalues. The first 4 values are approximated by the exponent, and the remaining ones are straight (the dotted line shows the 95% confidence interval), which according to the Cattel criterion corresponds to random estimation errors.

Table 1. The frequency of adjectives (taking into account the sign), which load the relevant factors.

Factor 1	Factor 2	Factor 3	Factor 4
Attraction (pleasant, kind, good)	Activity (fast, active, solid)	Dominance (large, expensive)	Score (light, fresh, smooth)
Nasty – nice 10	Fast – slow 17	Big – small 8	Light – heavy 7
Evil – good 8	Active – Passive 12	Expensive – cheap 6	Fresh – Rotten 6
Bad - good 6	Soft – hard 9	Smooth – rough 3	Smooth – rough 5
Dark – light 5	Joyful – sad 4	Weak – strong 2	Clean – dirty 5
Chaotic - ordered 4	Simple - complex 4	Simple - complex 2	Active – Passive 4
Expensive – cheap 4	Sharp - stupid 4	Wet – dry 2	Native - alien 4
Favorite – hated 4		Fresh – Rotten 2	
Fresh – Rotten 4			

Thus, the interpretation obtained is different from the classical one. Usually, the first factor is interpreted as "evaluation" (bad - good), but the table shows that it is rather an attraction (pleasant, kind, good). The assessment turned out to be highlighted in the new factor 4 (light, fresh, smooth). Interpretation of the second factor - activity - coincides with the traditional one. The third factor can be described as dominance (big, expensive).

3.3 Results of Comparison of Scales Obtained by Different Methods

Since the same stimulus objects were evaluated using different methods and the structure of factors according to SD and MDS turned out to be the same dimension, a direct comparison of results was carried out. The configuration of the stimulus points in the 4-dimensional model obtained according to the MDS data (for group 1) rotated to achieve the best fit with the model constructed using the SD data. The results are presented in Fig. 5.

In Fig. 5, it can be seen that the interpretation of factor 1 by SD really, corresponds to attraction, since the largest weight is taken by the passage Z141, marked by a triangle, i.e. the one that maximally loaded axis 3 in a spherical model of emotions. According to factor 2 SD, the stimuli Z129 - Z67 were located, which in the spherical model of emotions were located along axis 1 (sign evaluation), but the stimuli Z155-Z048 from axis 2 of the spherical model (surprise) were located similarly. The 3 SD factor combined the 1st and 3rd axes of the spherical model of emotions, and the 4 SD factor was connected with the 4 axis (rejection) of the spherical model.

Table 2 presents the data of the correlation analysis of the comparison of all selected SD axes, MDS axes rotated to them, and MAS scales.

Table 2 shows that the PXLab MAS scale 1 is more strongly associated with factor 4 SD (assessment), but also with factor 3. Scale 2 correlates with two factors SD - 2 (activity), and 3 (dominance). However, the PXLab MAS scale 3 (dominance) generally does not correlate well with the SD factors, however, it is associated with axis 2 (surprise) of the MDS space of group 2.

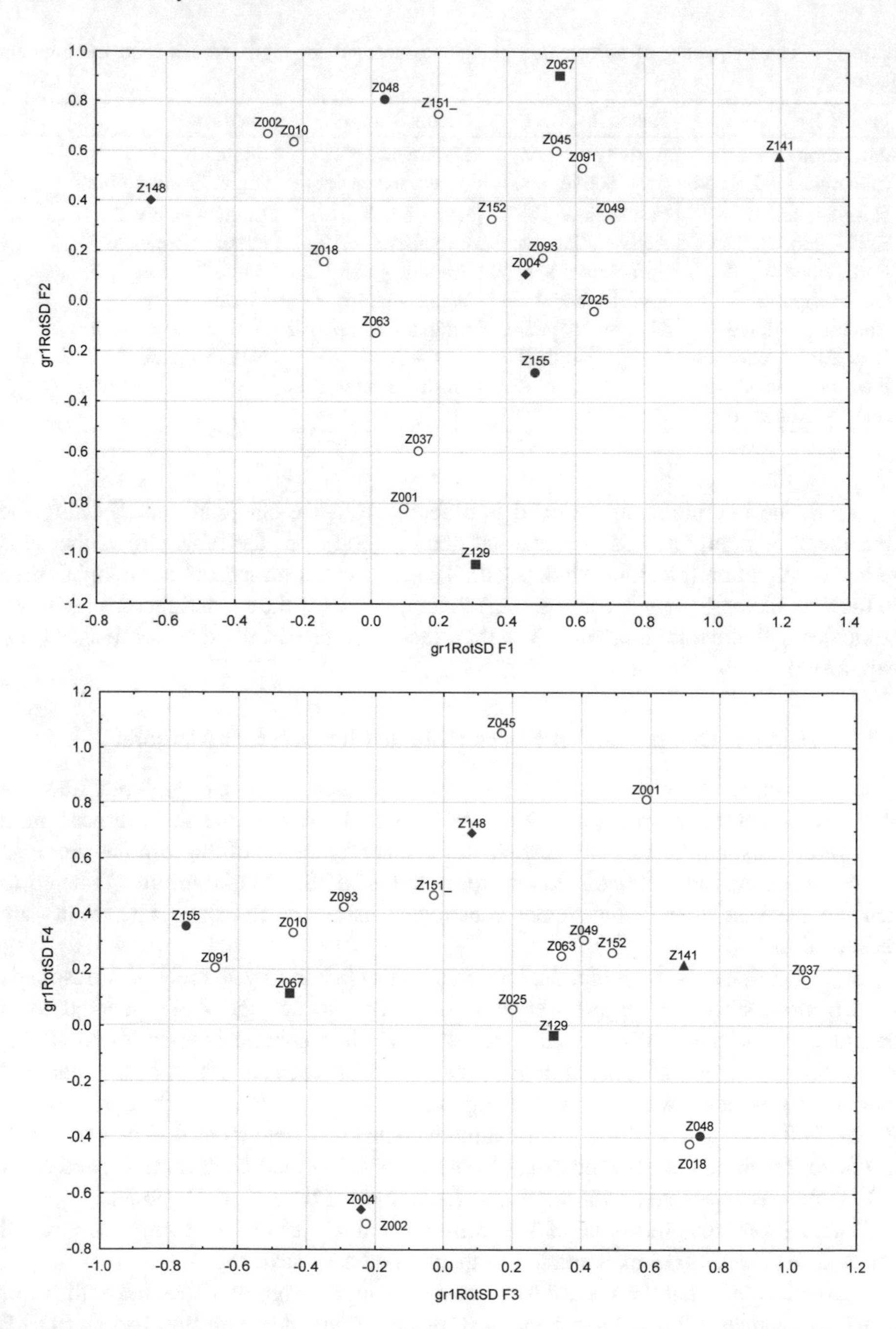

Fig. 5. The distribution of the studied stimuli - sound samples in the space of signs, rotated to best match the data SD. Above there is a projection on the plane of the axes 1–2, below there is a projection on the plane 3–4. Verification of interpretation is possible according to the stimuli selected by the marker, implemented for the axes in the MDS space (in accordance with the 4-dimensional spherical model of emotions): axis 1, shaded circle - along axis 2, triangle - along axis 3, diamonds - along axis 4.

Table 2. The results of the correlation analysis of the relationships between scales in models obtained by different methods.

	SD F1	SD F2	SD F3	SD F4	PxLab 1	PxLab 2	PxLab 3	gr1Rot SD F1	gr1Rot SD F2	gr1Rot SD F3	gr1Rot SD F4	gr2Rot SD F1	r2Rot SD F2	r2Rot SD F3	r2Rot SD F4
Attraction (pleasant, kind, good)	1.00	−0.28	−0.28	−0.24	0.39	−0.02	−0.35	*0.61*	−0.03	−0.14	0.01	*0.58*	0.07	−0.13	−0.29
Activity (fast, active, solid)	−0.28	1.00	−0.27	−0.37	−0.33	*0.75*	0.40	−0.16	*0.67*	−0.40	−0.28	−0.15	*0.64*	*−0.45*	−0.26
Dominance (large, expensive)	−0.28	−0.27	1.00	−0.37	*−0.71*	*−0.54*	0.30	−0.32	−0.43	*0.53*	−0.23	0.03	−0.22	*0.64*	0.21
Score (light, fresh, smooth)	−0.24	−0.37	−0.37	1.00	*0.57*	−0.03	−0.25	−0.08	−0.26	−0.17	*0.45*	*−0.52*	−0.31	−0.05	0.39
Valence PxLab 1	0.39	−0.33	*−0.71*	*0.57*	1.00	0.07	−0.37	0.28	−0.04	0.00	0.35	0.18	−0.25	−0.13	0.12
Arousal PxLab 2	−0.02	*0.75*	*−0.54*	−0.03	0.07	1.00	*0.45*	−0.12	*0.69*	*−0.62*	−0.10	−0.12	*0.66*	*−0.57*	−0.25
Dominance PxLab 3	−0.35	0.40	0.30	−0.25	−0.37	*0.45*	1.00	−0.40	0.25	−0.13	*−0.48*	−0.02	*0.54*	0.07	0.26
gr1RotSD F1	*0.61*	−0.16	−0.32	−0.08	0.28	−0.12	−0.40	1.00	0.02	−0.04	0.11	0.30	−0.03	−0.30	−0.09
gr1RotSD F2	−0.03	*0.67*	−0.43	−0.26	−0.04	*0.69*	0.25	0.02	1.00	−0.30	−0.09	−0.11	0.39	*−0.48*	*−0.60*
gr1RotSD F3	−0.14	−0.40	*0.53*	−0.17	0.00	*−0.62*	−0.13	−0.04	−0.30	1.00	−0.02	0.31	*−0.66*	*0.69*	0.16
gr1RotSD F4	0.01	−0.28	−0.23	*0.45*	0.35	−0.10	*−0.48*	0.11	−0.09	−0.02	1.00	−0.29	−0.40	−0.29	−0.15
gr2RotSD F1	*0.58*	−0.15	0.03	*−0.52*	0.18	−0.12	−0.02	0.30	−0.11	0.31	−0.29	1.00	−0.01	0.09	0.08
gr2RotSD F2	0.07	*0.64*	−0.22	−0.31	−0.25	*0.66*	*0.54*	−0.03	0.39	*−0.66*	−0.40	−0.01	1.00	−0.33	−0.16
gr2RotSD F3	−0.13	*−0.45*	*0.64*	−0.05	−0.13	*−0.57*	0.07	−0.30	*−0.48*	*0.69*	−0.29	0.09	−0.33	1.00	0.24
gr2RotSD F4	−0.29	−0.26	0.21	0.39	0.12	−0.25	0.26	−0.09	*−0.60*	0.16	−0.15	0.08	−0.16	0.24	1.00

Correlations
Marked correlations are significant at $p < .05000$
N = 20 (Casewise deletion of missing data)

As a result, the interrelation of all used scales is shown and a way of their integration into a four- dimensional spherical model of emotions is presented.

Acknowledgments. This work was supported by the Russian Science Foundation, Grant № 18-11-00336.

References

1. Samsonovich AV (2009) The constructor metacognitive architecture. In: The AAAI fall symposium-technical report, pp 124–134. AAAI Press: FS-09-01, Menlo Park
2. Samsonovich AV, Goldin RF, Ascoli GA (2010) Toward a semantic general theory of everything. Complexity 15(4):12–18. https://doi.org/10.1002/cplx.20293
3. Eidlin AA, Eidlina MA, Samsonovich AV (2018) Analyzing weak semantic map of word senses. Procedia Comput Sci 123:140–148
4. GarageBand for iOS https://www.apple.com/ru/ios/garageband/
5. Torgerson NS (1958) Theory and methods of dcaling. Wiley, New York
6. VartanovA V, Vartanova II (2018) Four-dimensional spherical model of emotions. Procedia Comput Sci 145:604–610. https://doi.org/10.1016/j.procs.2018.11.096
7. Osgood CE, Suci GJ, Tannenbaum PH (1957) The measurement of meaning. University of Illinois Press, Urbana
8. Petrenko VF (1983) Vvedenie v eksperimental'nuyu psikhosemantiku. Issledovanie form reprezentatsii v obydennom soznanii, 175 p. [Introduction to experimental psychosemantics. Study forms of representation in ordinary consciousness]. MGU Publ, Moscow. (In Russ.)
9. Bradley MM, Lang PJ (1994) Measuring emotion: the self-assessment manikin and the semantic differential. J Behav Ther Exp Psychiatry 25(1):49–59. https://doi.org/10.1016/0005-7916(94)90063-9

Factographic Information Retrieval for Biological Objects

Sergey D. Kulik[1,2(✉)]

[1] National Research Nuclear University MEPHI, Kashirskoe shosse 31, Moscow 115409, Russia
sedmik@mail.ru
[2] Moscow State University of Psychology and Education (MSUPE), Sretenka st. 29, Moscow 127051, Russia

Abstract. This paper describes the results of work to develop an automated factographic information retrieval system for biological forensic objects. The effective factographic information retrieval problem has been investigated in the paper. This factographic information retrieval includes a pattern recognition algorithm, and they are implemented for retrieval only one image among the variety of similar images of biological forensic objects. The automated factographic information retrieval system includes special retrieval block and human operator. Analytical formula was obtained to evaluate the effectiveness of factographic information retrieval using indicator. This formula is presented by effectiveness indicator: average length of the recommendatory list provided by the retrieval block enabling the human operator to take the final decision. The paper describes the structure of the algorithm for factographic information retrieval. Properties of the important indicator of effectiveness – the average length of the recommendatory list for the human operator were explored.

Keywords: Factographic information retrieval · Information system · Forensic science · Neural networks · Biological object

1 Introduction

In practice, a factographic information retrieval (FIR) is used in question-answering systems [1] for detailed analisys of biological and forensic objects. Automated Factographic Information Retrieval System (AFIRS) is a special case of question-answer systems (QA systems) [2]. This Automated Factographic Information Retrieval System [2] in its structure contains a special block of recognition. To develop this block we used neural networks. Robots [3], intelligent cobots [4], convolutional neural networks [5], become increasingly popular today. Human-oriented social robots, or cobots [6], will work with humans side by side. These robots contain a Human-like artificial emotional intelligence [4]. Some Intelligent agents, co-robots and systems agent-based search [7] include AFIRS and block of recognition for computer vision.

It is necessary to give a short review of the research field. A lot of papers and monographs have been published on topics related to the subject of this paper, such as neural networks and factographic information retrieval. Among them several recent

A. V. Samsonovich (Ed.): BICA 2019, AISC 948, pp. 277–282, 2020.
https://doi.org/10.1007/978-3-030-25719-4_35

works were found to be helpful during the research papers [8–11]. The papers [9, 11] overviews the algorithms and applications for forensic handwriting examination and factographic information retrieval. Paper [9] describes a forensic handwriting examination and a number of methods to reduce the impact of human factors on the results of handwriting examination. Another paper [11] deals with factographic information retrieval for semiconductor physics, micro- and nanosystems. As factographic information retrieval often deals with big amounts of data, it's also useful to understand core concepts of big data which are described in paper [10].

2 Factographic Information Retrieval

There are important scientific areas of applications for FIR, artificial neural networks, information technology and artificial intelligence methods:

- information security [8],
- association rules mining [16],
- big data [10] and neural network [20],
- medicine (biomedical engineering) [18],
- cognitive competence of graduates [13],
- regular agent technologies [19],
- gaze-based control [12],
- semiconductor physics [11],
- communication in multicultural society [2],
- Eye-Brain-Computer Interface [14],
- retrieval system [11],
- robots [3], intelligent cobots [4],
- forensic experts [9] and criminology [15]
- automated factographic information retrieval system [17]

and many other areas.

System structure for AFIRS was developed in [2]. For instance, the task of information retrieval is to find biological objects or Image which Is Identical (ImII) to the enquiry among the archive system documents (images) with the help of their descriptions in the search factographic database (FDB) and, in case of this document detection, to give necessary factographic information from this image (document).

The automated factographic information retrieval system includes six blocks for example: the block 1 of image indexing; the block 2 of retrieval; the block 3 of Recommendatory List (RL) processing; the block 4 – the searching FDB which includes image descriptions in the form of Searching Images Patterns (SIP); the block 5 of recognition; the block 6 (archive of documents).

It is supposed that there are N images (documents) which are stored in the archive and every enquiry have Searching Enquiry Pattern (SEP). Each image (document) of the archive is located in the place which is uniquely determined by the registration number. Only one image (document) can be located in only one place. All SIPs are stored in the searching array of the factographic database in the form of a consecutive linear list.

Request for the search (or SEP) includes the description of the image which can be stored or can be absent in the images array. It is supposed that the request with Q probability can be a description of the image which is identical to one of the archive images. Information in SIP and SEP can be misrepresented because of different hindrances (noises) or errors during the image indexation. Comparison of SEP and SIP by the AFIRS is realised using patterns recognition algorithm and is characterised by U, V probabilities where [2]:

- U – probability of the correct comparison of two identical images based on their descriptions (determines the target mission probability);
- V – probability of the correct comparison of two non-identical images based on their descriptions (determines the false alarm probability).

Analytical formulas were obtained to evaluate the effectiveness of factographic information retrieval for biological objects using indicator:

$$Z = Q \cdot E + (1 - Q) \cdot T, \tag{1}$$

where:

Z – average RL length generated by the search block of AFIRS for making a final decision by a human operator (for instance forensic expert).

E – average length of the RL during information retrieval in the area which includes ImII.

T – average length of RL during information retrieval in the area which does not include ImII.

It is supposed that L is maximum length of the RL. If $Q \approx 0$, the $Z \approx T$. The following analytical expressions for Z were obtained to evaluate the effectiveness FIR [2]:

$$Z \approx \left[\sum_{m=0}^{L-1} \{ m \cdot S_m \cdot D_m^N \} + \sum_{m=L}^{N} \{ L \cdot S_m \cdot D_m^N \} \right], \tag{2}$$

where:

$$S_m = V^{N-m} [1 - V]^m,$$

$$D_m^N = \frac{N!}{m!(N - m)!}.$$

3 Results

As a result of researches average length of RL, it was set that Z is changed during the changes of L, N, and V. A small part of these researches for different L, N and V we can see in the Table 1.

Table 1. Example of effectiveness FIR ($N = 15000$ and $V = 0.95$).

Maximum length of the RL (L)	Average RL length ($Z \approx$)
3	3
85	85
95	95
400	400
…	…
700	700
740	733
750	740
760	744
790	749
900	750
…	…
2500	750
5000	750
6000	750
…	…
15000	750

According to Table 1, if $V = 0.95$, $N = 15000$ (N – the number of SIP), and L 760 (L – maximum length of the RL), the effectiveness FIR is $Z \approx 744$. Analogously, if $V = 0.95$, $N = 15000$ and $L = 5000$, the effectiveness FIR is $Z \approx 750$.

4 Conclusion

Thus, as a result of research, the analytical formula for Z allowing the evaluation of the RL average length was developed. Properties of the important indicator of effectiveness – the RL average length for the human operator were explored. It allows reasonable analysis of the effectiveness of the factographic information retrieval which is implemented by the AFIRS. Necessary software, allowing the evaluation of the retrieval's effectiveness, was created for the developer of the automated factographic information retrieval system. This factographic information retrieval can be applied to full images of the biological forensic objects.

We also plan to apply some statistical methods for the developing of factographic information retrieval, which are more resistant to the dependence of the biological features.

We believe that our results are useful for scientists and researches in the field of cognitive technology for example robotic wheelchair [21].

Acknowledgements. This work was supported by Competitiveness Growth Program of the Federal Autonomous Educational Institution of Higher Education National Research Nuclear University MEPhI (Moscow Engineering Physics Institute). The topics presented in this paper have been discussed with a number of people in various conferences. I thank all them for the received comments.

References

1. Ko J, Si L, Nyberg E, Mitamura T (2010) Probabilistic models for answer-ranking in multilingual question-answering. ACM Trans Inf Syst (TOIS) 28(3):37. Article 16
2. Kulik S (2016) Factographic information retrieval for communication in multicultural society. In: Procedia - social and behavioral sciences (International conference on communication in multicultural society, CMSC 2015, 6–8 December 2015, Moscow, Russian Federation), vol 236, pp 29–33
3. Gridnev AA, Voznenko TI, Chepin EV (2018) The decision-making system for a multi-channel robotic device control. Procedia Comput Sci 123:149–154
4. Samsonovich AV (2013) Emotional biologically inspired cognitive architecture. Biol Inspired Cogn Arch 6:109–125. https://doi.org/10.1016/j.bica.2013.07.009
5. Chistyakov IS, Chepin EV (2019) Gesture recognition system based on convolutional neural networks. In: 2019 IOP conference series: materials science and engineering, vol 498, p 012023
6. Samsonovich AV (2018) On semantic map as a key component in socially-emotional BICA. Biol Inspired Cogn. Arch 23:1–6. https://doi.org/10.1016/j.bica.2017.12.002
7. Artamonov AA, Ionkina KV, Kirichenko AV, Lopatina EO, Tretyakov ES, Cherkasskiy AI (2018) Agent-based search in social networks. Int J Civ Eng Technol 9(13):28–35
8. Miloslavskaya N, Tolstoy A (2016) State-level views on professional competencies in the field of IoT and cloud information security. In: 2016 IEEE 4th international conference on future internet of things and cloud workshops (FiCloudW), Vienna, pp 83–90
9. Kulik S, Nikonets D (2016) Forensic handwriting examination and human factors: improving the practice through automation and expert training. In: Proceedings of the third international conference on digital information processing, data mining, and wireless communications, DIPDMWC 2016, 06–08 July, Moscow, Russia, pp 221–226
10. Miloslavskaya N, Tolstoy A (2016) Big data, fast data and data lake concepts. Procedia Comput Sci 88:300–305
11. Kulik, S.D.: Factographic information retrieval for semiconductor physics, micro - and nanosystems. AMNST 2017. In: IOP conference series: materials science and engineering, vol 498, 012026 (2019)
12. Shishkin S, Nuzhdin Y, Svirin E, Trofimov A, Fedorova A, Kozyrskiy B, Velichkovsky B (2016) EEG negativity in fixations used for gaze-based control: toward converting intentions into actions with an eye-brain-computer interface. Front Neurosci 10:1–20
13. Kireev V, Silenko A, Guseva A (2017) Cognitive competence of graduates, oriented to work in the knowledge management system in the state corporation “ROSATOM”. J Phys: Conf Ser 781(1):012060
14. Nuzhdin YO, Shishkin SL, Fedorova AA, Trofimov AG, Svirin EP, Kozyrskiy BL, Medyntsev AA, Dubynin IA, Velichkovsky BM (2017) The expectation based eye-brain-computer interface: an attempt of online test. In: Proceedings of the 2017 ACM workshop on an application-oriented approach to BCI out of the laboratory. ACM, pp 39–42

15. Yasnitsky LN, Vauleva SV, Safonova DN, Cherepanov FM (2015) The use of artificial intelligence methods in the analysis of serial killers' personal characteristics. Criminol J Baikal Natl Univ Econ Law 9(3):423–430
16. Kireev VS, Guseva AI, Bochkaryov PV, Kuznetsov IA, Filippov SA (2019) Association rules mining for predictive analytics in IoT cloud system. In: Samsonovich A (ed) Biologically inspired cognitive architectures 2018, vol 848. Advances in intelligent systems and computing. Springer, Cham, pp 107–112
17. Kulik S (2016) Factographic information retrieval for competences forming. In: Proceedings of the third international conference on digital information processing, data mining, and wireless communications, DIPDMWC 2016, 06–08 July, Moscow, Russian Federation, pp 245–250
18. Yasnitsky LN, Dumler AA, Bogdanov KV, Poleschuk AN, Cherepanov FM, Makurina TV, Chugaynov SV (2013) Diagnosis and prognosis of cardiovascular diseases on the basis of neural networks. Biomed Eng 47(3):160–163
19. Artamonov A, Onykiy B, Ananieva A, Ionkina K, Kshnyakov D, Danilova V, Korotkov M (2016) Regular agent technologies for the formation of dynamic profile. Procedia Comput Sci 88:482–486
20. Verbitsky NS, Chepin EV, Gridnev AA (2018) Experimental studies of a convolutional neural network for application in the navigation system of a mobile robot. Procedia Comput Sci 145:611–616
21. Voznenko TI, Chepin EV, Urvanov GA (2018) The control system based on extended BCI for a robotic wheelchair. Procedia Comput Sci 123:522–527

Recognition Algorithm for Biological and Criminalistics Objects

Sergey D. Kulik[1,2(✉)] and Alexander N. Shtanko[1]

[1] National Research Nuclear University MEPHI, Kashirskoe shosse 31, Moscow 115409, Russia
sedmik@mail.ru, shtanko-mephi@yandex.ru

[2] Moscow State University of Psychology and Education (MSUPE), Sretenka st. 29, Moscow 127051, Russia

Abstract. This paper describes the results of a work to develop an algorithm for analyzing images of embossed impressions in paper documents under oblique lighting. The described algorithm could also be used for recognition of similarly-structured objects, for example, some of biological structures. This type of analysis is necessary during forensic analysis of certain security features of paper documents. Part of this analysis is determining to which category new, uncategorized impression belongs to. This research explores the potential for automation of this task using neural networks. The core element of the algorithm is a neural network which determines the similarity between two embossed impressions. The paper describes the structure of the algorithm, a method for creating an image database for training and testing, as well as testing results for proposed algorithm.

Keywords: Computer vision · Forensic science · Neural networks · Pattern recognition

1 Introduction

In the modern world complex systems, such as agent-based search systems [1] and intelligent co-robots [2], are often used. These complex systems may rely on computer vision to intelligently process visual input data. Areas that require visual data analysis include biology and forensic science (criminalistics). Modern computer technologies allow us to automate many tasks which only experts could do in the past. One of the areas in which human experts are not yet replaced by machines is visual image analysis. But even in this area technologies are moving forward often due to developments in machine learning algorithms.

Computer vision is a science that deals with processing and analyzing of digital images, its applications vary from filters for enhancing photographs on social media to automated intelligent quality control systems. One of the promising areas for computer vision application is criminalistics (forensic science).

Forensic science studies collection and analysis of evidence. Tasks that may involve computer vision include automated handwriting examination, restoration of destroyed documents, fingerprint matching and many others.

A. V. Samsonovich (Ed.): BICA 2019, AISC 948, pp. 283–294, 2020.
https://doi.org/10.1007/978-3-030-25719-4_36

The object of this work is the analysis of embossed impressions under oblique lights. Embossing can be used as a security feature by itself and as method to forge complex security features like watermarks. The goal of this work is to create an algorithm with neural network at its core that could determine whether two embossed impressions are made with the same die or not from their images under oblique lighting. Such an algorithm could also be used to implement search of new impression in database of known samples.

1.1 Forensic Science

Forensic science or criminalistics is a science that studies factors of crimes, methods of acquiring information about them and analyzing evidence in criminal investigations. The general goal of forensic science is to solve crimes using modern science and technologies. One of the important subgoals of forensic science is developing and improvement of technical and forensic tools and methods of paper document analysis, in particular their security features [3].

There're many examination methods for establishing the authenticity of document or analyzing a forgery. Because there're so many examinations to be conducted on a single document a complete analysis is a laborious process and requires as much automation as possible.

This works focuses on developing a tool that could assist in con-ducting an analysis of a document in oblique light.

Often when examining security features forged by embossment (for example, forged watermarks or metallographic printing) expert has to conclude whether two impressions are made with the same die or not. In addition, sometimes expert has to find examined impression in database of known samples. This paper belongs to a series of research to automate this task. The final system is sup-posed to be an assistant program that stores known samples and can suggest close matching samples for any given new impression. To minimize number of mistakes the expert would make the final decision, but list of suggestions would reduce the search time significantly. This work explores the possibility of solving this task with neural network.

It is necessary to give a short review of the research field. A lot of papers and monographs have been published on topics related to the subject of this paper, such as neural networks and information retrieval. Among them several recent works were found to be helpful during the research: book [4] and papers [5, 6]. The book [7] overviews the most prominent algorithms and applications for computer vision and image processing. Paper [5] describes a forensic handwriting examination and a number of methods to reduce the impact of human factors on the results of handwriting examination. Another paper [6] deals with factographic information retrieval for semiconductor physics, micro- and nanosystems. As machine learning often deals with big amounts of data, it's also useful to understand core concepts of big data which are described in paper [8].

1.2 Computer Vision

There are important scientific areas of applications for computer vision [7], artificial neural networks and information technology:

- cognitive competence of graduates [18];
- regular agent technologies [20];
- big data, fast data [8];
- factographic information retrieval [16];
- Eye-Brain-Computer Interface [15];
- semiconductor physics [6];
- data analysis [13];
- medicine (biomedical engineering) [23];
- association rules mining [19];
- police, criminology [21] and forensic experts [5];
- digital image processing [4];
- retrieval system [6];
- Gaze-Based Control [14];
- communication in multicultural society [17];
- housing market [22];
- information security [9];

and many other areas. Computer vision is a very powerful tool for different tasks and areas of applications. For instance, a computer vision task of criminalistics objects and biological structures recognition.

2 Algorithm for Recognition

2.1 Neural Networks

Computer vision is a set of technologies and algorithms for automatic image and video analysis by computers in order to obtain knowledge about the environment [7].

One of the tools for image processing is neural networks. Artificial neural network is a very powerful computation model inspired by neural cells in animals' brains. Neural networks are used in various areas, such as medicine, engineering, economics and many others. The core idea of neural networks lies in optimization theory. This expresses itself as finding optimal values for network weights to minimize error.

Neural network consists of a set of artificial neurons connected to each other. Each connection has a weight value assigned to it. Weights determine how much connected neurons affect each other.

Usually neurons are divided into layers. Neurons of lower level are connected to neurons of immediate higher level. In simplest networks each neuron of level M is connected to every neuron of $M - 1$ level.

Fully connected networks can be applied to images; however, this type of networks completely ignores the structure of input pixels. Input values can be fed in any fixed order with the same result, while images have strong two-dimensional structure: spatially close variables are correlated. Local correlation of variables is the reason why extraction and combination of local features is so advantageous for following recognition of objects in an image. Configurations of neighboring pixels can be classified into small set of categories, such as edges and corners. Many image processing algorithms make use of feature extraction. In neural networks there're special convolutional neural networks designed to process images. Convolutional neural networks extract local features by restricting the perception field of neurons.

There're tree main architectural ideas in convolutional neural networks that provide a certain degree of invariance to image shifting, scaling and distortion: the locality of perception fields, shared connection weights and spatial pooling.

Input neurons receive pixel values from image. Each neuron of convolution layer receives values of neurons of previous layers that are located in a small area of neighboring neurons. So each convolution neuron has a restricted perception field. This idea allows neurons to learn and extract basic features of images, such as edges and corners. After that, these features can be combined by the following layers to form high-level features. As was said before, distortions and shifts can change the position of features. In addition to that basic feature detectors that are useful in one part of the image are likely to be useful over the whole image. To implement this knowledge into network, we may restrict weights of neurons over-seeing different parts of the image to be exactly the same. Neurons with the same set of weights are organized into planes. Neuron values in a plane are called feature map. Neurons of the same feature map perform the same operation in different parts of the image. Usually convolution layer consists of multiple feature maps (with different set of weights) so network could learn multiple features. Calculating neuron values of convolution layer equivalent to convolving image then adding a bias and applying activation function. In this interpretation neuron weights are convolution kernel. One of the properties of convolutional layer is that shifting of input image results in exact shift in output of the layer (without changing the values themselves). This property allows networks to be robust against shifts and distortions.

As soon as a feature is detected its exact position becomes less relevant and can even harm network accuracy, because exact positions change from image to image. The easiest way to mitigate this effect is to reduce the size of a feature map. This can be achieved by pooling layers. Pooling layer performs local averaging and map's size reduction, lowering network's sensitivity to shifts and distortions. Perception fields of pooling neurons don't overlap so if each neuron is connected to 2 by 2 square of previous layers the resulting feature map will be two times smaller.

Usually in convolutional neural networks one or two convolution and pooling layers followed by several fully connected layers [10].

In this work neural network is used to determine whether or not two impressions are created using the same die. Pair of analyzed images is passed to network as input. If they are made with the same die the network is supposed to return "1" (positive result) otherwise "0" (negative result). In other words, this neural network is a binary classifier.

To evaluate performance of binary classification algorithm there're multiple metrics.

Number of true positives *TP* – number of items correctly labeled by classifier as positive.

Number of false negatives FN – number of items incorrectly classified as negative.
Number of false positives FP – number of items incorrectly classified as positive.
Number of true negatives TN – number of correctly identified negative items.
True positive rate, that is the proportion of positive items in testing set that were correctly labeled as such [11]:

$$tp\_rate = \frac{TP}{TP + FN}. \quad (1)$$

False positive rate, that is the proportion of incorrectly labeled negative items [11]:

$$fp\_rate = \frac{FP}{FP + TN}. \quad (2)$$

Accuracy, that is the proportion of correctly labeled samples in testing dataset [11]:

$$accuracy = \frac{TP + TN}{TP + FP + FN + TN}. \quad (3)$$

Precision, that is the proportion of correctly labeled items among all positively labeled samples [11]:

$$precision = pr = \frac{TP}{TP + FP}. \quad (4)$$

Recall or sensitivity is the fraction of correctly labeled items among all positive samples (exactly matches *tp_rate*) [11]:

$$recall = re = \frac{TP}{TP + FN}. \quad (5)$$

F-measure (F-score) represents the balance between precision and recall [11]:

$$F = \frac{2 \cdot pr \cdot re}{pr + re}. \quad (6)$$

Also to visualize classifiers' performance ROC-curves are often used. ROC-curve is a two-dimensional graph in which *fp_rate* is plotted on the X axis and *tp_rate* is plotted on the Y axis. So ROC-curve shows a trade-off between benefits (true positives) and costs (false positives). The better classifier performs classification task, the closer its curve to upper-left corner (0, 1) of the graph. Curve of perfect classifier passes through (0, 1) point which represents correct classification of all samples [11].

2.2 General Structure of the Algorithm

The general structure of the algorithm is presented in Fig. 1.

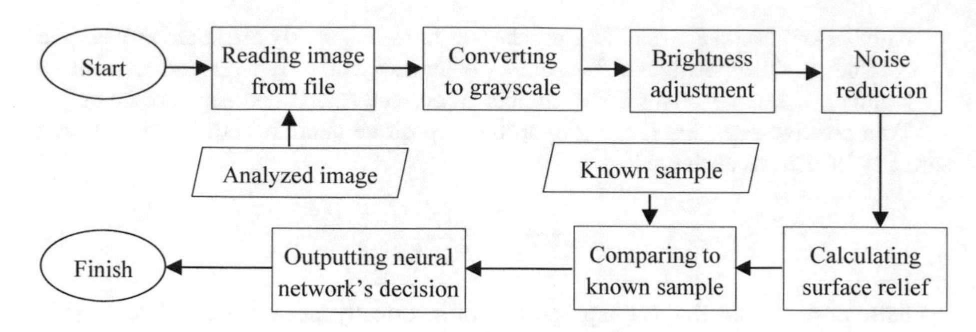

Fig. 1. General structure of the algorithm.

As can be seen from the figure the core element of the algorithm is a neural network which compares analyzed impression to known sample. Another part of algorithm is several pre-processing steps. Let's look at the stages of the algorithm in more details.

2.3 Pre-processing

Since color doesn't carry any useful information about embossed impressions the first step after acquiring an image is to convert it to grayscale using the following expression [12]:

$$g(i,j) = 0.299 f_r(i,j) + 0.587 f_g(i,j) + 0.114 f_b(i,j), \tag{7}$$

where $g(i,j)$ is intensity of pixel (i,j) of converted image; $f_r(i,j), f_g(i,j), f_b(i,j)$ are intensities of red, green and blue channels of original image respectively.

Images of impressions can vary in general level of brightness, as well as in brightness of individual regions. Because of that it's necessary to perform local averaging of brightness. Specifically, intensity of output pixel assigned a value exceeding 125 (half of maximal intensity) by how much original pixel exceeded average of intensities in its local region:

$$g^k(i,j) = \frac{I_{max}}{2} + f(i,j) - s^k(i,j), \tag{8}$$

where $g^k(i,j)$ is intensity of pixel (i,j) of output image, I_{max} is maximum possible intensity value, $f(i,j)$ is intensity of input image (grayscale). $s^k(i,j)$ is average intensity in neighborhood of a given pixel:

$$g^k(i,j) = \frac{I_{max}}{2} + f(i,j) - s^k(i,j), \tag{9}$$

$$s^k(i,j) = \frac{\sum_{\substack{|i'-i| \le k,\, i' \ge 0,\, i' < n_1 \\ |j'-j| \le k,\, j' \ge 0,\, j' < n_2}} f(i',j')}{ab}, \tag{10}$$

$$a = \min(i+k+1, n_1) - \max(i-k, 0), \qquad b = \min(j+k+1, n_2) - \max(j-k, 0),$$

where n_1 and n_2 are height and width of the image, k is argument of this operation which determines the size of a region considered for calculating single output pixel. Value k was chosen to be 25 for images with resolution 500 × 500 pixels.

After this operation noise reduction is performed according to the following formula:

$$G = \frac{(F \circ K) \cdot K + (F \cdot K) \circ K}{2}, \tag{11}$$

where F is input image, G is output image, $\circ$ is operation of morphological opening [4], $\cdot$ is morphological closing [4], K is operation kernel, for example:

$$K = \begin{pmatrix} 1 & 1 & 1 \\ 1 & 1 & 1 \\ 1 & 1 & 1 \end{pmatrix}. \tag{12}$$

Last stage of pre-processing is a custom transformation. This operation converts image so that image represents a map of indentations in paper surface, that is so the darker regions correspond to indents in paper and the lighter – to protrusions. Because documents are photographed in side-light, shadows and bright spots signal change in surface level. For example, if oblique lights shine from right to left, then while traversing image in the direction of lighting, shadows mean drop in surface level while bright spots mean rise in surface level. Based on this the following transformation will give us surface level map:

$$g(i,j) = \frac{1}{2r+1} \sum_{k=i-r}^{i+r} g(k, j+1) + p(i,j), \; \begin{matrix} 0 \le i < n_1 \\ 0 \le j < n_2 \end{matrix}, \tag{13}$$

$$g(i, n_2) = \frac{I_{max}}{2},$$

$$p(i,j) = \begin{cases} a(f(i,j) - mean), & f(i,j) \ge mean, \\ b(f(i,j) - mean), & f(i,j) < mean, \end{cases}$$

where $mean$ is average intensity of input image, r is parameter responsible for averaging adjacent rows to prevent distortions caused by image noise. $r = 1$ was enough to give good results. a and b are found from the following equation:

$$\sum_{i,j} p(i,j)(f(i,j) - mean) = 0. \tag{14}$$

$p(i,j)$ can be found through the following operations expressed in Python language:

```
import numpy as np
img = read_img()
m = np.median(img)
```

```
dev = img – m
pos = dev[dev > 0].sum()
neg = -dev[dev < 0].sum()
k = neg/float(pos)
dev[dev > 0] = dev[dev > 0] * k
result = dev
```

After these transformations intensity values can be higher than I_{max} or lower than 0. For convenience they are normalized:

$$g(i,j) = I_{max}\left(\frac{f(i,j) - mean}{6 \cdot std} + \frac{1}{2}\right), \tag{15}$$

where *std* is standard deviation. After this conversion values are clipped between 0 and I_{max}. Example of image pre-processing is shown in Fig. 2. All images of training and testing sets are stored in their pre-processed form.

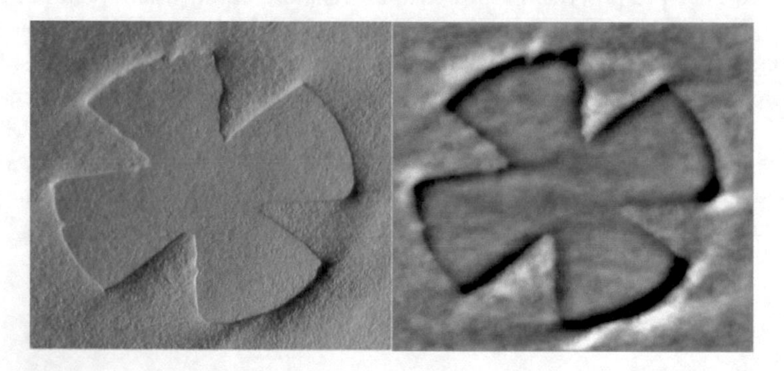

Fig. 2. Pre-processing: original (left) and pre-processed (right) images.

2.4 Image Database

An appropriate image database is required for neural network training and testing. To build the database several embossing dies were manufactured. Several impressions were embossed with each die. These impressions were photographed in oblique lighting and pre-processed.

Because the network accepts pairs of images (to determine whether they were made with the same die or not) images should be paired with each other. To create a large number of samples from a limited number of full images, each image was randomly cropped many times and scaled to 30 × 30 pixels. In other words, each image is used many times in database but with different crop offsets and a scale factor. For positive pairs (images made with the same die) images within pair always have the same crop offsets and a scale factor so impressions are aligned to each other.

2.5 Neural Network Architecture

Neural network used in this research has the following structure:

1. Input layer: two images 30 × 30 pixels organized in two layers. Intensity values are normalized into interval [0;1]. In other words, input layer is a 2 × 30 × 30 array of [0;1] values.
2. Convolution layer: five 5 × 5 filters.
3. Pooling 2 × 2.
4. Fully connected layer consisting of 10 neurons.
5. Output layer: 2 neurons corresponding to each output class.

3 Results

The algorithm was implemented using programming language Python. OpenCV library was used to perform operations with images. Theano, lasagne, nolearn stack of libraries was used to implement neural network and related machine learning algorithms.

Seven dies were manufactured for image database, four impressions were embossed for each die. For training set 40 000 image pairs were produced. For testing set 10 000 image pairs were produced for every pair of dies.

Testing results are shown in Table 1. Confidence level is $\alpha = 95\%$.

ROC-curve for testing set is shown in Fig. 3.

Table 1. Testing results (confidence level 0.95).

Metric	Value
Number of positive samples	70000
Number of negative samples	420000
True positives	69697
False negatives	303
False positives	1358
True negatives	418642
True positive rate	0.9957 ± 0.0005
False positive rate	0.00323 ± 0.00018
Accuracy	0.99661 ± 0.00017
Precision	0.9809 ± 0.0010
Recall	0.9957 ± 0.0005
F-measure	0.9882 ± 0.0006

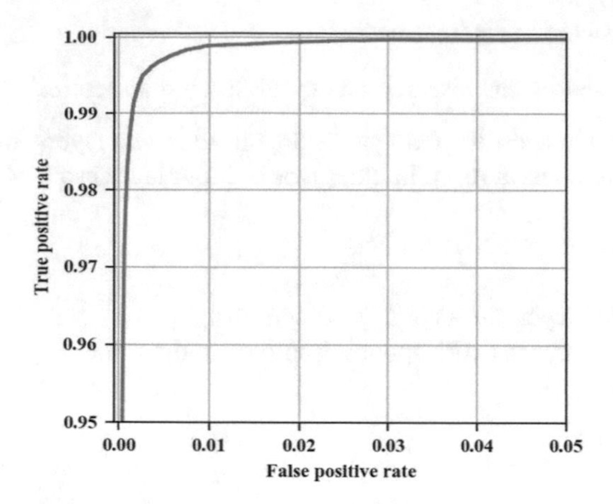

Fig. 3. Algorithm's ROC-curve for testing set.

4 Conclusion

Developed algorithm proved its effectiveness on a given testing set and showed accuracy around 99%, which shows the potential of using convolutional neural networks in examination of paper document security features. This algorithm can be applied to full images of the embossed impressions as well as to its parts which is useful in cases where part of the impression was destroyed by the elements.

The plan for future research is to expand these results to impressions of complex arrangements and adjust algorithm's sensitivity to work with finer differences in embossment dies. Additionally, in our further work we're planning to create a database for different biological objects. Object retrieval from this database will be based on features extracted using neural network approach.

Acknowledgement. This work was supported by Competitiveness Growth Program of the Federal Autonomous Educational Institution of Higher Education National Research Nuclear University MEPhI (Moscow Engineering Physics Institute). We would like to express our gratitude to students of NRNU MEPHI for their valuable discussion and suggestions.

References

1. Artamonov AA, Ionkina KV, Kirichenko AV, Lopatina EO, Tretyakov ES, Cherkasskiy AI (2018) Agent-based search in social networks. Int J Civ Eng Technol 9(13):28–35
2. Samsonovich AV, Kuznetsova K (2018) Semantic-map-based analysis of insight problem solving. Biol Inspir Cogn Arc 25:37–42
3. Belozerova II (1997) Forensic science. Oneworld, Moscow
4. Gonzales RC, Woods RE (2008) Digital image processing, 3rd edn. Pearson Education, Prentice Hall, Upper Saddle River

5. Kulik S, Nikonets D (2016) Forensic handwriting examination and human factors: improving the practice through automation and expert training. In: The third international conference on digital information processing, data mining, and wireless communications (DIPDMWC 2016), Proceedings, 06–08 July, Moscow, Russia, pp 221–226
6. Kulik SD (2019) Factographic information retrieval for semiconductor physics, micro - and nanosystems. In: AMNST 2017, IOP conference series: materials science and engineering, vol 498, p 012026
7. Szeliski R (2010) Computer vision: algorithms and applications. Springer, London
8. Miloslavskaya N, Tolstoy A (2016) Big data, fast data and data lake concepts. Procedia Comput Sci 88:300–305
9. Miloslavskaya, N., Tolstoy, A.: State-level views on professional competencies in the field of IoT and cloud information security. In: 2016 IEEE 4th international conference on future internet of things and cloud workshops (FiCloudW), Vienna, pp 83–90 (2016)
10. LeCun Y (1998) Gradient-based learning applied to document recognition. Proc IEEE 86 (11):2278–2324
11. Fawcett T (2006) An introduction to ROC analysis. Pattern Recogn Lett 27(8):861–874
12. OpenCV documentation. https://docs.opencv.org. Accessed 20 Jan 2018
13. Skiteva L, Trofimov A, Ushakov V, Malakhov D, Velichkovsky B (2016) MEG data analysis using the empirical mode decomposition. In: Samsonovich A, Klimov V, Rybina G (eds) Biologically inspired cognitive architectures (BICA) for young scientists., vol 449. Method in advances in intelligent systems and computing. Springer, Cham, pp 135–140
14. Shishkin S, Nuzhdin Y, Svirin E, Trofimov A, Fedorova A, Kozyrskiy B, Velichkovsky B (2016) EEG negativity in fixations used for gaze-based control: toward converting intentions into actions with an eye-brain-computer interface. Front Neurosci 10:1–20
15. Nuzhdin YO, Shishkin SL, Fedorova AA, Trofimov AG, Svirin EP, Kozyrskiy BL, Medyntsev AA, Dubynin IA, Velichkovsky BM (2017) The expectation based eye-brain-computer interface: an attempt of online test. In: Proceedings of the 2017 ACM workshop on an application-oriented approach to BCI out of the laboratory. ACM, pp. 39–42
16. Kulik S (2016) Factographic information retrieval for competences forming. In: The third international conference on digital information processing, data mining, and wireless communications, DIPDMWC 2016, Proceedings, 06–08 July, Moscow, Russian Federation, pp 245–250
17. Kulik S (2016) Factographic information retrieval for communication in multicultural society. In: Procedia - social and behavioral sciences, International conference on communication in multicultural society, CMSC 2015, 6–8 December 2015, Moscow, Russian Federation, vol 236, pp 29–33
18. Kireev V, Silenko A, Guseva A (2017) Cognitive competence of graduates, oriented to work in the knowledge management system in the state corporation “ROSATOM”. J Phys: Conf Ser 781(1):012060
19. Kireev VS, Guseva AI, Bochkaryov PV, Kuznetsov IA, Filippov SA (2018) Association rules mining for predictive analytics in IoT cloud system. In: Samsonovich A (ed) Biologically inspired cognitive architectures 2018, vol 848. Advances in intelligent systems and computing. Springer, Cham, pp 107–112
20. Artamonov A, Onykiy B, Ananieva A, Ionkina K, Kshnyakov D, Danilova V, Korotkov M (2016) Regular agent technologies for the formation of dynamic profile. Procedia Comput Sci 88:482–486

21. Yasnitsky LN, Vauleva SV, Safonova DN, Cherepanov FM (2015) The use of artificial intelligence methods in the analysis of serial killers' personal characteristics. Criminol J Baikal Natl Univ Econ Law 9(3):423–430
22. Yasnitsky LN, Yasnitsky VL (2016) Technique of design of integrated economic and mathematical model of mass appraisal of real estate property by the example of Yekaterinburg housing market. J Appl Econ Sci XI(8(46)):1519–1530 Winter
23. Yasnitsky LN, Dumler AA, Bogdanov KV, Poleschuk AN, Cherepanov FM, Makurina TV, Chugaynov SV (2013) Diagnosis and prognosis of cardiovascular diseases on the basis of neural networks. Biomed Eng 47(3):160–163

Cognitive Maps for Risk Estimation in Software Development Projects

Anna V. Lebedeva and Anna I. Guseva(✉)

National Research Nuclear University MEPhI (Moscow Engineering Physics Institute), 31 Kashirskoye Shosse, Moscow, Russia
lebedevaa28@gmail.com, AIGuseva@mephi.ru

Abstract. This article is the result of research in terms of risk management in the implementation of software development projects. In the conditions of active digitalization of the economy and the growth of both the number and budgets of projects in the software industry, this study is relevant. The article provides a detailed comparative analysis of existing methods and tools for project risk management, the authors define the comparison criteria and scale. To take into account the industry features of it projects, especially in the early stages of implementation, the authors propose a comprehensive approach, which includes an assessment of the cost and complexity of projects. There was also built a cognitive map of risks of software development projects. Twenty projects of different scale and type, including projects on development of technical documentation, on completion of already ready solutions were evaluated. The most accurate and significant results were achieved in the framework of projects for the development of reporting documentation. At the same time, the complexity of the project Manager involved in the calculation of risks and cost and complexity, decreased by 3 times. Author acknowledges support from the MEPhI Academic Excellence Project (Contract No. 02.a03.21.0005).

Keywords: Instrumental digital platforms · Digitalization · Risk management · Cost and labor input estimation · Project management methodologies · Cognitive maps · Software development projects

1 Introduction

This work is devoted to solving an important scientific and practical problem of project management, namely, risk assessment and forecasting. On the basis of the information obtained in the framework of the analysis, protective measures are developed to reduce the risk of risks. There is almost no information about the choice of ways to protect against various types of risk in software projects in the literature. To develop countermeasures, it is necessary to know at what stage of the software development life cycle and for what reason there is a risk. Such research is extremely limited.

A. V. Samsonovich (Ed.): BICA 2019, AISC 948, pp. 295–304, 2020.
https://doi.org/10.1007/978-3-030-25719-4_37

2 Main Approaches to the Problem

2.1 Risk Management Process

Only 29% of projects are successfully completed, while one of the main factors of the project success is competent project management (13.9%) and clear planning (9.6%) according to the annual review of the The Standish Group [1]. Despite the active development of information technology, the percentage of successful completion of such projects remains low. Figure 1 shows the statistics published by The Standish Group for the period from 2011–2015, which shows that on average, half of software projects are completed in excess of the existing restrictions [1, 2].

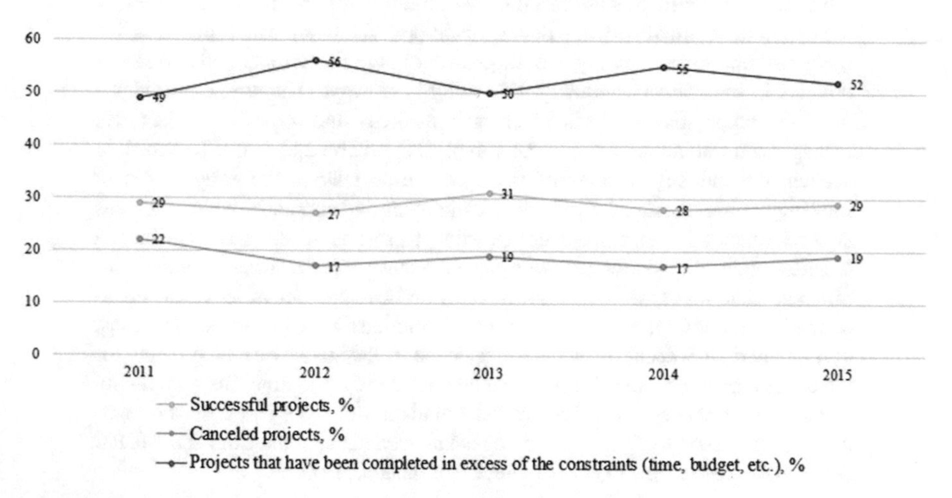

Fig. 1. Statistics of the success of information technology projects (according to The Standish Group from 2011–2015)

Work related to the change, implementation and creation of information technology should be integrated into a software project. The term "software project" should be understood as a task or a set of tasks for the implementation of software in the enterprise with predefined initial data (terms, budget, pre-functional and(or) non-functional requirements). Within the framework of this study, the software project should be understood as work in the field of software, hardware, network technologies, data storage technologies.

Due to the fact that the implemented information technology can have a significant impact on the future activities of both the individual business process and the enterprise as a whole, before the start of the software project, a comprehensive analysis of the need for such a project should be carried out, as well as the calculation of the effectiveness of the implemented solutions, including cost and labor The success of software project depends on many factors, but the main ones are: timing, quality, resources. Therefore, the main task of the software project manager can be formulated as follows – to meet the allocated budget within the specified time frame while ensuring the required level of quality. In some cases, project decisions are made on the basis of intuition, own experience and recommendations of consulting companies, which entails the adoption of significant risk. This once again confirms the relevance of developing a software tool for risk assessment and forecasting based on machine learning methods.

Many definitions of risk are known. In our opinion, the most suitable one for software risk is "Risk is the probability of occurrence of an event that can lead to danger or negative consequences, such as a lack of profit, a decrease in the efficiency of processes and quality of activity, security threats, the occurrence of losses, damages" [3].

In the process of management and in the decision-making process in particular must also carry out risk management. In general, the risk management process includes the following stages: risk identification, risk assessment, risk management, risk monitoring.

At the identification stage, it is expected to work to identify potential threats that may hinder the achievement of the project objectives. Then, at the next stage, there is a determination of which of the detected risks are the greatest danger, in other words, a risk assessment is made- the probability and possible consequences are estimated. Next, plans are made for the development of events in the conditions of implementation of the most dangerous risks and provides for measures to neutralize their impact. That is, strategies are being developed to protect the project from the most critical risks. At the final stage, it remains to maintain the formed project plan and the list of risks up to date.

Risk management in software development is associated with the need to eliminate various threats to which the created software may be exposed due to some minor or non-obvious errors in the project or development process. In fact, all stages of software development are potential sources of software risks, as they include hardware, software, technology, people, cost and schedule. One of the important aspects of risk management in software development is risk identification.

Thus, the problems of software project risk management are the lack of standardized and formalized risk reference for software development projects. While technology is being actively developed, which can contribute to the emergence of new risks, it is nevertheless possible to identify a typical set of risks inherent in such projects for use in project evaluation and management. Also, most of the decisions are made on the basis of intuition and experience of the project Manager, which entails the adoption of significant risk, in this regard, a software tool based on machine learning methods is necessary, which will aggregate the accumulated experience into a more objective and convenient form.

2.2 Comparative Analysis of Risk Assessment Methods

In this study, the methods were grouped into 4 risk groups (statistical; probabilistic-statistical; theoretical-probabilistic approach; expert), and the criteria of comparison were identified.

The result of the comparative analysis of the methods is presented in Table 1.

Table 1. Results of comparison of risk assessment methods

Methods/criteria	Statistical evaluation methods	Probabilistic evaluation methods	Probability-theoretic methods	Expert evaluation methods
Q_1	1	1	1	0
Q_2	0	0	0	1
S	2	1	0	0
V	0	0	1	1
A	1	1	1	0
P	0	1	0	1
Total	4	4	3	3

The following scale of criteria was used for comparison:

- Quantitative (Q_1) and qualitative (Q_2) evaluation criteria. $Q_1 = \{0, 1\}$, $Q_2 = \{0, 1\}$, where the 0 – method can be applied when the specific classes to which the objects to be evaluated should be assigned are pre-determined, 1 – method is suitable for quantitative/qualitative evaluation.
- Criterion of complexity of the method (S). $S = \{0, 1, 2\}$, where 0 – a complex method that requires the involvement of highly qualified experts or the use of complex mathematical apparatus; 1 – the method requires additional information; 2 – intuitive method.
- Criterion for the amount of statistical data (V). $V = \{0, 1\}$, where 0 – method requires a large number of statistics, 1 – method does not require a large number of statistics.
- The criterion of accuracy of estimation (A). $A = \{0, 1\}$, where 0 is an insufficiently accurate method; 1 the results of the method are sufficiently accurate.
- Criterion of possibility of application in software development projects (P). $P = \{0, 1\}$, where 0 – method is inconvenient and labor-intensive enough for use in software development projects, 1 – method is convenient to use in software development projects.

In practice, the quantitative method is always used in conjunction with the qualitative. The choice of a particular method and the setting of appropriate scales is individual for each project and the company as a whole.

When applying quantitative methods, it should be taken into account that the scope of application of such methods, depending on the available statistical data and the possibility of constructing theoretical models, can be limited. Speaking about the field of software development, it is worth noting that in the vast majority of cases, there is no necessary and sufficient data and statistics on past projects, so the use of the statistical method will be very difficult. Also, it is not recommended to apply a complex, time-consuming theoretical and probabilistic approach, which has proven itself in assessing the risks of rare phenomena, but has low accuracy and the costs are not justified for regular software development projects. At the same time, probabilistic-statistical and expert methods can give quite good results with a certain level of reliability.

The combination of the use of such methods of evaluation can give the most accurate result with a relatively small amount of work-bone application.

Once the probability of materialization of risks and losses due to the occurrence of various risks have been assessed, it is necessary to proceed to the next step – the allocation of priority problems and the formation of management strategy. During the software development process, you can choose different risk management strategies depending on the degree of risk impact. Based on the magnitude of the impact of risk and the probability of its occurrence in the project of software development, risk strategies can be divided into three classes: cautious, typical and flexible. Typically, a prudent risk management strategy is designed for new and inexperienced organizations whose development projects involve new and untested technologies. A typical risk management strategy is clearly defined as support for existing organizations with experience in software development projects and technologies used, but whose projects carry a sufficient number of risks. A flexible risk management strategy is implemented in experienced software development organizations, whose development projects are officially defined and based on proven technologies.

2.3 Comparative Analysis of Risk Assessment and Analysis Tools

In this study, the most popular tools that support risk assessment processes were considered: CRAMM, RiskWatch, COBRA and MethodWare.

The results of the comparative analysis are presented in Table 2.

Table 2. Results of comparison of risk assessment and analysis tools

Tools/criteria	CRAMM	RiskWatch	COBRA	Method-Ware
K_1	0	0	1	0
K_2	0	0	1	1
K_3	0	1	0	1
K_4	1	1	0	1
K_5	1	0	1	1
K_6	1	1	0	1
K_7	0	1	1	1
Total	3	4	4	6

The following scale of criteria was used for comparison:

- Software cost criterion (K1). K1 = {0, 1, 2}, where 0 – paid software, 1 – there is a free demo version of the software, 2 – free software.
- Operational complexity criterion (K2). K2 = {0, 1}, where 0 – for operation requires either highly qualified specialists or special training 1 – for the operation of the software does not require additional training.
- The criterion of convenience of the interface (K3). K3 = {0, 1}, where 0 – is an outdated or not-user-friendly/complex interface 1 – is an intuitive, modern interface.

- Criterion of possibility of customization (K4). K4 = {0, 1}, where 0 – there is no possibility to customize the software, 1 – there is a possibility to customize the software.
- Criteria for the use of quantitative (K5) and qualitative (K6) risk assessment methods. K5 = {0, 1}, K6 = {0, 1}, where 0 – software does not allow to use the appropriate risk assessment method, 1 – in software implemented the possibility of using the appropriate risk assessment method.
- Criterion for the existence of a logical inference function (K7). K7 = {0, 1}, where 0 is not supports the function of logical inference, 1 – in function of logical inference.

Based on this comparison and the general picture of the use of such tools, it can be concluded that these tools mainly use quantitative or mixed methods of risk analysis. However, none of the systems is without drawbacks, whether it is the impossibility of customization for the needs of a particular enterprise or the restriction in the use of certain methods of evaluation. For example, the COBRA system does not use quantitative assessment, RiskWatch does not provide the user with a deep qualitative assessment, and the CRAMM does not allow to assess the possibility of consequences for risks on the system "What if?", thereby reducing the completeness of forecasting.

3 Proposed Approach

In the framework of this study, an approach was proposed that allows us to consider all the resources in the relationship, use it at any stage of the project, as well as reduce the complexity of its use.

The proposed approach adopted the following definitions:

Resource – a decomposed stage of work, which is carried out within the framework of project activities.
Threat – impact on the resource, because of which there may be negative consequences on the project.
Weakness – lack of resource, using which it is possible to implement threats.

The risk assessment process consists of 5 steps:

Step 1. The first step is to calculate the threat level of Th weakness based on the criticality and probability of threat realization through this weakness. Threat level shows how critical is the impact of this threat on the resource, taking into account the probability of its implementation.

$$Th = \frac{ER_{cia}}{100} * \frac{P_{cia}(V)}{100} \tag{1}$$

where ER_{cia} is the severity of the threat (specified in %); $P_{cia}(V)$ is the probability of the threat being implemented through the weakness (specified in %). You have one or three values depending on the number of basic threats. Get the value of the threat level of weakness in the range from 0 to 1.

Step 2. In the second step, the threat level is calculated for all CTh weakness through which this threat can be implemented on the resource. The threat levels obtained through specific weakness are then summarized.

For each type of threat for all weakness:

$$CTh_{cia} = 1 - \prod\nolimits_{i=1}^{n} (1 - Th_i) \tag{2}$$

The value of level of threat on all weakness is obtained in the interval from 0 to 1.

Step 3. In the third step, the overall threat level for the CThR resource is calculated in the same way (taking into account all threats affecting the resource):

$$CThR_{cia} = 1 - \prod\nolimits_{i=1}^{n} (1 - CTh_{cia_i}) \tag{3}$$

The value of the overall threat level is obtained in the range from 0 to 1.

Step 4. The fourth step is to calculate the risk of works R from several information resources:

$$R_{cia} = 1 - \prod\nolimits_{i=1}^{n} (1 - CThR_{cia_i}) * D \tag{4}$$

where D is the criticality of the product or business process in the range 0–1.

Step 5. At the fifth step, the calculation of the total risk for works from several works is carried out:

$$CR_{cia} = 1 - \prod\nolimits_{i=1}^{n} (1 - R_{cia_i}). \tag{5}$$

From the author's point of view, the generalized model of risk assessment looks as shown in Fig. 2 [4]:

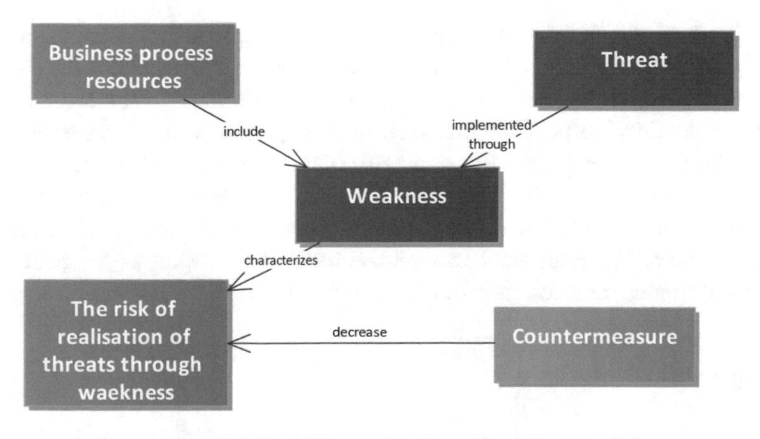

Fig. 2. Generalized risk assessment model

Previously, the authors identified groups of risk factors [5]. Modern managing processes require taking decisions within weakly-structured dynamic systems where interconnected parts ties cannot be described functionally. In order to carry out business-modeling of such systems that show qualitative nature of interconnected parts tires, analysts use special system models called cognitive maps [6].

On the basis of the identified factors, the cognitive map presented in Fig. 3 was constructed. Special software was used to build a cognitive map. Software for automated modeling based on cognitive maps is called cognitive mapper (cognitive cartographer) [7–9]. This cognitive map reflects the relationship and interdependence of risk factors among themselves.

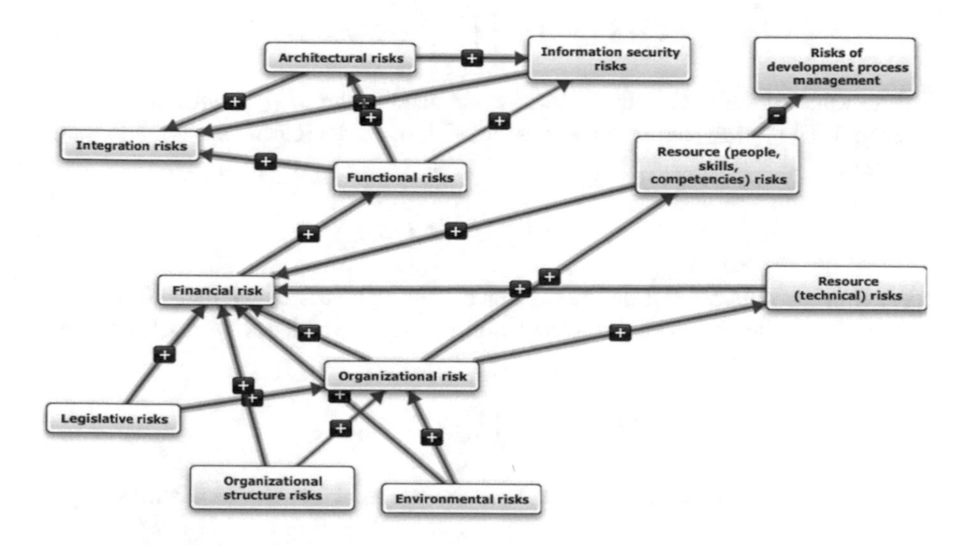

Fig. 3. Cognitive map of risk factor groups

The plus sign indicated on the arcs means that the increase in the cause-factor increases the effect-factor, and the minus sign means that the decrease in the value of the effect-factor as a consequence of the increase in the value of the cause-factor. With the use of appropriate web Analytics, it is possible to track how changing the value of one risk factor will affect several other risk factors at once [10]. Timely assessment of such impact will facilitate the adoption of timely countermeasures on the software development project.

For each such group of risk factors, there is a list of threats that can be implemented through weakness. For each weakness on the basis of an expert assessment the probability of occurrence, and for each threat the criticality of implementation is put down.

4 Results

During the use of this approach, 20 projects of various scale and type were evaluated, including projects for the development of technical software documentation, for the development of concepts of information systems, for the development of information systems, for the finalization of ready-made solutions.

The most accurate and significant results were achieved in the framework of projects for the development of reporting documentation.

It should also be noted that the complexity of the project manager involved in the calculation of risks and cost and complexity, decreased by 3 times. Thus, the project manager can take part in a much larger number of pre-sale works and carry out a more rapid calculation for the provision of commercial proposals to potential customers.

5 Summary

As a result of this work, the literature sources devoted to risk assessment in software development projects, as well as tools for risk assessment were analyzed.

The problem of risk assessment and preventive measures to prevent the failure of the project is relevant.

Existing methods, models and tools do not allow to consider all resources in interrelation, and also to control and manage the project at any stage. In this regard, it becomes relevant to develop an approach that allows the process of modeling the assessment taking into account a large number of parameters and limitations.

It should be noted that the developed approach has already shown positive results after implementation at the enterprise. The time spent by the project manager to evaluate the project decreased by 3 times.

In the future, it is planned to use this approach in conjunction with machine learning tools [11], which will achieve even more accurate results through the accumulation of historical data on projects, as well as reduce the complexity of use through the introduction of automated tools.

Acknowledgements. This work was supported by the MEPhI Academic Excellence Project (Contract No. 02.a03.21.0005).

References

1. Standish Group 2015 chaos report. http://blog.standishgroup.com/post/50. Accessed 03 Dec 2018
2. CHAOS summary 2009 report. https://www.classes.cs.uchicago.edu/archive/2014/fall/51210-1/required.reading/Standish.Group.Chaos.2009.pdf. Accessed 03 Dec 2018
3. Avdoshin SM (2011) Business informatization. Risk management. DMKPress, Moscow
4. Lebedeva AV (2017) Methods of risk identification in software projects. In: New information technologies in scientific research: materials of the XXII all-Russian scientific and technical conference of students, young scientists and specialists. Ryazan State Radio Engineering University, Ryazan, pp 42–44
5. Lebedeva AV (2017) Risk factors in software development projects. In: Education and science: based on the materials of the international scientific-practical conference №11-5 (27), pp 72–74. Journal of scientific conferences, Tambov
6. Samsonovich AV, Kuznetsova K (2018) Semantic-map-based analysis of insight problem solving. Biol Inspired Cogn Arch 25:37–42
7. Kulinich AA (2010) Computer cognitive map modeling systems. Control Sci 3:2–16

8. Gray SA et al (2013) Mental modeler: a fuzzy-logic cognitive mapping modeling tool for adaptive environmental management. In: 2013 46th Hawaii international conference on System sciences (HICSS). IEEE, pp 965–973
9. Carvalho JP, Tomè JAB (1999) Rule based fuzzy cognitive maps-fuzzy causal relations. In: Mohammadian M (ed) Computational intelligence for modelling, control and automation
10. Kireev VS, Smirnov IS, Tyunyakov VS (2018) Automatic fuzzy cognitive map building. Procedia Comput Sci 123:228–233
11. Lebedeva AV (2018) Algorithm of risk response formation in software development projects based on neural network. In: New information technologies in scientific research: materials of the XXIII all-Russian scientific and technical conference of students, young scientists and specialists. Ryazan State Radio Engineering University, Ryazan, pp 236–238

A Typicality-Based Knowledge Generation Framework

Antonio Lieto[1,2](✉), Federico Perrone[1], Gian Luca Pozzato[1], and Eleonora Chiodino[1]

[1] Dipartimento di Informatica, Universitá di Torino, Turin, Italy
{antonio.lieto,federico.perrone,gianluca.pozzato,
eleonora.chiodino}@unito.it
[2] ICAR-CNR, Palermo, Italy
http://di.unito.it/GOCCIOLA

Abstract. This short paper is an abridged version of [1], where we introduce a framework for the dynamic generation of novel knowledge obtained by exploiting a recently introduced extension of a Description Logic of typicality able to combine prototypical descriptions of concepts. Given a goal expressed as a set of properties, in case an intelligent agent cannot find a concept in its initial knowledge base able to fulfill all these properties, our system exploits the reasoning services of the Description Logic $\mathbf{T}^{\mathrm{CL}}$ in order to find two concepts whose creative combination satisfies the goal.

Keywords: Cognitive architectures · Knowledge generation · Concept combination

1 Introduction

A challenging problem in Artificial Intelligence concerns the capability of an intelligent agent to achieve its goals when its knowledge base does not contain enough information to do that. In this line of research, existing goal-directed systems usually implement a re-planning strategy in order to tackle the problem. This is systematically performed by either an external injection of novel knowledge or as the result of a communication with another intelligent agent [2]. Here, we describe an alternative approach, consisting in a dynamic and automatic generation of novel knowledge obtained through a process of commonsense reasoning. The idea is as follows: given an intelligent agent and a set of *goals*, if it is not able to achieve them from an initial knowledge base, then it tries to dynamically generate new knowledge by *combining* available information. Novel information will be then used to extend the initial knowledge base in order to achieve the goals. As an example, suppose that an intelligent agent is aware of the facts that, normally, coffee contains caffeine and is a hot beverage, that the chocolate with cream is normally sweet and has a taste of milk, whereas Limoncello is not a hot beverage (normally, it is served chilled). Both coffee and Limoncello are after meal drinks. Cold winters in Turin suggest to have a hot after-meal drink, also being sweet and having taste of milk. None of the concepts in the knowledge base of the agent are able to achieve the

A. V. Samsonovich (Ed.): BICA 2019, AISC 948, pp. 305–307, 2020.
https://doi.org/10.1007/978-3-030-25719-4_38

goal on their own, however, the combination between coffee and chocolate with cream provides a solution.

In this short paper we describe a framework following this approach in the context of Description Logics (for short, DLs) by exploiting the logic $\mathbf{T}^{\textsf{CL}}$, recently introduced in order to account for the phenomenon of concept combination of prototypical concepts [3,4]. The logic $\mathbf{T}^{\textsf{CL}}$ relies on the logic of typicality $\mathcal{ALC} + \mathbf{T}_{\mathbf{R}}^{RaCl}$ [5], whose semantics is based on the notion of rational closure, as well as on the DISPONTE semantics of probabilistic DLs [6], and is equipped with a cognitive heuristic used by humans for concept composition. In this logic, typicality inclusions of the form $p :: \mathbf{T}(C) \sqsubseteq D$ are used to formalize that "we believe with degree p about the fact that typical Cs are Ds". As in the distributed semantics, this allows us to consider different scenarios containing only some typicality inclusions, each one having a suitable probability. Such scenarios are then used to ascribe typical properties to a concept C obtained as the combination of two concepts, revising the initial knowledge base with the addiction of typical properties of C. In the example, the revised knowledge base provided by the logic $\mathbf{T}^{\textsf{CL}}$ contains typical properties of the combination of coffee and chocolate with cream, which suggests to consider a beverage corresponding to the famous Turin drink known as *Bicerín* (little glass), made by coffee, chocolate and cream.

2 Generating New Knowledge by Concept Combination in $\mathbf{T}^{\textsf{CL}}$

Given a knowledge base $\mathcal{K}$ in the Description Logic $\mathbf{T}^{\textsf{CL}}$, an intelligent agent has to achieve a goal $\mathcal{G}$ intended as a set of concepts $\{D_1, D_2, \ldots, D_n\}$. More precisely, the agent has to find a *solution* for the goal, namely a concept C such that, for all properties D_i, it holds that either $\mathcal{K} \models C \sqsubseteq D_i$ or $\mathcal{K} \models \mathbf{T}(C) \sqsubseteq D_i$ in the logic of typicality $\mathcal{ALC} + \mathbf{T}_{\mathbf{R}}^{RaCl}$. If $\mathcal{K}$ does not contain any solution for the goal, then the agent tries to generate a *new concept* by combining two existing ones C_1 and C_2 by means of the logic $\mathbf{T}^{\textsf{CL}}$: C is then considered a solution for the goal if, considering the $(C_1 \sqcap C_2)$-revised knowledge base $\mathcal{K}_C$ extending $\mathcal{K}$, we have that, for all properties D_i, it holds that either $\mathcal{K}_C \models C \sqsubseteq D_i$ or $\mathcal{K}_C \models \mathbf{T}(C) \sqsubseteq D_i$ in the logic of typicality $\mathcal{ALC} + \mathbf{T}_{\mathbf{R}}^{RaCl}$.

Definition 1. *Given a knowledge base $\mathcal{K}$ in the logic $\mathbf{T}^{\textsf{CL}}$, let $\mathcal{G}$ be a set of concepts $\{D_1, D_2, \ldots, D_n\}$ called* goal. *We say that a concept C is a* solution *to the goal $\mathcal{G}$ if either (i) for all $D_i \in \mathcal{G}$, either $\mathcal{K} \models C \sqsubseteq D_i$ or $\mathcal{K}' \models \mathbf{T}(C) \sqsubseteq D_i$ in the logic $\mathbf{T}^{\textsf{CL}}$ or (ii) C corresponds to the combination of two concepts C_1 and C_2 occurring in $\mathcal{K}$, i.e. $C \equiv C_1 \sqcap C_2$, and the C-revised knowledge base $\mathcal{K}_C$ provided by the logic $\mathbf{T}^{\textsf{CL}}$ is such that, for all $D_i \in \mathcal{G}$, either $\mathcal{K}_C \models C \sqsubseteq D_i$ or $\mathcal{K}_C \models \mathbf{T}(C) \sqsubseteq D_i$ in the logic $\mathbf{T}^{\textsf{CL}}$.*

Consider the example of the Introduction and suppose that $\mathcal{K}$ contains the information that, normally, coffee contains caffeine and is a hot beverage; moreover, we have that the chocolate with is normally sweet and has a taste of milk, whereas Limoncello is not a hot beverage (normally, it is served chilled). Both coffee and Limoncello are after meal drinks. We can represent these information in the logic $\mathbf{T}^{\textsf{CL}}$ as follows:

$0.9 :: \mathbf{T}(\mathit{Coffee}) \sqsubseteq \mathit{AfterMealDrink}$
$0.8 :: \mathbf{T}(\mathit{Coffee}) \sqsubseteq \mathit{WithCaffeine}$
$0.85 :: \mathbf{T}(\mathit{Coffee}) \sqsubseteq \mathit{HotBeverage}$
$\mathit{Limoncello} \sqsubseteq \mathit{AfterMealDrink}$
$0.9 :: \mathbf{T}(\mathit{Limoncello}) \sqsubseteq \neg \mathit{HotBeverage}$
$0.65 :: \mathbf{T}(\mathit{ChocolateWithCream}) \sqsubseteq \mathit{Sweet}$
$0.95 :: \mathbf{T}(\mathit{ChocolateWithCream}) \sqsubseteq \mathit{TasteOfMilk}$

Cold winters in Turin suggest to have a hot after-meal drink, also being sweet and having taste of milk. We can then define a goal $\mathcal{G}$ as

$$\mathcal{G} = \{\mathit{AfterMealDrink}, \mathit{HotBeverage}, \mathit{Sweet}, \mathit{TasteOfMilk}\}.$$

None of the concepts in the knowledge base represent a solution for the problem. However, the combination between the concepts *Coffee* and *ChocolateWithCream* represents a solution. Indeed, the revised knowledge base obtained by exploiting the logic $\mathbf{T}^{\text{CL}}$ to combine these concepts allows the agent to extend its knowledge with the following typicality inclusions:

$0.9 :: \mathbf{T}(\mathit{Coffee} \sqcap \mathit{ChocolateWithCream}) \sqsubseteq \mathit{AfterMealDrink}$
$0.85 :: \mathbf{T}(\mathit{Coffee} \sqcap \mathit{ChocolateWithCream}) \sqsubseteq \mathit{HotBeverage}$
$0.65 :: \mathbf{T}(\mathit{Coffee} \sqcap \mathit{ChocolateWithCream}) \sqsubseteq \mathit{Sweet}$
$0.95 :: \mathbf{T}(\mathit{Coffee} \sqcap \mathit{ChocolateWithCream}) \sqsubseteq \mathit{TasteOfMilk}$

providing a solution to the goal corresponding to the famous Turin drink known as *Bicerín* (little glass).

We have implemented our framework (http://di.unito.it/gocciola) and we have tested it in the task of concept combination by comparing the obtained results with other systems as well with human responses. The description of the implementation and the experimental results are fully reported in [1].

References

1. Lieto A, Perrone F, Pozzato GL, Chiodino E (2019, to appear) Beyond subgoaling: a dynamic knowledge generation framework for creative problem solving in cognitive architectures. Cogn Syst Res 1–20
2. Aha DW (2018) Goal reasoning: foundations, emerging applications, and prospects. AI Mag 39(2):3–24
3. Lieto A, Pozzato GL (2018) A description logic of typicality for conceptual combination. In: Proceedings of the 24th international symposium on methodologies for intelligent systems, ISMIS 2018
4. Lieto A, Pozzato GL (2019) A description logic framework for commonsense conceptual combination integrating typicality, probabilities and cognitive heuristics. arXiv preprint. arXiv:1811.02366
5. Giordano L, Gliozzi V, Olivetti N, Pozzato GL (2015) Semantic characterization of rational closure: from propositional logic to description logics. Artif Intell 226:1–33
6. Riguzzi F, Bellodi E, Lamma E, Zese R (2015) Reasoning with probabilistic ontologies. In: Yang Q, Wooldridge M, (eds) Proceedings of the twenty-fourth international joint conference on artificial intelligence, IJCAI 2015, Buenos Aires, Argentina, 25–31 July 2015, pp 4310–4316. AAAI Press

Application of Variable-Valued Logic to Correct Pattern Recognition Algorithms

Larisa A. Lyutikova(✉) and Eleva V. Shmatova(✉)

Institute of Applied Mathematics and Automation of Kabardin-Balkar Scientific Centre of RAS (IAMA KBSC RAS), St. Shortanova 89 a, Nalchik 360000, KBR, Russia
{lylarisa, lenavsh}@yandex.ru

Abstract. In this work the logical analysis is carried out for the predetermined subject domain, that represents an object and its describing characteristics in terms of variable-valued logic, besides the set of algorithms operating at the given domain is studied. The study developed logical procedure for creation of the correct algorithms that analyze subject domain, model the knowledge base for the set objects, minimize it and allocate a unique feature set for each object.

Keywords: Algorithm · Training set · Set of data · Knowledge base · Subject domain · Variable-valued logic · Disjuncts

1 Problem Formal Statement

We find that the logical approach can be the basis in building of the synthesis theory for the recognition correct algorithms with the help of the existing algorithm families [1, 5]. These methods, despite the lack of adequate mathematical models of the studied dependences between an image and its properties, incompleteness and discrepancy of data allow creating of the algorithms which produce expert's reasonings [6].

In this paper we study logical approach to theoretical justification for constructing correct algorithms which expand area of solutions obtained on the base of the existing algorithms.

On the subject domain consisting of objects and their characteristics a number of recognition problem algorithms $A_1, A_2, \ldots, A_n$, are considered.

Suppose $X = \{x_1, x_2, \ldots, x_m\}$, is variable of $x_i \in \{0, 1, \ldots, k_r - 1\}$, где where $k_r \in [2, \ldots, N], N \in Z$ is the set of features considered within the variable-valued logic; $X_i = \{x_1(y_i), x_2(y_i), \ldots, x_m(y_i)\}, i = 1, \ldots, l$ is the feature characterizing vector, $y_i \in Y$, $Y = \{y_1, y_2, \ldots, y_l\}$ is the set of objects; $A = \{A_1, A_2, \ldots, A_n\}$ is the set of algorithms, $a_j(X_i, y_i) \in \{0, 1\}$; $i = 1, 2, \ldots, l$; $j = 1, 2, \ldots, n$ - the performance quality of the algorithm on a given set $X_i = \{x_1(y_i), x_2(y_i), \ldots, x_m(y_i)\}, i = 1, 2, \ldots, l$: formulated as follows:

This work is supported by the RFBR grant 19-01-00648.

A. V. Samsonovich (Ed.): BICA 2019, AISC 948, pp. 308–314, 2020.
https://doi.org/10.1007/978-3-030-25719-4_39

$$a_j(y_i) = \begin{cases} 1, & A_j(X_i) = y_i \\ 0, & A_j(X_i) \neq y_i \end{cases}, \; i = 1, 2, \ldots, l, j = 1, 2, \ldots, n,$$

i.e. the algorithm operation with the given characteristics set is evaluated by Boolean algebra:

1 - algorithm recognizes object y_i by given characteristics of x_i,
0 - algorithm A_i does not recognize object y_i by given characteristics of x_i.

The set of recorded data can be represented in a two-dimensional matrix of the following form (Table 1):

Table 1. Input and evaluation functions algorithms

x_1	x_2	…	x_m	Y	A'_1	A'_2	…	A'_n
$x_1(y_1)$	$x_2(y_1)$	…	$x_m(y_1)$	y_1	$a_1(y_1)$	$a_2(y_1)$	…	$a_n(y_1)$
$x_1(y_2)$	$x_2(y_2)$	…	$x_m(y_2)$	y_2	$a_1(y_2)$	$a_2(y_2)$	…	$a_n(y_2)$
…	…	…	…	…	…	…	…	…
$x_1(y_2)$	$x_2(y_1)$	…	$x_m(y_1)$	y_1	$a_1(y_1)$	$a_2(y_1)$	…	$a_n(y_1)$

$A'_i = \{a_i(y_1), a_i(y_2)), \ldots, a_i(y_l)\}, i = 1, 2, \ldots, n.$ is a vector provided by a column of assessment values for the algorithm A_i performance quality.

Some of the given objects in the training sample remain unrecognized by any of the study algorithms. We can write it as follows:

$$\exists y_i \in Y | A_1(X_i) \neq y_i, A_2(X_i) \neq y_i, \ldots, A_n(X_i) \neq y_i, \; i = 1, 2, \ldots, l, j = 1, \ldots, n.$$

It is necessary to construct algorithm on the basis of the given algorithms which allows detection of all objects defined in the domain $A_{n+1}(X_i) | A_{n+1}(X_i) = y_i$ and $A_{n+1}(X) | A_{n+1}(X) = Y$.

Definition. We say that the algorithm is correct on the set of objects Y defined by the set of characteristics X если when $\forall y_i \in Y: \; a_j(X_i, y_i) = 1, i = 1, 2, \ldots, l;$ $j = 1, \ldots, n$. In other words, the algorithm is correct for that set of objects which it identifies correctly.

For the analysis of the subject domain we use algebra of variable valued logic [3, 4] which provides indicative coding of heterogeneous information, since each separate characteristic of $x_i \in \{0, 1, \ldots, k_r - 1\}$ can be encoded by a predicate of any suitable for this characteristic value.

The apparatus of variable-valued logic is good for simple and indicative coding and decoding of properties of the researched objects. It simplifies fuzzification and defuzzification procedures which are necessary in case with fuzzy logic. And also significantly simplifies creation of the logical constructions that reveal compliance of the researched objects and their properties. Within the offered approach these logical design is presented in the form of production rules.

1.1 Decision Rules and Response Quality Function

Definition: Let's call the statement:

$\&_{j=1}^{m} x_j(y_i,) \rightarrow y_i, \qquad i = 1,\ldots,l, x_j(y_i) \in \{0,1,\ldots,k_i - 1\}, k_i \in [2,\ldots,N], N \in Z$

decision rule.

In this case, the decision rule is a production rule whose logic interpretation says that a definite object follows from the set of definite characteristics (this or that characteristics).

Supposing there are n algorithms $\{A_1, A_2, \ldots, A_n\}$ that partially recognize the specified domain. For each given feature set X_i we build operation quality functions of each algorithm and obtain a set of vectors $A'_j = \{a_j(y_1), a_j(y_2)), \ldots, a_j(y_l)\}$, $j = 1, 2, \ldots, n$, presented in a matrix in a column A'_j. We get the algorithm results on every predetermined row corresponding to the object y_i, to the same object there corresponds the production rule

$$\&_{s=1}^{m} x_s(y_i) \rightarrow y_i x_s(y_i) \in \{0,1,\ldots,k_r - 1\},\ i = 1,\ldots,l, s = 1,\ldots,m.$$

The resulting column can be assumed as a partially defined Boolean function on $\{X, Y\}$.

1.2 Algorithm Design for Solutions Area Expanding

While data processing the choice of algorithm with: $a_j(X_i, y_i) = 1$ is a proper thing. In case when at least one algorithm has found the solution we have: $A_j(X_i) = y_i$, then $\vee_{j=1}^{n} a_j(y_i) = 1$. If none of the study algorithms does not recognize the object then y_i, то $\vee_{j=1}^{n} a_j(y_i) = 0$.

Suppose all the training sample are decision rules:

$$\&_{s=1}^{m} x_s(y_i) \rightarrow y_i,\ i = 1,\ldots,l, x_s(y_i) \in \{0,1,\ldots,k_r - 1\}, k_r \in [2,\ldots,N], N \in Z.$$

For each algorithm we select decision rules to recognize objects if $\exists a_j(y_i) = 1$, then $\&_{s=1}^{m} x_s(y_i) \rightarrow y_i$, $i = 1,\ldots,l, x_s(y_i,) \in \{0,1,\ldots,k_r - 1\}, k_r \in [2,\ldots,N], N \in Z$.

We create a function which is the conjunction of decision rules for the given algorithm. Being guided by the following logical reasonings: algorithm A_j recognizes the object y_i and algorithm A_j recognizes the object y_p and all remaining objects.

$$F_j(X_i) = \&_{a_{j(y_i)}=1} \left(\&_{s=1}^{m} x_s(y_i) \rightarrow y_i\right) = \&_{a_{j(y_i)}=1} \left(\vee_{s=1}^{m} \overline{x_s(y_i)} \vee y_i\right).$$

Further it is possible to apply reduction algorithm adapted for multiple-valued logics. If DNF has some single-valued disjunct x_i^j, then we remove all $x_i^j \& \ldots$, disjuncts (absorption law).

As a result, for a given algorithm A_j, we obtain F_j, corresponding to those decision rules recognized by the predetermined algorithm. This function has a number of features [2] and nearly builds up the knowledge base for this algorithm, breaking solution area into all possible classes.

Theorem. Necessary and sufficient condition, generated by the characteristic set $\{X_j\}$ for K_r is the equality: $F_j(X_i) = K_r$.

Having designed appropriate functions F_j, $j = 1, 2, \ldots, n$ for each algorithm we obtain a set of functions $F_1, \ldots, F_n$. Adhering to these arguments we construct the generalizing function which is a conjunction for $F_1, \ldots, F_n$: $F = \&_{i=1}^{n} F_i$. Carry out computation and conversions and obtain:

$$F(X, Y) = f_1(X) \vee f_2(X, Y),$$

where $f_1(X)$ is a function that holds only X_s variables $f_1(X)$; is a setup function; and function disjuncts are tuning elements which do not matter in object identification but matter for new algorithm design on the base of objects unrecognized earlier; $f_2(X, Y)$ is a function with features and objects conjunction, intended to determine individual features of the predetermined objects.

For the new correct algorithm design on the datasets unrecognized by previous algorithms it is enough to use function $f_1(X)$. A new algorithm is a conjunction of $f_1(X)$ and the object decision rule unrecognized by other algorithms. The result is a unique object characteristic and its features combination that do not belong to any of the previously recognized objects

$$A_{n+1} = f_1(X) \& (\&_{s=1}^{m} x_s^j \rightarrow y_j) \vee_{j=1}^{n} A_j$$

Example 1. Let $X = \{x_1, x_2, x_3\}$ be a feature set; the value of each characteristic is encoded within the three-valued logic system $x_s \in \{0, 1, 2\}$, s = 1, 2, 3.

The input data ratio (objects features), objects and recognition algorithms results are provided by the following matrix (Table 2).

Table 2. Example 1.

x_1	x_2	x_3	Y	A'_1	A'_2	A'_3
0	1	1	a	1	0	1
1	2	2	b	0	1	0
0	1	2	c	1	1	0
1	0	0	d	0	0	0

On the basis of the given ratios we can write:

$A_1 : F_1 = (x_1^0 \& x_2^1 \& x_3^1 \rightarrow a) \& (x_1^0 \& x_2^1 \& x_3^2 \rightarrow c)$ (A_1 recognizes object a and c)

$A_2 : F_2 = (x_1^1 \& x_2^2 \& x_3^2 \rightarrow b) \& (x_1^0 \& x_2^1 \& x_3^2 \rightarrow c)$

$A_3 : F_3 = (x_1^0 \& x_2^1 \& x_3^1 \rightarrow a)$

$F = F_1 \& F_2 \& F_3 = f_1(X) \vee f_1(X, Y)$

$f_1(X) = x_1^2 \vee x_2^0 \vee x_3^0 \vee x_1^1 x_2^1 \vee x_1^1 x_3^1 \vee x_2^2 x_3^1$

$f_2(X, Y) = b x_1^1 \vee b x_2^2 \vee a x_3^1 \vee c x_1^0 x_3^2 \vee c x_2^1 x_3^2 \vee b c x_3^2 \vee a x_1^0 x_3^1 \vee a x_2^1 x_3^1 \vee ab$

$A_4 = f_1(X) \& (x_1^1 \& x_2^0 \& x_3^0 \rightarrow d)$

$= x_1^0 x_2^0 \vee x_2^0 x_3^1 \vee x_2^0 x_3^2 \vee x_1^0 x_3^0 \vee x_2^1 x_3^0 \vee x_2^2 x_3^0 \vee x_1^1 x_2^1 \vee x_1^1 x_3^1 \vee x_2^2 x_3^1 \vee d x_2^0 \vee d x_3^0$

Algorithm A_4 объекта identifies individual features of the object d, namely $x_2 = 0$ и $x_3 = 0$. Algorithm A_4 in a disjunction with the earlier set algorithms presents the entire solution area in the predetermined data domain.

2 Logical Approach to Correct Algorithm Design on the Given Data Domain

When in the previous matrix we add correctness requirements to algorithm $A_{n+1}(X)$ we obtain the following matrix (Table 3):

Table 3. Baseline, estimators algorithms correct algorithm evaluation function

x_1	x_2	…	x_m	Y	A'_1	A'_2	…	A'_n	A'_{n+1}
$x_1(y_1)$	$x_2(y_1)$	…	$x_m(y_1)$	y_1	$a_1(y_1)$	$a_2(y_1)$	…	$a_n(y_1)$	1
$x_1(y_2)$	$x_2(y_2)$	…	$x_m(y_2)$	y_2	$a_1(y_2)$	$a_2(y_2)$	…	$a_n(y_2)$	1
…	…	…	….	…	…	…	…	…	…
$x_1(y_2)$	$x_2(y_1)$	…	$x_m(y_1)$	y_1	$a_1(y_1)$	$a_2(y_1)$	…	$a_n(y_1)$	1

I.e. for $A_{n+1}(X)$ all values are $a_{n+1}(y_i) = 1,\ i = 1,2,\ldots,l.$

Since we can consider $a_j(y_i)$ as a Boolean variable then $A'_{n+1}(A'_1, A'_2 \ldots A'_n)$ is the Boolean function with a value of 1 in all specified sets of the domain $(A'_1, A'_2 \ldots A'_n)$. And we can write:

$$A'_{n+1}(A'_1, A'_2 \ldots A'_n) = \vee_{i=1}^{l} \&_{j=1}^{n} A_j^{\sigma'}(y_i),\ i = 1,2,\ldots,l, \quad j = 1,2,\ldots,n.$$

$$A_j^{\sigma\prime}(y_i) = \begin{cases} A'_j, a_j(y_i) = 1 \\ \overline{A'_j}, a_j(y_i) = 0 \end{cases},$$

We assume that A'_j, is a set of decision rules recognized by the algorithm, $\overline{A'_j}$ is a set of decision rules unrecognized by this algorithm.

$$A'_j = \&_{i=1}^{l}\left(\&_{s=1}^{m} x_s(y_i) \to y_i\right) = \&_{i=1}^{l}\left(\vee_{s=1}^{m} \overline{x_s(y_i,)} \vee y_i\right) \text{ when } a_j(y_i) = 1,$$

$$\overline{A'_j} = \overline{\&_{i=1}^{l}\left(\&_{s=1}^{m} x_s(y_i) \to y_i\right)} = \&_{i=1}^{l}\left(\&_{s=1}^{m} x(y_i,)\&\overline{y_i}\right) \text{ when } a_j(y_i) = 0,$$

The whole study data domain can be presented as decision rules:

$$\&_{s=1}^{m} x_s(y_i) \to y_i,\ i = 1,\ldots,l, x_s(y_i) \in \{0,1,\ldots,k_r - 1\}, k_r \in [2,\ldots,N], N \in Z. \quad (1)$$

Theorem: We define a set of decision rules of the form

$$\&_{j=1}^{m} x_s(y_i) \rightarrow y_i,\ i = 1,\ldots,l, x_j(y_i) \in \{0,1,\ldots,k_r-1\}, k_r \in [2,\ldots,N], N \in Z,$$

that represents a certain subject domain under study

$$A'_{n+1}(A'_1, A'_2 \ldots A'_n) = \vee_{i=1}^{l} \&_{j=1}^{n} A^{\sigma\prime}{}_j(y_i) = 1,\ i = 1,2,\ldots,l, j = 1,2,\ldots,n.$$

Each algorithm enters the proposed disjunction and the one or more conjunctions as A'_j, and also one or more conjunctions as $\overline{A'_j}$, otherwise it is a universal algorithm, for which all $a_j(y_i) = 1,\ i = 1,2,\ldots,l,$ or not operating algorithm $a_j(y_i) = 0,\ i = 1,2,\ldots,l$. Since A'_j is the set of decision rules recognized by algorithm A_j, and $\overline{A'_j}$ is the set of decision rules unrecognized by this algorithm so the disjunction of these rules provides full description of the study domain for each algorithm. In case of DNF creation

$$A'_{n+1}(A'_1, A'_2 \ldots A'_n) = \vee_{i=1}^{l} \&_{j=1}^{n} A^{\sigma\prime}{}_j(y_i)$$

it can be reduced to a deadlock DNF by known methods. Further when A J ^ is replaced by the decision rules it is possible to apply a reduction algorithm adapted for multiple-valued logics.

If some variable enters DNF with one value in all disjuncts, we delete all disjuncts containing this variable; (this variable isn't informative)

If DNF has a single-value disjunct x_i^j, то применяем правило поглощения дизъюнкта then we apply the rule of absorption of a disjunct.

As a result, each disjunct get minimized knowledge base relevant to the set of rules described by this disjunct. Such disjuncts have a number of properties [2]. They break solutions domain into all possible within it classes. By combining these domains we minimize the knowledge base for the entire predetermined area (Table 4).

Table 4. Example

x_1	x_2	x_3	Y	A'_1	A'_2	A'_3	A'_4
0	0	1	a	1	0	1	0
0	2	1	b	0	0	1	1
2	1	2	c	0	1	0	1
1	2	0	d	0	0	0	0

Example 2. Let $X = \{x_1, x_2, x_3\},\ x_i \in \{0,1,2\}$.
We build a disjunction in lines

$$F = A'_{n+1}(A'_1, A'_2 \ldots A'_n) = \vee_{i=1}^{l} \&_{j=1}^{n} A^{\sigma\prime}{}_j(y_i)$$

$$F = A_1 \& \overline{A_2} \& A_3 \& \overline{A_4} \vee \overline{A_1} \& \overline{A_2} \& A_3 \& A_4 \vee \overline{A_1} \& \overline{A_3} \& A_2 \& A_4 \vee \overline{A_1} \& \overline{A_2} \& \overline{A_3} \& \overline{A_4}$$

and then we write algorithms through decision rules, transform them and obtain the following expression:

$$\begin{aligned} A_5 &= (x_1^0 \& x_2^0 \& x_3^1 \to a) \& (x_1^0 \& x_2^2 \& x_3^1 \to b) \& (x_1^1 \& x_2^2 \& x_3^0 \to d) \\ &= x_1^2 \vee x_3^2 \vee x_2^1 \vee x_1^1 x_2^0 \vee x_1^1 x_3^1 \vee x_1^0 x_3^0 \vee x_2^0 x_3^0 \\ &\vee x_3^0 d \vee b x_1^0 x_2^2 \vee b x_2^2 x_3^1 \vee b d x_2^2 \vee a x_1^0 x_2^0 \vee a x_2^0 \vee a x_2^0 x_3^1 \vee x_1^1 d \end{aligned}$$

The algorithm A_5 selects personal features of object d.

Inference

The results of the logical analysis of the predetermined domain and decision rules that describe objects make clear that complexity of the obtained algorithm depends on the algorithms quality that has already been given and regularities that are hidden in the subject domain. The proposed logic synthesis method allows building of a correct algorithm on the entire data area, simulating the knowledge base, minimizing it, and selecting unique set of features for each object.

References

1. Zhuravljov Ju. I (1978) Ob algebraicheskom podhode k resheniju zadach raspoznavanija ili klassifikacii. Problemy kibernetiki 33:5–68
2. Shibzukhov ZM (2014) Correct aggregation operations with algorithms. Pattern Recognit Image Anal 24(3):377–382
3. Timofeev AV, Lyutikova LA (2005) Razvitie i primenenie mnogoznachnyh logik i setevyh potokov v intellektual'nyh sistemah. Trudy SPII RAN 2:114–126
4. Ljutikova LA (2006) Modelirovanie i minimizacija baz znanij v terminah mnogoznachnoj logiki predikatov. Preprint – Nal'chik, NII PMA KBNC RAN, 33 s
5. Voroncov KV (2000) Optimizacionnye metody linejnoj i monotonnoj korrekcii v algebraicheskom podhode k probleme raspoznavanija. Zhurnal vychislitel'noj matematiki i matematicheskoj fiziki 40(1):166–176
6. Zhuravljov Ju. I (1987) Rudakov K. V. Ob algebraicheskoj korrekcii procedur obrabotki (preobrazovanija) informacii. Problemy prikladnoj matematiki i informatiki 187–198

Brain-Computer Interfaces for Controlling Unmanned Aerial Vehicles: Computational Tools for Cognitive Training

Sonia López[1], José-Antonio Cervantes[1(✉)], Salvador Cervantes[1], Jahaziel Molina[1], and Francisco Cervantes[2]

[1] Centro Universitario de los Valles, 46600 Ameca, Mexico
{sonia.lopez,antoniocervantes,salvador.cervantes, jahaziel.molina}@valles.udg.mx

[2] Instituto Tecnológico y de Estudios Superiores de Occidente, 45604 Tlaquepaque, Mexico
fcervantes@iteso.mx

Abstract. Attention deficit-hyperactivity disorder is considered a mental disease that affects a significant number of the world's youth population. Brain-computer interfaces have been used to study and treat this mental disease. In this paper, we present the current state of unmanned aerial vehicles controlled by mental commands. We hope this study can be useful to guide future research focused to develop brain-computer interfaces able of controlling unmanned aerial vehicles for therapeutic purposes.

Keywords: Brian-computer interface · Unmanned aerial vehicles · Mental health · Cognitive training · ADHD

1 Introduction

Mental illness is a relevant issue because it affects the wellbeing and life-style of both patients and their relatives. Studies show that neuropsychiatric conditions are the leading cause of disability in young people in all regions. Children with mental disorders face major challenges with stigma, isolation, and discrimination, as well as lack of access to health care and education facilities [1–3]. The research community on mental health care has been working on new therapies and mechanisms to close the treatment gap for mental illnesses. The use of computational technology has been an option for developing alternative strategies and therapies [4–6].

We are aware of opportunities that technology offers and how it is constantly changing the way we approach tasks both in our daily-life and at work, and the health field is not an exception to these changes. For example, new concepts and approaches based on computational models have been useful to simulate and analyze people's responses after traumatizing events, including their development and recovery [7]. Moreover, advances in technology and falling costs have led to novel uses of unmanned aerial vehicles (UAVs), for example, to improve the mobility of people how

A. V. Samsonovich (Ed.): BICA 2019, AISC 948, pp. 315–320, 2020.
https://doi.org/10.1007/978-3-030-25719-4_40

have much limited movement [8], to deliver medical goods [9], or to offer emergency medical services by way of a drone ambulance [10].

This paper presents a plausibility study of the use of UAVs as computational tools for therapeutic purposes through an analysis of the current state of UAVs and how they have been coupled to brain-computer interfaces (BCIs) in order to control them through mental commands.

2 Attention Deficit-Hyperactivity Disorder

Attention deficit-hyperactivity disorder (ADHD) is considered a neurodevelopmental disease that affects an estimated 5% of the world's population. It has been seen as more common in childhood, but its prevalence in adolescence and adulthood has also been reported [11]. Symptoms associated to this neurobehavioral disorder are commonly presented before the age of 12 years, and in at least two contexts, such as home and school.

Electroencephalographic activity has been used to describe the functional characterization of patients with ADHD. Additionally, electroencephalographic activity has been used as a tool for intervention, showing the subjects real-time feedback from their brainwave activity, which has to be modulated in order to reach the objective of the task; this technique is called Neurofeedback [12]. Neurofeedback Training (NFT) has been employed to treat a number of diseases, such as cerebral infarctions [13], traumatic brain injuries [14], learning disabilities [15] as well as ADHD [16]. Studies have reported relevant reductions in symptom severity in patients with ADHD after they have engaged in game-based therapy combining both cognitive training and NFT [17]. We believe that BCIs for controlling a UAV can implement NFT in order to develop a computational tool for therapeutic purposes. In the next section, the current state of UAVs controlled by BCIs is presented.

3 Brain-Controlled Systems for UAVs

Unmanned aerial vehicles (UAVs) more popularly known as drones, were initially developed for military purposes, but nowadays, advances in technology and falling costs of UAVs have allowed civilians to work and conduct experiments with UAVs. These aerial robots have been widely used to offer new solutions in different areas such as search and rescue missions [18], delivery of medical goods [9], among other tasks [8, 10].

Moreover, novel brain-computer interfaces (BCIs) are being developed at a rapid pace for different purposes, including entertainment, communication, education, rehabilitation, among others [19]. Basically, a BCI is a communication system that enable users to send commands to a computer by using only their brain activity. These devices can be grouped into two categories, which are invasive and non-invasive BCIs.

Works described in this section have been classified as conventional or hybrid BCI systems. Whereas a conventional BCI has only one input modality commonly based on measuring electrical brain activity, a hybrid BCI has more than one input modality.

3.1 Conventional BCI Systems

- Yu and colleagues [8] proposed a BCI with an EEG to control a UAV for the entertainment of handicapped subjects. The system allows the subjects to interact with the UAV by sending wireless control commands through a commercial EEG to control the UAV's basic moves, send stream video to a computer and take pictures. The EEG sends commands according to the detection of the activation and the activation strength of three motor imagery brain activities: think left, think right, and think push. Additionally, EEG signals associated with eye blinking and tooth clenching are detected to add control commands; these signals can introduce artifacts into EEG signals, which can also be detected and converted into specific control commands.
- Kosmyna and colleagues [20] proposed an architecture to reduce the long training phases of BCI systems based on a machine learning approach in order to obtain a much more enjoyable interaction modality for recreation applications. The architecture was implemented using a signal amplifier with 16 electrodes that allow commands to be sent to a UAV. This proposal uses a bidirectional feedback between the user and the system, where the signals classified by the system can be confirmed or contradicted by the user. The signals detected are associated with imaginary movements (hands, arms, legs, etc.). The imaginary movements are determined by the system using an automatic classification of the signals and the corresponding commands are sent to the UAV to perform tasks such as taking off, going forward and landing.
- Shi and colleagues [21] proposed the use of a BCI based on the analysis of motor imagery (MI) signals obtained with an EEG in order to achieve an easy-to-use and stable control of a UAV independent of the manual operating level. The BCI system allows a low-speed UAV to be controlled in indoor scenarios using three imaginary motor signals: left-hand imagination movement to turn left, right-hand imagination movement to turn right, and idle to fly forward. A cross-correlation method is used for the feature extraction of the EEG signal and a logistic regression method is used for the signal classification. Additionally, the UAV uses a semi-autonomous navigation subsystem based on scanning lasers to avoid obstacles automatically and provide feasible directions. The combination of a BCI system and the semi-autonomous navigation subsystem offers low computational cost and high control efficiency of a UAV for the indoor target searching task.
- Lin and Jiang [22] implemented a BCI to control a UAV by processing EEG signals that were represented as feature vectors. These vectors were reduced by using Principal Component Analysis (PCA) to recognize facial expressions, which were converted into control commands for the UAV. The commands to fly left/right, forward, backward, take-off and landing were generated by the recognition of the facial expressions left/right smile, raise/frown brow, and left/right wink respectively. The objective of this research was the creation of an easy-to-use interface for human-computer interaction.

3.2 Hybrid BCI Systems

- Kim and colleagues [23] proposed a wearable hybrid interface for controlling a quadcopter. This hybrid BCI uses information from an eye tracking system and an EEG to extend the number of control commands. In this way, the hybrid interface enables a remote subject to control a quadcopter in a three-dimensional physical environment. Also, a front-view camera on the quadcopter is used to provide visual feedback to the remote subject on a laptop display. According to [23], the hybrid interface was tested by five human subjects who participated in a series of experiments and the results were validated by comparing them with results from a keyboard-based interface. The purpose of this study was to demonstrate the applicability of the hybrid interface to interact with a quadcopter.
- Khan and colleagues [24] proposed a hybrid BCI based on hybrid electroencephalography with near-infrared spectroscopy (EEG-NIRS). This hybrid BCI uses two commands to operate a quadcopter. The hybrid BCI decodes a command by an active brain signal coming from the user's own will through the use of a motor imaginary task. The second command is decoded when a reactive brain signal is generated by a visual flickering of light. Tasks performed by the quadcopter might look like trivial tasks. However, results presented in this research indicate that the proposed scheme can be suitable for BCI control applications.
- Khan and Hong [25] proposed hybrid electroencephalography with near-infrared spectroscopy (EEG-NIRS) to decode brain commands from the frontal brain region for a hybrid BCI. Four commands were decoded by a NIRS and four other commands were decoded by an EEG. Thus, the hybrid BCI was able to decode a total of eight brain activity commands. The hybrid BCI was tested using a quadcopter in an open space. Results show an average accuracy of 75.6% in decoding commands with the NIRS and 86% in decoding commands with the EEG [25].

4 Conclusion

This research focused on showing the plausibility of using UAVs for clinical purposes through a review of the current state of UAVs and how they have been coupled to non-invasive BCIs in order to control them through mental commands. As a result of this study, we consider that current technology is robust and has proven its effectivity in a wide range of tasks, which are outstanding innovations for recording and translating brain signals into control commands for UAVs. We consider that the next step in this study field could be the application of requirements engineering process in order to identify specific requirements for developing a BCI focused on cognitive training of children with ADHD.

References

1. Shidhaye R, Lund C, Chisholm D (2015) Closing the treatment gap for mental, neurological and substance use disorders by strengthening existing health care platforms: strategies for delivery and integration of evidence-based interventions. Int J Ment Health Syst 9(1):1–11. https://doi.org/10.1186/s13033-015-0031-9
2. Patel V, Xiao S, Chen H, Hanna F, Jotheeswaran A, Luo D, Parikh R, Sharma E, Usmani S, Yu Y (2016) The magnitude of and health system responses to the mental health treatment gap in adults in India and China. Lancet 388(10063):3074–3084. https://doi.org/10.1016/S0140-6736(16)00160-4
3. Kazdin AE (2017) Addressing the treatment gap: a key challenge for extending evidence-based psychosocial interventions. Behav Res Ther 88:7–18. https://doi.org/10.1016/j.brat.2016.06.004
4. Spruijt-Metz D, Hekler E, Saranummi N, Intille S, Korhonen I, Nilsen W, Rivera DE, Spring B, Michie S, Asch DA (2015) Building new computational models to support health behavior change and maintenance: new opportunities in behavioral research. Transl Behav Med 5(3):335–346. https://doi.org/10.1007/s13142-015-0324-1
5. Silva BM, Rodrigues JJ, de la Torre Díez I, López-Coronado M, Saleem K (2015) Mobile-health: a review of current state in 2015. J Biomed Inform 56:265–272. https://doi.org/10.1016/j.jbi.2015.06.003
6. Naslund JA, Aschbrenner KA, Barre LK, Bartels SJ (2015) Feasibility of popular m-health technologies for activity tracking among individuals with serious mental illness. Telemed E-Health 21(3):213–216. https://doi.org/10.1089/tmj.2014.0105
7. Formolo D, Van Ments L, Treur J (2017) A computational model to simulate development and recovery of traumatised patients. Biol Inspired Cogn Arch 21:26–36. https://doi.org/10.1016/j.bica.2017.07.002
8. Yu Y, He D, Hua W, Li S, Qi Y, Wang Y, Pan G (2012) Flying-Buddy2: a brain-controlled assistant for the handicapped. In: UbiComp. ACM, pp 669–670
9. Scott JE, Scott CH (2018) Models for drone delivery of medications and other healthcare items. Int J Healthc Inf Syst Inform (IJHISI) 13(3):20–34. https://doi.org/10.4018/IJHISI.2018070102
10. Van de Voorde P, Gautama S, Momont A, Ionescu C, De Paepe P, Fraeyman N (2017) The drone ambulance [A-UAS]: golden bullet or just a blank? Resuscitation 116:46–48. https://doi.org/10.1016/j.resuscitation.2017.04.037
11. Polanczyk G, De Lima MS, Horta BL, Biederman J, Rohde LA (2007) The world-wide prevalence of ADHD: a systematic review and metaregression analysis. Am J Psychiatry 164:942–948
12. Hammond DC (2011) What is neurofeedback: an update. J Neurother 15(4):305–336. https://doi.org/10.1080/10874208.2011.623090
13. Bearden TS, Cassisi JE, Pineda M (2003) Neurofeedback training for a patient with thalamic and cortical infarctions. Appl Psychophysiol Biofeedback 28(3):241–253. https://doi.org/10.1023/A:1024689315563
14. Tinius TP, Tinius KA (2000) Changes after EEG biofeedback and cognitive retraining in adults with mild traumatic brain injury and attention deficit hyperactivity disorder. J Neurother 4(2):27–44. https://doi.org/10.1300/J184v04n02_05
15. Fernández T, Bosch-Bayard J, Harmony T, Caballero MI, Díaz-Comas L, Galán L, Ricardo-Garcell J, Aubert E, Otero-Ojeda G (2016) Neurofeedback in learning disabled children: visual versus auditory reinforcement. Appl Psychophysiol Biofeedback 41(1):27–37. https://doi.org/10.1007/s10484-015-9309-6

16. Van Doren J, Arns M, Heinrich H, Vollebregt MA, Strehl U, Loo SK (2018) Sustained effects of neurofeedback in ADHD: a systematic review and meta-analysis. Eur Child Adolesc Psychiatry 1–13. https://doi.org/10.1007/s00787-018-1121-4
17. Johnstone SJ, Roodenrys SJ, Johnson K, Bonfield R, Bennett SJ (2017) Game-based combined cognitive and neurofeedback training using focus pocus reduces symptom severity in children with diagnosed AD/HD and subclinical AD/HD. Int J Psychophysiol 116:32–44. https://doi.org/10.1016/j.ijpsycho.2017.02.015
18. Silvagni M, Tonoli A, Zenerino E, Chiaberge M (2017) Multipurpose UAV for search and rescue operations in mountain avalanche events. Geomat Nat Hazards Risk 8:18–33. https://doi.org/10.1080/19475705.2016.1238852
19. Choi I, Rhiu I, Lee Y, Yun MH, Nam CS (2017) A systematic review of hybrid brain-computer interfaces: Taxonomy and usability perspectives. PLoS ONE 12:e0176674. https://doi.org/10.1371/journal.pone.0176674
20. Kosmyna N, Tarpin-Bernard F, Rivet B (2015) Towards brain computer interfaces for recreational activities: piloting a drone. In: Human-computer interaction. Springer, pp 506–522. https://doi.org/10.1007/978-3-319-22701-6_37
21. Shi T, Wang H, Zhang C (2015) Brain computer interface system based on indoor semi-autonomous navigation and motor imagery for unmanned aerial vehicle control. Expert Syst Appl 42:4196–4206. https://doi.org/10.1016/j.eswa.2015.01.031
22. Lin JS, Jiang ZY (2015) Implementing remote presence using quadcopter control by a non-invasive BCI device. Comput Sci Inf Technol 3:122–126. https://doi.org/10.13189/csit.2015.030405
23. Kim BH, Kim M, Jo S (2014) Quadcopter flight control using a low-cost hybrid interface with EEG-based classification and eye tracking. Comput Biol Med 51:82–92. https://doi.org/10.1016/j.compbiomed.2014.04.020
24. Khan MJ, Hong KS, Naseer N, Bhutta MR (2015) Hybrid EEG-NIRS based BCI for quadcopter control. In: 54th annual conference of the society of instrument and control engineers of Japan (SICE). IEEE, Hangzhou, pp 1177–1182. https://doi.org/10.1109/sice.2015.7285434
25. Khan MJ, Hong KS (2017) Hybrid EEG-fNIRS-based eight-command decoding for BCI: application to quadcopter control. Front Neurorobotics 11:1–13. https://doi.org/10.3389/fnbot.2017.00006

Sparsified and Twisted Residual Autoencoders

Mapping Cartesian Factors to the Entorhinal-Hippocampal Complex

András Lőrincz(✉)

Department of Artificial Intelligence, Eötvös Loránd University, Budapest, Hungary
lorincz@inf.elte.hu

Abstract. Previously, we have put forth the concept of Cartesian abstraction and argued that it can yield 'cognitive maps'. We suggested a general mechanism and presented deep learning based numerical simulations: an observed factor (head direction) was non-linearly projected to form a discretized representation (head direction cells). That representation, in turn, enabled the development of a complementing factor (place cells) from high dimensional (visual) inputs. It has been shown that a related metric, in the form of oriented hexagonal grids, may also be derived. Elements of the algorithms were connected to the entorhinal-hippocampal complex (EHC loop). Here, we make one step further in the mapping to the neural substrate. We consider (i) the features of signals arriving at deep and superficial CA1 pyramidal cells, (ii) the interplay between lateral and medial entorhinal cortex efferents, and the nature of 'instructive' input timing-dependent plasticity, a feature of the loop. We suggest that the circuitry corresponds to a special form of Residual Networks that we call Sparsified and Twisted Residual Autoencoder (ST-RAE). We argue that ST-RAEs can learn Cartesian Factors and fit the structure and the working of the entorhinal-hippocampal complex to a reasonable extent, including certain oscillatory properties. We put forth the idea that the factor learning architecture of ST-RAEs has a double role in serving goal-oriented behavior, such as (a) the lowering the dimensionality of the task and (b) the mitigation of the problem of partial observation.

Keywords: Factor learning · Entorhinal-hippocampal loop · Residual networks · Skip connections · Sparsification

1 Introduction

Ever since the famous case of patient H.M. was found [1], the puzzle of algorithmic functioning of the entorhinal-hippocampal complex (EHC loop) has become an enigmatic challenge in neuroscience. Patient H.M. suffered anterograde amnesia following removal of the EHC loop and, possibly, neighboring regions. Questions emerge, such as what this loop is doing, compared to the neocortical areas? Is it the key structure to form episodic and declarative memory [2]? Or, along the discovery of place cells in the hippocampus [3], is it having a leading role in the creation of the representation of

A. V. Samsonovich (Ed.): BICA 2019, AISC 948, pp. 321–332, 2020.
https://doi.org/10.1007/978-3-030-25719-4_41

space serving navigation? How is it possible that upon lesioning, category formation remains intact, while episodic memory is severely spoiled [4]?

The set of hippocampal place cells in rats are known as the "cognitive map", but how is it used for cognition? What other functions it may have – in rats and in primates – beyond developing such place cells? How come that upon early lesioning the mentioned functionalities are spared to some extent [5]? What are the critical components of information processing in the EHC?

Answers to some of the questions have been proposed upon the discovery of grid cells [6]: one can think of the role of the hippocampus in the loop that it projects to and discretizes a low dimensional space, being two dimensional for rats and three dimensional for bats in the case of navigation [7], whereas the entorhinal cortex represents the metric in that space, making dead reckoning and planning possible. But then, how can we account for the diverse responses in neocortical areas? How come that such hexagonally symmetric gridlike code may appear across the brain in diverse cognitive tasks [8]? We are to address some of these issues.

In our recent work [9], the concept of Cartesian Factors was put forth. We demonstrated that head direction based semi-supervision on visual information may emerge both place cells and oriented grid cells in a sparsified deep neural network [9, 10]. Another work [11] showed that reinforcement learning and recurrent neural networks can produce direction independent grid cells. This may elucidate the spatial navigation view, namely, that a particular multi-layered factor representation is formed both for space and for spatial metric. It is then a falsifying issue, if the algorithms can be mapped onto the EHC loop.

Here, we make an effort to do so and map the algorithms to the EHC loop. We base our proposal on Cartesian Factors and on recent achievements concerning a modification of Residual Networks [12, 13] that changes the role of the skip connections. We extend these networks with the concept of sparsification. The new network is called Sparsified and Twisted Residual Autoencoder (ST-RAE).

The paper is organized as follows. In the next section (Sect. 2) we describe the classical view of the EHC loop, a base model, its shortcomings and recent findings about the neuronal substrate. In Sect. 3 the ST-RAE model is built and we argue that it can form Cartesian Factors. We explain the advantages of sparsification and map the artificial neural network structure to the EHC loop. The architecture is discussed in Sect. 4. We consider the generality of the algorithm, its temporally modulated potential neuronal manifestations, and reinforcement learning in the context of partially observed Markovian decision making. We conclude and consider future work in the last section (Sect. 5).

2 Features and Models of the Neural Substrate

2.1 'Classical' Description of the EHC Loop

Neural layers, the main input and output connections of the entorhinal hippocampal complex, the superficial and deep layers of the entorhinal cortex (EC), the subiculum (SUB), and the components of the hippocampus, including the CA3 and CA1 subfields, and the dentate gyrus (DG) formed by the fascia dentata (FD) and the hilus (H) are depicted in Fig. 1. In this structure, the CA3 subfield and the deep layers (EC V-VI) have recurrent collaterals.

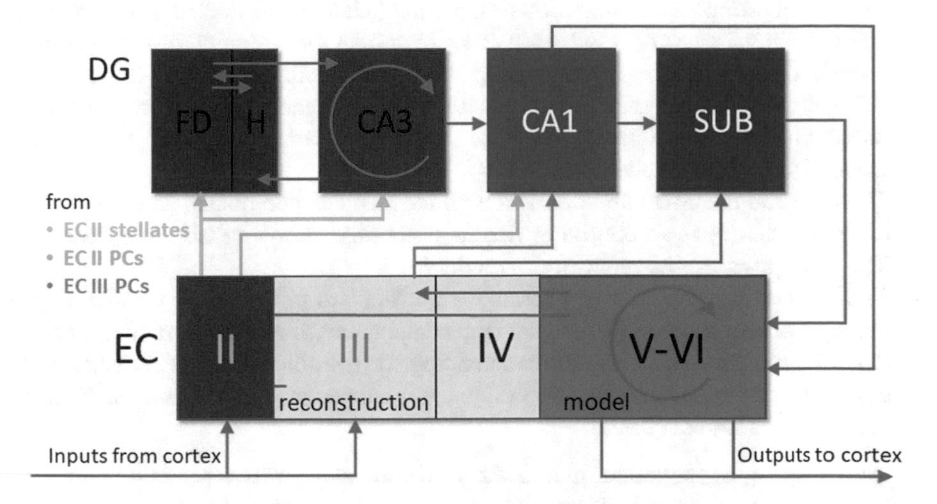

Fig. 1. Sketch of the loop formed by entorhinal cortex (EC) and the hippocampal complex, the EHC loop. Main internal connections are shown. Superficial (deep) layers of the EC receive inputs (provide outputs) from (to) neocortical areas. Main stops of the information flow in the loop are the deep (V–VI) and superficial (II and III) layers of the EC, the dentate gyrus (DG), the CA3 and CA1 subfields and the subiculum. The loop in the DG connects the granule cells of the fascia dentata (FD) and the mossy cells of the hilus (H). The EC provides input to the CA3 subfield (perforant path and mossy fibers) and sends information to the CA1 subfield. Input also arises from EC III. Recurrent connections are present in EC V–VI and in CA3. PC: pyramidal cell. For thc meaning of the words 'error', 'reconstruction', and 'model', see text.

2.2 The Base Model

We start by inspecting two *reconstruction network* models of this loop [14–16] and [21], respectively. We follow this line of reverse engineering in order to clarify the *algorithmic problems* arising from the nature of the input-output signals and the structure of the architecture to be matched. Relevant additional references can be found in the above works as well as in other works that we cite.

The starting point of this reverse engineering effort is the development of internal models that can replay and predict episodes occurring in the environment and serve the learning of behaviors.

For the sake of forming internal representations and for predicting future inputs, reconstruction networks make a natural choice. The simplest reconstruction networks are made of three layers, operate as autoencoders, but represent no dynamics. A more complex structure was suggested in [14, 15] for the EHC loop. It included a model network that we describe below:

1. The network has a dynamical model layer in deep entorhinal layers (EC V-VI)
2. The model produces an estimation of the input (a.k.a. reconstructed input) in EC III and computes its error (in EC II) in order to correct the operation of the dynamical model. The error also serves training. Processing delays that may differ for the different sensory components are (a) estimated by a putative blind source separation method in the dentate gyrus. They are (i) compensated and then (ii) the synchronized signals are modelled by the loop.
3. The CA3 and the CA1 subfields deal with independent component analysis that can discretize the input space giving rise to place fields upon suitable preprocessing. The CA1 provides the input to the model layer
4. The network operates in two phases. One of the two phases is responsible for the learning of the bottom-up independent components, the internal model, and can leave learning traces for top-down encoding for the other phase. This other phase encodes long-term memories into the synapses connecting the dynamic model to the reconstructed input.

An extension of the model placed the whitening stage of independent component analysis into the entorhinal cortex and thus the model produced grid cells, too [16]. Biologically sound parameter tuning was suggested that can represent causal relationships and can define events.

The dynamic model enables behavior optimization of the 'model', since it can act as a dynamical reservoir [17, 18]. Such dynamical reservoirs can overcome problems originating from non-Markovian observations as shown in [19].

2.3 Shortcomings

The model is not sufficient in many ways. We list the important ones and add potential resolutions to some of them:

(a) In case of a perfect model, the error becomes zero and this is not the case for the neural substrate. However, the concept of reconstruction network can be saved if we consider oscillations. Then, we have a driving signal and 'zero' is represented by an out-of-phase version giving rise to 'zero' upon integration with the input signal. Such oscillations support (a) phase-locked loop synchronization and control [20], offering a control view for the EHC loop that has been sketched in [21] and that allows for sinusoidal modulation. The phase-locked loop view emerged early in the literature [31] and it was suggested to contribute to learning. Spike-timing dependent plasticity (STDP) [22] harmonizes with the idea, since they change sign and provide training signals unless the reconstruction signal is out-of-phase.

(b) Input dimensions are very large. Discretization of high dimensional spaces are not feasible and low-dimensional embedding is not solved in the model except – maybe – through the whitening stage [16]. Topographical constraints could lead to such low-dimensional projections, but the discretizing subfields, CA3 and CA1, do not seem to have such constraints. This problem was circumvented to some extent by the proposal on semi-supervision based on known features of the loop [10] and subject to the condition of strong sparsification of the representation. Semi-supervision means that one modality (one Cartesian Factor) supervises the formation of another complementary factor. The algorithm is promising, supporting arguments exist, but the mapping to the neuronal substrate was left open.

(c) There are many factors that the neuronal EHC loop can emerge and they all may be crucial for behavioral optimization. The model should be able to explain their existence, too. A general mechanism for semi-supervision would be attractive for this specific model type.

(d) The EHC loops is much more complex than the classical description provided earlier. Novel findings may serve model construction.

In the next part (Subsect. 2.4) we review some of the intriguing features of the EHC loop that were discovered recently. They come to our rescue in the quest for developing algorithms for forming factors *and* satisfying neuronal constraints, while still *serving* behavioral optimization. Fast progress of deep learning technology offers a novel route that we describe in Sect. 3.

2.4 Novel Findings in Neuroscience

The entorhinal cortex has two major divisions, the lateral EC (LEC) and the medial EC (MEC). The EC layers, the neurons and the connectivity structure have been studied and many special properties have been found. Out of those that have relevance for us, we list the most prominent ones. A more comprehensive list and the original publications can be found in the (review) papers that we cite [23–28].

1. Semi-supervision requires supervisory mechanisms. There are ‘instructive’ supervisory signals working through long-term heterosynaptic plasticity. They don’t require somatic spiking and target perforant path dendrites of CA1 neurons giving rise to long-term potentiation at Schaffer collateral synapses, provided that the supervisory information arrives 10–20 ms earlier. The phenomenon was termed input-timing-dependent plasticity (ITDP) [23].
2. The mechanism is complex; it exploits long-range inhibitory projections (LRIPs) from both MEC and LEC to CA1 inhibitory neurons giving rise to a disinhibitory mechanism suppressing feedforward inhibition. The mechanism can enhance contextual specificity and increases the precision of long-term memory in the hippocampus [24].
3. There are robust differences in CA1 pyramidal cell population that go beyond the proximal and distal division of the CA1 subfield targeted mainly by MEC and LEC efferents, respectively [26]. For example, superficial and deep layer neurons of this subfield behave differently [25]: firing from these layers segregate along theta and gamma cycles, giving rise to distinct options depending on if CA1 pyramidal cells target cortical neurons together or differentially.
4. EC II stellate cells and EC II and EC III pyramidal cells act in diverse ways: they can associate discontinuous sub-episodes, may have context dependencies, can interact in many ways spatially, including connections to CA2 pyramidal cells that target CA1 neurons, among other possibilities [26].

5. It has been suggested that pyramidal cells of different subfields, layers and sublayers, interneurons of many different kinds, and stellate cells form four distinct sub-circuits of the EHC loop [26] that can support diverse functions.
6. Short time-scale changes without changes in spatial specificity, called rate-remapping occur in the hippocampus. This feature is another source of context modulation. Rate remapping in the CA1 subfield depends on the task and occurs during the first half of the theta cycle. The other half of the theta cycle exhibits the well-known phase precession without task modulations [27].
7. Context may be represented as a modulated rate code that also has spatial dependencies as suggested for the case of tastes [28].

Now wc turn to the architectural description and note that points 1-to-7 indicate that there can be diverse ways for representing factors. The diversity of factor formation and the capability of encoding behaviorally relevant factors into cortical regions are highly desired features of an architecture that serves dimension reduction (e.g., 2D space or taste), discretization of the reduced dimensions (e.g., place cells), the development of a metric in that space (e.g., grid cells), and similar capabilities for the different non-spatial factors, such as discrimination capabilities and spatial representation for taste. All of these can be relevant for the optimization of behaviors. In turn, factor formation may have alternative mechanisms beyond the architecture and its functioning to be presented below. We tried to use a minimal set of assumptions that can fit the experimental findings.

3 Factor Learning with a Sparsified Residual Network

We review the concept of Cartesian Factors and a prototype architecture that appeared recently [12, 13]. The prototype is shown in Fig. 2a. It is dealing with depth estimation from monocular images and it is to be trained on stereo recordings. We consider (a) how to generalize the architecture for input pairs beyond left and right images and (b) how to extend it with a discretization that could emerge place cells [9, 10].

3.1 Autoencoding and Sparsification

Assume a specific autoencoder network with a twisted reconstruction task: the input of the network is from one modality and the role of the network is to (re)construct or predict another, a complementing one. This other information source is the instructive, or supervisory information. This task alone is easy, provided that the inputs and the desired outputs are available. However, we want to satisfy two additional constraints: we would like to (a) develop the factor being responsible for the input-to-output mapping and (b) discretize this factor in order to develop a metric.

For the prototype example assume that we have a stereo camera and plan to learn the mapping from the left images to the right ones. Deep learning approaches are very efficient, so they could do that to a certain extent. Our constraints, however, do not allow such brute force mapping. The idea is the following: develop a minimal representation from the input and combine it with the input itself to produce the desired output with a relatively simple, possibly linear network. Furthermore, try to eliminate most of the information about the input from the minimal representation itself hoping that a new factor, in our case the depth, will be represented (Fig. 2a).

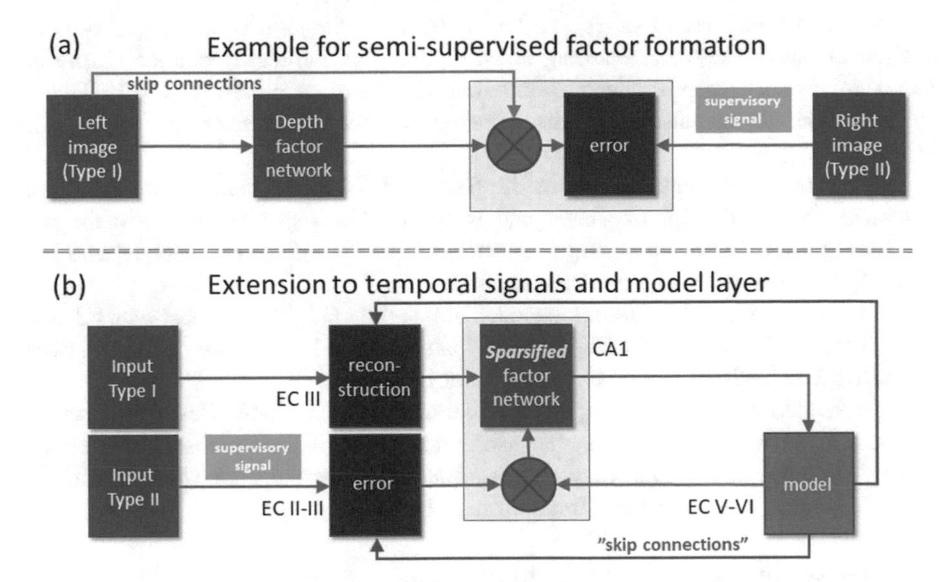

Fig. 2. Sparse residual network with skip connections for factor learning.

(a) Depth learning network. Left image enters a network. This input and the output of the depth learning part are joined by means of skip connections. Supervision is provided by the right image, i.e., by the complementary information. The error between the output of the network and the supervisory signal serves training via error backpropagation. Training results in the Cartesian factor, the depth, related to the disparity, in this case [12, 13].
(b) Mapping to the EHC loop: the Sparsified and Twisted Reconstruction Autoencoder (ST-RAE) model. The EC efferents to the CA1 subfield can induce input-timing-dependent plasticity (ITDP) on the Schaffer collateral EPSPs. In turn, signals through these EC efferents may act as supervisory inputs. These instructive signals can be induced without recruiting dendritic spikes [23]. CA1 representation is sparsified for the discretization of Cartesian factors of different kinds and they feed the internal model. It is the internal model that provides skip connection like associations, instead of the complementary input. These connections produce both the reconstructed input and its error. Reconstruction based error signal generation fits the Hebbian learning paradigm as opposed to error backpropagation. For more details, see text.

Sparsification is not included directly into the cited works [12] and [13], most probably, because convolutional neural networks (CNNs) were utilized. Such networks are overcomplete and may become sparse by themselves. However, CNNs exploit translational invariance and neither CA3 nor CA1 show translational such invariances. In turn, in our model the development of minimized representations may need sparsification in order to eliminate as much input information as possible, but still being able to generate the other modality in the twisted architecture.

In the case of stereo images, we know that an estimation of the disparity is the desired output of the factor learning network, being an indirect (reciprocal) representation of distance (depth). Such estimations can be very precise [12, 13]. Due to sparsification, we may also have a discretized estimation of the factor, i.e., the disparity in this case.

Skip connections make an interesting feature of the architecture. Alike to residual networks, the input skips one layer and is added – in a specific way – to the next processing layer. In our case, skip connections form part of the autoencoder and they simplify the estimation of the other modality.

In sum, the network is an autoencoder-like architecture that reconstructs different signals from other ones, while exploiting sparsification. In turn, we call the network Sparsified and Twisted Residual Autoencoder (ST-RAE).

We should mention that – in general – individual pixels are insufficient for disparity estimation. Such estimation, i.e., the estimation of the value of the factor requires contextual information. One may suspect that the more the contextual information about the input, the better the estimation may become.

3.2 Reconstruction Model of the EHC Loop

The stereo example of the previous subsection depicted in Fig. 2a and the architecture described in [10] that could produce place cells by semi-supervision provided by direction-oriented signal make good starting points to develop a new model for the EHC loop. This time, it is only a verbal model. Computational studies are to follow later.

In the case of the stereo example, error computation was the task of two signals, the supervisory signal and the sum of the factor learning network and the signal arriving via the skip connections. This procedure is not a necessity if we include time and aim to learn an internal dynamical model of the external signals, since the internal model can produce the estimated input and then the reconstruction error becomes approximately zero. In a phase-locked loop system phase may indicate that the model is precise, so the learning of the model may stop and factor learning can start.

This modification, namely that the estimated input can replace the true one is included into the model architecture (Fig. 2b) in order to match the structure of the EHC loop. Due to this change, the name skip connections has been put between quotation marks. Concerning the factor learning architecture, "skip connections" make two steps. First, the reconstruction error is computed, setting the error free output out-of-phase. This is the method that circumvents the lack of backpropagation capabilities. We shall come back to this issue in Sect. 4.

Here, we note that in deep learning, backpropagation is the method of weight tuning and this option does not seem to fit the neurobiological system. Instead, the error should be delivered to the dendrites of the neurons. The problem may be overcome in an autoencoding architecture with model-based synchronization between the input and its reconstructed version. The ITDP mechanism suits the needs, since it works in the absence of somatic spiking and thus it can wire in *or* reuse certain parts of the architecture when novel information is to be learned.

4 Discussion

4.1 Modified Architecture of the EHC Complex

Given our model of semi-supervised learning of place cells [10] and the recent advances in deep neural networks plus the novel findings concerning the entorhinal-hippocampal complex, we propose to modify the model slightly and we map the modified model to the entorhinal-hippocampal loop.

The basis of the architecture is an autoencoder equipped with a dynamical internal model layer that estimates the actual input. Due to the delays in sensory processing, the estimation works by model-based prediction in the loop. The new model takes into consideration brain waves: the reconstruction error of the autoencoder is zero when the error is the out-of-phase version of the input signal.

The principle of semi-supervision seems to match peculiar features of the neural substrate, including the rich variety of phase relations, possibly both in the gamma and the theta waves. Furthermore, recently discovered connections in the EHC loop [23–28] and their specific roles including the existence of 'instructive' signals all support the idea that this loop develops Cartesian Factors of diverse kinds.

4.2 The Cognitive Advantage: Mitigating Combinatorial Explosion

The main advantage of the development of many Cartesian factors is that by selecting a few of them, they may be sufficient for the characterization of an episode, the prediction of the episodic outcomes. In turn, decision making may involve a relatively small cognitive space (that may change from episode to episode) as opposed to the tremendous size of the space of sensory information. These factors are hidden variables and may serve the discoveries of higher order ones. Cognition then becomes possible, being impossible in the space of 10^6—10^8 dimensions of sensory observations.

Such reduction of the input space serves reinforcement learning by turning it into a factored one and thus limiting combinatorial explosion [29, 30]. This feature is of relevance if the number of learning samples is limited.

4.3 The Behavioral Advantage: Overcoming Problems Arising from Partial Observation

Another advantage is the following: assume that part of the information is missing. For example, if one lens of the stereo camera is covered, the estimation of depth is still feasible. This is what Cartesian Factors can do: in case of missing information, relevant quantities can be estimated from these learned and computable factors, which in our example is disparity being proportional to the inverse of the depth. In turn, the problem of partial observation, a key obstacle in reinforcement learning is mitigated.

4.4 The Need for Sparsification

Deep convolutional neural network searches for Cartesian Factors. Each factor needs its own metric for the purpose of control and for goal-oriented decision making. The hint from the neuronal substrate is that the metric can be developed via discretization [3, 7] and the global grid-like representation [6]. Our numerical simulations [10] point to sparsification as the key for solving both tasks, i.e., local and global representations together.

It has been found that a similar mechanism works for metric learning in higher order cognitive tasks. In particular, a two-dimensional decision-making problem (length of neck and legs) produced hexagonal grid-like activity patterns in the brain [8]. Such findings point to general mechanisms shared by the EHC loop and neocortical regions.

4.5 Features of the Neuronal Substrate

It is interesting that early lesioning of the hippocampus does not corrupt spatial relational learning [5]. It is then reasonable to assume that cortical regions can replace the lesioned part, at least to some extent and until connectivity structures allow considerable reorganization.

Model learning and model-based reconstruction could be one of the methods to encode long-term memories into cortical areas in a similar way to the encoding of the deep-to-superficial connections, i.e., the decoder of the autoencoding architecture.

Our model offers solutions to certain constraints of the neural substrate, including Hebbian learning. Firstly, the reconstruction method enables instructive training that can work like error-backpropagation. Secondly, brain waves can serve both bottom-up learning of structures and top-down supervisory training for processing signals.

5 Conclusions and Future Works

We sketched an architecture for the development of Cartesian Factors. These factors are relevant for

(a) overcoming combinatorial explosions in behavior optimization and
(b) diminishing the problem of partial observation in reinforcement learning
(c) supporting the creation of higher order factors based on prewired or learned ones
(d) controlling and planning in a much lower dimensional space than the sensory one

We proposed a mapping of the architecture to the neural substrate, namely to the EHC loop, being responsible, e.g., for the learning of the representation of space, including its discretized representation and its metric. It also offers a solution to supervisory 'instructive' learning, necessary for semi-supervised learning of Cartesian Factors.

The suggested mapping offers experimental predictions concerning the interplay between working, phase control, and learning in the loop and possibly the mechanisms of encoding long-term memories into the neocortical areas.

We put forth arguments that the algorithms can work in oscillatory networks. This assumption should be the subject of further numerical investigations in order to uncover the strengths or the weaknesses of the architecture.

Acknowledgments. The research has been supported by the European Union, co-financed by the European Social Fund (EFOP-3.6.3-VEKOP-16-2017-00002) and by the ELTE Institutional Excellence Program (1783-3/2018/FEKUTSRAT) supported by the Hungarian Ministry of Human Capacities.

References

1. Scoville WB, Milner B (1957) Loss of recent memory after bilateral hippocampal lesions. J Neurol Neurosurg Psychiatry 20:11–21
2. Cohen NJ, Squire LR (1980) Preserved learning and retention of pattern analyzing skill in amnesia: dissociation of knowing how and knowing that. Science 210:207–210
3. O'Keefe J, Dostrovsky J (1971) The hippocampus as a spatial map. Preliminary evidence from unit activity in the freely-moving rat. Brain Res 34:171–175
4. Knowlton BJ, Squire LR (1993) The learning of categories: parallel brain systems for item memory and category knowledge. Science 262(5140):1747–1749
5. Lavenex P, Lavenex PB, Amaral DG (2007) Spatial relational learning persists following neonatal hippocampal lesions in macaque monkeys. Nat Neurosci 10(2):234
6. Moser EI, Kropff E, Moser MB (2008) Place cells, grid cells, and the brain's spatial representation system. Annu Rev Neurosci 31:69–89
7. Yartsev MM, Ulanovsky N (2013) Representation of three-dimensional space in the hippocampus of flying bats. Science 340(6130):367–372
8. Constantinescu AO, O'Reilly JX, Behrens TE (2016) Organizing conceptual knowledge in humans with a gridlike code. Science 352(6292):1464–1468
9. Lőrincz A (2016) Cartesian abstraction can yield 'cognitive maps'. Procedia Comput Sci 88:259–271
10. Lőrincz A, Sárkány A (2017) Semi-supervised learning of cartesian factors: a top-down model of the entorhinal hippocampal complex. Front Psychol 8:215
11. Banino A, Barry C, Uria B, Blundell C, Lillicrap T, Mirowski P, Wayne G, Pritzel A, Chadwick MJ, Degris T, Modayil J, Wayne G, Soyer H, Viola F, Zhang B, Goroshin N, Rabinowitz N, Pascanu R, Beattie C, Petersen S, Sadik A, Gaffney S, King H, Kavukcuoglu K, Hassabis D, Hadsell R, Kumaran D (2018) Vector-based navigation using grid-like representations in artificial agents. Nature 557(7705):429–433
12. Garg R, Kumar VBG, Carneiro G, Reid I (2016) Unsupervised CNN for single view depth estimation: Geometry to the rescue. In: European conference on computer vision. Springer, Cham, pp 740–756
13. Godard C, Mac Aodha O, Brostow GJ (2017) Unsupervised monocular depth estimation with left-right consistency. In: Proceedings of the IEEE conference on computer vision and pattern recognition, pp 270–279
14. Lőrincz A, Buzsáki G (2000) Two-phase computational model training long-term memories in the entorhinal-hippocampal region. Ann N Y Acad Sci 911(1):83–111
15. Chrobak JJ, Lőrincz A, Buzsáki G (2000) Physiological patterns in the hippocampo-entorhinal cortex system. Hippocampus 10(4):457–465
16. Lőrincz A, Szirtes G (2009) Here and now: how time segments may become events in the hippocampus. Neural Netw 22(5–6):738–747

17. Jaeger H (2002) Tutorial on training recurrent neural networks, covering BPPT, RTRL, EKF and the "echo state network" approach, vol 5. GMD-Forschungszentrum Informationstechnik, Bonn
18. Maass W, Natschläger T, Markram H (2002) Real-time computing without stable states: a new framework for neural computation based on perturbations. Neural Comput 14 (11):2531–2560
19. Szita I, Gyenes V, Lőrincz A (2006) Reinforcement learning with echo state networks. In: International conference on artificial neural networks. Springer, Heidelberg, pp 830–839
20. Johnson MG, Hudson EL (1988) A variable delay line PLL for CPU-coprocessor synchronization. IEEE J Solid-State Circuits 23(5):1218–1223
21. Lőrincz A (1998) Forming independent components via temporal locking of reconstruction architectures: a functional model of the hippocampus. Biol Cybern 79(3):263–275
22. Markram H, Gerstner W, Sjöström PJ (2012) Spike-timing-dependent plasticity: a comprehensive overview. Front Synaptic Neurosci 4:2
23. Dudman JT, Tsay D, Siegelbaum SA (2007) A role for synaptic inputs at distal dendrites: instructive signals for hippocampal long-term plasticity. Neuron 56(5):866–879
24. Basu J, Zaremba JD, Cheung SK, Hitti FL, Zemelman BV, Losonczy A, Siegelbaum SA (2016) Gating of hippocampal activity, plasticity and memory by entorhinal cortex long-range inhibition. Science 351(6269):aaa5694
25. Mizuseki K, Diba K, Pastalkova E, Buzsáki G (2011) Hippocampal CA1 pyramidal cells form functionally distinct sublayers. Nat Neurosci 14(9):1174
26. Valero M, de la Prida LM (2018) The hippocampus in depth: a sublayer-specific perspective of entorhinal–hippocampal function. Curr Opin Neurobiol 52:107–114
27. Sanders H, Ji D, Sasaki T, Leutgeb JK, Wilson MA, Lisman JE (2019) Temporal coding and rate remapping: representation of nonspatial information in the hippocampus. Hippocampus 29(2):111–127
28. Herzog LE, Pascual LM, Scott SJ, Mathieson ER, Katz DB, Jadhav SP (2019) Interaction of taste and place coding in the hippocampus. J Neurosci 39(16):3057–3069
29. Szita I, Lorincz A (2008) Factored value iteration converges. Acta Cybern 18(4):615–635
30. Szita I, Lőrincz A (2009) Optimistic initialization and greediness lead to polynomial time learning in factored MDPs. In: Proceedings of the 26th annual int. conf on machine learning. ACM, pp 1001–1008
31. Miller R (1989) Cortico-hippocampal interplay: self-organizing phase-locked loops for indexing memory. Psychobiology 17(2):115–128

Developing Ontology of Properties and Its Methods of Use

Nikolay V. Maksimov[1(✉)], Anastasia S. Gavrilkina[2], Viktoryi A. Kuzmina[1], and Valeriy I. Shirokov[1]

[1] National Research Nuclear University MEPhI, Kashirskoe shosse, 31, Moscow, Russia
nvmaks@yandex.ru, vikulya_kuzmina@inbox.ru, v.shirokov@ase-ec.ru
[2] JSC ASE, Dmitrovskoe shosse, 2/1, Moscow, Russia
asgavrilkina@yandex.ru

Abstract. Developing ontology of properties and units of measurement includes fundamental properties links and units of measurement, as well as links that show the formation of concepts. General accepted metrology definitions used as the basis of ontology construction. Semiotic ontology nature taken into account: ontology includes the signs system – appellations, designations, linguistic constructions that represent information in a communicative form, as well as concepts, measures, relations that allow modeling the reality. Examples of ontology applying in the consistent representation of objects parameters analysis are shown and their using in automatic properties and values extractions from texts.

Keywords: Ontology · Properties identification · Units of measurement · Text processing · Digital twins

1 Introduction

Properties, quantity and their measure are the essential parts of reasonable human activity and, first of all, a component of knowledge. Tasks like automatic extraction of factography properties from texts, identification and comparison of various parameters, stability ensuring problems of digital twins functioning, require well-formalized representations of relevant knowledge.

Ontology, which is an appropriate tool for solving such problems, is considered according to [1–4] as *a specification of conceptualization at the level of explicit knowledge, depending on the subject area or task for which it is intended, and focused on joint multiple use.*

Representing properties and units of measurement are considered in different projects.

For instance, quantities, units, dimensions and data types standardization is the basic part of the NASA information architecture concept.

Ontology "Quantities, Units, Dimensions and Types" (QUDT) [5] is a high-level formalization of metrology concepts. It aims to manage the creation and updating of

A. V. Samsonovich (Ed.): BICA 2019, AISC 948, pp. 333–339, 2020.
https://doi.org/10.1007/978-3-030-25719-4_42

units and quantities and describe their use in a standardized way. The QUDT ontology includes components like quantity, units of measurement, prefix, system of units, system of quantity, quantity kind, and dimension. The QUDT ontology provides the link between units of measurement of different systems of units (e.g. CGS) with the SI system (International System of Units).

Ontology of Units of Measure is intended for use in scientific and engineering practice [6]. The ontology includes the following components: quantity, measurement scale, dimension, prefix, unit of measurement, application area, system of units, quantity kinds. Application area determines using of different units of measurement of one quantity, and the kind of quantities shows a hierarchy of quantities. Components "unofficial labels" and "unofficial abbreviation" hold misuse of values or incorrect values records.

The Units Ontology is intended to describe qualitative and quantitative observations in biology. The ontology includes components like quantities, units of measurement, prefixes [7].

Units Markup Language (UnitsML) is a specification language for coding scientific units of measurement. UnitsML consists of different sets of elements describing various aspects: units of measurement, quantities, dimensions, items and prefixes [8]. A database contains information on units of measurement—both included and not included in the SI system, and is maintained.

The research shows that ontologies of properties are developed for specific application areas for solving different problems. This paper presents developing of ontology of properties and units of measurement, based on the analysis of entities and relationships.

Examples of ontology applying in the consistent representation of objects parameters analysis are shown and their using in automatic properties and values extractions from texts.

2 Developing of Ontology of Properties

Properties have different nature and different representations in different application areas. Measurement methods depend on the environment in which the quantities are measured. Considering this, ontology supposed to be presented as fundamental links of quantity, properties, units, etc., and hold links that show the relationships of these concepts. In addition, it is ought to take into account the semiotic nature of ontology: besides the objects and reality relationships, as well as concepts and relationships that modeling this reality, ontology has a linguistic component – a sign system that allows consistent and sufficiently flexible information presentation in a form suitable for recording and transmission. The presence of a sign component provides the text selection by the automatic analysis of quantities and units of measurement presented differently in texts. This approach makes possible to build a systematic representation of the semantics of measurement space and eventually provide greater completeness and accuracy of information identification.

The definitions adopted in metrology will be used as a logical-semantic basis for the construction of the ontology of properties and units of measurement. The conceptual scheme of basic abstractions of the ontology of properties is based on the category named *Cognizable Realizable* [9] that allows to consider the information image (in particular, the value of the measurement quantity) as a mapping of a particular object (some quantities of reality) to a certain model (subject of knowledge).

Figure 1 shows the developed semantic graph of ontology of properties and units of measurement[1], built on this basis.

Under the object we understand not just any part of objective or subjective reality but only one to which the attention of the subject is directed, which is involved in the activities of the subject and becomes the subject of its theoretical or practical activities —the application area. The application area corresponds to a property and a quantity. The subject or "subject area" is the set of object's properties being selected by cognitive or practical means.

Property is a philosophical category that expresses such a kind of a subject (object, phenomenon, process) that determines its difference or commonality to the other objects and in its relations to them [10]. The property is manifested by the objects' interaction, which, in turn, are determined by the subject area. We can say the property is the projection of the "object" to the "subject area". Property has name, definition and symbol.

«Quantity is a property of a phenomenon, body, or substance, where the property has a magnitude that can be expressed as a number and a reference» [11]. The quantity is manifested by the interaction (e.g. with the measured instrument) as a measure that is determined by the property and corresponding subject area. I.e. the quantity is defined as the object's property that can be estimate (perceived, measured, calculated, etc.).

«The system of quantities is a set of quantities together with a set of non-contradictory equations linking these quantities» [11]. The system of quantities includes basic and derived quantities. The latter are determined via the main.

«Material measure is a measuring instrument reproducing or supplying, in a permanent manner during its use, quantities of one or more given kinds[2], each with an assigned quantity value» [11]. A measure has its unit of measurement. Generally, the measure shows the relationship of qualitative and quantitative characteristics: «the measure shows a borderline where change in quantity causes change in quality» [12]. I.e. the quantity is reproduced by the measure, and the property is determined by the measure.

«Quantity dimension is the quantity dependence on basic quantities of the system of quantities as a product of powers of factors corresponding to the base quantities, omitting any numerical factor» [11]. The basic units determine the quantity dimension.

«Unit of measurement is a real scalar quantity, defined and adopted by convention. Any other quantity of the same kind can be compared to express the ratio of the two quantities as a number» [11]. The unit of measurement is adopted for the quantitative expression of homogeneous values with it. In combination with units, multipliers

---

[1] The figure does not show the linguistic component – a sign system of ontology, which represents for each unit of measurement a possible form of its recording in the form of an enumeration of linguistic constructions or regular expressions.

[2] «Kind of quantity is an aspect common to mutually comparable quantities» [11], showing which class (on the basis of property) the quantity belongs.

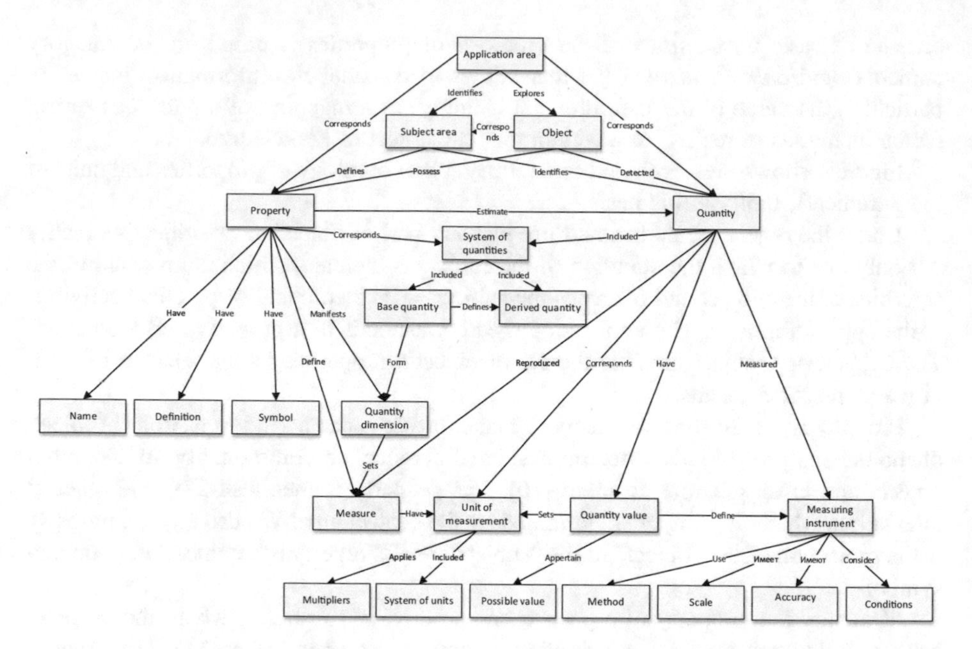

Fig. 1. Ontology of properties and units of measurement

(name prefixes) can be used to reduce the expression of numeric values (for example, 1 μm instead of 0.000001 m).

«System of units is a set of base units and derived units with their multiples and submultiples, defined in accordance to given rules, for a given system of quantities» [11].

«Quantity value is a value number and reference together expressing magnitude of a quantity» [11]. Quantity value is defined by the measurement instrument. The values may be limited to the range of possible values determined by the nature of the quantity or by the nature of the application area.

According to [11] «Measuring instrument is a device used for making measurements, alone or in conjunction with one or more supplementary devices», the measurement interval can be specified by technical conditions. Measurement instruments have scales and measurement accuracy, and use a specific measurement method. «Scale is a part measurement instrument consisting of an ordered set of marks with associated quantity values». «Accuracy of measurement is a closeness of agreement between measured quantity value and true quantity of a value being measured». «Method of measurement is a generic description of logical organization used in measurement». When measuring the value is also taken into account in what conditions the measurement takes place.

To implement the ontology of properties, a database containing the corresponding attributes has been developed. The database includes 1014 records for properties related to the subject areas: space and time, mechanics, thermodynamics, acoustics, electricity, magnetism, electromagnetism, ionizing radiation, electrical engineering, economics, information technology, cooking, astronomy, pharmaceuticals, printing, forestry, jewelry industry, oil industry, textile industry, construction, music, marine navigation, colorimetric method, transport.

3 Applying Ontology in Extracting Properties from Text

Extraction of quantities, their values and units of measurement from the text as separate elements leads to the emergence of disambiguation. Representation of some units of measurement can be the same for different quantities, and vice versa, the quantity can be assigned to several units of measurement. For an unambiguous identification of properties by taking into account the context of their use, an ontology of properties is applied. Comparison of semantically significant elements of the text, extracted within the boundaries of one or several sentences and forming a semantic neighborhood of property with the corresponding components of the ontology of properties allows restoring missing semantic fragments or identifying contradictions, in particular, to find errors in text or discrepancies in notation. I.e. if the measuring instrument is known, the name of quantity and quantity dimension can be determined via ontology. In case the name of quantity, dimension and system of quantities is known the unit of measurement can be synthesized.

Let us consider an example of processing[3] the following fragment of an article [13] on atomic energy: «Channel-type uranium-graphite reactors make it possible to produce 1,000 MW or more from one unit. There is an important issue from the point of view of electricity supply, namely, that NPPs produce power evenly, and the metallurgical plant has power surges reaching 60% or more».

Tokens "1", "000", "MW" were extracted from the first sentence, and then the token "MW" was identified as a unit of measurement using the ontology of properties. The sequence of tokens consisting of the numbers preceding the unit of measurement was combined and identified as the quantity value. At the same time, the unit of measurement "MW" in ontology is associated with properties related to different subject areas: "power" in mechanics, "sound energy flow" in acoustics, and "radiation flow" in electromagnetism "active power" in electricity, etc. Since the next sentence in text refers to power and NPPs (application area), it is possible to identify it as "active power".

[3] Text processing to extract meaning and build a formalized conceptual structure for its presentation includes following steps: (1) Tokenization – separation of characters sequences according to the assigned delimiters; (2) Preliminary identification of tokens based on the analysis of graphematic signs, regular expressions, the use of dictionaries of names and units, in particular, abbreviations, proper nouns, units of measurement, numerical values, etc.; (3) Identifying semantic units (concepts, properties, nominal groups, etc.) by combining tokens based on an analysis of their morphological characteristics, order, group structures, dictionaries, thesauri, etc.

4 Applying Ontology in the Analysis of Parameters

Consider an example of controlling pumping equipment at a nuclear power plant. The pumps operating in the system belong to different types, differ in parameters and properties and work in different systems (water supply, cooling of steam generators, heating), but they perform a common function – changing the energy of the working environment flow.

Consider the specifications of two types of pumps provided by different suppliers [14, 15]: the HF40-25-160(D) pump designed for the heat supply system and the XKJ-1100I pump for water supply.

Their technical data analysis showed that they differ in the main functional property: for the HF40-25-160(D) it is the "Maximum lift height", for the XKJ-1100I it is the "Pressure", and the application area is different – the heat supply system and water supply system, respectively.

Based on the definitions and units of measurement presented in the ontology, the maximum lift height, and the pressure measured in meters of the working (pumped) fluid column, we can say these values are identical. Since in both cases there is working fluid and they have a common subject area (hydraulics), it can be concluded that the fields of application are also close.

Thus, using ontology components is possible to identify differences in properties that are necessary for a more consistent operation of equipment.

5 Conclusion

The fundamental provision adopted in the development of ontology is its semiotic nature. Besides the real objects, quantities, measuring instruments, as well as concepts, measures, relations, allowing modeling this reality, ontology includes a sign system – appellations, designation, and linguistic constructions, representing information in a communicative form. The ontology of properties and units of measurement uses both fundamental connections of properties, quantities and measures, and connections which show formation of concepts. As a conceptual and terminological basis for the ontology construction, the definitions adopted in metrology are used.

This ontology can be used to extract factual data from texts during the design and operation of large projects to create and use digital twins of objects.

Property identification technology is based on comparison of semantically significant elements of the text, extracted within the boundaries of one or several sentences and forming a semantic neighborhood of property, with the corresponding components of the ontology of properties thus allowing restore missing parts of meaning or identify differences in the notation.

The database of ontology includes 1014 records for properties related to the subject areas: space and time, mechanics, thermodynamics, acoustics, electricity, magnetism etc. was developed.

In general, the use of ontology allows to ensure the completeness and accuracy of identifying factual information.

References

1. Gruber Th (2009) What is an ontology. http://www-ksl.stanford.edu/kst/what-is-an-ontology.html
2. Guarino N (1994) Understanding, building, and using ontologies. http://ksi.cpsc.ucalgary.ca/KAW/KAW96/guarino/guarino.html
3. Van Heijst G, Wielinga BJ (1996) Using: explicit ontologies in KBS development. http://ksi.cpsc.ucalgary.ca/KAW/KAW96/borst/node16.html
4. Golitsyna OL et al (2012) The ontological approach to the identification of information in tasks of document retrieval. Autom Doc Math Linguist 46(3):125–132
5. Hodgson R, Keller P (2014) QUDT - quantities, units, dimensions and data types in OWL and XML. http://www.qudt.org
6. Rijgersberg H, van Assem M, Top J (2013) Ontology of units of measure and related concepts. Semant Web 4(1):3–13
7. Gkoutos GV, Schofield PN, Hoehndorf R (2012) The units ontology: a tool for integrating units of measurement in science. In: Database
8. Celebi I, Dragoset RA, Olsen KJ, Schaefer R, Kramer GW (2010) Improving interoperability by incorporating UnitsML into markup languages. J Res Natl Inst Stand Technol 115(1):15–22
9. Maksimov NV (2018) The methodological basis of ontological documentary information modeling. Autom Doc Math Linguist 52(3):57–72
10. Ilichev LF, Fedoseev PN, Kovalev SM, Panov VG (1983) Philosophical encyclopedic dictionary. Soviet encyclopedia, 597. (in Russian)
11. JCGM 200 (2008) International vocabulary of metrology. Basic and general concepts and associated terms
12. Hegel GVF (1970) Science of logic 3(1):428. (in Russian)
13. Kudrin BI, Kosharnaya YuV (2017) History of decisions on creation of a nuclear metallurgical complex on the basis of the Kola nuclear power plant. Vestnik MGTU 4. (in Russian)
14. Hepbasli A, Kalinci Y (2009) A review of heat pump water heating systems. Renew Sustain Energy Rev 13(6–7):1211–1229
15. Drive W (2006) Product catalogue

Development of Graph-Theoretical Model and Operations on Graph Representations of Ontologies as Applied to Information Retrieval Tasks

Anastasia Maksimova, Valentin Klimov, and Nikita Antonov(✉)

National Research Nuclear University MEPhI (Moscow Engineering Physics Institute), Moscow, Russia
{mks-anastasiya, nikit-antono2011}@yandex.ru

Abstract. This paper presents a graph-theoretical model of ontologies of subject domains. An ontology represented as a weighted graph. At the vertices of the graph are concepts. The edges of the graph marked the relationship between concepts. In addition, the basic operations on the ontology graph representations introduced for ontology editing in the semantic search system.

Keywords: Ontology · Graph · Information retrieval · Semantic search

1 Introduction

Sense formalization with respect to the properties of a single subject area is a complex and ambiguous task. Ambiguity is a consequence of the fact that several ways and forms existing of describing knowledge that have their advantages and disadvantages.

Today exists and developing lot of new ontologies reflect certain aspects or levels of abstraction of each subject area (SA). At the same time, the most similar conceptual systems, in terms of terminology, can describe different subject areas [1].

In theory, it is customary to classify ontologies according to the degree of dependence on the tasks or the applied field, by language of ontological knowledge and its expressive capabilities, and other parameters [2].

Thus, one of the main tasks is the ability to manipulate all this variety of conceptual systems on the example of graph representations, namely:

- find commonalities in the ontology (the intersection);
- build a common conceptual system (Association);
- from general to particular, find knowledge reflecting a certain aspect of the subject area (aspect projection);
- build the required abstraction level (scaling).

Either these operations allow creating the maximum universal ontology for the subject area, or the most highly specialized, specific conceptual system, depending on the task set by the user [3]. However, these operations do not solve all the problems associated with the representation of knowledge [4].

A. V. Samsonovich (Ed.): BICA 2019, AISC 948, pp. 340–345, 2020.
https://doi.org/10.1007/978-3-030-25719-4_43

2 Proposed Approach

2.1 Problem Statement

The initial data are:

1. Ontology

$$O = \langle Sf, Sc, St, \equiv \rangle, \quad (1)$$

where:

- Sf - functional system;
- Sc - conceptual system;
- St - terminological system.

2. Conceptual system

$$Sc = \langle Mc, Ac, Rc, Zc \rangle, \quad (2)$$

where:

- Mc - a set of symbolic descriptions of SA concepts;
- Ac - set of characteristic attributes of symbolic descriptions;
- Rc – generic and associative relations;
- Zc – composition law, according to which particular system chosen basis {Mc, Ac, Rc}.

3. A graph

$$G = \langle V, E \rangle, \quad (3)$$

where:

- $V = \{v1, v2, \ldots vn\}$ is the set of vertices
- $E = \{(u, v) | u \in V, v \in V\}$ is the set of edges of the graph.

4. The structure of the ontology is given as a directed weighted graph

$$G = \langle V, E \rangle, \quad (4)$$

where:

The vertices $V = \{vi\}$ are concepts, and the arcs $X = \{xn\}$, where $xn = (vi, vj, wn)$ such that vi is the vertex of the beginning of the arc, vj is the vertex of the end of the arc, and wn is the weight of the arc, in our case, the name of the relation between concepts [5].

2.2 Definition of Graph-Theoretic Model

Ontology of the subject area is represented as a graph

$$G = \langle V, E, T, R \rangle, \tag{5}$$

where:

- V – the set of meta-vertices, such that;
- $\forall \mathrm{k} \in \overline{0, n}, \forall v_i \in V^k v_i = \left\langle N, k, V_i^{k+1} \right\rangle : V^0 = V, v_i \in V^k, v_i^k = \bigcup_{l=1}^{m} v_{il}^{k+1}, V_i^{k+1} \subset V^k / v_i$;
- E – the set of edges, such that;
- $\exists e_i \in E e_i = \langle v_k, v_l, t_i \rangle : v_k, v_l \in V, t_i \in T$ (T – set types of relationships);
- R – rules of marking of the graph G (subgraph selections);
- $\forall R_i \in R R_i : f(E) = E_i \subset E$.

2.3 Mathematical Models of Operations on Graph Representations of Ontologies in Relation to Information Retrieval Problems

In this paper, we consider four operations on ontologies.

- Binary:
  - Intersection operation.
  - Union operation.
- Unary:
  - Aspect projection operation
  - Scaling operation.

To implement the operations it is proposed to use the thesaurus [6] as an auxiliary element that complements the ontology to one global domain [7]. It is assumed that the thesaurus, similar to ontologies, is presented in the form of a graph, which allows us to bring operations on ontologies to a set of operations on several graphs.

To store and represent the thesaurus routes (links) between the vertices of the graphs, a matrix is used, in the first row of which the vertices of the first graph are written, and in the first column – the second. If there is a route Lij between concepts i and j in the thesaurus, the route is written at the intersection of the corresponding row and column in the matrix, and if there is no route, zero is written.

Ontology Intersection Algorithm. Let two ontologies A, B and thesaurus T given.

$$\text{If} Va \cap Vb \neq \emptyset, i.e. Va \cap Vb = \{a1, a2 \ldots ai\}, \tag{6}$$

then:

- perform the intersection operation according to the standard algorithm of graph theory. As a result, we get a new graph:

$$G = Va \cap Vb \tag{7}$$

$$\text{If } A \cap B = \varnothing \text{ and, or } A \cap T = \varnothing, \text{ or } B \cap T = \varnothing, \tag{8}$$

that:

- end. A and B cannot be crossed through thesaurus
- Let $A \cap T = Vat = \{a1, a2...ai\}, B \cap T = Vbt = \{b1, b2...bj\}$.
- Build all kinds of routes *Lij* из *Vat* в *Vbt* (and, versa, from *Vbt* to *Vat*).

Leave only those routes that meet the following conditions:

- *Lij* does not contain more than one associative link (since the associative link is not transitive, the transition to associative links reduces the syntactic connectedness of concepts) [9].
- $S(i,j) > 0$ (or greater than the specified value).
- Of the two routes L_{ij} and L^*_{ij} selects the one for which the measure $S(i,j)$ takes the maximum value or one that does not contain associative arcs.
- Of the two routes L_{im} и L_{in} selects the one whose measure is greater, i.e. $S(i,m) > S(i,n) \mapsto Lim$, or $S(i,m) < S(i,n) \mapsto Lin$.
- Graph $L = \{L11, L12...Lmn\}$, combining the routes of point 5, combined with the intersection of ontologies A and B from point 1, i.e. $L \cup G$ - this is the intersection graph of ontologies A and B [10].

Algorithm of Ontology Combining Operation. Let two ontologies A, B and thesaurus T given.

$$\text{If } V_a \cap V_b \neq \emptyset, \ i.e. \ V_a \cap V_b = \{a_1, a_2...ai\}, \tag{9}$$

then:

- perform the join operation according to the standard algorithm of graph theory, described in Sect. 1.
- As a result, we get a new graph $G = A \cup B$.

$$\text{If } Va \cap Vb = \varnothing, \tag{10}$$

then:

- graph G = A ∪ B is a disconnected graph of two connected components.
- We carry out steps 2–5 of the graph intersection algorithm described above. As a result, we get routes:

$$\mathrm{L} = \{\mathrm{L}_{11}, \mathrm{L}_{12}...\mathrm{L}_{mn}\} \tag{11}$$

- To the graph G add routes from paragraph 2. The resulting graph L ∪ G - this is a graph combining ontologies A and B.

Aspect Projection. Let the ontology B, aspect A and the thesaurus T are given. Aspect ontology A can be of three types:

1. consist only of concepts (terms describing concepts), that is $A = <V_a>$;
2. consist only of relations (connections) between concepts, that is, $A = <R_a>$;
3. include both concepts and relationships between them, that is, $A = <V_a, R_a>$.

The following restrictions apply to an aspect ontology:

1. the concepts of an aspect ontology must exist in the ontology under study;
2. the relationship between concepts should exist ontology or/ and studied in the thesaurus.

Accordingly, there are three algorithms for constructing aspect ontologies, depending on their type [11].

Scaling Operation. Let ontology A and thesaurus T are given.

Detailing:

- $V_a \cap V_t = V_{at} = \{a_1, a_2 ... ai\}$;
- Compose a square matr looking for routes *Lij* in thesaurus T between concepts *ai* and a_j such that $i \neq j$ and in ontology A there is a path L^*_{ij} such that $|L^*_{ij}| = 1$;
- choose routes $|Lij| > 1$, satisfying the conditions of the intersection/union algorithm except for the fourth;
- combine original ontology with found routes:

$$A \cup \{L_{11}, L_{12} ... L_{mm}\} \tag{12}$$

Enlargement. Let ontology A and thesaurus T are given.

- $Va \cap Vt = Vat = \{a_1, a_2 ... a_i\}$;
- looking for routes L_{ij} in thesaurus T between concepts *ai* and a_j such that $i \neq j$ and in ontology A there is a path L^*_{ij} such that $|L^*_{ij}|\rangle 1$.
- let $k = 2$, choose routes $|L_{ij}| < \left|L^*_{ij}\right| = k$, satisfying the conditions of the intersection/union algorithm, with the exception of the fourth;
- if the routes are not found, then proceed to the next step;
- if $\forall L^*_{ij} < k$, then end. The resulting graph is an enlargement of the original ontology A, otherwise next step;
- if $\exists L^*_{ij} > k$, then $k = k + 1$ and fulfilled step 3;
- replace the pathways in the original ontology L^*_{ij} to *Lij*;
- remove the rows and columns corresponding to the removed vertices from the matrix. Perform step 3.

3 Conclusion

In this research work, considered the actual topic of the description and presentation of software ontologies. A graph-theoretical model for the representation of an SA developed, axiomatic operations on the ontology graph representations described in relation to information retrieval tasks [12]. In addition, the paper analyzed the methods of presenting an SA, provided an overview of the mathematical descriptions of an SA in the form of ontologies, and presented the basic concepts of graph theory and operations on graphs.

Acknowledgements. This work was supported by the Council on grants of the President of the Russian Federation, Grant of the President of the Russian Federation for the state support of young Russian scientists - candidates of sciences MK-6888.2018.9. The funding for this research provided by the RSF Grant № 18-11-00336.

References

1. Chernyshov A, Balandina A, Klimov V (2018) Intelligent processing of natural language search queries using semantic mapping for user intention extracting. In: Biologically inspired cognitive architectures meeting. Springer, Cham
2. Balandina AI, Chernyshov AA et al (2018) Dependency parsing of natural Russian language with usage of semantic mapping approach. Procedia Comput Sci 145:77–83
3. Golitsyna OL, Maksimov NV (2011) Information retrieval models in the context of retrieval tasks. Autom Doc Math Linguist 45(1):20–32
4. Balandina, A, et al (2016) Usage of language particularities for semantic map construction: affixes in Russian language. In: International symposium on neural networks. Springer, Cham
5. Golitsyna OL, Maksimov NV (2011) Information retrieval models in the context of retrieval tasks. Autom Doc Math Linguist 45(1):20–32
6. Glava M, Malakhov V (2018) Information Systems Reengineering Approach Based on the Model of Information Systems Domains. Int J Softw Eng Comput Syst (IJSECS) 4(1):95–105
7. Levashova TV (2002) Principles of ontology management used in the knowledge integration environment. Tr SPIIRAN 1(2):51–68
8. Chernyshov A et al (2017) Intelligent search system for huge non-structured data storages with domain-based natural language interface. In: First international early research career enhancement school on biologically inspired cognitive architectures. Springer, Cham
9. Meenachi NM, Sai Baba M (2017) Matrix rank-based ontology matching: an extension of string equality matching. Int J Nucl Knowl Manag 7(1):1–11
10. Meenachi NM, Sai Baba M (2014) Development of semantic web-based knowledge management for nuclear reactor (KMNuR) porta. DESIDOC J Libr Inf Technol 34(5)
11. Meenachi NM, Sai Baba M (2012) A survey on usage of ontology in different domains. Int J Appl Inf Sys 9:46–55
12. Klimov V et al (2016) A new approach for semantic cognitive maps creation and evaluation based on affix relations. In: Biologically inspired cognitive architectures (BICA) for young scientists. Springer, Cham, pp 99–105

An Automatic System for Learning and Dialogue Based on Assertions

Umberto Maniscalco(✉), Antonio Messina, and Pietro Storniolo

Istituto di Calcolo e Reti ad Alte Prestazioni - CNR,
Via Ugo La Malfa, 153, 90146 Palermo, Italy
umberto.maniscalco,antonio.messina,pietro.storniolo@cnr.it
https://www.icar.cnr.it

Abstract. When someone says a statement about a particular subject, we memorize the assertion and, implicitly, we can construct all the possible questions that have as a right answer to the assertion just heard. This means that, in this specific case, our learning process based on assertions subsists. When we read a book, we do nothing but learn through a succession of assertions. In this article, we present a system for automatically constructing a conversational agent, which uses only assertions to build the dialog engine. The whole architecture is based on the "Robot Operating System" (ROS), and the experiments were conducted using a humanoid robot.

Keywords: Learning · Robotics · Knowledge base · Conversational agent

1 Introduction

A conversational agent is a software, based on artificial intelligence algorithms, able to dialogue with human beings mainly using a natural language. The first attempts to build an automatic dialogue system date back to the early seventies, and it were part of the famous DARPA project. However, even earlier, in the sixties, Joseph Weizenbaum published ELIZA, a software, as the title of the article says, "for the study of natural language communication between man and machine"[1].

In recent years, even the cognitive multinationals have invested billions in the field of conversational agents. This is the case of IBM with "Watson"[1], Google with "Google Home"[2] or Amazon with "Alexa"[3].

Most studies and projects concerning conversational systems have the ultimate goal of passing the Touring test, that is, to demonstrate that the artificial intelligence that animates the system is indistinguishable from the human one.

[1] https://www.ibm.com/watson.
[2] https://en.wikipedia.org/wiki/Google_Home.
[3] https://en.wikipedia.org/wiki/Amazon_Alexa.

A. V. Samsonovich (Ed.): BICA 2019, AISC 948, pp. 346–351, 2020.
https://doi.org/10.1007/978-3-030-25719-4_44

This goal is perfectly acceptable, but it may not even be the goal of all conversational systems.

Such an ambitious goal involves a great effort to build a knowledge base, even on a limited domain, to analyze all the answer question pairs, even in prototype form, to process the knowledge base to train the system, often of a neural type, which is the basis of the recognition of the intent of the question, etc.

Therefore, in some cases, a less ambitious goal of passing the Touring test could be placed first, favoring the speed or simplicity of creating and using knowledge. In the article presented here, the goal is not to create a perfect dialogue system. Instead, the goal is to design a system able to generate a conversational agent in a fast and unsupervised way.

The idea is to try to replace that human learning process based exclusively on assertions. When we hear or read a statement like "*Leonardo da Vinci was born in Anchiano in 1452*", we simultaneously memorize two notions: the year and the birthplace of Leonardo da Vinci. Therefore, listening to the statement mentioned above, a human will be able to answer all the questions that concern both the year of birth and the birthplace of Leonardo da Vinci.

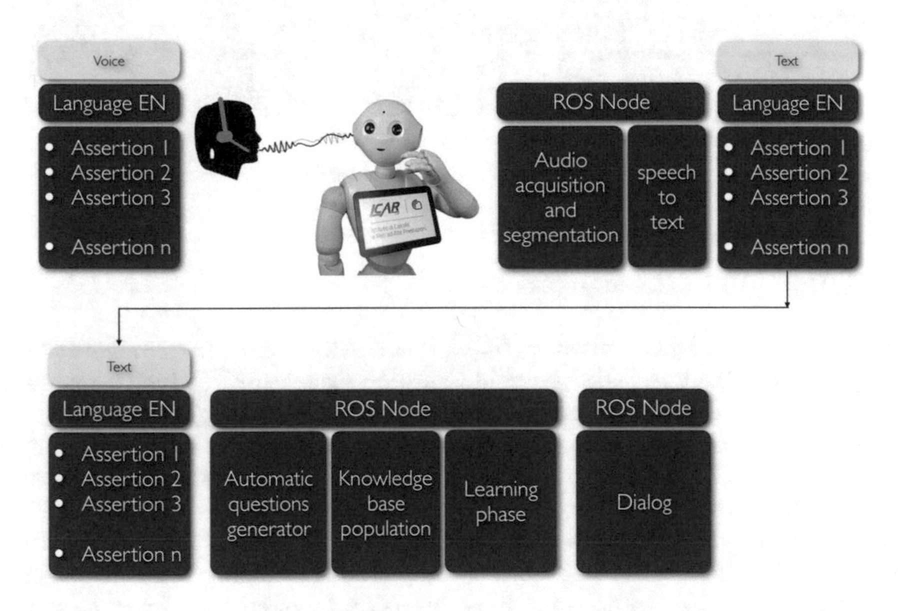

Fig. 1. The scenario in which a robot learns to answer simple questions about a specific domain by listening only to specific statements.

More in detail, in the scenario described in Fig. 1, a humanoid robot listens to some statements from a human regarding a specific domain of knowledge. These statements are automatically acquired and the audio transformed into text using a "speech to text" service (i.e. Google Cloud Speech-to-Text)[4]. Then, these

[4] https://cloud.google.com/speech-to-text/docs/.

statements are processed to get a set of questions compatible with them using an automatic question generator. All possible pairs of questions and answers are stored in a knowledge base. Now, a contextual conversational agent can be trained using, for example, an open source framework similar to RASA[5]. As Fig. 1 shows, the entire architecture is implemented using "Robot Operating System" (ROS)[6] nodes to maintain the independence from the specific robot.

We want to emphasize that, in this context, we do not want to achieve a perfect conversational agent. We want to experience the possibility of creating and transferring knowledge to a humanoid robot simply by instructing it with statements. Therefore, the proposed architecture is evaluated by measuring how much the user that interacts with the robot perceives to have an acquaintance, even limited, of the topic object of the dialogue.

Next sections will explain the different parts of the architecture and how they work in more detail (Fig. 2).

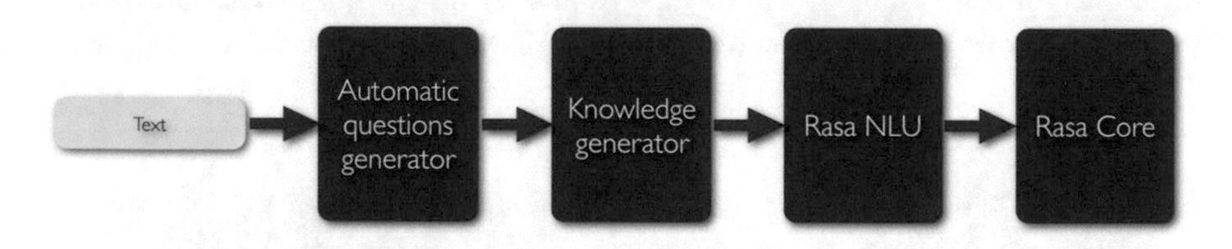

Fig. 2. Steps of learning process

2 Questions Generation

The automatic question generator (AQG) is currently derived from a slightly modified version of the tool "NLP Factoid Question Generator"[7], which implements the idea of an overgenerate-and-rank framework for question generation [2].

Starting from a given paragraph or article, the tool is able to generate simple basic what, when, who, where kind-of questions. To mitigate the problem of generation of non-sense or poor quality questions, they are overgenerated and ranked making use of some information metrics. In other words, the tool can extract a simplified statement from a compound statement. Then, it performs some procedures like subject-auxiliary inversion followed by answer to phrase mapping and finally WH-movement[8] to arrive at a simple question from that simplified statement.

[5] https://rasa.com/docs/.

[6] http://www.ros.org/about-ros/.

[7] https://github.com/tarungavara/nlp-factoid-questions-generator.

[8] In linguistics, *WH-movement* is about rules of syntax involving the placement of interrogative words, that is an asymmetry between the syntactical arrangement of words (*morphemes*) in a question and the form of answers to that question.

For example, using as input the statement "Leonardo da Vinci was born in Anchiano in 1452", the AQG will generate the following questions:

- *Who was born in Anchiano in 1452?*
- *Where was Leonardo da Vinci born in 1452?*
- *When was Leonardo da Vinci born in Anchiano?*

The AQG is implemented around some well-known human language technology tools, such as Stanford CoreNLP [3], NLTK [4], WordNet [5] and Spacy [6].

3 Knowledge Base Population and Learning

The starting statement and the automatically generated questions are the essential building blocks to populate the knowledge base. After the learning phase, the system will be able to answer the questions asked. The following subsections describe the conversational system we chose, how the knowledge base is represented and populated, and the learning phase.

Understanding natural language (NLU) or interpreting natural language (NLI) is a subset of natural language processing in artificial intelligence. Understanding natural language is considered a difficult problem for artificial intelligence. There is considerable interest in the sector due to its possible application in various contexts such as automated reasoning, automatic translation, answering questions, gathering news, categorizing texts, voice activation, archiving and large-scale content analysis.

A particularly popular area is that of AI assistants. These assistants, based on conversational systems, have two primary tasks to perform: understanding the user and giving correct answers or adequate actions.

Typically, three steps are needed to implement a conversational system:

- Recognition of the intent: how to understand what users want
- Extraction of entities - how to determine the entities that characterize the intent in a specific domain
- Determination of hyperparameters: how to select and optimize the parameters that contextualize the intent.

For this purpose we used the AI RASA framework. The Rasa Stack deals with these activities with the *Rasa NLU* natural language comprehension component and the dialogue management component with *Rasa Core*.

3.1 Building the Knowledge Base

The RASA Stack works on the basis of the content of three files:

- *nlu_data.md*, used by RASA NLU, contains the definition of intents with the relative statements that characterise them;
- *domain.yml*, contains the list of the recognized intents and all the utterances, that is the set of the possible answers;
- *stories.md*, defines the various possible dialogue paths.

Therefore, building the knowledge base for our system consists in write contents into those files following a proper syntax. This task is done by some Python scripts that keep the textual information generated by the AQG, generate unique identifiers for intents and utterances and, also thanks to the Chatette tool[9], build the training datasets.

3.2 The Learning Phase

The training dataset is used by RASA to train two neural networks: the first for intents recognition, the second for handle the dialog phase. Supervised classification techniques handled by TensorFlow (Keras)[10] and machine learning-based dialogue management allow the system to predict the best action based on the input from NLU, the conversation history and the training data.

First, we train RASA NLU with a command like

```
python -m rasa_nlu.train -c nlu_config.yml --data nlu.md -o models \
      --fixed_model_name nlu --project current --verbose
```

Then, we train Rasa Core with a command like

```
python -m rasa_core.train -d domain.yml -s stories.md -o models/dialogue
```

Of course, those trainings are executed programmatically.

4 Dialog

The integration of the conversational system with the humanoid robot is carried out using the ROS infrastructure. A specific ROS node has been set up, which deals with interfacing with RASA through querying the JSON service on port 5005 (http://rasa.server:5005/webhooks/rest/webhook).

An example of the dialogue presented in Fig. 3, regarding the assertion "Leonardo da Vinci was born in Anchiano in 1452", is shown below.

```
- Rasa Core server is up and running on http://rasa.server:5005
Bot loaded. Type a message and press enter (use '/stop' to exit):
Your input ->  Who was born in Anchiano in 1452?
Leonardo da Vinci was born in Anchiano in 1452
Your input ->  Where was Leonardo da Vinci born in 1452?
Leonardo da Vinci was born in Anchiano in 1452
Your input ->  When was Leonardo da Vinci born in Anchiano?
Leonardo da Vinci was born in Anchiano in 1452
Your input ->  Who was born in 1452?
Leonardo da Vinci was born in Anchiano in 1452
Your input ->  When was Leonardo born?
Leonardo da Vinci was born in Anchiano in 1452
Your input ->  /stop
```

[9] https://github.com/SimGus/Chatette.

[10] https://www.tensorflow.org/guide/keras.

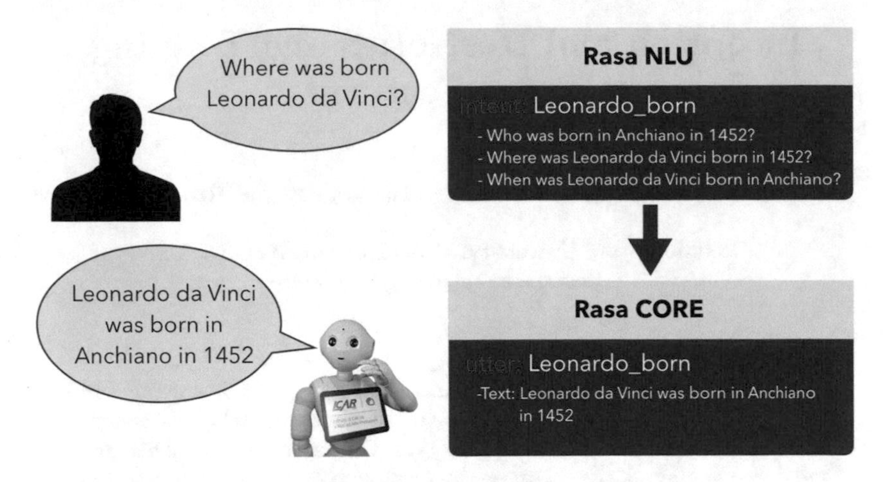

Fig. 3. A dialogue example

5 Conclusion and Future Works

In this work we wanted to present an architecture to create an automatic system for learning and dialogue based on assertions. The architecture is entirely based on open-source components and implemented in ROS. The first results obtained, although on limited knowledge domains seem to be good. The next works will concern the improvement of performance and the introduction of a module for the management of other languages besides English.

Acknowledgements. This research was partially supported by the project AMICO - Assistenza Medicale In COntextual Awareness, with funding from the National Programs of the Italian Ministry of Education, Universities and Research (code: ARS01_00900).

References

1. Weizenbaum J (1966) ELIZA - a computer program for the study of natural language communication between man and machine. Commun ACM 9(1):36–45
2. Heilman M (2011) Automatic factual question generation from text. Carnegie Mellon University, Pittsburgh. ISBN =978-1-267-58224-9
3. Manning CD, Surdeanu M, Bauer J, Finkel J, Bethard SJ, McClosky D (2014) The stanford CoreNLP natural language processing toolkit. In: Proceedings of the 52nd annual meeting of the association for computational linguistics: system demonstrations, pp 55-60
4. Loper E, Bird S (2002) NLTK: the natural language toolkit. In: Proceedings of the ACL-02 workshop on effective tools and methodologies for teaching natural language processing and computational linguistics, vol 1. Association for Computational Linguistics, Stroudsburg, PA, USA, pp 63–70
5. Fellbaum C (1998) WordNet: an electronic lexical database. MIT Press, Cambridge. ISBN =9780262061971
6. Honnibal M, Montani I (2017) spaCy 2: Natural language understanding with Bloom embeddings, convolutional neural networks and incremental parsing. https://spacy.io

Inception and ResNet: Same Training, Same Features

David G. McNeely-White(✉), J. Ross Beveridge, and Bruce A. Draper

Colorado State University, Fort Collins, CO 80523, USA
david.white.679@gmail.com

Abstract. Deep convolutional neural networks (CNNs) are the dominant technology in computer vision today. Unfortunately, it's not clear how different from each other the best CNNs really are. This paper measures the similarity between two well-known CNNs, Inception and ResNet, in terms of the features they extract from images. We find that Inception's features can be well approximated as an affine transformation of ResNet's features and vice-versa.

The similarity between Inception and ResNet features is surprising. Convolutional neural networks learn complex non-linear features of images, and the architectural differences between the systems suggest that these functions should take different forms. Instead, they seem to have converged on similar solutions. This suggests that the selection of the training set may be more important than the selection of the convolutional architecture.

Keywords: ResNet · Inception · CNN · Feature mapping

1 Introduction

Deep convolutional neural networks (CNNs) are the dominant technology in computer vision today, and much of the recent computer vision literature can be thought of as a competition to find the best vision architecture within the deep convolutional framework. Despite all this effort, however, it's not clear how different from each other the best architectures really are. The best systems correctly classify images at approximately the same rate. Are they extracting the same information from images, or just equally discriminating information?

This paper compares two sophisticated CNNs, Inception and ResNet, in terms of the features they extract from images. Both systems perform similarly on the ILSVRC2012 image recognition challenge [8]. ResNet-v2 152 [4] labels 78.9% of ILSVRC2012 test images correctly, while Inception-v4 [10] labels 80.2% correctly. On the surface, however, Inception and Resnet appear quite different. Inception [11] divides processing by scale, merges the results, and repeats. ResNet [3] has a simpler, single-scale processing unit with data pass-through connections. Inception produces 1,536 features per image, while ResNet produces 2,048. They also have disjoint pedigrees: Inception was developed by Google

A. V. Samsonovich (Ed.): BICA 2019, AISC 948, pp. 352–357, 2020.
https://doi.org/10.1007/978-3-030-25719-4_45

[5,10–12]; ResNet was developed at Microsoft in 2016 [3,4]. Nonetheless, we find that the features extracted by Inception are very similar to the properties extracted by ResNet, in the sense that affine mappings of one predict the other. These mappings are accurate enough that mapped features can be used to label images without retraining the underlying classifier. This suggests that Inception and ResNet, despite their structural differences, extract essentially the same properties from images.

The finding that Inception and ResNet features are linked by affine transformations is surprising. Convolutional neural networks have revolutionized the field of computer vision precisely because they learn complex non-linear features of images, and the architectural differences between the systems suggest that these non-linear functions take different forms. Yet, Inception and ResNet seem to extract similar properties from images, since their features are (almost) linear transformations of each other. In essence, their training algorithms seem to hill-climb in totally different spaces and yet find similar local optima. This may explain why the two systems perform similarly, but we believe it also has broader implications. It suggests that the features extracted by Inception and ResNet are driven less by the details of convolutional architectures, and more by the content of the training images. If this is true, one would expect many sophisticated CNNs to perform at similar levels.

2 Related Work

Comparison of task performance (e.g. classification accuracy) on a common dataset is the most prevalent method for comparing deep learning systems [8]. Visualization techniques are used to peer into internal model representations, for example network inversion [2] and dissection [1]. Transfer learning can also be used to compare networks [6].

The closest work to this paper is Lenc et al. [7]. They also find affine mappings between CNNs (AlexNet, VGG-16 and ResNet-50). However, they learn mappings between convolutional layers and train their mappings using supervision in the form of image labels. We learn unsupervised mappings between final convolutional layers and fully-connected classifiers. As a result, the notion of feature equivalence established here is independent of manually-assigned image labels.

3 Methodology

Our goal is to compare convolutional neural networks, not in terms of their recognition rates, but in terms of how similar their features are. To do this, we describe CNNs as the composition of two functions. The first function, F(), extracts features from input using convolutional layers. In this paper the inputs are images, but in other domains they could be videos or audio signals or any

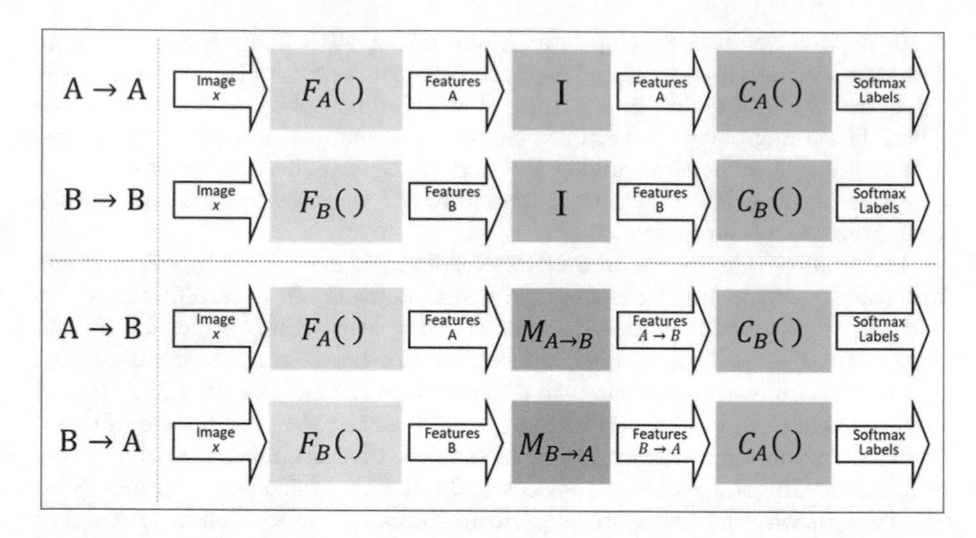

Fig. 1. Illustration of how a pair of standard CNNs can be used to create two alternative CNNs where the features from one are affine mapped to the classifier of another. The first two rows show CNNs A and B without alteration; the mapping from $F()$ to $C()$ is the identity mapping I. The next two illustrate the swapping of classifiers and the introduction of affine mappings $M_{A \to B}$ and $M_{B \to A}$.

other complex input. The second function, C(), is the classifier, usually implemented through fully-connected layers. Classifiers map feature vectors to labels. Thus a typical CNN is written as C(F()).

Throughout the evaluation process, we never retrain or otherwise modify any of the classification functions C() or feature extraction functions F(). We do, however, fit affine matrices $M_{A \to B}$ and $M_{B \to A}$ which create linear mappings between F_A and F_B. In other words

$$\tilde{F_A}() = M_{B \to A} F_B() \qquad \text{and} \qquad \tilde{F_B}() = M_{A \to B} F_A()$$

where $\tilde{F_A}$ is the best approximation to F_A and $\tilde{F_B}$ is the best approximation to F_B across the training data set.

Our goal is to determine if $F_A()$ and $F_B()$ extract the same properties from the input. The key insight is that $C_A()$ was trained to classify samples based on the information in $F_A()$. If the same information is in $F_B()$, then the mapping $M_{B \to A}$ should preserve it, and $C_A()$ should be able to label images as well from $M_{B \to A} F_B()$ as from $F_A()$. If, on the other hand, $F_A()$ contains information not in $F_B()$, then that information will be missing in $M_{B \to A} F_B()$.

We therefore define a directional loss function between feature extractors. Let $Acc(C())$ be the accuracy of a classifier $C()$ across a test data set. Then

$$Loss_{A,B} = Acc(C_A(F_A())) - Acc(C_A(M_{B \to A} F_B()))$$

is the loss in discriminatory power that results from replacing A with B. For example, if system A classifies 90% of all inputs correctly using its own features $F_A()$, but only classifies 80% of samples using $\tilde{F}_A() = M_{A \to B} F_B()$, then $Loss_{A,B} = 10\%$, because 10% of the discriminatory power in $F_A()$ was missing from $F_B()$. Note that if A and B are both good systems but extract and rely on different information, both $Loss_{A,B}$ and $Loss_{B,A}$ may be high. It is similar to a divergence in this sense.

It is possible for the loss function defined above to be negative (in which case we refer to it as a gain). It implies that system A's classifier, which was trained on system A's features, nonetheless performs better on the mapped version of system B's features. Since $C_A()$ is not retrained, the information gain cannot be in the form of a new property extracted by B but ignored by A, since the information would have been dropped in the mapping process (and A's classifier wouldn't know what to do with it anyway). Instead, it implies that some property detected by A is detected more robustly by B, so that $M_{B \to A} F_B()$ is a more reliable predictor of the property than $F_A()$ is.

The experiments in this paper use the Inception-v4 network found in the Tensorflow-Slim GitHub repository [9] and described in [10], and the ResNet-v2 network found in the same repository and described in [4]. These networks were trained "internally at Google" using identical preprocessing on all 1.1 million ILSVRC2012 training samples, as described in Szegedy et al. [10]. After loading these pretrained parameters, we internally validated each network's top-1 single-crop classification accuracy on all 50k ILSVRC2012 validation set samples. This internal validation yielded 78.0% accuracy for ResNet, as opposed to the 78.9% reported in He et al. [4]. Our Inception-v4 model achieved 80.1%, as opposed to the 80.2% reported by Szegedy et al. [10].

Given source features from Inception $F_I() \in \mathbb{R}^{1536}$ and target features from ResNet $F_R() \in \mathbb{R}^{2048}$, we solve for an affine mapping. Specifically, we define the affine mapping as:

$$\tilde{F}_R(T) = M_{I \to R} \begin{bmatrix} F_I(T) \\ \mathbf{1} \end{bmatrix} \quad (1)$$

The weights in the affine transformation are trained using the ILSVRC2012 1.1 million training images for T in Eq. 1. The terms in the affine transformation are trained using gradient descent. The error function E is the euclidean distance between the unit-normalized true and predicted vectors, i.e.

$$E(x, y) = \left(\frac{x}{\|x\|} - \frac{y}{\|y\|} \right)^2$$

where x and y are feature vectors. We compute this function on random batches of predicted vectors $y \in \tilde{F}_R(T)$ and their corresponding true vectors $x \in F_R(T)$. This error function was chosen as it is similar to angular distance, but is not sensitive to computational issues surrounding zero vectors.

After training, features are once again generated as in the previous section. Using the ILSVRC2012 50,000 validation images V, we produce class predictions $C_R(M_{I \to R} F_I(V))$ for mapping $I \to R$ and similarly $C_I(M_{R \to I} F_R(V))$ for mapping $R \to I$. This is illustrated in Fig. 1. Evaluation is performed by simply measuring the amount of correctly classified samples for each mapping configuration.

4 Results

Section 3 presented the technical details of how we create mappings from one feature space to another. This section returns to the question of whether the features learned by Inception-v4 and ResNet-v2 are similar. We measure similarity over the validation images from the ILSVRC2012 image recognition challenge. We apply $F_I()$, Inception's convolutional network, to these images, producing $1,536$ features per image. We also apply $F_R()$, ResNet's convolution network, to the validation images, producing $2,048$ features per image. Finally, we apply the I→R and R→I mapping functions described above, thereby creating $\tilde{F}_I = M_{R\to I}F_R()$ and $\tilde{F}_R = M_{I\to R}F_I()$. By applying $C_I()$ to $F_I()$ and $\tilde{F}_I()$ and applying $C_R()$ to $F_R()$ and $\tilde{F}_R()$ we can measure the similarity between Inception's and ResNet's features.

Table 1. Comparison of mapped and unmapped network performance. The values indicate the number of correctly labelled samples out of 50,000 on the ILSVRC2012 validation dataset is shown along with the percent correct. For the bottom two with the affine mappings the loss defined in Sect. 3 is shown.

Description	Mapping	Correct	Percent	Loss
Inception-v4	(I → I)	40,037	80.2	
ResNet-v2 152	(R → R)	39,022	78.0	
Inception-v4 to ResNet-v2 152	(I → R)	39,686	79.4	−1.4
ResNet-v2 152 to Inception-v4	(R → I)	37,866	75.7	4.5

The central question of the paper is whether the features learned by Inception-v4 and ResNet-v2 are equivalent. Table 1 shows the experimental results for mapping Inception features onto ResNet features (I→R) and vice-versa (R→I). The Inception classifier correctly labels 40,037 validation images using its own features $F_I()$; using ResNet features mapped through R→I (i.e. $\tilde{F}_I()$) it correctly labels 37,866 images. This is a loss of 4.5%. In the other direction, the ResNet classifier correctly labels 39,022 validation images using its own features $F_R()$ and 39,686 images using Inception's features mapped through I→R (i.e. $\tilde{F}_R()$). This is a gain rather than loss of 1.4%.

This result strongly suggests that all the information in ResNet's 2,048 features is also contained in Inception's 1,536 features. Otherwise, I→R could not have learned a mapping with no loss. The result in the other direction is weaker. The performance of Inception's classifier drops by 4.5% when it is given the mapped version of ResNet's features. This suggests that Inception-v4, which is the higher performing of the two systems, may capture some information in its features that ResNet's features miss.

5 Conclusion

We looked for equivalence between ResNet and Inception features to see whether they capture different but roughly equally discriminative information, or whether they actually extract the same information, just using a different architecture and encoding to do it. The findings suggests that they are, in fact, extracting qualitatively the same features, while Inception does it slightly more robustly. We speculate that this may have implications for other convolutional nets as well. As nets grow more complex, there may be a law of diminishing returns if they all end up extracting similar features, but that may be what happens because that is what the training data supports.

References

1. Bau D, Zhou B, Khosla A, Oliva A, Torralba A (2017) Network dissection: quantifying interpretability of deep visual representations. In: Proceedings of the IEEE conference on computer vision and pattern recognition, pp 6541–6549
2. Dosovitskiy A, Brox T (2016) Inverting visual representations with convolutional networks. In: Proceedings of the IEEE conference on computer vision and pattern recognition, pp 4829–4837
3. He K, Zhang X, Ren S, Sun J (2016) Deep residual learning for image recognition. In: Proceedings of the IEEE conference on computer vision and pattern recognition, pp 770–778
4. He K, Zhang X, Ren S, Sun J (2016) Identity mappings in deep residual networks. In: European conference on computer vision. Springer, pp 630–645 (2016)
5. Ioffe S, Szegedy C (2015) Batch normalization: accelerating deep network training by reducing internal covariate shift. arXiv preprint arXiv:1502.03167
6. Kornblith S, Shlens, J, Le QV (2018) Do better imagenet models transfer better? arXiv preprint arXiv:1805.08974
7. Lenc K, Vedaldi A (2019) Understanding image representations by measuring their equivariance and equivalence. Int J Comput Vis 127(5):456–476. https://doi.org/10.1007/s11263-018-1098-y
8. Russakovsky O, Deng J, Su H, Krause J, Satheesh S, Ma S, Huang Z, Karpathy A, Khosla A, Bernstein M et al (2015) Imagenet large scale visual recognition challenge. Int J Comput Vis 115(3):211–252
9. Silberman N, Guadarrama S (2016) Tensorflow-slim image classification model library. https://github.com/tensorflow/models/tree/master/research/slim
10. Szegedy C, Ioffe S, Vanhoucke V, Alemi AA (2017) Inception-v4, inception-resnet and the impact of residual connections on learning. In: Thirty-first AAAI conference on artificial intelligence (2017)
11. Szegedy C, Liu W, Jia Y, Sermanet P, Reed S, Anguelov D, Erhan D, Vanhoucke V, Rabinovich A (2015) Going deeper with convolutions. In: Proceedings of the IEEE conference on computer vision and pattern recognition, pp 1–9
12. Szegedy C, Vanhoucke V, Ioffe S, Shlens J, Wojna Z (2016) Rethinking the inception architecture for computer vision. In: Proceedings of the IEEE conference on computer vision and pattern recognition, pp 2818–2826

Multicriteria Approach to Control a Population of Robots to Find the Best Solutions

Sergey Yu. Misyurin[1,2], Andrey P. Nelyubin[2(✉)], and Mikhail A. Potapov[3]

[1] National Research Nuclear University MEPhI, 31 Kashirskoe Shosse, Moscow, Russia
[2] Mechanical Engineering Research Institute RAS, 4 Malyi Kharitonievski Pereulok, Moscow, Russia
nelubin@gmail.com
[3] Institute of Computer Aided Design RAS, 18/19 2nd Brestskaya, Moscow, Russia

Abstract. A bio-inspired and multicriteria approach to the design of promising robotic systems consisting of groups of robots and able to rebuild and adapt to changing goals and operating conditions is proposed.

Keywords: Robotics · Design problem · Multicriteria optimization · Criteria importance theory

1 Introduction

Robotics are increasingly beginning to be used not only to automate typical repetitive operations, but also to solve more complex tasks, including search, monitoring, navigation, control, and automatic decision making [1–6]. Examples of such tasks are research, explore, search and rescue mobile robots in hard-to-reach areas, in space, deep under water, under extreme conditions regarding temperature, pressure, radiation etc. [1, 7, 8]. Promising micro- and nano-robots are being developed for the diagnosis and treatment of internal organs [9]. Future military operations will be carried out mostly by autonomous or remote-controlled robots. Modern technologies of security systems and smart home include a distributed sensor system for tracking changing conditions and an automatic response system [10, 11].

In the course of performing the above tasks by robotic systems, the goals and external conditions may change. Moreover, it is possible to talk about the accomplishment of missions consisting of various successively performed tasks. Therefore, robotic systems should be quite versatile. The complexity of the arising problems requires the construction of efficient robotic systems and their control algorithms. To increase the efficiency of finding the best solutions, it often makes sense to use distributed systems of groups of robots in which individual robots can simultaneously explore different areas of solutions, exchanging signals between themselves and with

A. V. Samsonovich (Ed.): BICA 2019, AISC 948, pp. 358–363, 2020.
https://doi.org/10.1007/978-3-030-25719-4_46

the control center [1–3, 12]. This is especially true when performing search and monitoring tasks.

Similar systems of parallel search and group cooperation exist in nature, for example, a swarm of bees, a flock of ants, and the immune system of the body. The ways of organizing such natural systems can suggest a number of ideas for creating robotic systems. One of these ideas is to maintain the diversity of individuals in the population. At the same time, individuals can vary greatly in their characteristics and even have different specializations. Maintaining biodiversity allows the entire population to quickly adapt to changing environmental conditions. To perform a robotic system of complex missions, consisting in solving various tasks under different conditions, it can also be useful to have robots differing in characteristics and functionality within the whole group. The specialization of robots is especially effective in conditions of limited resources, when the use of universal robots is either impossible or expensive.

In this paper, we develop an approach to the design of robotic systems consisting of groups of robots, based on the principles of organization of populations of living organisms, using methods of multicriteria analysis [13–15].

2 The Problem of a Robotic System Design

Consider the problem of creating a separate robot as a multicriteria design problem. Target characteristics of robots are considered as criteria. For example, speed, accuracy, maneuverability, reliability, dimensions. Technical or resource constraints make it impossible to achieve the best values for all criteria at the same time. Solving a multicriteria optimization problem, one can obtain a set of Pareto-efficient solutions (robot projects) [16]. Selection of any one project among this set will correspond to fixed preferences regarding the criteria relative importance. However, these preferences should be different for robots with various specializations. Moreover, preferences regarding the importance of criteria may change over time and depend on the external conditions of operation of robots. Therefore, in this paper we propose to save and maintain all or part of Pareto-efficient solutions in a group of robots, based on the idea of maintaining the diversity of individuals in a population.

By controlling the population of robots, we mean the real-time control of the number and composition of a group of robots with different characteristics. The mechanism for implementing such a control may vary depending on the particular system: robots can be generated or disabled; withdrawn from the base or return to the base; robots can be rebuilt (reconfigured), changing their characteristics.

The robots population can be controlled by an operator, a set of rules, or algorithmically. A combination of these methods is also possible. The control based on a rule set may be suitable when the best robots are known in advance for certain operating conditions. However, the most interesting cases involve finding solutions during the process of the robotic system operation. As a rule, there is a limit in time and in the number of tests in such situations.

When the designed robotic system turns out to be in new, unfamiliar conditions, it should as soon as possible and at the lowest cost reconfigure itself into a new mode of functioning, the most effective in the new conditions. The system must be adaptable to repeated changes in conditions. To improve the efficiency of finding solutions to such problems, we propose an approach based on methods of multicriteria analysis and the criteria importance theory [17].

3 Multicriteria Approach

Finding the best solution regarding the mode of operation of the robotic system in the new conditions means determining the characteristics of robots that better cope with the fulfillment of the system goals in these conditions. That is, the problem of finding the best solution can be reduced to a multicriteria optimization problem.

A set of Pareto-efficient solutions of a multicriteria optimization problem can be obtained (or approximated) in advance and incorporated into the robotic system at the design stage [13]. Which of these solutions will be the best in the new conditions depends on the relationship between the importance of individual criteria in these conditions. For example, for ground mobile robots the criterion of speed of movement is more important on a flat road, whereas the criterion of passability is more important in a swampy area. In contrast to this example, in the general case, determining the relative importance of criteria in new conditions may not be easier than solving the original problem of finding the best solution. However, there are methods of multicriteria analysis that can improve the efficiency of solving this problem.

The mathematical theory of criteria importance, created by prof. V.V. Podinovski and currently being developed [17–19] contains methods for determining information about the relative importance of criteria and the scale of criteria and correctly using it for comparing alternatives by preference. Both qualitative and quantitative information on the preferences of the decision maker is used. Special iterative procedures to solve the choice problem have been developed within the framework of the criteria importance theory, which consistently refine the information on the preferences and reduce the set of potentially best alternatives.

To describe the proposed method, we present the necessary information from the criteria importance theory. Consider the following mathematical model:

$$M = <X, K_1, \ldots, K_m, Z_0, R> ,$$

here X is the set of alternatives (robot designs), K_1, …, K_m are individual criteria ($m \geq 2$), i.e. functions K_i: $X \rightarrow Z_0$, $Z_0 = \{1, \ldots, q\}$ is the range of criteria values (set of scale estimates) ($q \geq 2$), and R is the preference model. It is assumed that each of the criteria is independent in preference from the others [16] and its larger values are preferable to smaller ones. Each alternative x from the set X is characterized by its vector estimate $y(x) = K(x) = (K_1(x), \ldots, K_m(x))$. The set of all vector estimates is $Z = Z_0^m$.

In the criteria importance theory, preferences on the set of vector estimates of Z are modeled using a nonstrict preference relation R: expression yRz means that the vector estimate y is no less preferable than z. The relation R is a partial quasi-order and produces the relations of indifference I and strict preference P on Z:

$$yIz \Leftrightarrow yRz \wedge zRy; \quad yPz \Leftrightarrow yRz \wedge \neg zRy.$$

Since preferences increase along the criteria scale Z_0, the Pareto relation $R^{\varnothing}$ is defined on the set of vector estimates Z:

$$yR^{\emptyset}z \Leftrightarrow y_i \geq z_i, \quad i = 1, \ldots, m.$$

As a rule, one can't obtain a solution to the multicriteria choice problem only with the help of the Pareto relationship. Therefore, it needs to be extended, involving additional information about preferences of the decision maker. In the criteria importance theory, information on the relative importance of criteria is used, which is formally introduced as follows [18]. Denote by y^{ij} the vector estimate obtained from the vector estimate $y = (y_1, \ldots, y_m)$ by permuting its components y_i and y_j.

The criteria K_i and K_j are *equally important*, or equally important (such a message is denoted by $i \sim j$) when the vector estimates of y and y^{ij} are the same in terms of preference. The message $i \sim j$ sets the indifference relation on the set Z:

$$yI^{i\sim j}z \Leftrightarrow z = y^{ij}, \; y_i \neq y_j. \tag{1}$$

The criterion K_i is *more important* than the criterion K_j (such a message is denoted by $i \succ j$) when the vector estimate y, in which $y_i > y_j$, is more preferable than y^{ij}. The message $i \succ j$ sets the preference relation on the set Z:

$$yP^{i\succ j}z \Leftrightarrow z = y^{ij}, \; y_i > y_j. \tag{2}$$

Qualitative information Ω is a set of messages of the form $i \sim j$ or $i \succ j$.

The relation R^{Ω}, generated on Z by qualitative information Ω about the criteria importance, is defined as the smallest transitive relation containing the Pareto relation $R^{\varnothing}$ and the relation R^{ω} for all messages $\omega \in \Omega$:

$$R^{\Omega} = \mathrm{TrCl}\left[\left(\cup_{\omega\in\Omega}R^{\omega}\right)\bigcup R^{\emptyset}\right], \tag{3}$$

here TrCl is the transitive closure binary relation operator, $R^{\omega} = I^{i\sim j}$ for $\omega = i \sim j$, and $R^{\omega} = P^{i \succ j}$ for $\omega = i \succ j$. There are efficient algorithms for checking the relation R^{Ω} for arbitrary vector estimates [18, 19]. Thus, a comparison of the preferences of specially selected alternatives allows to automatically draw conclusions about the preference of other alternatives. This is the effect of the proposed method.

In the criteria importance theory, information about the relative importance of criteria is obtained in the course of a special procedure in which the decision maker interacts with an analyst or computer system [17]. To determine the relative importance of the criteria K_i and K_j, the decision maker compares by preference the pairs of vector estimates y and y^{ij} presented to him/her. In practice, only a limited set of specially selected vector estimates are used. First, we take vector estimates that have the best or the worst values for the rest of the criteria, except for K_i and K_j. If the results of the comparisons lead to a definite conclusion, $i \sim j$ or $i \succ j$, then further, to confirm this result, we can compare vector estimates that have middle values for the other criteria.

Let us return to the problem of finding the best solution for a robotic system, which turned out to be in new conditions. The procedure described above for obtaining information on the relative importance of criteria can be implemented either with the participation of the operator as a decision maker, or automatically. For this purpose, a method of direct pairwise comparison of specially selected alternatives is necessary. In new unknown conditions, modeling capabilities are often limited. Therefore, the only available method remains the experimental run of robots with different characteristics in real conditions. A comparison of robots by preference should be made by organizing a competition that reflects the essence of the mission being carried out. Robots better coping with the main task are considered to be more preferable. It is the analogue of natural selection among living organisms in nature.

The first come to mind solution is to run the whole group of Pareto-efficient robots and identify the winner among them. But this method is either too expensive or impossible. The proposed procedure for determining the relative importance of criteria reduces the number of experimental runs of robots. This procedure is flexible in the sense that the number of comparisons of alternatives can vary depending on the required accuracy of the information and time constraints. The result of each additional comparison of alternatives adds a new piece of information that may reduce the set of potentially best alternatives.

The proposed multicriteria approach can also be applied in the process of continuous adaptation of the robot system to changing conditions. For this purpose, a certain variety of characteristics must be maintained in the group (population) of operating robots. Robots that are most adapted to current conditions, i.e. the best in accordance with current information about the relative importance of the criteria, should prevail quantitatively in the population. But periodically, comparisons should be made of the operating efficiency of specially selected pairs of robots in the population in accordance with the definitions (1–2). So information about the importance of criteria can be updated. As a result, robots with other characteristics will be recognized as the best in the new conditions and their number in the population should be increased.

4 Conclusion

We have considered a rather extensive area of application of promising developments of robotic systems, which should automatically adapt to new unknown operating conditions. Our approach is based on the principle of maintaining diversity of solutions, similar to the principle of maintaining biological diversity in a population of living organisms in order to adapt species. This principle can be naturally implemented in robotic systems consisting of groups of robots with different characteristics.

To improve the efficiency of finding the best solutions during the adaptation of the system, we use multicriteria analysis methods. First, it is proposed to consider only Pareto-efficient solutions, which already limits the diversity of solutions in the population. Secondly, special methods should be used to determine information on the relative importance of criteria, which reduces the total number of experimental runs of robots in real time.

The study was carried out within the framework of the State Program of the ICAD RAS during the research in 2016–2019 and was financially supported by the RFBR project No. 18-29-10072.

References

1. Burgard W, et al. (2000) Collaborative multi-robot exploration. In: IEEE international conference on robotics and automation (IRCA), 476–481
2. Kalyaev IA, Gaiduk AR, Kapustyan SG (2009) Models and algorithms of the collective control in groups of robots. Fizmatlit, Moscow
3. Wuhui C et al (2018) A study of robotic cooperation in cloud robotics: architecture and challenges. IEEE Access 6:36662–36682
4. Mirus F et al (2018) Neuromorphic sensorimotor adaptation for robotic mobile manipulation: From sensing to behavior. Cognitive Syst Res 50:52–66
5. Gridnev AA, Voznenko TI, Chepin EV (2018) The decision-making system for a multi-channel robotic device control. Procedia Comput Sci 123:149–154
6. Demin A, Vityaev E (2018) Adaptive control of multiped robot. Procedia Comput Sci 145:629–634
7. Lopatina NB, Frolov DV (2017) Analysis of prospects for the development of ground-based robotic systems for conducting radiation, chemical and biological intelligence in the United States. Robot Tech Cybern 4(17):3–5
8. Vasiliev IA (2017) The use of mobile robots for rescue missions in the sea. Robot Tech Cybern 4(17):6–9
9. Dolenko T et al (2018) Application of wavelet neural networks for monitoring of extraction of carbon multi-functional medical nano-agents from the body. Procedia Comput Sci 145:177–183
10. Wilson G et al (2019) Robot-enabled support of daily activities in smart home environments. Cognitive Syst Res 54:258–272
11. Alaparthy VT, Amouri A, Morgera SD (2018) A study on the adaptability of immune models for wireless sensor network security. Procedia Comput Sci 145:13–19
12. Rekleitis I, Dudek G, Milios E (1998) Accurate mapping of an unknown world and online landmark positioning. Proceedings of Vision Interface 6:455–461
13. Misyurin SYu, Nelyubin AP, Ivlev VI (2017) Multicriteria adaptation of robotic groups to dynamically changing conditions. IOP Conf Series: J Phys 788(1):012027
14. Misyurin SYu, Nelyubin AP (2017) Multicriteria adaptation principle on example of groups of mobile robots. IOP Conf Series: J Phys 937(1):012034
15. Misyurin SYu, Nelyubin AP, Potapov MA (2019) Applying partial domination in organizing the control of the heterogeneous robot group. IOP Conf Series: J Phys 1203(1):012068
16. Keeney RL, Raiffa H (1976) Decisions with multiple objectives: preferences and value tradeoffs. John Wiley & Sons, New York
17. Podinovski VV (2007) Introduction into the theory of criteria importance in multi-criteria problems of decision making. Fizmatlit, Moscow
18. Podinovski VV (1976) Multicriterial problems with importance-ordered criteria. Automation Remote Control 37(11, part 2):1728–1736
19. Nelyubin AP, Podinovski VV, Potapov MA (2018) Methods of criteria importance theory and their software implementation. Springer Proc Math & Stat 247:189–196

Multi-agent Algorithm Imitating Formation of Phonemic Awareness

Zalimkhan Nagoev, Irina Gurtueva(✉), Danil Malyshev, and Zaurbek Sundukov

The Federal State Institution of Science Federal Scientific Center Kabardino-Balkarian Scientific Center of Russian Academy of Sciences, I. Armand Street, 37-a, 360000 Nalchik, Russia
gurtueva-i@yandex.ru

Abstract. This paper proposes the cognitive speech perception model necessary as a theoretical basis for the development of universal automatic speech recognition systems that are highly effective in conditions of high noise and cocktail party situations. A formal description of the general structure of the act of speech perception and the main elements of the structural dynamics of the speech recognition process has been developed. The necessity of using the articulation event as a minimal basic pattern of sound image recognition has been proved. Using articulation event gives an opportunity to analyze such aspects of speech message as extra-linguistic components and intonation means of expression. Multi-agent systems are chosen as the formal means of implementation. An algorithm for supervised machine learning with an imitation of the mechanism of the formation of a human's phonemic awareness is developed. It gives the possibility to create speech systems that are resistant to the diversity of accents and individual characteristics of the user.

Keywords: Speech recognition · Artificial intelligence · Cognitive models · Multi-agent systems

1 Introduction

The rapid development of ambient intelligence imposes stringent requirements for speech systems. The main problems appear when speech applications use in noisy conditions or in the analysis of the so-called cocktail party situations [1, 2]. The good solution to the problem of acoustic variation is yet to be found [3, 4]. The automatic speech recognition systems offered by modern IT technologies are not effective enough in solving the mentioned problems with satisfactory accuracy and cannot be admitted as universal [5]. In our opinion, a successful solution of this problem is possible as consequence of solving problems of artificial intelligence, since it is necessary to use an internal semantic model built on the basis of cognitive functions that a person uses when decoding audio messages. In [6] the approach was proposed to formalize the semantics of rational thinking using the cognitive modeling based on the concept of a

The work was supported by RFBR grants № 18-01-00658, 19-01-00648.

A. V. Samsonovich (Ed.): BICA 2019, AISC 948, pp. 364–369, 2020.
https://doi.org/10.1007/978-3-030-25719-4_47

recursive cognitive architecture and the hypothesis of an invariant of the organizational and functional structure of intellectual decision-making based on cognitive functions. The new approach was built on the base of cognitive sciences [7–12] and multi-agent modeling paradigm [13, 14].

This article offers basic elements of the cognitive model of speech recognition as a theoretical basis for developing a new approach to the study of these problems.

The relevance of the research is that the lack of reliable speech recognition systems that is in many ways a deterrent to the development of intelligent robotics. The object of research of this work is speech recognition in real environments based on cognitive modeling. The subject of the research is the structural and functional organization of the auditory analyzer of an intelligent agent based on neurocognitive architectures.

2 The Cognitive Model of Speech Perception Mechanism Based on Multi-agent Systems

The act of speech perception, as a stable phase sequence, regularly reproduced by an individual in speech activity, can be structured as a consequent processing of audio information at the following stages: preliminary recognition, subconscious recognition, conscious recognition, level of situations. The structure and functions of each level are determined by a set of agents, actors and their contract-based interactions [5].

The first stage, pre-recognition, is the registration of the acoustic parameters of the signal by the human auditory receptors and phoneme clusterization. [5, 15]. A human tries to establish connection between the designate (phoneme) and the denotatum (letter). Technically, it is implemented on the base of the machine-learning algorithm, described in detail later in this paper and in [16].

At the next stage, called subconscious recognition, the signatures of the previous layer are grouped around significant objects and actions. We presume that for reliable speech recognition it is necessary to establish a connection between the spectral characteristics of the signal and the "articulatory event" underlying it. The articulatory event is a couple of agents, one of which identifies the object, and the other the action, as well as their contract, that is, the dynamic link with the source of the sound. At the training stage, the expert comments the sound signals in natural language as follows: "The car drove", "Helen said", etc.

If we consider an articulatory event in terms of linguistics, then we can say that one agent is the topic (theme) of the utterance, and the other is the comment (rheme). We rely on the information structure of the message, as in contrast to the grammatical structure. The latter reflects the grammatical aspect of the sentence; the former is synonymous with logical and semantic structure. The information structure of the statement considers the message from the communicative and cognitive points of view.

The choice of the articulatory event and its connection with the spectral characteristics of the statement, as an independent object of analysis, is in good agreement with the main postulate of the motor theory of speech perception. Namely, a person identifies non-verbal sound information as an event. For example, for identifying a non-verbal signal, a person uses simple statements: "steps were heard," "a bird sang" [17]. Finally, neuropsychologists distinguish two subsystems in the structure of the auditory system: nonverbal hearing and phonemic awareness. According to the results of clinical studies, it is known that right-handers lose the ability to differentiate and analyze verbal messages (phonemic or speech hearing) when the left temporal lobe of the cerebral cortex is affected, and nonverbal – right [18]. In the most severe cases of auditory agnosia, patients cannot determine the meaning of the simplest everyday sounds (the creaking of doors, for example).

Phonemic hearing, in turn, is heterogeneous. It includes phonemic awareness and intonation perception. Adequate processing of intonation content of speech as well as non-verbal messages is violated in cases of right-sided localization of the lesion (for right-handers). The relative autonomy of the articulation event as an independent object of analysis makes it possible to imitate the differentiation of the processing of verbal and non-verbal sound images by different hemispheres of the brain [18].

Thus, from the point of view of neuropsychology, the introduction of such a concept as an articulatory event is necessary for building the correct mechanism of speech perception. The articulatory event allows to bring into the framework of the analysis a large amount of information transmitted by intonation means, it makes possible to formalize the occurrence and construction of the context, to include in the scope of analysis communicative intentions.

At the third phase of conscious recognition, a human determines significant current priority events, using the so-called cogniton of emotional evaluation [6] - a functional node of multi-agent recursive intelligence, containing a priori and learned information about the significance of events for the realization of the objective function of an intelligent agent.

At the fourth level, a “situation” is formed. Situation is a set of events with their emotional values. Dynamic connection between events and emotional evaluations is the basis for prediction [6].

Thus, the proposed model of the cognitive mechanism for speech perception allows to take into account all the aspects of speech message in the signal analysis procedure, including the extra-linguistic component expressed in this approach in terms of an event and a situation.

3 Multi-agent Algorithm Imitating the Formation of Phonemic Awareness

At the stage of preliminary processing, the signal is converted into a set of signatures and the matrix of features is created. The matrix characterizes acoustic features of the signal [5]. Its information (the location of the sound source, amplitude, frequency and duration of signal sound) is fed to the first level of the architecture – preliminary recognition [5].

The learning process starts when the system receives signatures from the actors of the afferent tract. The agent of special type, the so-called factory [19], determines the type of a signature and, creates the agent-processer, which is located in the predefined area of spatial localization of the multi-agent cognitive architecture. To create such an agent, a special procedure is used that reads agent’s genome — the initial set of production rules created by the developers [6, 19]. According the rules, in particular, such an agent acquires the ability to recognize the very signature immediately after its creation. The signature arrival served as a trigger for the creation of the agent by the neuro-factory.

Thus, the developers form a set of neuro-factories necessary to create agents of all types involved in the cognitive architecture, and then the agents are created by the factories dynamically on demand in the process of functioning and learning the system. Figure 1 illustrates the described process. The factories are shown in double contours and the corresponding agents are represented by geometric figures in single contours.

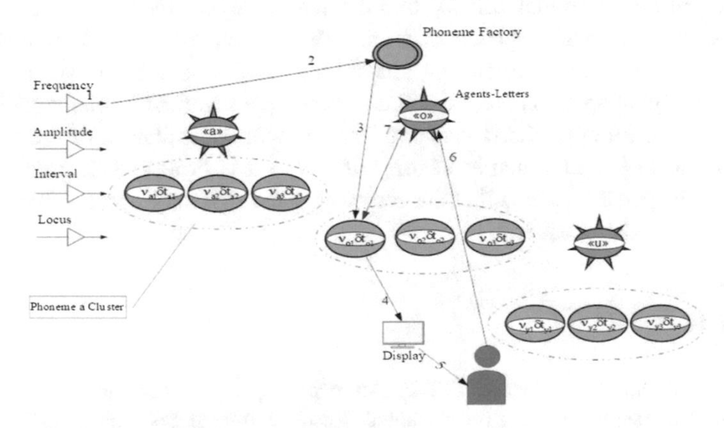

Fig. 1. Machine learning procedure imitating the process of phonemic awareness of a human while mastering native language.

After formation, the agent responsible for a certain phoneme, realizing the behavior determined by the rules recorded in its own genome, in order to find the class to which it belongs, appeals to an expert with the question. The question is formulated in natural language (for instance, "what letter means the sound?"). Then it is displayed on the screen. On the base of expert's answer it connects with the letter-agent characterizing the phoneme of the class. The contract-based interaction between the actor and the letter-agent arises after the expert's response, that is, it is implemented as supervised machine-learning [20, 21]. The training process of the robot involves the presentation in the field of its "hearing" a complete set of training material, followed by the above procedure. Unknown information is classified using the method of k-nearest neighbors [20].

The purpose of the functioning of the first level of the developed architecture is the formation of sets of agents corresponding to each minimal speech unit of a language. The proposed algorithm allows reducing the high variability of speech with respect to the variety of accents, the effect of co-articulation and speaker's individual characteristics. By selecting this or that training material, one can trace the mechanism of the formation of human hearing patterns [22, 23]. At subsequent levels, phonological and grammatical limitations are planned to accelerate the recognition process. The introduction of feedback is also planned to correct and refine the decoding results.

Thus, based on the analysis of experimental data of behavioral studies and model representations about the mechanisms of speech recognition from the point of view of psycholinguistic knowledge, a method of machine learning was developed with an imitation of the formation of a human phonemic awareness.

4 Conclusion

General structure and main elements for a cognitive model of speech perception are developed. Proposed model allows to take into analysis the linguistic and extra-linguistic components of a speech message. The articulatory event is selected as the basic pattern of speech recognition. The refusal of the search for an invariant are critical. Multi-agent systems are chosen as a technical mean of implementation.

An algorithm for supervised machine learning with an imitation of the mechanism of the formation of the phonemic hearing of a person is proposed. It gives the possibility to create speech systems that are resistant to the diversity accents and individual characteristics of the speaker.

References

1. Marti A, Cobos M, Lopez J (2012) Automatic speech recognition in cocktail-party situations: a specific training for separated speech. J Acoust Soc Am 131(2):1529–1535. https://doi.org/10.1121/1.3675001
2. Zion Golumbic EM, Ding N, Bickel S et al (2013) Mechanisms underlying selective neuronal tracking of attended speech at a "cocktail party". Neuron 77(5):980–991. https://doi.org/10.1016/j.neuron.2012.12.037
3. Jurafsky D, Martin J (2008) Speech and language processing: an introduction to natural language processing, computational linguistics, and speech recognition, 2nd edn. Prentice Hall, New Jersey
4. Waibel A, Lee K-F (1990) Readings in Speech Recognition. Morgan Kaufman, Burlington
5. Nagoev Z, Lyutikova L, Gurtueva I: Model for Automatic Speech Recognition Using Multi-Agent Recursive Cognitive Architecture, Annual International Conference on Biologically Inspired Cognitive Architectures BICA, Prague, Chech Republic http://doi.org/10.1016/j.procs.2018.11.089
6. Nagoev ZV (2013) Intellectics or thinking in living and artificial systems. Publishing House KBSC RAS, Nalchik [Nagoev, Z. V.: Intellektika ili myshleniye v zhyvych i iskusstvennych sistemach. Izdatel'stvo KBNC, Nal'chik (2013)]
7. Chomsky NA (1967) A review of skinner's verbal behavior. In: Jakobovits LA, Miron MS (eds) Readings in the psychology of language. Prentice-Hall, New Jersey
8. Gazzaniga M (2009) Conversations in the cognitive neuroscience. The MIT Press, Cambridge
9. Minsky M (1988) The Society of Mind. Simon and Shuster, New York
10. Haikonen P (2003) The cognitive approach to conscious machines. Imprint Academic, Exeter
11. Newell A (1990) Unified Theories of Cognition. Harvard University Press, Cambridge
12. Schunk DH (2011) Learning theories: an educational perspective. Pearson Merrill Prentice Hall, New York

13. Wooldridge M (2009) An introduction to multi-agent systems. Wiley, Hoboken
14. Kotseruba Iu, Tsotsos J K, A Review of 40 Years of Cognitive Architecture Research: Core Cognitive Abilities and Practical Applications. arxiv.org/abs/1610.08602
15. Nagoev ZV, Nagoeva OV (2015) Knowledge Extraction from Multimodal Streams of Unstructured Data on the Base of Self-Organization of Multi-Agent Cognitive Architecture for Mobile Robot. News of KBSC of RAS 6(68):73–85 [Nagoev Z V, Nagoeva O V: Izvlechenie znanii iz mnogomodal'nyh potokov nestrukturirovannyh dannyh na osnove samoorganizatsii mul'tiagentnoi kognitivnoi arhitektury mobil'nogo robota. Izvestia KBNC RAN 6(68), 73–85 (2015)]
16. Nagoev ZV, Denisenko VA, Lyutikova LA (2018) Learning system of autonomous agricultural robot for static images recognition on the base of multi-agent cognitive architectures. Sustainable Dev Mountain Territ 2:289–297 [Nagoev, Z. V., Denisenko, V. A., Lyutikova, L. A.: Sistema obucheniya avtonomnogo sel'skohozyaistvennogo robota raspoznavaniyu staticheskih izobrazhenii na osnove multiagentnyh kognitivnyh arhitektur. Ustoichivoie razvitie gornyh territoii 2, 289-297 (2018).]
17. Sorokin VN (2007) Motor speech perception theory and inner model theory. Inf Process 7 (1):1–12 [Sorokin, V. N.: Motornaya teoriya vospriyatia rechi i teoriya vnutrennei modeli. Informatsionniye protsessy 7(1), 1-12 (2007).]
18. Morozov VP, Vartanyan IA, Galunov VI (1988) Speech Perception: Problems of Functional Brain Asymmetry. Science, St. Petersburgh
19. Nagoev Z V, Nagoeva O V (2017) Visual analyzer of intellectual robot for unstructured data processing on the base of multi-agent neurocognitive architechture. In: Advanced systems and management tasks: proceedings of the 12th all-russian conference, pp. 457–467. Rostov-on-Don. [Nagoev, Z. V., Nagoeva, O. V.: Zritel'nyi analizator intellektual'nogo robota dlya obrabotki nestrukturirovannyh dannyh na osnove mul'tiagentnoi neirocognitivnoi arhitektury. In:Perspektivnye sistemy I zadachi upravleniya: Materialy vserossiiskoi nauchno-prakticheskoi konferencii, 457–467. Rostov-na-Donu (2017)]
20. Coates A, Ng AY (2012) Learning feature representations with K-means. In: Montavon G, Orr GB, Müller K-R (eds) Neural networks: tricks of the trade, vol 7700. LNCS. Springer, Heidelberg, pp 561–580. https://doi.org/10.1007/978-3-642-35289-8_30
21. Russel S, Norvig P (2009) Artificial intelligence: a modern approach, 3rd edn. Pearson, London
22. Weber A, Scharenborg O (2012) Models of spoken-word recognition. WIREs Cogn Sci 3(3):387–401. https://doi.org/10.1002/wcs.1178
23. Strange W (1995) Speech perception and linguistic experience: issues in cross-language research. York Press, Baltimore

A Simulation Model for the Cognitive Function of Static Objects Recognition Based on Machine-Learning Multi-agent Architectures

Zalimkhan Nagoev, Inna Pshenokova, Irina Gurtueva(✉), and Kantemir Bzhikhatlov

The Federal State Institution of Science Federal Scientific Center, Kabardino-Balkarian Scientific Center of Russian Academy of Sciences, I. Armand Street, 37-a, 360000 Nalchik, Russia
gurtueva-i@yandex.ru

Abstract. The purpose of the research is the development of the theoretical foundations and algorithms for the unstructured data flows recognition based on multi-agent neurocognitive architectures in artificial intelligence systems. The task of the study is to develop a simulation model of the cognitive function of static objects recognition according to the video stream.

The article developed a simulation model of the cognitive function of static objects recognition based on multi-agent architectures and a software system that demonstrates its work in geometric shapes recognition. The simulation model based on multi-agent neurocognitive architectures allows to create concepts and categories in the autonomous mode according to the data of multimodal input information (event occurred). Through interaction with the user, the system can expand these concepts and categories, and correct the links between them.

The system presented in the paper is autonomous and self-learning. It can be used in autonomous artificial intelligence systems, such as Smart Systems, robotic complexes, etc., to recognize unstructured data streams.

Keywords: Multi-agent systems · Neurocognitive architecture · Simulation model · Image recognition · Artificial intelligence systems

1 Introduction

Simulation modeling is expedient to use in solving the problems of technical systems development for image recognition. Particularly when it is necessary to structure big streams of multimodal input data.

One of the promising directions in simulation modeling is multi-agent systems (MAS). The basics of multi-agent modeling are presented in [1–5]. The first implementations of the MAS are associated with the names of Lesser [6], Hewitt [7],

The work was supported by RFBR grants №18-01-00658, 19-01-00648.

A. V. Samsonovich (Ed.): BICA 2019, AISC 948, pp. 370–378, 2020.
https://doi.org/10.1007/978-3-030-25719-4_48

Lenat [8] and Smith [9]. Currently, MASs find their applications to resolve problems in different spheres: control systems of complex processes in industry and economics, medicine, search engines in the Internet, telecommunication systems, energy networks, transport and pedestrian traffic, computer games and film industry, etc. [10–13].

The named tasks are associated with large flows of unstructured input data processing. That is why the structuring of unstructured processes in artificial intelligence systems is at the center of the unsolved problems of modeling. The approach based on the computational abstraction of multi-agent neurocognitive systems that illustrates architectural conformity to self-organizing neurocognitive networks of the brain [14], enables to develop the model capable to learn independently, recognize and understand data flows using existing knowledge, context and experience.

The purpose of the study is to develop the theoretical foundations and algorithms for the unstructured data flows recognition based on multi-agent neurocognitive architectures in artificial intelligence systems.

The task of the study is to develop a simulation model of the recognition function of static objects shown on the camera.

2 The Multi-agent Neurocognitive Architecture

The multi-agent neurocognitive architecture (MARI) consists of rational software neuron agents. Their objective function is to maximize energy in the limiting conditions imposed by the environment [15]. The agents act proactively [16]; therefore, at each step of discrete time, they solve the problem of synthesizing a behavior plan few steps ahead up to the planning horizon. Agents interact through contracts. The execution of contracts helps to reach the system-wide goal and to receive additional energy. Energy is considered here as a measure of the activity of an agent in a medium. It is a dimensionless quantity. A contract is a dependence arising and developing when agents enter into contractual obligations on the terms of a mutually beneficial exchange of energy for information through sending messages. Contractual link is similar to a mathematical function that sets the model of the interconnection between abstract values x and y. In our case, abstract values are messages in the knowledge base of agents, i.e. there is always an agent which sends a message for request of another agent and returns its message as a value. Therefore, a functional connection between agents is called either a multi-agent existential function (MAEF), or ע-function (Ayin-function) [17].

The agent receives information through sensory subsystem consisting of receptors that register external (exteroceptors) and internal (interoceptors) parameters of different modalities and build a system of afferent signals of the agent. Agents realize their functions on the basis of internal knowledge bases consisting of production rules. The conditional part of production rules defines the initial and final situation, and the core defines the action that transfers the agent from the initial situation to the final one. At the same time, each incoming situation, formed on the basis of the sensor subsystem, is compared with the energy estimation that the agent acquires, or loses, having fallen into this situation.

Situation is a set of events united by emotional evaluation. The event, in its turn, is information that system receives from interoceptive and/or exteroceptive receptors, which is formed by agents through a multi-agent existential mapping.

There are several cognitive nodes in the neurocognitive architecture differentiated as peculiar parts as they perform heterogeneous functions. But cognitive nodes are related to each other by data. The following interrelated cognitive nodes can be distinguished: input image recognition, emotional evaluation, goal setting, plan synthesis, modeling the effects of plan implementation, and management of plan implementation [18].

3 Simulation Model of the Cognitive Function of Static Objects Recognition Based on Machine-Learning of Multi-agent Architectures

This paper considers the functioning of the cognitive recognition node of input images by the example of static objects recognition shown on the camera.

The minimal architecture of a multi-agent neurocognitive system for image recognition (see Fig. 1) should consist of exteroceptive agents that receive and process external signals (these are agents: Video-color, Video-shape, Keyboard); exteroeffector agents that transmit internal signals to the user (these are: Display and Factory-agents which create agents of a certain type to fix a new event).

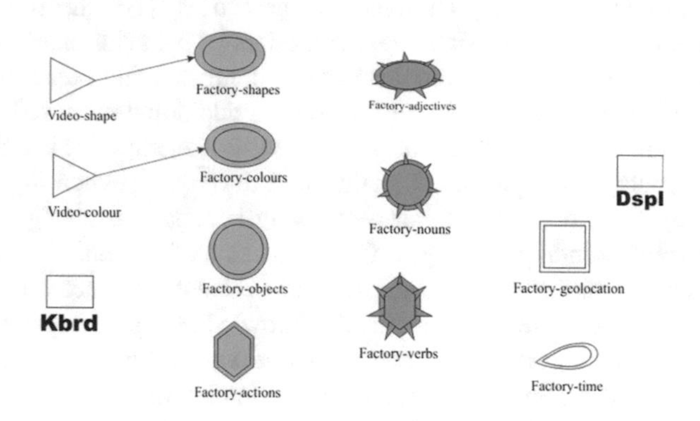

Fig. 1. Minimal architecture of multi-agent neurocognitive.

Agents-Factories, namely, *Factory-shapes, Factory-colors, Factory-objects, Factory-actions, Factory-geolocation, Factory-time* create units of knowledge. They are conceptual agents such as concept-attribute, concept-object, concept-action, concept-time, and concept-place. Since any conceptual notion must be supported by verbal representation [19], Agents-Factories, namely, *Factory-Adjectives, Factory-Nouns* and *Factory-Verbs*, create for each conceptual notion agents of the concept-word type (adjective, noun, verb respectively). Agents build interconnections in the form of energy exchange contracts to construct an event.

Then a certain event $s^{j}_{i\tau_\kappa}$, which occurred at some discrete point of time τ_κ, $k = 1, \ldots N$, where N - is the number of agents in the system, may be represented as functional dependencies between conceptual agents $\aleph^{j}_{i}$(i - is the name of the agent, j - is the type agent in the multi-agent neurocognitive architecture) formed on a contractual basis (see Fig. 2). The arrows indicate the multi-agent existential mapping (ע-function).

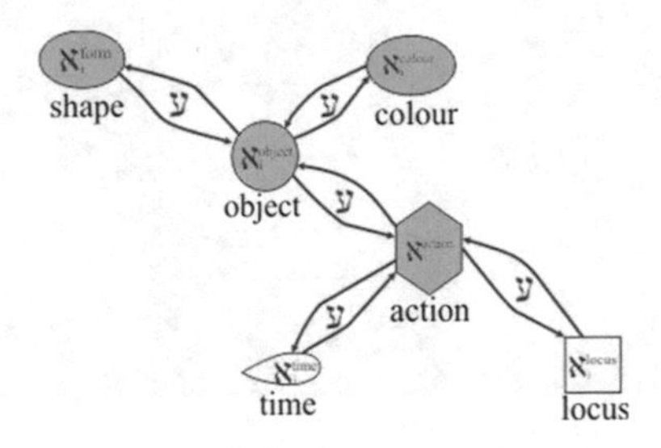

Fig. 2. Event composition on the base of conceptual agents of different types.

Formally, the event received at the input of the system can be written in the form:

$$s^{j}_{i\tau_k}=\left\{\aleph^{j}_{i} : \aleph^{j}_{i} = \cup^{j_k}_{i_k}\{ע(\aleph^{j+1}_{i+1})\}\right\}.$$

Therefore to recognize and understand an event, agents need to form links between concepts of such types as object, attribute, action, locus and time. The procedure for finding partners and concluding contracts is defined in the so-called agents' *genome*. *Genome* is a set of specific production rules for agents of certain type. An agent receives corresponding genome at the moment of creation. The rules contain information necessary for an agent to operate in a multi-agent system (see Fig. 3).

Let us consider the event when the multi-agent neurocognitive architecture received a static picture with the image of geometric figures in it. The system begins to analyze the event: a cube, a sphere and a pyramid were shown. Messages with signatures of the shape of the figure and of color come along the afferent path to the exteroceptive agents. If a signature enters the system for the first time, then the system sends it at the input of agents-factories. In our example, these are the Factory-shapes and the Factory-colors, which create new agents, namely, concept-shape and concept-color, respectively.

As the information is new, the system addresses to the user through the Display with the questions in natural language "What color is this?", "What shape is this?" User through the Keyboard inputs the names that are assigned to agents. And it is the start for the special procedure to define multi-agent contracts and create new agents such as concept-object, concept-action, concept-time, and concept-locus. For each conceptual agent, concept-word agents are created with the appropriate factories.

```
"AgentType" : "colour",
"knowledgebase" : [
    {
        "currentCondition" : [
            {
                "From" : "Status",
                "Message" : "New"
            }
        ],
        "desiredCondition" : [
            {
                "From" : "Stomach",
                "Message" : "Energy increased"
            }
        ],
        "ruleAction" : [
            {
                "Message" : "What colour is it?",
                "To" : "Display"
            },
            {
                "Message" : "Waiting for the answer",
                "To" : "Status"
            }
        ],
        "ruleStatus" : 1
```

Fig. 3. Fragment of the genome for conceptual agent-color.

In accordance with the data from genomes, agents fill their valences and conclude energy exchange contracts sending messages to agents of corresponding types. Agents make a newsletter to potential buyers of information that answers the questions: “What color?”, “What shape?”, “What object is it?”, “What does it do?”, “When?”, “Where?”.

Agents interested in increasing their energy, in exchange for the information they have, respond to the request. The implementation of the concluded contracts allows agents to increase their life-span and exclude terminal states, i.e. the agent’s death [20].

A multi-agent algorithm based on interactive training is presented in the Fig. 4. It shows creation of conceptual and concept-word agents by factories and establishing contract links between them through sending messages for recognizing and understanding geometric shapes formed by the MARI system. The composition of the event through multi-agent contracts is indicated by bold lines. For simplicity, the Fig. 4 shows the recognition of the only geometric shape - a cube.

The procedure for entering into energy-exchange contracts can be formally written as follows.

For concept agents of the *object* type:

$$\aleph_i^j \to \aleph_{i+1}^{j+1^*} : \mathrm{m}(\text{"What is it?"}). \tag{1}$$

Messages of the form (1) are sent to all agents of *action*, *color* and *shape* types, as well as agent-word of the *noun* type.

For concept agents of the *action* type:

$$\aleph_i^j \to \aleph_{i+1}^{j+1^*} : \mathrm{m}(\text{"What does it do? "}). \tag{2}$$

Messages of the form (2) are sent to all agents of *object*, *locus* and *time* types, as well as the agent-word of the *verb* type.

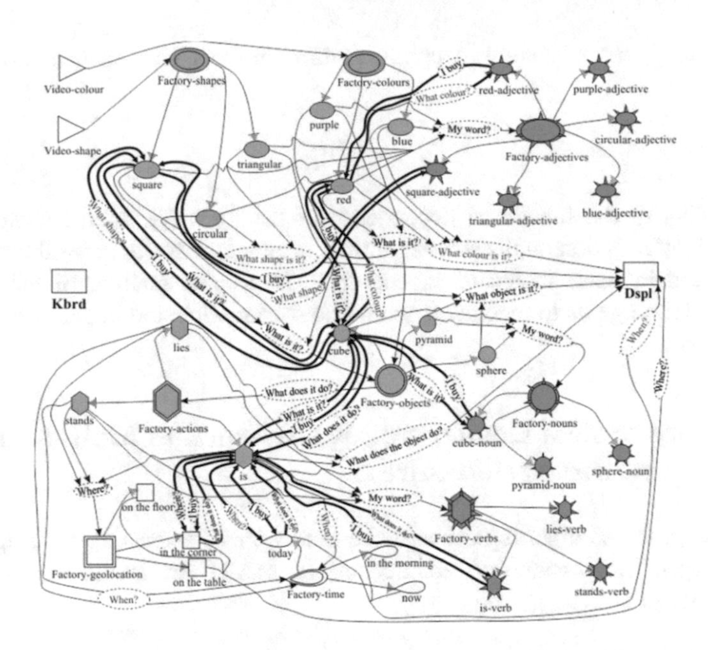

Fig. 4. The algorithm of the self-learning multi-agent neurocognitive architecture for recognizing geometric shapes.

For agents of concepts of *color* and *shape* types:

$$\aleph_i^j \rightarrow \aleph_{i+1}^{j+1^*} : \mathrm{m}(\text{"What colour/shape? "}). \tag{3}$$

Messages of the form (3) are sent to all agents of the *object* type, as well as to the agent-word of the *adjective* type.

For concept agents of the *locus* type:

$$\aleph_i^j \rightarrow \aleph_{i+1}^{j+1^*} : \mathrm{m}(\text{"Where? "}). \tag{4}$$

Messages of the form (4) are sent to all agents of the *action* type.

For concept agents of the *time* type:

$$\aleph_i^j \rightarrow \aleph_{i+1}^{j+1^*} : \mathrm{m}(\text{"When? "}). \tag{5}$$

Messages of the form (5) are sent to all agents of the *action* type.

Agents respond to such requests uniformly as follows:

$$\aleph_i^j \rightarrow \aleph_{i+1}^{j+1^*} : \mathrm{m}(\text{"I"}). \tag{6}$$

In (1–6), the * symbol next to the agent type indicates that the message is sent to every agent of the specified type in the system.

The multi-agent existential mapping resulting from such an exchange of messages can be written as:

$$\aleph_i^j = ע\left(\aleph_{i+1}^{j+1}\right). \tag{7}$$

According to the functional dependence of the form (7), agents always seek to spread their information on the request of their contractors to maximize the total energy rewarded and increase their life expectancy. The implementation of the concluded contracts allows agents to recognize geometric shapes when the image is displayed on camera again.

4 Software System Demonstrating the Work of Multi-agent Neurocognitive Architecture

To test the presented algorithm, a software system that demonstrates the operation of the multi-agent neurocognitive architecture of MARI for image recognition was developed (see Fig. 5).

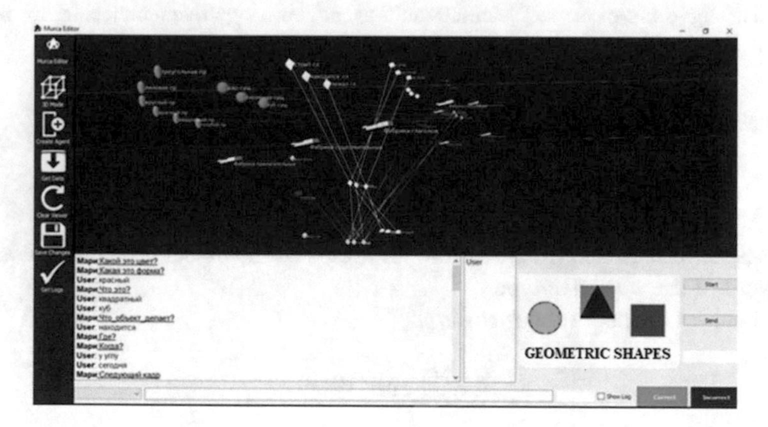

Fig. 5. Visual editor of the multi-agent neurocognitive architecture of MARI, for creating conceptual and word agents for recognizing geomctric shapes

Figure 5 shows the implementation of the algorithm for creating conceptual and concept-word agents by factories and concluding energy-exchange contracts through sending messages for recognizing geometric shapes: a cube, a pyramid and a sphere, and constructing the corresponding event.

Similar to the Fig. 4, contracts concluded by conceptual agents and word agents are showed with lines. For clarity, agents of different types are shown in different shapes and colors.

5 Conclusion

In the article a simulation model of the cognitive function for recognizing static objects, based on multi-agent architectures, and a software system that demonstrates its work in recognizing geometric shapes are presented.

The developed simulation model based on multi-agent neurocognitive architectures allows automatic creation of concepts and categories according to the data of multimodal input information (event occurred). Through interactive user interaction, the system can expand these concepts and categories, and clarify the links between them.

Due to the fact that the presented system is autonomous and self-learning, it can be used in autonomous artificial intelligence systems such as Smart Systems, robotic complexes, etc. to recognize unstructured data streams.

References

1. Tarasov VB (2002) From multi-agent systems to intellectual organizations: philosophy, psychology, computer science. [Ot mnogoagentnykh sistem k intellektual'nym organizatsiyam: filosofiya, psikhologiya, informatika.]. Editorial URSS, Moscow, 352 p
2. Russell S, Norvig P (2006) Artificial intelligence. Modern approach, 2nd edn. Williams Publishing House, Moscow, 1408 p
3. Wooldridge M (2009) An introduction to multiagent systems, 2nd edn. Wiley, Hoboken 484 p, ISBN: 0470519460
4. Weiss G (ed) (1999) Multiagent systems: a modern approach to distributed artificial intelligence. The MIT Press, Cambridge 643 p, ISBN: 0262232030
5. Shoham Y, LeytonBrown K (2008) Multiagent systems: algorithmic, game theoretic, and logical foundations. Cambridge University Press, Cambridge 504 p, ISBN: 0521899435
6. Lesser VR, Erman LD (1980) Distributed interpretation: a model and experiment. IEEE Trans Comput 29(12):1144–1163
7. Hewitt C (1977) Viewing control structures as patterns of message passing. Artif Intell 8 (3):323–364
8. Lenat D (1975) BEINGS: knowledge as interacting experts. In: Proceedings of the 1975 IJCAI conference, pp 126–133
9. Smith RG (1980) The contract net protocol: high level communication and control in a distributed problem solver. IEEE Trans Comput 29(12):1104–1111
10. Rzevski G (2012) Modelling large complex systems using multi-agent technology. In: Proceedings of 13th ACIS international conference on software engineering, artificial intelligence, networking, and parallel/distributed computing (SNPD 2012), Kyoto, Japan, 8–10 August, pp 434–437
11. Chen Y et al (2013) Multi-agent systems with dynamical topologies: consensus and applications. IEEE Circ Syst Mag 13(3):21–34
12. Chen M, Athanasiadis D, Al Faiya B, McArthur S, Kockar I, Lu H, De Leon F (2017) Design of a multi-agent system for distributed voltage regulation. In: 19th international conference on intelligent systems application to power systems (ISAP), 19 October 2017. IEEE, Piscataway, 6 p
13. Granichin O, Khantuleva T, Amelina N (2017) Adaptation of aircraft's wings elements in turbulent flows by local voting protocol. In: IFAC proceedings

14. Nagoev ZV (2012) Multiagent recursive cognitive architecture. In: Biologically inspired cognitive architectures 2012, Proceedings of the third annual meeting of the BICA Society. Advances in intelligent systems and computing series, pp 247–248. Springer, Berlin
15. Nagoev ZV (2013) Intellect, or thinking in living and artificial systems [Intellekt ili myshleniye v zhivykh i iskusstvennykh sistemakh]. Publishing House KBNTS RAS, Nalchik 211 p
16. Anokhin PK (1974) System analysis of the integrative activity of a neuron [Sistemnyy analiz integrativnoy deyatel'nosti neyrona]. Successes fiziol Sci 5(2):5–92
17. Nagoev ZV (2013) Multiagent existential mappings and functions [Mul'tiagentnyye ekzistentsial'nyye otobrazheniya i funktsii], Izvestiya KBNTS RAS, no 4(54), pp 64–71. Publishing KBNC RAS, Nalchik
18. Ivanov P, Nagoev Z, Pshenokova I, Tokmakova D (2015) Forming the multi-modal situation context in ambient intelligence systems on the basis of self-organizing cognitive architectures. In: 5th world congress on information and communication technologies (WICT), Morocco, 14–16 December
19. Reformatsky AA (2007) Introduction to linguistics [Vvedeniye v yazykoznaniye]. Aspect Press, Moscow
20. Nagoev Z, Nagoeva O, Tokmakova D (2016) System essence of intelligence and multi-agent existential mappings. In: Abraham A et al (eds) 15th international conference on hybrid intelligent systems (HIS 2015), Seoul, South Korea. Advances in intelligent systems and computing, vol 420, pp 67–76. Springer International Publishing Switzerland, Cham. https://doi.org/10.1007/978-3-319-27221-4_6

Multi-agent Algorithms for Building Semantic Representations of Spatial Information in a Framework of Neurocognitive Architecture

Zalimkhan Nagoev, Olga Nagoeva, Irina Gurtueva(✉), and Vladimir Denisenko

The Federal State Institution of Science Federal Scientific Center Kabardino-Balkarian, Scientific Center of Russian Academy of Sciences, I. Armand Street, 37-a, 360000 Nalchik, Russia
gurtueva-i@yandex.ru

Abstract. The paper outlines the conceptual basics of multi-agent modeling of the semantics of subjective reflexive mapping of the interaction between real objects, space and time. A holistic view of the intellectual agent about the environment is needed to develop a learning system for internal representation of the events localization space to realize orientation and navigation of autonomous mobile systems. The paper proves that the multi-agent neurocognitive architecture is an effective formalism for describing the semantics of the spatial localization of events. Main theoretical foundations have been developed for the simulation of spatial relations using the so-called multi-agent facts, consisting of software agents-concepts, reflecting semantic categories corresponding to parts of speech. It is shown that locative software agents that describe the spatial location of objects and events, forming homogeneous connections, compose the so-called *field locations*.

Keywords: Multi-agent systems · Neurocognitive architecture · Spatial localization · Artificial intelligence systems

1 Introduction

The ways of describing space in human's irrational picture of the world are based on correlative procedures. Different objects are the parameters in the correlative procedures, and the relations are oriented to the presentation of the features of the mutual arrangement of these objects.

Neuro-morphologically, the system of relationships is justified by the presence of specified brain neurons responsible for the localization of objects [1, 11]. In this paper, we use the concept of so-called locative neurons, which are used to conclude contracts with neurons that represent concepts. The essence of these contracts comes down to the establishment of two significant facts: (i) definition and reflection of points in space in which objects are located; (ii) description of the spatial relationship between these points.

A. V. Samsonovich (Ed.): BICA 2019, AISC 948, pp. 379–386, 2020.
https://doi.org/10.1007/978-3-030-25719-4_49

Of course, the properties of the so-called functional-semantic field have not been established yet and are debatable question [2, 9, 10].

In [2, 10], the types of spatial relationships are highlighted. They are expressed by language means as a multifaceted sphere of values, the central content of which is event localization. The main type of relationship in a given semantics, its dominant, is the designation of an action, an event flow within a space. The second type is an expression of dynamic spatial relationships associated with the situation of movement, where locality characterizes the starting, final point of movement, the route, etc. The third type is associated with the designation of the position of objects relative to each other. In this case, one object serves as a guideline, according to which the location of another object is determined. A significant sphere of spatial relationships has a pronounced anthropocentric content, since it is associated with the location and perception of the face. The position of the speaker in the situation of speech is most consistently manifested in deictic nominations and forms a subject-oriented localization - the fourth semantic type of spatial relationships. The fifth type is the parametric characteristics of objects and phenomena of reality, which are a necessary component of this range of values, as they reflect the three-dimensionality of space as one of its most important properties.

However, according to most researchers, the main dominant type of semantics of spatial relationships is the localization of action, event. This paper focuses on the modeling of this type of semantics of spatial relations.

We are interested in the semantics of spatial relations from a constructivist position, from the point of view of the possibilities of its formalization for subsequent use in artificial intelligence systems and intelligent systems (software systems and robots).

Event localization is necessary to perform functional tasks in intelligence systems.

The *object* of research in this paper is the semantics of the event localization space.

The *subject* of research is the possibility to create multi-agent models of the semantics for locative relations.

The *purpose* of the study is to develop a learning system for internal representation of the event localization space, orientation and navigation of autonomous mobile systems.

The *task* of the research is the development of simulation models of the semantics of the event localization space based on multi-agent neurocognitive architectures.

2 Neurocognitive Multi-agent Location Models

In [3], agents-concepts were described that perform the function of the representation of objects, phenomena, and their attributes in the cognitive architecture. In particular, it was shown that agents-concepts are classified according to the types of objects being represented. These can be conceptual agents-objects, representing objects of the real world, conceptual agents-actions, describing any actions, agents-attributes of objects, and agents-adverbials of actions, modifying, respectively, attributes of objects and manners of action, indicating place or time [4–8].

An agent of one of these types possesses a certain set of valences characteristic only for its type and uses them to conclude contracts with other agents. Agents are generated during system functioning on demand based on unstructured data input streams generated by intelligent agent sensor systems. The essence of structural reduction realizing by agent-concepts, is in replacement afferent data sequences on a conceptual agent in the process of interiorization. The agent-concept is activated whenever it recognizes its input signature in the afferent flow and performs an action aimed at the internal representation of the real-world object, which allows it to build reasoning about this object using the constructed representation. At the same time, the structuring of input data flows proceeds in the direction of schematization - the formation of agents-concepts of different types and the definition of relations between them.

The use of a system of conceptual agents generated on demand in a situationally determined context allows, after appropriate training, to recognize input events and build their description based on a multi-agent presentation of some fact about objects of the real environment. In [3], a description of the agents is given, the contracts between which form a multi-agent presentation of such a fact. In particular, it is shown that the semantic structure of such a construction corresponds to the types of relations in a sentence in a natural language (see Fig. 1).

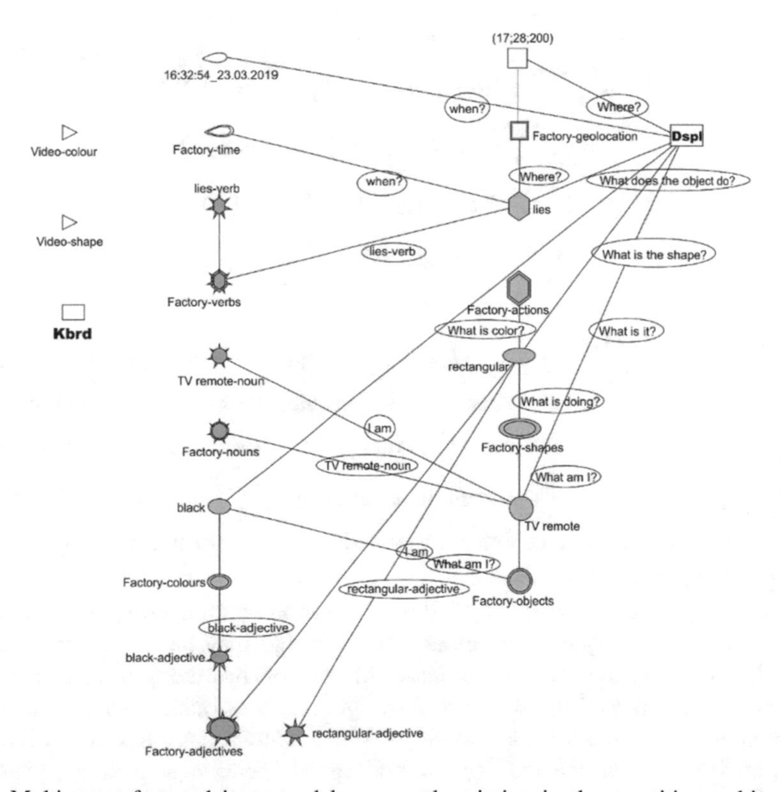

Fig. 1. Multi-agent fact and its natural language description in the cognitive architecture of intelligent agent (on the example of TV remote control recognition and its location determining).

In the Fig. 1 geometric figures depict agents-concepts (subject, action, object, attribute, manner of action, locus, time), and geometric figures inscribed in polygons represent the corresponding agents-words that perform certain syntactic functions in the sentence (subject, predicate, object, modifiers of object and action, adverbial modifiers of place and time). The Fig. 1 shows the relations between agents corresponding to their valences. A multi-agent fact is complete comprised if there are no empty valences for the agents included in its composition. Of fundamental importance is the structure of links between agents, which is formed on the basis of these valences. For example, as follows from the Fig. 1, links between attributes and objects, actions and manners of actions are formed according to the type of semantic relations in the noun group and the adverbial phrase, respectively.

As shown in [3, 8], the types of relations defined by valences correspond to the algorithms for the exchange of information and energy between agents based on a specific interaction protocol. Agents seeking to receive energetic rewards on the basis of certain protocol form proposals for the “sale” of information have known them in a message format, which is written as follows:

$$\aleph_{iRN}^{(\text{ל}-2)\text{ד}j}: name,\ e_i^{j1} = \text{ע}(?)$$

$$\aleph_{iRN}^{(\text{ל}-2)\text{ד}j}: \text{"}what\ is\ doing?\text{"},\ e_i^{j2} = \text{ע}(?)$$

$$\aleph_{iRN}^{(\text{ל}-2)\text{ד}j}: \text{"}which\ one?\text{"},\ e_i^{j3} = \text{ע}(?)$$

$$\aleph_{iRN}^{(\text{ל}-2)\text{ד}j}: \text{"}where?\text{"},\ e_i^{j4} = \text{ע}(?)$$

$$\aleph_{iRN}^{(\text{ל}-2)\text{ד}j}: \text{"}when?\text{"},\ e_i^{j5} = \text{ע}(?)$$

The record: $\aleph_{iRN}^{(\text{ל}-2)\text{ד}j}: name,\ e_i^j = \text{ע}(?)$ means that the agent $\aleph_{iRN}^{(\text{ל}-2)\text{ד}j}$ is looking for an agent to acquire a contract that will acquire the information contained in the message $\aleph_{iRN}^{(\text{ל}-2)\text{ד}j}: name$ in exchange for the fee e_i^j. The record: $\aleph_{kRV}^{(\text{ל}-2)\text{ד}}: \text{"}who?\text{"},\ e_k^j = \text{ע}(?)$ means that the agent $\aleph_{kRV}^{(\text{ל}-2)\text{ד}}$ is looking for an agent-seller to enter into a contract that will provide information that answers “*who*?” in exchange for the reward e_i^j.

Accordingly, another agent “buys” this information by transferring a certain portion of energy to the first agent. The initiator of exchange may be the buyer-agent (customer). In this case, the offer to purchase information has the type of question of a certain type. The types of questions asked by agents to each other are determined by the types of agents and the set of their valences. When multi-agent interaction arises in the process of forming “on demand” agents, conceptual agents representing certain properties of reality, connecting with each other in accordance with similar types of semantic relations, describe fragments of reality in a multi-agent fact format. Agents-words form contractual relations with concept-agents that provide relationships between these agents that are equivalent to the relationship between the vertices of the Frege triangle, representing designate and concept.

Thus, natural language statements interpreted by cognitive architecture are provided by multi-agent semantic models. Agents of noun phrase form the theme of a multi-agent fact and agents of predicate and adverbial modifiers build a statement/sentence. Theme and rheme of the fact and their dynamic connection constitute an event. Objects, attributes and phenomena of the event form in afferent flows generated by the intelligent agent sensory system input signatures of different types of concept-agents previously created in the cognitive architecture of an intelligent agent in the learning process. Signatures are recognized by the concept-agents, which then form contractual relationships, in accordance with the structure of their valence, thereby providing an internal representation of the observed event in a multi-agent fact format.

The Fig. 2 shows the so-called locative agent (agent-concept of place). It is depicted as a square. There are two contracts between the agent-concept of subject and agent-concept of action. The first one corresponds to the case when the agent-concept of subject "buys" information about the action that it performs, the second one - to the case when the agent-concept of action "buys" information about who exactly performs the action they represent.

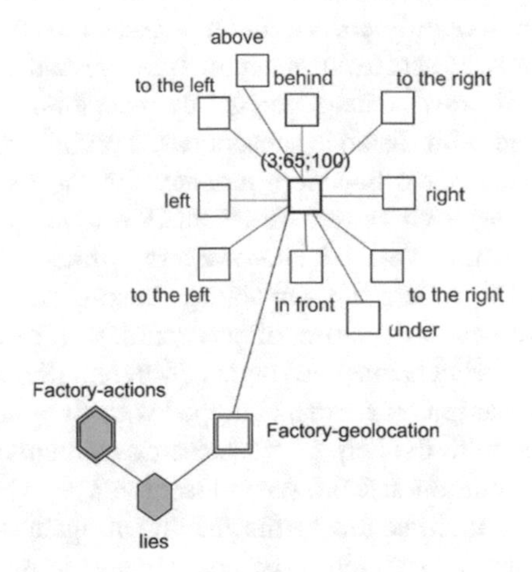

Fig. 2. Locative sensor structure of a cognitive architecture.

There is the only one contract between agent-concept of action and agent-concept of locus. This means that the semantic function of the agent-concept of locus is not related to the need to manage complex contexts formed on the basis of other agents-concepts, but is limited to identifying the location of objects involved in real-world events described by a multi-agent fact.

It also shows the so-called time agents. They are connected with agent- concept of action by a contract of only direction and, accordingly, do not provide energy, but provide just information in response to requests from the agents-concepts of action. By forming contracts with agent-concept of object, agent of time and a locative agent,

agent-concept of action marks the fact of the presence of an object at a certain point of space in a certain moment of time. Later, when the object's position changes, for example, when it is moved to a nearby location, another multi-agent fact is automatically generated in the neuro-cognitive architecture, marking the absence of this object at given point in corresponding moment of time.

Thus, an experience is formed that describes events of the real world in space-time reference, which is of fundamental importance for the possibility of recalling past events and the synthesis of multi-agent facts representing locative relationships between imaginary future events that have not yet occurred.

Let us describe the location of several objects that are not involved in the same event. Let the autonomous mobile robot need to investigate an unfamiliar room, discover all the familiar objects and examine (having learned recognition and natural language nomination) all unfamiliar, having made an internal map of the location of these objects. The map should allow the robot to build a route from one object to another and move between these objects by a command in natural language. In addition, it is necessary to teach the robot to determine the spatial relationships between objects and to answer questions in natural language regarding the absolute and relative position of these objects.

Afferent flows for the intelligent agent that performs the functions of control the robot and represents multi-agent neurocognitive architecture provide multimodal sensors consisting of a hardware-software receptor (sensor) and a software actor implementing preprocessing of "raw" data generated by receptors.

The robot is equipped with distance sensors based on infrared and ultrasonic range finders, two video cameras, two encoders mounted on the axes of the wheels, pain sensor (collision detection), and sensor of satisfaction (energy). Sensors of pain and satisfaction are based on the work of interoceptors, which record the values of the internal parameters of the robot; the remaining sensors are based on the work of exteroceptors, which record the values of environmental parameters. The robot is equipped with two motor wheels that can rotate in different directions and two movable wheels. It also has two manipulator arms, equipped with five-finger brushes with cable drives. Hands are sensed with the help of piezoelectric elements (touch receptors), Hall effect sensors, angle of rotation sensors (proprioceptors).

In the neurocognitive architecture of the intelligent agent of the robot, the possibility of generating agents of different types on demand is realized with the help of special actors, conventionally called factories. The behavior algorithm of a robot aimed at exploring a room consists of the following basic steps.

1. Determination of the starting point from which the robot begins learning. To realize it, the neurocognitive architecture of the intelligent agent of the robot generate a group of agents-concepts describing events, the meaning of which is that the robot has begun to study the new room (location). The event, accordingly, is documented by a multi-agent fact that includes a locative agent (see Fig. 1).
2. Further, the robot, using the image recognition system, determines objects familiar to it, located in the room and within its field of view, focuses on the largest of them and moves towards it, up to the triggering of the logical pain sensor (collision detection), which is triggered at a certain predetermined distance.

3. The robot registers the event describing the presence of this object at a given point in a room (location) by forming a multi-agent group (fact), similar to a multi-agent fact formed earlier for the starting point of the robot's stay in a given room (location).
4. Locative agents have a set of 11 valences. These are the only connection with the agent-concept of action, and 10 links with the locative agents closest to the agent known at the moment or in the future. To be exact, these are one from above, one from the below, four in the directions of the cardinal directions in the horizontal plane and four in directions that are 45° from the cardinal points, also in the horizontal plane. After creation the first locative agent, the rest 10 valences require their filling. However, it is not possible to fill them as other locative agents are absent in the system. After the second agent creation, also equipped with 11 valences, the agents have the opportunity to form bidirectional contractual relations with each other, filling one valence for each agent in accordance with the relative directions of the agents relative to each other. The essence of the contractual relationship is the purchase and sale of the mutual location information (distance, direction) of the locative agents. This information is generated automatically on the basis of data obtained from displacement sensors and angle of rotation, built on the basis of measurements made by encoders.
5. Next, the robot again looks around and moves to a new known object. The procedure of forming a multi-agent fact, the generation of a locative agent and the formation of its connections with previously formed locative agents, is repeated. The logical pain sensor (collision detection) is implemented on the basis of software processing of data from distance receptors. The statement of multi-agent fact describing events of a certain object in a certain place of a room is evaluated by a neuro-cognitive architecture as a positive event, bringing additional energy, which leads to an automatic addition of energy (value programmed generated) and the fact that an energy sensor increases the sensor of satisfaction. Thus, the robot is encouraged to study objects located in a given location and memorize their location.
6. Having described the location and the spatial position between them of all known objects, the robot proceeds to the study of unknown objects. Here, the described procedure, in general, is repeated, except that to describe these new, previously unknown objects and features, new agent-concepts are generated, and using the interactive system, users provide information about the natural language description of the established multi-agent fact, representing the event of the presence of this object at this point. Having completed the study of the room, the robot not only builds an internal spatial map of the room, but also gains knowledge of what objects are in certain parts of the room, and also learns the interpretation of natural language commands describing the tasks of moving from one object (locus) to another.

3 Conclusion

Thus, the multi-agent neurocognitive architecture is an effective formalism to describe the semantics of spatial localization of events. The theoretical foundations for the spatial relationships simulation were developed using the so-called multi-agent facts, consisting of program agents-concepts, reflecting semantic categories corresponding to parts of speech. It is shown that locative agents that describe the spatial location of objects and events, forming homogeneous connections, compose so-called field of locations, describing a holistic view of the intellectual agent about the environment.

Conceptual foundations for multi-agent modeling of reflection semantics of the interaction of real objects, space and time are developed.

The algorithm for forming triple connections between concept agents describing events, their localization and the time of occurrence has been developed and implemented.

Acknowledgements. The research was supported by the RFBR, grants № 18-01-00658 A, 19-01-00648 A.

References

1. Hafting T, Fyhn M, Bonnevie T, Moser M-B, Moser EI (2008) Hippocampus-independent phase precession in entorhinal grid cells. Nature 453:1248–1252
2. http://www.dissercat.com/content/prostranstvennye-otnosheniya-v-sovremennom-russkom-yazyke-semantika-i-sredstva-vyrazheniya
3. Nagoev ZV (2013) Decision-making and control methods in unstructured tasks based on self-organizing multi-agent recursive cognitive architectures. Thesis for the competition. Degrees of Doctor of Technical Sciences, Nalchik
4. Nagoev ZV (2013) Multiagent recursive cognitive architecture. In: Chella A, Pirrone R, Sorbello R, Jóhannsdóttir K (eds) Biologically inspired cognitive architectures 2012. Advances in intelligent systems and computing. Springer, Heidelberg, pp 247–248
5. Nagoev ZV (2013) Intellect, or thinking in living and artificial systems. Publishing house KBNTS RAS, Nalchik
6. Nagoev ZV, Nagoeva OV (2017) Building a formalism to describe the semantics of natural language statements and the process of understanding these statements based on self-organizing cognitive architectures. News of the Kabardino-Balkarian Scientific Center of the Russian Academy of Sciences, No 4 (78)
7. Nagoev ZV, Nagoeva OV: Modeling the semantics of phrases with attributive adjectives based on multi-agent recursive cognitive architecture. News of the Kabardino-Balkarian Scientific Center of the Russian Academy of Sciences №3 (83), 11–20 (2018)
8. Nagoev ZV, Nagoeva OV, Pshenokova IA (2017) Formal model of semantics of natural language statements based on multi-agent recursive cognitive architectures. News of the Kabardino-Balkarian Scientific Center of the Russian Academy of Sciences. №4 (78), 19–31
9. Ramachandran VS (2011) The tell-tale brain: a neuroscientist's quest for what makes us human. W. W. Norton&Company, NewYork, p 357. ISBN 978-0-393-07782-7
10. Renzhong GUO (1998) Spatial objects and spatial relationships. Geo-spatial Inf Sci 1(1):38–42. https://doi.org/10.1080/10095020.1998.10553282
11. Solstad T, Boccara CN, Kropff E, Moser M-B, Moser EI (2008) Representation of geometric borders in the entorhinal cortex. Science 322:1865–1868

Emotional and Moral Impressions Associated with Buddhist Religious Terms in Japanese Blogs-Preliminary Analysis

Jagna Nieuważny[1(✉)], Fumito Masui[1], Michal Ptaszynski[1], Kenji Araki[2], Rafal Rzepka[2], and Karol Nowakowski[1]

[1] Department of Computer Science, Kitami Institute of Technology, Koencho 165, Kitami, Hokkaido 090-8507, Japan
jagnajagna86@ialab.cs.kitami-it.ac.jp

[2] Faculty of Information Science and Technology, Hokkaido University, Kita 14, Nishi 9, Kita-Ku, Sapporo, Hokkaido 060-0814, Japan

Abstract. This paper is an attempt at analyzing how much religious vocabulary (in this case Buddhist vocabulary taken from a large scale dictionary of Buddhist terms available online) is present in everyday Japanese social space (in this case in a repository of blog entries form the Ameba blog service) and thus in the consciousness of people. We also investigate and what associations (positive or negative) it generates, thus indicating the connotations associated with several Buddhist terms – whether expressions containing Buddhist vocabulary are considered proper or not from a moral point of view – as well as the emotional response of Internet users to Buddhist terminology.

Keywords: Buddhist terminology · Automated moral reasoning · Machine ethics

1 Introduction

Religion in Japan is dominated by two main doctrines – Buddhism and Shintō. According to a survey[1] carried out in 2014, less than 40% of the population of Japan identifies with an organized religion and around 45% of those are Buddhists. However, a trait characteristic for Japan is, in contrast to Western countries, a low level of identification with an organized religion paired with a high level of actual participation in religious rituals of both main religions (or even more than two religions) –the total number of people estimated to participate in some form of religious ritual according to the Japanese Ministry of Education, Culture, Sports, Science and Technology topped 190 million people, which is more than the actual population of Japan (127,171,854 people as of July 17, 2018).

It has not been yet measured to what degree is Buddhism present in the consciousness of the masses, or whether does what is considered ethical behavior according to Buddhist religious principles correspond to what the general population considers to be ethically proper or improper behavior. In the present paper we will

[1] http://www.bunka.go.jp/tokei_hakusho_shuppan/tokeichosa/shumu/pdf/h26kekka.pdf.

A. V. Samsonovich (Ed.): BICA 2019, AISC 948, pp. 387–392, 2020.
https://doi.org/10.1007/978-3-030-25719-4_50

analyze on a number of particular examples to what degree is religious terminology present in the vocabulary of a usual Japanese Internet user – thus what could be considered a general consciousness, and what associations does it generate, from the point of view of emotional life and moral implications.

The outline of this paper is as follows. We introduce the resources used in this study in Sect. 2, describe the method used to analyze the data in Sect. 3, the experiment for defining connotations of Buddhist terms with ethical criteria and its results are presented in Sect. 4 and the paper is concluded in Sect. 5.

2 Resources Applied in this Study

The Digital Dictionary of Buddhism[2] is the largest dictionary of Buddhist terms available online. It is a lexicon of Chinese ideograph-based terms including titles of texts, names of temples, schools, persons, etc. found in Buddhist canonical sources. As of July 20, 2018, the dictionary contains 70,814 entries.

Yet Another Corpus of Internet Sentences (YACIS, Ptaszynski et al. 2012) is a large-scale corpus of Japanese, based on blog entries from Ameba blog service.

Automatic Moral Judgement Agent Based on Wisdom of WebCrowd and Emotions (Rzepka and Araki 2005; Komuda et al. 2010) is an automatic reasoning agent applying a dynamic algorithm for moral reasoning. The model takes a sentence as an input and searches in a specified source (such as the Internet or blog corpus as the one mentioned above) for emotion types and morality-related concepts (i.e.social consequences) associating with the sentence contents. The modified phrases are queried in the corpus (e.g. the Internet) for a set specified number of snippets for one modified phrase (default is 300 per phrase). This way a large number of snippets for each queried phrase is extracted from the Web and cross-referenced with the emotive and moral lexicons. The higher hit-rate an expression had in the Web search, the stronger was the association of the original phrase to the emotion and moral types[3].

3 Initial Analysis

In the analysis we first verified how many of the headwords from the Digital Dictionary of Buddhism are present in the YACIS corpus. 22,397 terms appeared at least once in YACIS (see Fig. 1 for results). Among the terms appearing, the most popular one appeared in the corpus a total of 27,884 times, while on the other hand 3,616 least frequent terms appeared only once. A total number of 15,087,745 blog posts contained Buddhism-related terms.

It must be noted that among the terms appearing in YACIS, the majority were not strictly related to the Buddhist religion, but rather – like the term with the largest number of hits: 麻布 (jap. *azabu*, "linen", 27,884 hits) – expressions that are part of

[2] http://www.buddhism-dict.net/ddb/.

[3] For example, the phrase "thank you" is likely to associate with gratitude, relief and joy and "killing a person" is more likely to associate with such moral associations as "going to jail", or "condemn".

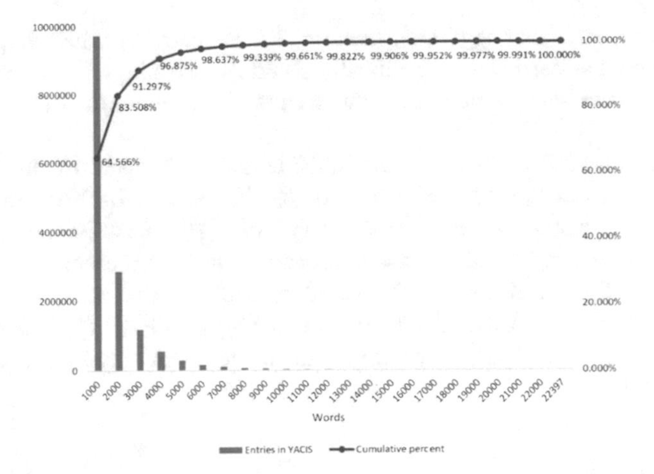

Fig. 1. 7.5% of the obtained vocabulary amounts for 83.5% of the total number of posts (the number of posts is indicated by the vertical axis)

Table 1. Top 10 terms from the Digital Dictionary of Buddhism with the greatest numbers of occurrences in YACIS corpus

Term	Reading	Meaning	Occurrences in YACIS
麻布	*azabu*	linen	27 884
時起	*Jiki*	monk name	25 391
株	*shu*	stump	24 404
胡麻	*goma*	sesame	23 713
日日	*nichi-nichi*	everyday	23 553
白	*shiro*	white	23 170
入門	*nyūmon*	to become a disciple	23 109
起動	*kidō*	move, awakening	23 015
右	*migi*	right (direction)	22 808
海水	*kaisui*	sea water	22 635

regular Japanese vocabulary, and also appear in Buddhist scriptures. Ten words with the biggest number of hits are shown in Table 1.

3.1 Phrase Filtering for Retaining Buddhist Context

After having specified which Buddhist terms appear in blogs, we needed to specify the seed phrases, which typically appear in Buddhist context, and not as lexicalized expressions. We selected the words covering 83.5% of all results, which only amounted to 7.5% of the total vocabulary found in blogs (Fig. 1). This means that the top 1,727 words generated an output of 12,070,196 posts.

We then divided the terms into those composed of one, two, three and four characters and noticed that most of the vocabulary composed of one character covered common words, such as "egg" (卵, *tamago*, 17,842 entries), which appeared in the Buddhist canon but carried no specifically Buddhist meanings in everyday life. We decided to remove such words from the results (it accounts to a loss of only 269 words).

Proper names, such as names of Buddhist temples (in Japan usually composed of three characters, as in 吉祥寺 – *Kichijōji*, or 金閣寺 – *Kinkakuji*) or names of monks (composed of two characters, as in 一休 – *Ikkyū*, or 空海 – *Kūkai*) often appear among 2- and 3-character long words in the dictionary, but their presence in the extracted sample was negligible. Among the vocabulary composed of 4 characters, all compounds had a Buddhist meaning, one example being 自業自得 (*jigō jitoku,* "the outcome of one's own karma") – however, there were only five such words in the extracted sample.

4 Experiment

4.1 Extraction of Word N-Grams for the Investigation Emotional and Moral Associations

In the next step, we used the 12,070,196 blog posts from YACIS containing any of the 7.5% most frequent terms. The procedure was as follows: first we tokenized the sentences and lemmatized them, using MeCab[4]. Then, we extracted 3-grams and 4-grams from the sentences and retained only those n-grams that actually contained the Buddhist term. Finally we applied an additional filter to retain only those n-grams, which contained not only the word in question but also grammatical particles[5], in order to increase the proportion of samples where it appears as part of a meaningful sentence. The total number of n-grams left to consider upon performing this preliminary cleaning was 19,305,978.

4.2 Examining Ethical and Emotional Associations of Buddhist Terms

From the pool of n-grams containing Buddhist terms obtained in the previous step, we selected several representative examples and investigated their emotional and ethical associations using the Moral Judgment Agent. As a proof of concept, below we present the results for three n-gram phrases:

1. 健康を祈願する (*kenkō wo kigan suru*, "to wish/pray for health");
2. 果報を受ける (*kahō wo ukeru*, "to get what one deserves, reward or retribution");
3. 嘘をつく (*uso wo tsuku*, "to tell a lie").

[4] http://taku910.github.io/mecab/.

[5] Belonging to the categories of case markers (格助詞 *kaku-joshi*)-namely が *ga*, の *no*, を *wo*, に *ni*, へ *he*, と *to*, で *de*, から *kara* and より *yori*, adverbial particles (副助詞 *fuku-joshi*): ばかり *bakari*, まで made, だけ *dake*, ほど *hodo*, くらい *kurai*, など *nado*, なり *nari*, やら *yara*, binding particles (係助詞 *kakari-joshi*): は *wa*, も *mo*, こそ *koso*, でも *demo*, しか *shika*, さえ sae, だに *dani* and conjunctive particles (接続助詞 *setsuzoku-joshi*): ば *ba*, や *ya*, が *ga*, て *te*, のに *noni*, ので *node*, から *kara*, ところが *tokoroga*, けれども *keredomo*.

For the first term, there were only 14 results found, with social consequences annotated as "good" in all of them. However the emotional consequences (3 in total) were all "bad", with the verdict that praying for health is "an action that most people dislike" – which can be explained when presuming that prayers for health are typically done in the context of someone's illness. As a result however, the term was considered "unethical" with the confidence of 75%, while indicating that 100% of the responses found it to be "the right thing to do". At the same time, 0% of the results stipulated that it is an action "worth praise" or "in accordance with the law" – which seems contradictory.

For the second term, in the case of which connotations could be both negative and positive, a total of 253 examples were found. 100% of the 44 social and moral consequences found belonged to the positive category. There were however 25 emotional consequences, out of which only 60% were tagged as "good".

The total of both (social and emotional) factors was described using 69 emotional categories from Nakamura's dictionary; 71% of the modified phrases found indicated that this was an action worth praise and 3% further signaled that it was "the proper thing to do".

For the last term finally – one that should certainly evoke negative connotations from the standpoint of Buddhist ethics – there were 279 examples of usage found in the YACIS corpus. The total of social consequences was unexpectedly equally divided into 50% of good and bad repercussions, while the total of emotional consequences (142 instances) shifted towards "bad", with 73% versus 26%, indicating that telling a lie was considered an unethical act. In addition, 11% of the responses indicated that telling a lie was "forgivable" (vs. 3% of "unforgivable" results). There were 45 hits for telling a lie being a morally acceptable act versus 113 hits for it being unacceptable from an ethical point of view. 142 expressions of emotions were associated with telling a lie, most of which indicated it to be an act regarded as "unpleasant". The algorithm concluded that telling a lie was a definitely immoral conduct.

5 Conclusions

In this paper we presented a preliminary analysis on whether and to what degree religious vocabulary is present in the everyday of Japanese Internet users (in this case in a repository of blog entries), what is the emotional response of bloggers to Buddhist terminology and what are the results of ethical evaluations of snippets of texts including Buddhist vocabulary. In order to do this, we checked which headwords from the Digital Dictionary of Buddhism appeared in YACIS, a large-scale corpus of Japanese language. We focused on terms covering roughly 80% of all results, dividing them into 1, 2, 3 and 4 character words. We tokenized and lemmatized the blog entries containing these terms, extracted n-grams (up to 4-grams, limited to those containing the term in question and a grammatical particle) and checked a few hand-selected representative examples of expressions containing Buddhist vocabulary, using an Automatic Moral Judgement Agent Based on Wisdom of WebCrowd and Emotions (Rzepka and Araki 2005; Komuda et al. 2010)

Although a lot of headwords from the Digital Dictionary of Buddhism appeared in YACIS, a majority of those were in fact words that are part of regular Japanese vocabulary and not specifically religious terms. Analysis of the data with the Moral Judgment Agent revealed that Buddhist terms were indeed not absent as a theme from Japanese blogs and generated a strong emotional response.

While the general reaction to several expressions using Buddhist terms was as expected, there were sometimes surprising twists in terms of social consequences. One such instance was the total of social consequences for the term uso wo tsuku ("to tell a lie"), which were unexpectedly equally divided into 50% of good and bad moral repercussions – while Buddhist ethics don't allow a concept such as a white or good lie.

As the next step in the research, we plan to develop heuristic rules in order to analyze how many words from the Digital Dictionary of Buddhism appearing in YACIS are strictly Buddhist religious terms.

References

Komuda R, Ptaszynski M, Momouchi Y, Rzepka R, Araki K (2010) Machine moral development: moral reasoning agent based on wisdom of web-crowd and emotions. Int J Comput Linguist Res 1(3):155–163

Nakamura A (1993) Kanjō hyōgen jiten [Dictionary of Emotive Expressions]. Tokyodo, Tokyo

Ptaszynski M, Dybala P, Rzepka R, Araki K, Momouchi Y. (2012) YACIS: a five-billion-word corpus of japanese blogs fully annotated with syntactic and affective information. In: Proceedings of the AISB/IACAP world congress 2012 in honour of Alan Turing, 2nd symposium on linguistic and cognitive approaches to dialog agents, LaCATODA 2012. University of Birmingham, Birmingham, pp 40–49

Rzepka R, Araki K (2005) What statistics could do for ethics?-the idea of common sense processing based safety valve. In: AAAI Fall Symposium on Machine Ethics, Technical Report FS-05-06. The AAAI Press, Menlo Park, California, pp 85–87

Application of Registration of Human Vegetative Reactions in the Process of Functional Magnetic Resonance Imaging

Vyacheslav A. Orlov[1], Yuri I. Kholodny[1,3(✉)], Sergey I. Kartashov[1], Denis G. Malakhov[1], Mikhail V. Kovalchuk[1], and Vadim L. Ushakov[1,2(✉)]

[1] NRC "Kurchatov Institute", Moscow, Russia
ptica89@bk.ru, kholodny@yandex.ru
[2] National Research Nuclear University MEPhI
(Moscow Engineering Physics Institute), Moscow, Russia
ushakov_vl@nrcki.ru
[3] Bauman Moscow State Technical University, Moscow, Russia

Abstract. This paper presents the first results of experiments on the joint use of MRI and MRI compatible polygraph in the study of neurocognitive processes, in particular - related to revealing of concealed information in humans. Experiments have shown the feasibility of using an MRI compatible polygraph in the process of fMRI studies. In the course of the study, the criterion for classifying subjects according to the dynamics of their vegetative reactions was discovered: the criterion allows for a more focused approach to the study of neurocognitive processes and may contribute to improving the quality of fMRI research for various purposes.

Keywords: MRI compatible polygraph, fMRI, Vegetative reactions · Concealed information · Neurocognitive processes

1 Introduction

In recent years, in order to in-depth study of brain activity during the implementation of certain mental phenomena, there has been a steady trend of joint use of several methods, including magnetic resonance imaging (hereinafter referred to as MRI). Researchers believe that "it is the combination of MRI with other methods ... will be the most effective strategy for understanding brain function" [1].

To conduct experimental studies on the neurocognitive mechanisms of identifying information concealed by a person, NRC "Kurchatov Institute", acting in line with the above-mentioned trend, developed a prototype of hardware and software complex: MRT compatible polygraph (hereinafter referred to as MRIcP), which provides high-quality registration of physiological reactions of a person during functional magnetic resonance imaging (hereinafter - fMRI).

A. V. Samsonovich (Ed.): BICA 2019, AISC 948, pp. 393–399, 2020.
https://doi.org/10.1007/978-3-030-25719-4_51

Being a complex technology, fMRI to obtain reliable results of the study requires control of movements of a person, dynamics of activity of his cardiovascular system (hereinafter CVS) and respiration [2,3]. However, the process of human perception of stimulus information during fMRI research remained out of the field of view of researchers. We assumed that the use of MRIcP will allow us to "see" the process of perception of human stimulus information in real time and better control the quality of the conducted fMRI research [4].

The creation of MRIcP opened up the possibility of using correctly and controllably the methodical means (i.e., tests), traditionally used in studies using a polygraph (hereinafter referred to as SUP), also in MRI environment.

In the framework of the planned research using fMRI, it was supposed to study the neurocognitive mechanisms of the processes that ensure during the SUP that a person reveals concealed information about past events. To carry out this study, the paradigm of revealing concealed information (hereinafter - the CI paradigm) [5] and, accordingly, tests based on it was chosen.

2 Materials and Methods

In order to carry out the installation series of experiments, the so-called "test with a concealed name" (hereinafter - TCN) was borrowed from the arsenal of forensic SUP, during which the person under study (hereinafter referred to as the subject) concealed his own name from the polygraph examiner along with five other names; the series of names were presented to the subject during the test five times. With the exception of one name, which stood under the number "0", all the others were presented in a random, unknown order to the subject with the phrase "Your passport name is ...". The names were set by the experimenter with an interval of about 20 s with the obligatory account of the current dynamics of the physiological parameters of the test, recorded using MRTcP. The cumulative graphical representation of the physiological parameters during the TCN was visualized on the computer screen in the form of a polygram. The dynamics of the electrical properties of the skin, i.e. galvanic skin reactions (hereinafter referred to as GSR), as well as reactions observed in the cardiovascular system and manifested in a change in heart rate (hereinafter referred to as heart rate) and narrowing of the blood vessels of the test subject's fingers (the so-called vascular spasm) were subject to analysis and evaluation.

Registered physiological reactions were subjected to expert evaluation on a 3-point scale widely used in SUP practice [6].

Experiments with the TCN was aimed at:

- during the experiments, evaluate the applied capabilities of MRTcS and determine its necessary refinements;
- detect specific brain areas involved in the process of concealing personally significant information for the subject (his own name);
- evaluate the sufficiency of MRI data recorded during the test for methodologically correct assessment of the activity of brain areas of research interest.

Experiments were performed on a homogeneous group of 20 healthy subjects (men aged 22–25 years).

Permission to conduct the experiment was obtained from the ethics committee of NRC "Kurchatov Institute". The experiment was conducted using MRI scanner Siemens Magnetom Verio 3 T based on NRC "Kurchatov Institute". To obtain anatomical MRI images, a three-dimensional T1-weighted sequence was used in the sagittal plane with high spatial resolution (176 slices, TR = 2530 ms, TE = 3.31 ms, thickness = 1 mm, angle = 7, inversion time = 1200 ms and FOV = 256 256 mm^2). Functional data was obtained using a standard echo-planar sequence (32 slices, TR = 2000 ms, TE = 24 ms and isotropic voxel 222 mm^3).

Pre-processing of MRI data was carried out on the basis of the freely distributed software package SPM8 [7], and specially adapted and developed terminal scripts of the MacOS system. Structural and functional data were brought to the center in the front commissure. Further, the calculation and correction of motion artifacts was made. With the help of magnetic field inhomogeneity maps recorded during the study, functional data was corrected in order to remove magnetic susceptibility artifacts. Structural and functional MRI volumes were normalized to template images in the MNI (Montreal Neurological Institute) space. In order to remove incidental emissions, a Gaussian based filter with a $6 \times 6 \times 6\,mm^3$ core was applied to the functional data. The preprocessing procedure was carried out according to the above scheme for each of the 20 subjects.

3 Results

Figure 1 shows the results of a group assessment of the active brain zones of 20 subjects when they conceal personally significant information (their own name) in comparison with the activity of names that are insignificant in TCN conditions. The figure shows large bilateral (on both sides) activation in the lower frontal and temporal gyri, the upper orbital gyrus; caudate nucleus and cerebellum (VI) on the left. It should be noted that the statistical analysis was carried out using T-statistics at $p < 0{,}001$.

Analysis of vegetative reactions (GSR, heart rate and vascular spasm), recorded using MRTPS, showed that the subjects did not constitute a homogeneous group and, according to the dynamics of these reactions, should be divided into two subgroups - high-reactive (see Fig. 2) and low-reactive subjects (see Fig. 3).

In the subgroup of high-reactive subjects (15 persons), the severity of GSR was in the range of 60–100% (that is, the subjects have according to the GSR in TCN from 6 to 10 points out of 10 possible). In low-reactive subjects (5 persons), the severity of GSR was 40% or less (i.e., subjects received 4 or less from 10 possible).

The response dynamics of high and low reactive subjects during the TCN is presented in polygrams (Figs. 2 and 3).

Based on the classification of the subjects, carried out using MRTcP and TCN, the data obtained during the MRI study were also divided into two subgroups and subjected to repeated group evaluation.

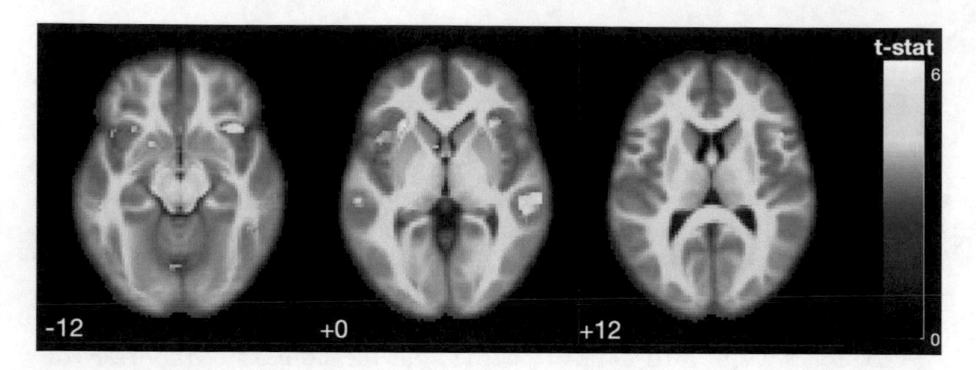

Fig. 1. The results of statistical evaluation for a group of 20 subjects. The figure shows a group statistical map printed on a high-resolution T1 image at levels z $= -12, 0, 12$ ($p < 0{,}001$)

The polygram (Fig. 2) shows the fourth (of five) presentation of TCN. Concealing of significant information (own name is Artur) causer maximal GSR (channel 6) to the subject, and also clear drop in heart rate (channel 7; moving lens shows 49 beats/min.) and constriction of the vessels of the fingers (channel 8).

The polygram (Fig. 3) shows the second (of five) presentation of TCN. Concealing of significant information (own name is Vadim) does not cause GSR to the subject (channel 6) and constriction of the vessels of the fingers (channel 8), but it is accompanied by a drop in heart rate (channel 7; moving lens shows 66 beats/min).

The analysis of MRI data for each of the subgroups showed a significant difference in the patterns of activation of the brain zones for high-reactive (see Fig. 4) and low-reactive subjects (see Fig. 5). Firstly, these subjects have a clear difference in the volume of involvement of the brain areas in the process of concealing information, and secondly, the group of highly reactive subjects has a prevalence in the activation of brain areas while concealing their own name against the background of other names, while for low-reactive subjects, the opposite is true.

In particular, in the presented sections for a group of high-reactive subjects (15 persons) in both hemispheres, activation is seen in the caudate nuclei, the inferior temporal, frontal, lingual and calcine gyri, the cerebellum (VI), the olfactory cortex, the insula and the thalamus on the left, and the signal drop is observed in the shell on the left.

On sections for a group of low-reactive subjects (5 persons), activation is present only in the putamen on the left. A drop in activation is observed in the middle occipital gyrus on the left and the calcine gyrus on the right.

The difference in the dynamics of the activation/deactivation of the brain areas during the concealment of information by high and low reactive subjects was discovered for the first time and is of indisputable scientific and applied

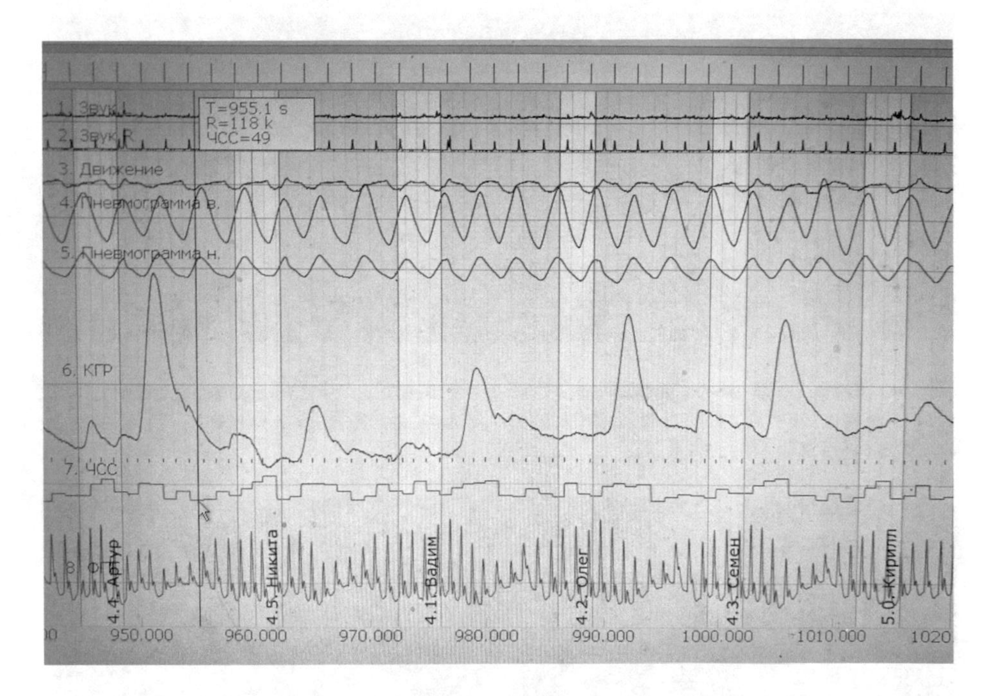

Fig. 2. TCN polygram of a high-reactive subject

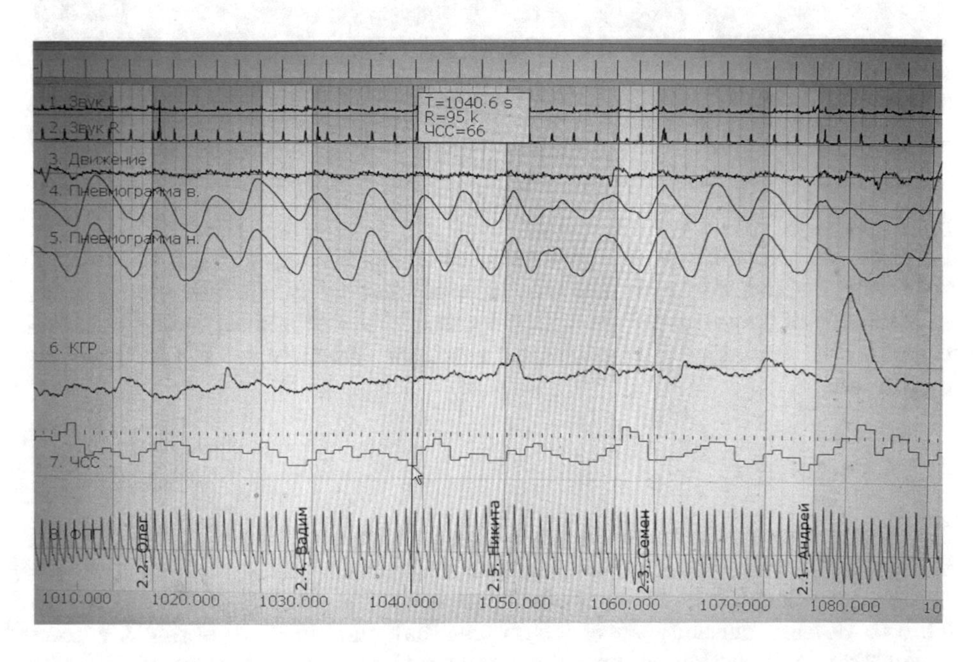

Fig. 3. TCN polygram of a low-reactive subject

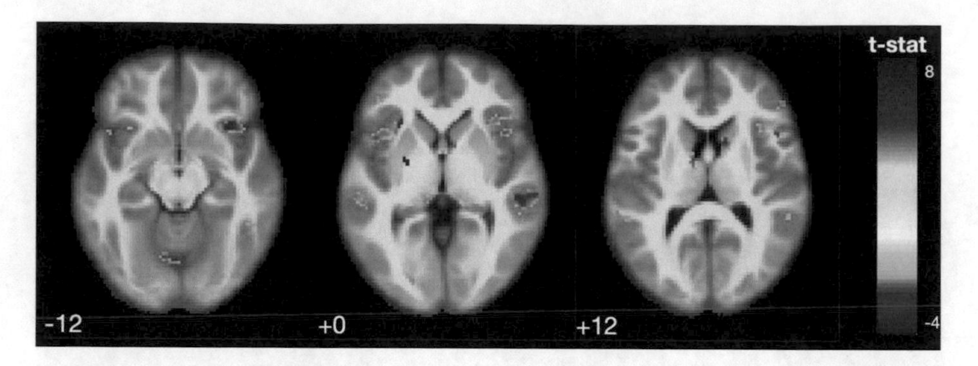

Fig. 4. The results of statistical evaluation for a group of 15 high-reactive subjects. The figure shows a group statistical map printed on a high-resolution T1 image at levels $z = -12, 0, 12$ ($p < 0{,}001$)

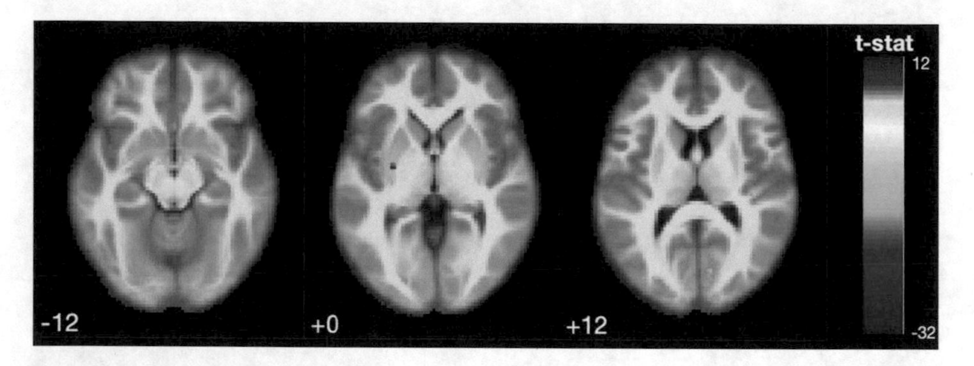

Fig. 5. The results of statistical evaluation for a group of 5 low-reactive subjects. The figure shows a group statistical map printed on a high-resolution T1 image at levels $z = -12, 0, 12$ ($p < 0{,}001$)

interest. However, due to the small size of the subgroups (15 and 5 persons), it is still premature to state the existence of such a pattern. The study of the relationship of the involvement of brain zones in the process of hiding information and the dynamics of physiological reactions (recorded by MRTcP) will be continued.

4 Conclusion

The experiments confirmed the promising prospects of the joint use of fMRI technology and SUP to study neurocognitive processes [4].

In particular, the integration of fMRI and SUP technologies [5] made it possible to develop methodological techniques that allowed us to obtain a correct assessment of the activity of neural structures, identify specific areas of the brain involved in the process of concealing personally significant information for the test (proper name) and detect (with the help of MRI), the different dynamics of the response of individual areas of the brain, due to the individual psychophysiological characteristics of the subject.

This work was in part supported by the National Research Center Kurchatov Institute (MR compatible polygraphy), by the Russian Science Foundation, grant 18-11-00336 (data preprocessing algorithms) and by the Russian Foundation for Basic Research, grants ofi-m 17-29-02518 (study of thinking levels). The authors are grateful to the MEPhI Academic Excellence Project for providing computing resources and facilities to perform experimental data processing.

References

1. Logothetis NK (2008) What we can do and what we cannot do with fMRI. Nature 12(453):870
2. Birn RM, Smith MA, Jones TB, Bandettini PA (2008) The respiration response function: the temporal dynamics of fMRI signal fluctuations related to changes in respiration. NeuroImage 40:644–654
3. Chang C, Cunningham JP, Glover GH (2009) Influence of heart rate on the BOLD signal: the cardiac response function. NeuroImage 44:857–869
4. Kovalchuk MV, Kholodny YI. Functional Magnetic Resonance Imaging augmented with Polygraph: new capabilities, In print
5. Ushakov VL, Malakhov DG, Orlov VA, Kartashov SI, Kholodny YI, Kovalchuk MV (2018) Research of neurocognitive mechanisms of revealing of the information concealing by the person. In: Biologically inspired cognitive architectures 2018. Proceedings of the ninth annual meeting of the BICA society. Advances in intelligent systems and computing, vol 848. Springer, pp 310–315
6. The accuracy and utility of polygraph testing (Department of Defense, DC) (1984) Polygraph 13:1–143
7. https://www.fil.ion.ucl.ac.uk/spm/software/spm8/

A Review of Method and Approaches for Resting State fMRI Analyses

Vyacheslav A. Orlov[1(✉)], Vadim L. Ushakov[1,2], Stanislav O. Kozlov[1,3], Irina M. Enyagina[1], and Alexey A. Poyda[1]

[1] National Research Centre “Kurchatov Institute”, Moscow, Russian Federation
ptica89@bk.ru, tiuq@yandex.ru

[2] National Research Nuclear University MEPhI (Moscow Engineering Physics Institute), Kashirskoe Highway 31, Moscow 115409, Russian Federation

[3] MIPT, Moscow, Russia

Abstract. Resting-state functional Magnetic Resonance Imaging (R-fMRI) measures spontaneous low-frequency oscillations of the BOLD signal in order to identify the functional architecture of the human brain. The analysis of such data allowed to identify resting state networks (RSN) and other areas of the brain that operate synchronously in time. Over the past few years, the interest of both scientists and clinicians to various methods of R-fMRI data analysis has greatly increased. In this article, we present a review and comparison of various methods and algorithms for analyzing the functional connectivity of the human brain in resting state, developed in the world, based on an analysis of the literature.

Keywords: Neuroimaging · fMRI · Static functional connectivity · Dynamic functional connectivity

1 Introduction

The interest in determining the connectivity of functional networks in the resting state of the brain is associated with the possibility of building a neural network architecture while operating a basic level of consciousness. The R-fMRI considers spontaneous low-frequency fluctuations (no more than 0.1 Hz) of the BOLD signal. For the first time in the work of Bisval [1], the functional significance of these fluctuations was shown. The subsequent work of other authors confirmed the presence of synchronization between the somatosensory regions of the brain at rest. One of the main discoveries in the analysis of R-fMRI became the resting state networks (RSN). RSN were first shown in the work of Raichle et al. [2]. During the analysis of positron emission tomography data of the state. It was found that certain parts of the brain were involved during a state of rest and reduced their activity when performing various cognitive tasks.

Functional connectivity is a matrix, in rows and columns, which hosts the nodes (regions of interest), and at the intersection is the strength of the functional connectivity (most often a correlation or partial correlation) between two corresponding nodes. In early works on functional connectivity, it was suggested that FC does not change in

A. V. Samsonovich (Ed.): BICA 2019, AISC 948, pp. 400–404, 2020.
https://doi.org/10.1007/978-3-030-25719-4_52

time and can be represented as a stationary matrix. In recent studies, researchers abandoned this assumption and began to apply various methods to assess non-stationary changes in the connectivity matrix during the entire scan period, which usually varies from 5 to 20 min.

2 Static Functional Connectivity Approaches

The most common method for assessing static functional connectivity is to calculate the Pearson correlation between the time series of regions of interest that reflects synchrony in cerebral blood flow and oxygenation between different regions.

Another popular method is the decomposition of the entire brain BOLD signal into statistically independent components. For example, applying the independent component method to a resting state data set, reveals different RSN. However, despite the small amount of a priori assumptions, most of the components received are noise and require expert evaluation to select the components associated with the signal. Despite the enormous number of differences in these two methods, Rosazza et al. showed that the results of seed base analysis and ICA are significantly similar in a group [4].

Graph method is an essential alternative to. This approach considers the regions of interest as nodes, and the correlation between them is represented as edges. Then, using the graph theory, we can estimate the connection characteristics. In graph theory, there are a large number of characteristics of the graph itself, which can then be calculated. For example, the coefficient of clusterization, which shows the connectivity of neighboring nodes and shows the possibility of the existence of subgraphs [5].

Also, clustering algorithms can be used to construct a functional connectivity matrix. When analyzing data, voxels can be combined into groups based on the similarity of some features. The most frequent metric for such an analysis is the Pearson correlation. Such an approach allows using the K-means clustering algorithm allows to build tree hierarchies based on thc analysis of distance metrics (for example, Pearson correlation or coherence). With this method, each voxel is assigned a membership in one of the clusters based on the distance of the voxel from the center of the cluster. Further, iteratively, changing the belonging of voxels to clusters and cluster centers, is achieved by dividing the brain into regions.

Yuan Zhong et al. proposed to consider the FCs combine the method of principal components (PCA) and regression analysis [6]. With the PCA, the authors proposed to select the most informative clusters. Next, using regression to the obtained components to build maps, applying to which the student's statistics can be left only statistically significant maps that can be represented as functional connectivity. There are several types of methods for assessing the time-varying fMRI patterns of data. Each of these methods can consider time intervals starting from a separate volume (1 time point) and ending with several minutes. Of course, the choice of the time scale is a compromise between the reliability and sensitivity of the method used, not to mention the fact that different neuronal processes can take place at different time.

3 Dynamic Functional Connectivity Approaches

Sliding Window Analyses. The first, simplest and most common method for evaluating dynamic functional connectivity is the sliding window analyses. The main idea of the method is to divide the entire time series into a group of shorter segments and calculate static functional connectivity within these segments [7]. However, the major drawbacks of this approach are the strong dependence of the results obtained, depending on the choice of window parameters. Theoretically, the choice of window should be determined by the required number of samples in order to get statistically significant results and cover at least 1 full cycle of the slowest wave present in the signal. However, the choice of an overly large window is most likely reduced to a blurring of switching times between different states of the brain. Thus, the question of choosing the optimal window parameters remains open and is the direction of future research.

Time-Frequency Analyses. Since the frequency characteristics of the BOLD signal are known (< 0.1 Hz), a number of recent studies have identified heterogeneous spectral characteristics within and between different quiescent networks [8]. Typically, a metric called coherence is used for this kind of analysis, which allows, based on the Hilbert transform, to calculate the phase shift between two time series. However, the main disadvantage of this method is the reduction of the signal-to-noise ratio with increasing frequency.

Dynamic Graph Analyses
The application of graph theory to a set of matrices obtained in the course of the analysis of dynamic connectivity can be used to calculate different properties of networks. Consider each matrix separately in terms of graph theory and calculate all the relevant characteristics of the graph and the division of the graph into subgraphs, for example, rich club structures [9].

Point Process Analysis
Recent studies showed that using paradigm free mapping or point process analysis can be used to find many brain voxels, that co-activate in meaningful patterns [10]. Liu and Duyn assumed that if you average only high-intensity events over time, you can get multiple repetitive, but different spatial patterns, which were named co-activation patterns [11]. Such patterns are obtained by decomposing all temporal fMRI volumes into clusters and averaging all the volumes included in one cluster. The authors have reliably shown that the resulting activity patterns are in very good agreement with the networks obtained using the method of independent components on the same data. However, the use of this method, unlike all previous ones, does not impose any restrictions on the data and does not require additional checks before use. Also, this method allows to evaluate various instantaneous states of the brain instantly (within one volume). However, this method has not yet received widespread use and is still under development.

4 Discussion and Conclusions

The approaches shown in the paper are not comprehensive, but consider the most interesting modern approaches to the analysis of neural network activity of the brain. One of the main problems of their use in practice is verification, obtained on the basis of the proposed results approaches. The only way to verify is to record the potentials of the brain directly from the cortex during intracranial operations, which is almost impossible in non-clinical conditions. An alternative to the authors of this work is creating a framework that allows to apply any of the proposed approaches to assess the connectivity of the neural networks of the brain, followed by testing the stability of the result obtained with different approaches, determining the sensitivity of the resulting connectivity matrices of choice for analyzing the volume of voxels within the functional area. This will allow in the near future to develop a methodology for building architectures of the brain based on an objective assessment of the result of computing neural network connectivity with subsequent practical application in the field of medicine, neurocognitive research.

This work was in part supported by the Russian Science Foundation, Grant № 18-11-00336 (data preprocessing algorithms) and by the Russian Foundation for Basic Research grants ofi-m 17-29-02518 (study of thinking levels), by the Russian Foundation of Basic Research, grant RFBR 18-29-23020 mk (method and approaches for fMRI analyses). The authors are grateful to the MEPhI Academic Excellence Project for providing computing resources and facilities to perform experimental data processing.

References

1. Biswal B, Yetkin FZ, Haughton VM, Hyde JS (1995) Functional connectivity in the motor cortex of resting human brain using echo-planar MRI. Magn Reson Med 34(4):537–541
2. Raichle ME, MacLeod AM, Snyder AZ, Powers WJ, Gusnard DA, Shulman GL (2001) A default mode of brain function. Proc Natl Acad Sci U S A. 98(2):676–682
3. Allen EA et al (2014) Tracking whole-brain connectivity dynamics in the resting state. Cereb Cortex 24:663–676. https://doi.org/10.1093/cercor/bhs352
4. Rosazza C, Minati L, Ghielmetti F (2012) Functional connectivity during resting-state functional MR imaging: study of the correspondence between independent component analysis and region-of-interest-based methods. AJNR Am J Neuroradiol 33:180–187
5. Behrens TE, Sporns O (2012) Human connectomics. Curr Opin Neurobiol 22(1):144–153
6. Zhong Y, Wang H, Lu G et al (2009) Detecting functional connectivity in fMRI using PCA and regression analysis. Brain Topogr 22:134. https://doi.org/10.1007/s10548-009-0095-4
7. Preti MG, Bolton TA, Van De Ville D (2017) The dynamic functional connectome: state-of-the-art and perspectives. Neuroimage 160:41–54
8. Gohel SR, Biswal BB (2015) Functional integration between brain regions at rest occurs in multiple-frequency bands. Brain Connect 5(1):23–34
9. Zalesky A, Fornito A, Cocchi L, Gollo LL, Breakspear M (2014) Time-resolved resting-state brain networks. Proc Natl Acad Sci U S A. 111(28):10341–10346

10. Gaudes CC, Petridou N, Dryden IL, Bai L, Francis ST, Gowland PA (2011) Detection and characterization of single-trial fMRI bold responses: paradigm free mapping. Hum Brain Mapp 32:1400–1418
11. Liu X, Duyn J (2013) Time-varying functional network information extracted from brief instances of spontaneous brain activity. Proc Natl Acad Sci USA 110(11):4392–4397

Use of Cognitive Maps in Learning Management System Vector

Daria Pavlenko(✉), Leonid Barykin, Eugeny Bezverhny, and Sergey Nemeshaev

National Research Nuclear University "MEPhI", 115409 Moscow, Russian Federation
pavlenko.da27@gmail.com

Abstract. The main objective of the study is to identify methods that can help improve the quality of learning using LMS. Through the use of a tag structure for a bank of questions and the use of cognitive maps, it becomes possible to more accurately construct individual learning paths. New methods can reduce the degree of subjectivity in the composition of tests and allow to identify gaps in the knowledge of a particular student.

Keywords: LMS · Cognitive map · Testing

1 Introduction

Learning is relevant throughout life, we gain new skills and knowledge, first in educational institutions, then as part of our work. Nowadays, an individual approach to learning based on AI becomes a particularly important advantage. Building individual learning paths allows you to make recommendations in accordance with the level of knowledge, psychological portrait, and interests of the employee. All this allows the most efficient use of the employee's work for the benefit of the company, giving him interesting tasks, growing up an employee professionally. Alghamdi and Bayaga found in their research that training on LMS must be organized based on individual courses [1]. In another study presented a method of test questions classification based on hybrid frame mixing semantic comprehension [2]. Analyzing the information from the studies of these authors, we conclude how important it is to introduce individual learning trajectories and at the same time to use for training questions that have passed a certain classification. To achieve the objectives of training, it is necessary to use a system that allows you to assess the knowledge of students, provide analytical reporting on learning outcomes. The implementation of this system (LMS Vector) has the potential to provide a universal software environment to optimize and modernize the educational process, stimulate students' interest and eliminate the possibilities of value judgments based on personal acceptance or rejection of students.

A. V. Samsonovich (Ed.): BICA 2019, AISC 948, pp. 405–410, 2020.
https://doi.org/10.1007/978-3-030-25719-4_53

2 LMS Vector and Organization of the Question Bank

2.1 LMS Vector

The Vector Learning Management System combines the functions of LMS and TMS, being a multifunctional software product designed to automate the process of integrated learning, student testing, statistical processing of results and talent development.

The system forms a single information space for all participants in the educational process. The system is built in the concept of a modular architecture and provides a wide range of opportunities to create test questions that can then be used to monitor students' knowledge. Teacher can create common standard types of questions - single choice, multiple choice, open answer, question with blocks, question for correspondence. In addition, the system allows you to create specialized types of questions - a calculated question (allows you to specify a formula according to which a specific task will be formed randomly when a student passes the test. The formula must contain variables, the values of which are also set when the question is created by intervals of values), the question of indicating the area at image. All questions are stored in a common database and can be used to generate any number of tests with different settings for different groups of students.

2.2 Hierarchical Tag Structure

Tag Cloud. On the page, you can create any tag, as well as relate it to a group of tags. For example, you can create a Java tag and tag groups Developers, Test engineers. In this case, the Java tag can be assigned to both Developers and Test engineers. Tags also have a language structure, so when creating a tag it is desirable to fill it with all language versions. The main language version is English and only this language version of the tag is allowed. If necessary, you can create Russian, Spanish, French versions. A tag group cannot include another tag group.

Multi-language Questions. When creating a question, you must add a language field that will show which language the question belongs to. The question can be created in any language, not necessarily in English. At the same time, after creating a question, it is possible to create a language version and select from the list a language in which the issue has not been created yet. After that, a copy of the question (all fields of the original question) must be created, which can be manually translated into the desired language. To the question you can attach tags only the corresponding language. In this case, tests can be constructed only from questions of one language.

By default, the standard language in the system is English. If the user chooses any other language, when passing the test, the system will search for the language version of the questions corresponding to the user's chosen language. If the system does not find necessary question's language version, then the user will see a question in English. Thus, different users who use different languages will see the same test in different languages. Of course, if all questions' language versions are filled in correctly. For a single user, when changing languages, the same test may also look different.

Question Quality Control. After creating a test question, an employee of the training department may assign an expert to check this question. For each question, there should be a discussion page on which the entire dialogue between the expert and the employee of the calving training is maintained. The expert can suggest how to change the wording, answer options, and assess the correctness of the question raised by writing a comment. At the same time, the expert cannot correct the question, answer options, tags and other information.

Question Bank Maintenance. In connection with the constant release of new releases on products issues may lose their relevance. It should be possible to include notifications that certain questions need to be checked for their relevance. Default value for question is "not to track obsolescence", but you can also choose a tracking period - month, quarter, year.

3 Tag Cognitive Map

3.1 Tag Cognitive Map Description

A cognitive map is a type of mental representation, which serves an individual to code, store, recall, and decode information about the relative locations and attributes of phenomena in their everyday or metaphorical spatial environment [3]. Cognitive maps have been studied in various fields, such as psychology, education, archaeology, planning, geography, cartography, architecture, landscape architecture, urban planning, management and history.

Cognitive map is a tool for quantitative representation of meaning. In particular, an affective space is usually understood as an abstract metric space, in which representations of various affects are allocated in such a way that the resulting geometric relations among them, such as proximity or orthogonality, such as similarity or independence [4].

As already noted, the LMS Vector uses an approach in which question tags can be combined into groups that form a stable hierarchy. The disadvantage of this approach is dependence on the teacher's subjective representation. At the same time, the influence of this factor can manifest itself both in the organization of the tag structure, and in attributing a question to a specific tag. In addition, there may be hidden links between topics that are not so easy to detect. In this regard, an approach that uses the tools of AI, namely cognitive maps, is proposed.

The concept, namely, the vertex in the cognitive map will be the question tag itself. There can be only one link between each pair of tags, and its weight and direction is determined dynamically based on the results of passing the tests. We will talk more about updating weights in the next section.

3.2 Cognitive Map Dynamic Update

Consider an example of a tag map dedicated to programming questions. For simplicity and clarity, we will consider only 4 tags - Java, Python, GC, Cycles. Suppose that a sufficient number of questions according to these tags have been added to the bank, and several of them have been included in the test. The dynamic change of relationships in the tag cognitive map will occur based on the answers that students give during the test.

For each arc, we will consider the factor f, a number that indicates how strongly these tags are related to the results of the test. However, the use of a linear quantity can deviate too strongly a measure in one direction or another, therefore, to determine the degree of connectivity as the weight of the arc W in the cognitive map, we will use the logarithm:

$$W = \ln(f) \tag{1}$$

Initially, for each arc f = 1, and W = 0, respectively. When passing the test, the value of the factor will change by one, and the weight:

$$W = \ln(f + k) \tag{2}$$

where f - current value of the factor. However, the value of the factor will not necessarily increase. For this is the coefficient k, taking values of 1 or −1. We present an algorithm for obtaining the value of the coefficient k, depending on the answer to the questions in the test.

Take an attempt to test, which turned out to be questions with the following tags:

- Question_A(Java, Python) +
- Question_B(Java, GC) −
- Question_C(Java, Cycles) +

The student answered correctly the first and third questions, and made a mistake in the second. Note that we have selected questions with a matching java tag. We will consider pairs in which both answers are valid for the matching tag, or one answer is correct, and the second one is wrong. We will not consider pairs of two incorrect answers, since they cannot guarantee the correct influence on our map (as an analogy, we can consider the logical operation of implication, which always returns the value "true" when a false message is made).

So, the first pair is Question_A and Question_C. For these questions, you can change the relationship between the tags Python and Cycles. The direction of the arc between these tags is determined by the weight of the arcs from the general concept of Java. As can be seen from Fig. 1, the weight of the Java-Cycles arc is greater than that of Java-GC, which means the weight will be determined for the Cycles-Python arc. Since both answers are correct, k will be equal to 1. If this arc does not exist and the reverse arc of Python-Cycles exists, then the sign of the coefficient k changes to the opposite. In the considered case, the arc exists, both answers are correct, therefore, the value of k = 1.

Now consider the Question_A and Question_B pair. In this case, a pair of tags - Python and GC. The calculation of the coefficient k is carried out in a similar way. Since W(Java-GC) > W(Java-Python), we will change the weight of the arc GC-Python. This arc also exists, but an incorrect answer was given to the question Question_B, then k takes the value −1.

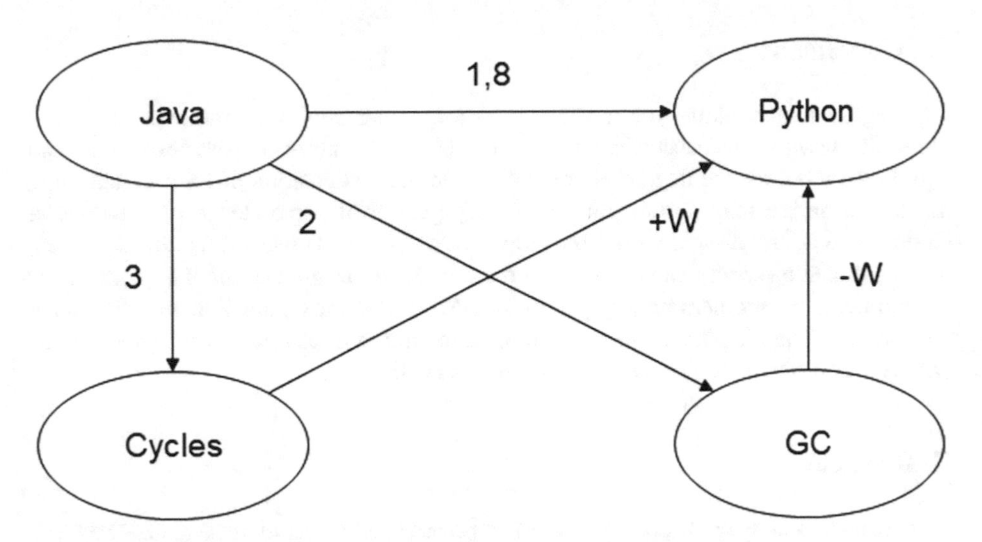

Fig. 1. Tag cognitive map example

As already noted, there can be only one arc from vertex to vertex. However, not only its weight can vary. If the situation W < 0 arises during the next change in the arc weight, then the direction of the arc changes to the opposite. The ability to change the direction of the arc makes the proposed model very flexible, and allows you to interpret the results presented on the tag cognitive map.

3.3 Interpretation

One of the most common approaches in cognitive maps analysis is loop analysis. In this method all the cycles are found, then the cycle sign is found by multiplying all the arc signs. When obtaining a positive value, the cycle is called amplifying, otherwise - stabilizing. In our model, it is impossible to obtain stabilizing cycles, as the links with negative weights change direction.

Receiving amplifying loops in the tag cognitive map will mean the presence of tag groups. Similar groups of tags are created with a hierarchical approach to the organization of the structure of storing questions. However, in this case, the subjectivity of the teacher and the presence of hidden links are excluded.

Another important area in which the tag cognitive map can be applied to is the formation of an individual student development trajectory. In this case, when building a tag cognitive map, you should consider the answers only of the student whose development trajectory we are building. Particularly interesting in this case will be the peaks, from which emanates at least 3 arcs. If there is an incoming arc for such a vertex, then the tag from which it originates will correspond to the area of knowledge in which the student most likely has a space. It is on her that you need to focus on the subsequent training.

4 Conclusion

Tag cognitive map allows you to group questions, abstracting from the experience of a particular teacher, and according to the results of students who pass tests. This can significantly reduce the degree of subjectivity in the composition of tests. In addition, the tag cognitive map allows you to identify gaps in the knowledge of a particular student and adjust his individual development trajectory. However, as already noted, the share of subjectivity can be made in the correlation of the tag and the question. At the moment, this problem may appear in the LMS Vector, in connection with which the tag cognitive map is planned to be developed in the future so that it can be determined whether a specific question corresponds to a specific tag.

References

1. Alghamdi SR, Bayaga A (2018) Suitable LMS content and format for training. In: 21st Saudi computer society national computer conference, NCC
2. Wang R, Cui X, Wang C (2017) Research of test questions classification based on hybrid frame mixing semantic comprehension and machine learning. In: ACM international conference proceeding series, pp 1–5
3. Tolman EC (1948) Cognitive maps in rats and men. Psychol Rev 55(4):189–208
4. Samsonovich AV (2018) On semantic map as a key component in socially-emotional BICA. Biol Inspir Cogn Arc 23:1–6

Automatically Estimating Meaning Ambiguity of Emoticons

Michal Ptaszynski(✉), Fumito Masui, and Naoto Ishii

Kitami Institute of Technology, Kitami, Japan
{ptaszynski,f-masui}@cs.kitami-it.ac.jp, ishii@ialab.cs.kitami-it.ac.jp

Abstract. Emoticons are one of the means of communication widely used by Internet users. Emoticons are used to express information that cannot be fully transmitted only by text, such as emotions or feelings. However, not all emoticons are easily understandable. In this research, we aimed at estimating such general level of comprehensibility, or "meaning ambiguity", defined by their correspondence with linguistic expressions, such as onomatopoeia. The method could help users apply more meaningful and understandable emoticons in their everyday communication and improve their online communication.

Keywords: Emoticons · Linguistic expressibility · Meaning ambiguity

1 Introduction

When it comes to communication between human users on the Internet, many means supplementary to the text-based communication have been developed to eradicate misunderstandings and facilitate the text based communication. One of such communication means include pictograms that appear in user interface during online conversation as simple pictures after entering specific words in such communication tools, like "stamps," "stickers," or "emoji" (high resolution pictograms) popularized in such communication tools like LINE or Facebook. Among such pictograms, emoticons have been widely applied since the times when it was still impossible to send high resolution images. Emoticons[1] are expressions that are made of letters and symbols including special characters and imitate facial expressions and postures to express emotions in online communication environment similarly to how they appear in human body language. In this sense, emoticons are used to convey nonverbal information. An example, could be "(≧∇≦)" or "(-_-;)". However, the meaning conveyed by emoticons could be ambiguous, causing misunderstandings due to difficulties in conveying nuances that are not sufficiently expressed only with pictures. Therefore, emoticons are often used together with linguistic expressions, such as onomatopoeia or exclamations such as, "(´･ω･`) ショボーン" or "m9(＾Д＾) プギャー". As was

[1] In this paper we focus especially on Eastern/Japanese emoticons as defined by Ptaszynski et al. (2010).

A. V. Samsonovich (Ed.): BICA 2019, AISC 948, pp. 411–416, 2020.
https://doi.org/10.1007/978-3-030-25719-4_54

shown in the analysis by Ptaszynski et al. (2012, 2013), these expressions may appear in sentences together with emoticons but are also often used as inherent parts of emoticons, and in cases where expressions or emotions cannot be represented with emoticons only, the additional linguistic expressions fulfill a complementary role of disambiguating the nuance.

In the following research we focus on linguistic expressions used in conjunction with emoticons, investigate the relationship between these two, and propose a method for quantifying whether an emoticon is ambiguous or not, based on linguistic expressions used in conjunction with the emoticon.

2 Automatic Estimation of Meaning Ambiguity of Emoticons

We propose a method to quantify the ambiguity of the meaning of emoticons and to perform automatic estimation of emoticon meaning ambiguity. Firstly, based on the questionnaire performed in our previous research (Ishii et al. 2017), we obtained the ratio of "I don't know" for each emoticon in the response to the questionnaire, and found that the ambiguity of the meaning differed depending on the symbols used to produce the emoticons. Therefore in the method each emoticon was divided into symbols, and each symbol in the emoticon was assigned the ratio of "I don't know" the same as the whole emoticon. Furthermore, for symbols used in multiple emoticons, the average of the ratio of "I don't know" in the answer was considered as the absolute linguistic expressibility of the symbol. However, because the same symbol may be used sometimes as eyes and sometimes as a mouth, absolute linguistic expressibility was calculated separately for each of such cases even if the used character was the same. Finally, ambiguity of meaning was calculated in the following three ways and performance of each version was verified.

Method 1: Matching Whole Emoticons. Ten emoticons used in the preliminary survey and their corresponding language expressibility score were saved in a database and when an emoticon was matched to an input sentence, the corresponding score was used as ambiguity score of the meaning of the found emoticon.

Method 2: Matching Based on Separate Characters. Ten emoticons used in the preliminary survey were divided into parts such as eyes, mouth, and other parts, and the ambiguity of each part was calculated. Thereafter, the language expressibility score of each part was calculated with respect to the emoticons matched to the input sentence, and the average of all parts was used as the ambiguity score of the detected emoticons.

Method 3: Stepwise Matching. First, matching is performed using method 1, and when method 1 does not detect a full emoticon, method 2 is used. Since Method 1 matches the entire emoticon, it can be predicted that the Recall rate would be lower because the detected emoticons are limited to those in the database. However, if matching succeeds, the score should be more accurate. On the other hand, in the case of method 2, it can be expected that the Recall rate would be higher because the ambiguity score can be calculated even if only one symbol can be matched, but it is expected that the Precision will be lower instead. Therefore method 3 can be predicted to obtain intermediate results.

2.1 Criteria for Ambiguity

Since the ambiguity score is calculated only for matched emoticons, it is 0 when it cannot be matched, and 0 or more when matching is successful. However, it is impossible to determine whether the score is high or low simply by calculating the linguistic expressibility. Therefore, it was necessary to set a threshold value based on certain objective conditions. In this research the average value of the absolute language expressibility score was used as the threshold. The average value of the language expressibility score of all emoticons in the database was calculated, and if the new score calculated for the new input emoticon was higher, the ambiguity was considered as high. Otherwise the ambiguity was set as low.

3 Experiment

In order to verify the performance of the above three methods, we prepared an experiment and we collected a gold standard data. Firstly, experiment settings and gold standard conditions were set as follows. Since ambiguity score is calculated for emoticons, matching was done not at a sentence level but at an emoticon level. Also, for gold standard data, Precision of each method (number of cases for which correct ambiguity was estimated within detected emoticon), and Recall (the number of cases for which correct ambiguity was estimated for in all emoticons included in the gold standard data), were calculated, and on its basis the harmonic mean of Precision and Recall (F-score) was calculated as follows.

$$F = \frac{2(P * R)}{(P + R)} \tag{1}$$

Moreover, gold standard data was collected as follows. First of all, the developed method was executed for 10,137 emoticons stored in the CAO system database (Ptaszynski et al. 2010). The CAO system is a system that extracts emoticons from input sentences and estimates the types of emotions they express. The number of emoticons covered by the CAO system is about 3 million. In addition to the conditions of the above experiment setting, coverage was obtained for each method. The coverage was defined as the number of emoticons that could be extracted from the original 10,137 emoticons divided by all emoticons in the database. As a result, method 1 succeeded in extracting three emoticons (coverage = 0.000296), and methods 2 and 3 successfully extracted 86 emoticons (coverage = 0.008484). Although the coverage was low, the number of emoticons used to make the method was also small (10 emoticons). In the case of exact matching, approximately one third (3/10) of emoticons used for database creation were also matched in the input. On the other hand, in the case of separate character-based matching, the coverage was about 30 times larger (0.008484/0.000296). From this, it can be expected that coverage can be improved by increasing emoticons in the database in the future. Emoticons (86) that could be extracted by all methods were used as the current gold standard data. In addition, a separate questionnaire survey was conducted for those 86 emoticons. In this questionnaire, the participants were first introduced to a list of emoticons and estimated whether or not the meaning of the emoticon is known to them or not. If they understood the meaning, they chose "understand" and wrote the meaning. If they didn't understand the meaning, they chose the option "don't understand". An example of a part of the designed questionnaire was shown in Fig. 1.

There was a total of 9 participants in the questionnaire. All nine participants were in their twenties. Each emoticon was annotated by three participants. The number of people who answered "I don't know" for each of the 86 emoticons was shown in Table 1. As a gold standard condition, when 0 people or 1 person answered "I don't know" for each emoticon, the ambiguity was set as low. Also, when 2 or 3 answered "I don't know", we set that ambiguity was high.

4 Results and Discussion

We examined and evaluated the number of emoticons whose estimation result of each ambiguity estimation method matched the questionnaire result. The numbers of matching for each method were represented in Tables 2, 3, and 4. In addition, evaluation measures such as the Precision, Recall, and balanced F-score of methods 1 and 2 were calculated and represented in Tables 5 and 6, respectively.

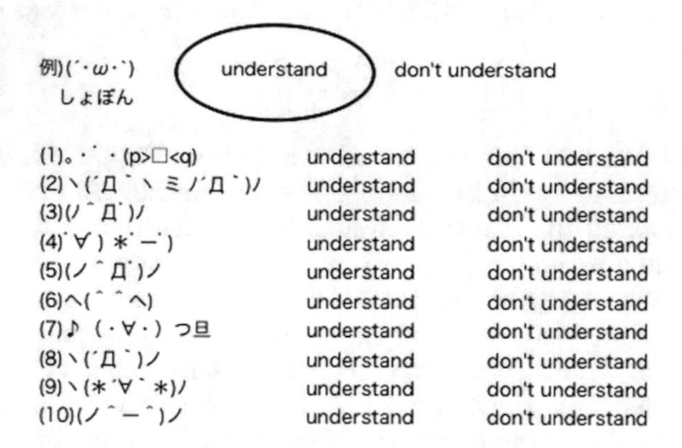

Fig. 1. Questionnaire inquiring about ambiguity of emoticons.

Better results were obtained for method 2 than for method 1 for all scores: Precision, Recall, and F score. This shows that by dividing emoticons by symbols, we can estimate the ambiguity of emoticons more accurately. Method 3

Table 1. Number of answers of "I do not know" for each emoticon.

# of answers	# of emoticon
3	15
2	29
1	23
0	19

Table 2. Method 1: result of matching the entire emoticon.

# of answers	# of emoticon	Matches
3	15	0
2	29	1
1	23	0
0	19	0
Total	86	1

Table 3. Method 2: result of separate-character based matching.

# of answers	# of emoticon	Matches
3	15	4
2	29	14
1	23	12
0	19	8
Total	86	38

Table 4. Method 3: result of stepwise matching.

# of answers	# of emoticon	Matches
3	15	4
2	29	15
1	23	11
0	19	8
Total	86	38

obtained the same result as method 2. Analysis of differences in results for Method 2 and Method 3 showed that, there was no difference in emoticon "♪(・∀・)つ旦". However, there was a difference between the emoticon "ヽ(°∀°)メ" and the emoticon "ヽ(°∀°)ノ". Looking at these two emoticons, in the questionnaire, the emoticon "ヽ(°∀°)メ" was more ambiguous and the emoticon "ヽ(°∀°)ノ" was less ambiguous. On the other hand, in Method 2, the ambiguity of both emoticons was low, while the output indicated high ambiguity for Method 3. Therefore, it was impossible to make a clear decision on whether Method 2 or Method 3 was better. From the above results we can clearly say that emoticons are ambiguous expressions, and it is difficult to estimate their ambiguity. A method created based on a preliminary survey (relationship between emoticons and linguistic expressions), when evaluated on gold standard data collected in a separate questionnaire, scored lower than assumed. In the future it is necessary to repeat the experiment by increasing the data, but we can conclude that calculating the ambiguity of the meaning of emoticons is a difficult task because correlation was not seen in this experiment. In recent

Table 5. Results for Method 1.

Precision	Recall	F score
0.33	0.01	0.02

Table 6. Results for Method 2.

Precision	Recall	F score
0.44	0.44	0.44

years, in most of the research on emoticons, there are many simple labels used such as "joy" "sadness" or "greetings" as the meaning of emoticons. However, as it was shown in this study, such top-down assumptions are not realistic, and research on affect and sentiment analysis of emoticons must also consider the ambiguity of emoticon meanings.

5 Conclusions

In this paper, based on the results of previous survey, in which we investigated the relationship between emoticons and various linguistic expressions, we proposed a method for automatic estimation of the meaning ambiguity of emoticons. As a result, we found out that the degree of comprehension differs for each emoticon. This also indicates that meaning of some emoticons is more context-dependent than others. Also, from the experiment with automatic estimation of the meaning ambiguity, we found out that emoticons are highly ambiguous expressions, and estimating this ambiguity is not a trivial task. However, we found out that it could be more effective to estimate this ambiguity by calculating it from separate characters, not from the whole emoticons.

As a future work, we plan to verify the ambiguity of emoticons when used in wider contexts, such as sentences. We also plan to apply other information-reducing measures, such as the one proposed by Dybala et al. (2011) for more accurate disambiguation of meaning ambiguity of emoticons.

References

Ptaszynski M, Jacek M, Pawel D, Rafal R, Kenji A (2010) CAO: a fully automatic emoticon analysis system based on theory of kinesics. IEEE Trans Affect Comput 1(1):46–59

Ptaszynski M, Dybala P, Rzepka R, Araki K, Momouchi Y (2012) YACIS: a five-billion- word corpus of Japanese blogs fully annotated with syntactic and affective information. In: Proceedings of the AISB/IACAP world congress 2012 in honour of Alan Turing, 2nd symposium on linguistic and cognitive approaches to dialog agents (LaCATODA 2012), 2–6 July 2012, pp 40–49

Ptaszynski M, Rzepka R, Araki K, Momouchi Y (2013) Automatically annotating a five-billion-word corpus of Japanese blogs for sentiment and affect analysis. Comput Speech Lang (CSL) 28(1):38–55

Dybala P, Ptaszynski M, Sayama K (2011) Reducing excessive amounts of data: multiple web queries for generation of pun candidates. Adv Artif Intell 2011. Article ID 107310

Ishii N, Masui F, Ptaszynski, M (2017) The analysis of the relation of emoticons and Japanese onomatopoeia. In: Proceedings of the 31st annual conference of the Japanese society for artificial intelligence (JSAI 2017), pp 1–4, Paper ID: 2O2-OS-22a-1 (in Japanese)

A VR-Based System and Architecture for Computational Modeling of Minds

Saty Raghavachary[1(✉)] and Lurong Lei[2]

[1] University of Southern California, Los Angeles, CA 90089, USA
saty@usc.edu

[2] Chongqing University of Arts and Science, Chongqing 402160, China
20030051@cqwu.edu.cn

Abstract. Computational modeling of natural cognition is a crucial step towards achieving the grand goal of human-level computational intelligence. Successful ideas from existing models, and possibly newer ones, could be assembled to create a unified computational framework (e.g. the Standard Model of the Mind, which attempts to unify three leading cognitive architectures) - this would be of great use in AI, robotics, neuroscience and cognitive science. This short position paper proposes the following: a VR-based system provides the most expedient, scalable and visually verifiable way to implement, test and refine a cognitive mind model (which would always embodied in a character in a virtual world). Such a setup is discussed in the paper, including advantages and drawbacks over alternative implementations.

Keywords: Virtual reality · Embodiment · Computational mind modeling · Cognitive architectures · Artificial general intelligence · Artificial intelligence

1 Introduction

In this paper, we describe a system for creating an artificial general intelligence (AGI) agent - one that would be capable of carrying out cognitive tasks such as perception and concept formation. There has been continued interest in developing such an agent capable of human-level intelligence, but there has been limited success in achieving it. We hope that the system we are proposing, can help the effort along.

Our key idea is this: an agent capable of achieving human-level intelligence, necessarily needs to inhabit a body, be it physical or virtual - it cannot be a disembodied presence (a "brain in a jar", as it were). Further, such an agent needs to be educated and trained from 'infancy', similar to how a human baby would be. The agent's concept acquisition, the most essential aspect of intelligence, needs to be grounded via immediate experience of reality, and tied to relevant contexts; again, embodiment is the most direct way to ensure this. The agent would need to learn continuously and incrementally, by interacting with its environment, which includes acquisition of basic skills and capabilities, leading up to language acquisition, and higher levels skills related to reasoning and problem-solving; once again, it is difficult to achieve this, without embodiment.

A. V. Samsonovich (Ed.): BICA 2019, AISC 948, pp. 417–425, 2020.
https://doi.org/10.1007/978-3-030-25719-4_55

Given the choice between physical and virtual embodiment, we are advocating for a virtual setup, for reasons we provide in an upcoming section of the paper.

In the following sections and subsections, we describe the system architecture, agent design, the agent's (virtual) environment, implementation, and what we might gain from creating such a system.

2 Architecture

We are about to describe a 'virtual reality operating system' (VR-OS) that would create both a virtual environment/world (VR-EN), as well as a virtually real agent (VR-AG) that would inhabit the environment. The VR-AG would learn and grow by interacting with its environment, and with a human 'teacher' who would appear to the VR-AG as another inhabitant of the VR-EN.

2.1 VR-OS

Much like a traditional OS, VR-OS is a platform for creating and executing multiple concurrent processes, scheduling tasks/events, managing resources (such as memory), providing I/O capabilities, etc. It would provide the following:

- a facility for creating and populating a 3D virtual world (VR-EN) - this includes being able to populate the world with objects that can exhibit behaviors, control the passage of time, schedule events (such as sunrise and sunset, growth of plants, weather events, etc.), include 'physics' behavior such as gravity and collision detection
- facility for creating a 3D virtual embodiment ('VR-AG') that would live in the world, and interact with it; the VR-AG would contain a 'mind' implementation which will let it perceive, learn, grow, reason, etc.
- a human 'teacher' agent who interacts with the VR-AG, by being embodied as an avatar in the VR-EN
- 'God-like' ability to modify the VR-EN from outside the environment, and modify the VR-AG's internal state
- time-stamping and recording facility for capturing activities, events, interactions, VR-AG behaviors, VR-AG's mental processes, etc.
- editing, and playback, of recorded sessions
- real-time observation of happenings in the VR-EN and inside the VR-AG, by a human observer outside the virtual world.

2.2 VR-AG

As mentioned earlier, the VR-AG is an agent with embodied cognition. Specifically, it would live in the VR-EN, experiencing it via its senses, embedded in a small human-like body with articulated limbs. The VR-AG would include a binocular first-person view of the world as seen through its eyes, binaural audio perception via microphones inside its ears, and be able to sense contact and touch via its skin. The agent would be able to observe itself, interact with the world (e.g. navigate, pick up small objects) and interact with its teacher as well (see, hear, get near, hold hands etc.). Most importantly, the VR-AG would start out ('be born') with an almost "tabula rasa" brain/mind (pre-programmed with some basic instincts and behaviors, explained in an upcoming section), and be able to store in its memory, experiences, concepts, etc. as it learns and grows. The VR-AG would need to nourish itself via an energy 'food' source when it runs low on energy. The VR-AG's mind would be totally transparent, visible to an observer outside the VR-OS. Every change to the mind's memory, every cognitive process (such as perception, thought and emotion) and every interaction with the VR-EN would be recorded with a timestamp, for purposes of analysis, playback, etc.

2.3 VR-EN

The virtual world the VR-AG lives in, would contain geometry (large objects such as buildings, medium objects such as chairs, and small objects such as balls and books), materials that are associated with the geometry, cameras (situated in the VR-AG, and elsewhere desired), and photoreal rendering. The world would also contain 'game physics' phenomena such as non-interpenetration, gravity, sunlight, fracturing, bouncing, etc. - such phenomena would be part of the VR-AG's 'qualia' and be recorded as experiences in its mind.

2.4 Implementation

We surveyed existing implementations of AGI-oriented as well as regular VR platforms, hoping to leverage the one closest to our proposed one. But to the best of our knowledge, such a system does not exist. What we found was too specific, or lacked functionality:

- Deepmind's Gym [1] is an environment for comparing reinforcement learning algorithms
- DeepMind Lab [2] is a first-person 3D game platform, for experimenting with deep reinforcement learning
- CogPrime [3] is a platform for embodied AGIs, much like what we are proposing - but it is not a VR system, and the underlying agent architecture is not easily replaceable with alternatives
- BabyX [4] is photoreal simulation of a baby's face, backed by biologically-accurate neural network to perform AGI-like functions such as perception and language acquisition - not open-source, and is not a VR environment
- there are several smaller implementations of specialized worlds (e.g. a visually-guided 2D agent [5] but they are not extensible, by design.

As a result, we are proposing building a system from the ground up. Here are some specifics regarding the implementation:

- the underlying VR-OS would be built using Node.js, a scalable, cross-platform, JavaScript-based runtime engine that can run code on multiple platforms (laptops, browsers, servers); it is a non-blocking system that runs asynchronous code on multiple threads (e.g. Pomelo [6] is a 3D game engine framework that runs on Node.js)

- WebGL, Three.js, BabylonJS and other graphics and game libraries (in JavaScript) could be used to create browser-based clients that can interface with VR-OS, making it possible to interact with the VR-AG via a phone browser attached to Cardboard VR, for example
- the VR-EN and VR-AG architectures would borrow heavily, features and design aspects from game engines such as Unreal and Unity
- the OS would provide dashboards with controls for monitoring and altering the VR-AG's internal states and configuration, for observing the VR-AG/VR-EN interaction as an invisible observer (so that the VR-AG is not aware of the extra presence), and for altering the VR-EN (e.g. causing an event, re-positioning objects, etc.)
- the OS would also provide an interface for exporting and importing JSON-based files that contain VR-AG/VR-EN interaction logs, the VR-AG's internal state (including its memory), the VR-EN's configuration, etc. - in short, everything for modifying or repeating world events, interactions and agent behavior.

2.5 Comparisons with Unembodied AGI and Physically-Embodied AGI

We strongly agree with the following characterization by Pfeifer and Scheier [7], that 'Intelligence cannot merely exist in the form of an abstract algorithm but requires a physical instantiation, a body.' Without being able to experience physical sensation, navigation, self-directed gaze and other body-based phenomena, an AGI is simply not going to be as relatable to human, as an embodied equivalent would be - in the real world, we do not have human intelligence that functions without a body.

Physical embodiment of an agent in the form of a robot, is certainly an alternative to a VR-AG. However, robots are expensive to create, are prone to mechanical failures, have safety issues when interacting with humans, and are not trivially upgradeable as they 'grow up'. Further, robots operate in the real world which can be unpredictable and not controllable (events cannot be isolated, prevented, or replayed, for example); extra actors and objects in the scene might interfere with an AGI's learning, memory formation and problem-solving.

2.6 Shortcomings of the Virtual AGI Approach

In the above sections, we laid out all the advantages of being able to interact with a virtual agent in a virtual world - these chiefly come down to lower cost, total controllability, configurability, and repeatability.

On the flip side, it might not be easy to replace an object with another, for example (we might not have the replacer object in our assets library, or if we had one, it might lack the functionality we desire, because that was not coded in during creation). Further, the environment's capabilities would be only as good as the simulation (e.g. we might not be able to tip over a paint can to have our VR-AG watch the paint ooze out, because fluid simulation is not part of our physics yet). In our opinion, such drawbacks are not insurmountable (and need to be addressed just once), so we do strongly prefer the virtual approach.

3 Usage

In this section, we itemize, under subsections, a wide variety of experiments related to teaching the VR-AG, studying the VR-AG's mental states and behavior, simulating animal physiology and animal behavior (including body-related ones), evolution simulation, comparative studies between VR-AGs, etc., all of which can be carried out in the proposed platform. While many of these can be done in a non-VR environment as well, the VR platform provides richer possibilities, namely first-person viewpoint and 'natural' interactions for the agents (to the extent that their anatomy and brain modeling permits), and for the participating humans, observability (of external behavior, and internal states) from outside the VR-EN, and interactability from within.

The possible experiments are listed in the form of a wish-list, with no details on how to set them up (since specifics of the physical and mental construction of the agents would determine the setup).

3.1 Instincts and Built-In Capabilities

In a manner similar to how human and animal/bird/fish/insect babies are born with pre-wired instincts, a VR-AG could be outfitted with them from 'birth' so that the capabilities and behaviors tied to those instincts could be carried intuitively and automatically, without needing thought or memory. Each instinct would be coded as a pair: what conditions would activate it, and what the resulting behavior should be. In addition to instincts, the VR-AG would possibly need some innate capabilities that are related to its survival. Possible instincts and built-ins include:

- self-preservation (no damage to self, mostly related to bodily harm)
- dealing with hunger
- face recognition
- avoiding discomfort (e.g. loud noise, bright light, shaking, too hot or cold exterior)
- fear of unfamiliar faces or situations
- motivation, curiosity to explore
- basic capacity for pleasure and pain
- tendency to seek reward (this could form the basis for reinforcement learning tasks)
- urge to socialize, and communicate.

3.2 Things that Could Be Taught to a 'toddler' VR-AG

A human toddler who attends pre-school, gets immersed in an environment that is conducive to learn a variety of foundational knowledge and gain basic experiences. A VR-AG would likewise benefit from learning and experiencing similar items, which include:

- spatial navigation
- object invariance with respect to head rotation
- object permanence (for older agents)

- a sense of time
- language
- things: labels (names), parts/structure
- simple physical phenomena
- concepts (such as big, outside)
- feelings
- motor responses
- opposites
- number sense, counting
- basic weather
- handedness
- simple mechanisms (e.g. via manipulatives such as toys)
- simple heuristics.

3.3 Ways to Train/Teach a VR-AG

Some of the basic ways by which to train or teach a VR-AG (identical to techniques employed on human children) include the following [note that they all presuppose language understanding and basic speech generation, on part of the VR-AG]:

- showing and naming (objects, pictures on board books) - repetition would cause reinforcement
- using manipulatives - learning occurs by grasping, turning, etc.
- performing simple experiments - for the VR-AG to watch, or repeat
- catechism - question and answer is a very effective technique that works on humans
- verbal exposition, including reading stories and singing nursery rhymes
- using reward and consequences, to reinforce and suppress behavior, respectively.

3.4 Physical and Mental States to Simulate and Observe

If our VR-AG is created with a need to consume energy ('eat'), and spend it by performing physical and mental tasks, we could invoke feelings of hunger, tiredness, sleepiness, and inattention by having the VR-AG deplete its stored energy.

Since we can directly observe the mental processes in action, we would be able to study the following aspects [if they are built into the VR-AG]:

- thinking - this would be the process of utilizing perceived inputs, memory-retrieved facts/skills/events, guided by expectation, influenced by feelings, to generate outputs ("thoughts") and actions
- feeling - e.g. the feeling of fear, or peace
- memory formation, including semantic, procedural and episodic memories
- concept formation - including generalization by induction, and deductively situating new concepts under existing ones
- memory recall, including partial recall
- mis-remembering
- memory modification
- forgetting.

3.5 Simulation of Animal/Bird Behavior

Our VR-AGs do not have to be humans, they could be other animals, birds, fish, etc., whose nervous systems have been sufficiently studied so their anatomy, physiology and behavior could be accurately modeled and simulated in the VR-EN. Following are some interesting possibilities:

- panoramic vision in birds (coupled with limited binocular vision)
- echolocation in bats
- gaze holding, and quick gaze shifting (saccadic motions)
- footedness in birds, e.g. parrots
- head bobbing motion in birds, e.g. pigeons
- pecking, in birds
- 'plunge diving' by waterfowl, to catch fish
- prey centering (foveated imaging), e.g. by eagles
- communication between bees, using 'waggle' dance [8]
- firefly flash synchronization, using a 'spiking' neuronal model for example [9]
- recreation of complete neural circuitry, for creatures whose entire neural atlas is known, e.g. C.elegans [10] - for the purposes for studying stimulus response behavior, for example.

3.6 More Fanciful Experiments

At the VR-AG becomes increasingly intelligent, with attendant brain mechanisms and structures modeled in, it would be interesting to indulge in more wishful pursuits such as these:

- teach the VR-AG to produce creative output: compose music, make art, generate poetry
- study the aesthetic response by having the VR-AG look at beautiful art or listen to a lovely tune
- have multiple VR-AGs interact, study socialization
- introduce multiple real-life avatars into the VR-EN, for the VR-AG(s) to interact with
- create a collective brain, by linking multiple VR-AG brains, study its behavior and capabilities
- graft parts of a VR-AG's brain on to another, and study the change in behavior of the target agent
- transplant a VR-AG brain on to a physical humanoid robot
- compare various cognitive architecture implementations [Alexei S], under identical VR-EN configurations
- using brain-computer interfaces (BCI), connect a human brain to a VR-AG's, study the effects on the human as well as the VR-AG
- provide additional senses to the VR-AG, e.g. extra pair of eyes on the back of the head, extra sensing capability (e.g. radio waves)
- study metacognition, and epistemology
- study the effects of brain damage and disease, on the VR-AG's brain.

While the above might sound outlandish at this point, it is worth remembering that the goal of BICA researchers is nothing less than full-scale AGI, achieving which would make these experiments become feasible.

4 Summary

In this paper, we have made the case for pursuing AGI development using a virtual platform (instead of a disembodied, or robot-based one) to set up agent/environment interactions.

We discussed the various components of the proposed system (VR-OS, VR-EN, VR-AG), implementation specifics, and mentioned the pros and cons of our approach; then we briefly touched on a number of items that could be testable on the platform (VR-AG's instincts, mental states, memory formation and memory access; non-human cognition; more fanciful sensor and perception experiments; body-based cognitive behaviors, animal behaviors, etc.).

5 Future Work

We note that the design we have discussed above, is at proposal stage; it has yet to be implemented, so that would constitute future work. That said, a 'version 2' rollout could expand the possibilities, via the enhancements listed below.

Future capabilities of VR-OS would include:

- access via the Internet, for remote 'teacher' (or teachers) participation
- capability to set up multiple, simultaneous human avatars that can populate the VR-EN and engage with the VR-AG
- a 'cloud VR' setup, where the VR-EN would be hosted as well as rendered on a cloud server, which would stream rendered video to its remote clients - clients would not need any setup or installation, they would simply participate via web browsers.

Future capabilities of VR-AG would include the following.
Body:

- more senses, such as being able to feel hot and cold, feel the wind, feel surface texture
- enhanced biomechanics, including a muscle and skin layer, bipedal gait synthesis.

Brain:

- proprioception
- depth perception.

And finally, here are new VR-EN-related features to add:

- more VR-AGs, for social interaction and communication
- more geometry, for a more interesting world
- more physics, e.g. weather phenomena
- more places, for the VR-AG(s) to visit.

References

1. Brockman G et al (2016) OpenAI Gym. https://arxiv.org/abs/1606.01540
2. Beattie C et al (2016) Deepmind Lab. https://arxiv.org/abs/1612.03801
3. Goertzel B (2012) CogPrime: an integrative architecture for embodied artificial general intelligence. https://wiki.opencog.org/w/CogPrime_Overview
4. Lawler-Dormer D (2013) BABY X: digital artificial intelligence, computational neuroscience and empathetic interaction. In: ISEA 2013 conference proceedings
5. Beer RD (1996) Toward the evolution of dynamical neural networks for minimally cognitive behavior. Animals Animats 4:421–429
6. Pomelo. https://github.com/NetEase/pomelo. Accessed 15 May 2019
7. Pfeifer R, Scheier C (1999) Understanding intelligence. MIT Press, Cambridge
8. Fernando S, Kumarasinghe N (2015) Modeling a honeybee using spiking neural network to simulate nectar reporting behavior. Int J Comput Appl 130(8):33–39. (0975–8887)
9. Kim D (2004) A spiking neuron model for synchronous flashing of fireflies. In: Bio systems 2004. https://doi.org/10.1016/j.biosystems.2004.05.035
10. WORMATLAS. https://www.wormatlas.org/neuronalwiring.html. Accessed 15 May 2019

A Bioinspired Model of Decision Making Considering Spatial Attention for Goal-Driven Behaviour

Raymundo Ramirez-Pedraza(✉), Natividad Vargas, Carlos Sandoval, Juan Luis del Valle-Padilla, and Félix Ramos

Department of Computer Science,
Center for Research and Advanced Studies of the National Polytechnic Institute (CINVESTAV IPN) Unidad Guadalajara, Guadalajara, Jalisco, Mexico
rramirez@gdl.cinvestav.mx

Abstract. Cognitive architectures (**CA**) are currently used to approach computer systems' behavior to human behavior and intelligence. Fundamental human capability is planning and decision-making. In that regard, numerous AI systems successfully exhibit human-like behavior but are limited to either achieving specific objectives or too heavily constrained environments, which makes them unsuitable in the presence of unforeseen situations where autonomy is required. In this work, we present a bioinspired computational model to undertake the autonomous navigation problem as a result of the interaction between planning and decision-making, spatial attention and the motor system. The proposed model is embedded in a greater cognitive architecture. In the case study developed, it is proposed and tested that the process of planning and decision-making plays an important role to carry out spatial navigation. In it, the agent must move through an unexplored maze from an initial point to a final point, which it accomplished successfully. The gathered results prompt us to continue working on the model that considers attentional information to guide the agent's behavior, which is strongly supported by the concise selection of neuroscientific evidence related to the cognitive functions we provided.

Keywords: Brain model · Decision-making · Planning · Spatial attention · Motor system · Goal-driven

1 Introduction

Cognitive science is an interdisciplinary research area whose objective is to understand and model human intelligence. A fundamental capability of our intelligence is planning and decision-making. In that regard, numerous AI systems successfully exhibit human-like behavior but are limited to either achieving specific objectives or to a heavily controlled environment, which makes them unsuitable in the presence of unforeseen situations where decisions must be made autonomously, such as in critical systems or videogames.

A. V. Samsonovich (Ed.): BICA 2019, AISC 948, pp. 426–431, 2020.
https://doi.org/10.1007/978-3-030-25719-4_56

This work describes a bio-inspired computational model that emphasizes the importance of the interaction cognitive functions of *planning and decision-making, motor system* and *attention* to provide a virtual creature with the ability to navigate in a two-dimensional maze environment similar to how a human would, traversing from a starting point to a destination point. The proposed model is based on different modules related to the brain areas that are involved in one or more *cognitive functions* which belong to the cognitive architecture proposed by our research group. To prove that the maze can be solved with the interaction between these cognitive functions -through the definition and modeling of the processes and information flows involved- we present a case study of a labyrinth, that consists of an entrance point, an exit point and obstacles (walls).

By reviewing the state of the art, different CAs were found that were taken into account because they share common elements required to achieve the specified goal and due to the nature of their architectures. *ACT-R* works with production rules that are stored in memory work together to select actions to execute, which satisfy the agent's goals [1]. On the other hand, the action selection stage in *LIDA* is a learning phase in which several processes operate in parallel. A copy of each action scheme creates an instance with its linked variables and sends it to action selection, where it competes to be the selected behavior for this cognitive cycle [2]. *iCub* is constituted by a distributed network of percepto-motor circuits, a modulation circuit for action selection that affects the anticipation through the simulation of perception-action [3].

However, none of this architectures specifies the way they carry out the long term planning [4]. That is, the ability of virtual creatures to imagine a plan in advance, before executing it. Some ACs, like CLARION that does not have an explicit attention function or ACT-R that does have attentional mechanisms but they function as perceptual filters through the modification of the G parameter of the utility function of the rules of production [5] or LIDA that bases its attention function on a series of predefined schemes that are enabled according

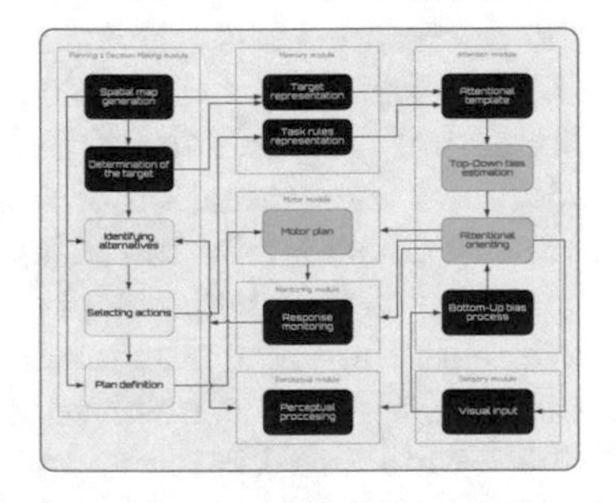

(a) Conceptual model: It shows the cognitive functions related to the navigation process and its connection with other identified functions.

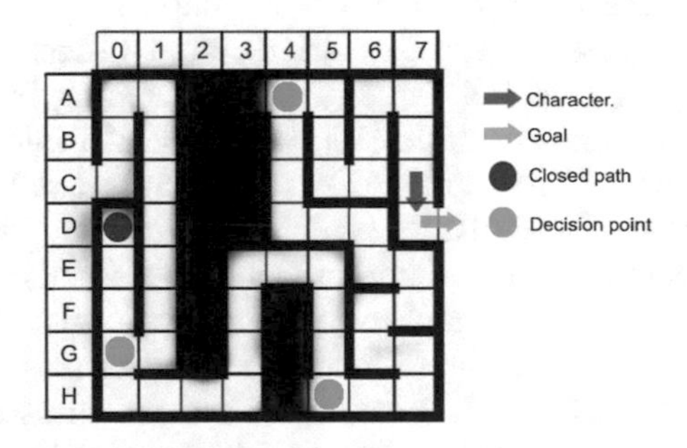

(b) Maze: It shows an example of a maze used like study case and the objects involved.

Fig. 1. Conceptual model and virtual environment

to the set of actions [6], requires attentional processes that function as top-down control mechanisms used to direct the actions determined by planning and decision-making.

To develop a computational model capable of emulating the cognitive processes of the human brain. We decided to base our model on neuroscientific evidence which will be described in detail in the Sect. 2. Therefore, we determined the main cerebral structures involved in the maze navigation process. It should be noted that some structures described here are also related to other cognitive functions currently in development, like memory, perception or visual sensory system, whose description is outside the scope of this work.

This article is structured as follows: In Sect. 2, we present our proposed model. After that, in Sect. 3 we present a case study used to validate the proposed model and the results obtained from the case study.

2 Bio-inspired Cognitive Model

In this section, we present a model to solve the spatial navigation problem in a virtual environment (Fig. 1b). Below we describe the information flow and briefly explains the processes performed by each module, according to the brain area they represent.

Based on the information gathered from neuroscientific evidence. We outline the processes involved the maze navigation: **Planning and decision-making** processes are used to solve a maze start from a given state and continue until reaching a final state. The three main planning and decision-making processes considered for the maze navigation task are *Identifying alternatives*, that is, once the objective is determined in **OFC**, it receives information from the perceptual system, about the current state, which is used to identify possible movement alternatives; then, **VTA** provides information related to novelty, uncertainty and saliency [7] that support the decision-making process [8]. *Selecting actions*, **dlPFC** keeps the objective and retains reward information for planning; **OFC** processes the information about objects in the environment [9]; then, **AMY** computes the value of stimuli coming from **OFC**; after that, **OFC** calculates the action's value [10], and then **DS** selects the action [11]. In *Plan definition*, **dlPFC** and **vmPFC** get the reward information and the evaluation of efforts, respectively [12]; then, **OFC** controls the voluntary behavior by initiating the required movements in **DS** [11]. **Attentional processes** that help the planning and decision making processes to guide behavior, as they are, *Top-down bias estimation* and *Attentional orientation* processes in which the involved areas like **OFC** that is used to generate an attentional template which must contain relevant characteristics that improve the detection of task-relevant stimuli. **LIP** help in the creation of a proto-object [13]. **FEF**), to orient the system towards the objective object's location. There are also **Motor System processes** that are important for the execution of the motor plan. The information travels to regions of the premotor cortex, where it is integrated with responses coming from the sensory system to create information about the object's location, and other object's properties such as grasping, holding, tearing and manipulating [14]. The information of the desired movement is added to the motor context to generate specific motions in the Primary motor cortex, after which it will travel to the pons where it will be combined with information from the cerebellum to generate patterns to be executed in the spinal cord [15].

As shows in Fig. 1a, there are processes that are belong to modules like memory, perception, visual sensory system and others, however, in this work we aim to evaluate only the interaction between the decision-making, attention and motor systems, and will not deepen in them.

3 Results and Conclusions

Our case study's goal was to analyze the behavior of the virtual character when trying to autonomously navigate the proposed maze. We focused on the validation of the cognitive processes involved in human decision-making behavior. We show a list of cases used to validate the behaviors explained before: Phase 1, Set the objective; Phase 2, Filter Stimuli information to detect possible options; Phase 3, Choose an option; Phase 4, Select an action; Phase 5, Perform control and execution of the motor action; Phase 6, Verify if the goal has been reached; Phase 7, Identify closed paths; Phase 8, Return to the last decision point; Phase 9, Finish the task when the objective has been reached.

Some of the assumptions and limitations considered for the implementation of the model: We assume the existence of cognitive functions like visual sensory system, perception and memory, which provide useful information about the agent's internal and external state. For simplicity, those functions will be considered as black box elements; In this moment, we do not consider a learning process, as the maze will be different in each trial; therefore it is not imperative that the virtual creature learns; We assume the agent knows the goal to reach; There is always a path from the starting point to the goal.

The implementation of our cognitive model was made for a virtual agent in a digital labyrinth (It was developed using the Unity game engine). The execution of the case study was carried out in a virtual environment, where the agent explores a maze from a starting location to reach the objective. The results we obtained from experimentation are mostly positive, in this simple test. The plan generated by planning and decision-making is sent to the motor system, which is responsible for carrying out the plans through its control and execution mechanisms. This process would be repeated until step n-1, in which the attention system would detect the exit of the maze and send the coordinates to decision-making processes, to notify that the objective has been reached and the execution of the motor plan must stop. Despite the promising results, some flaws became evident, such as when the virtual creature is at a decision point, its behavior could be different and appeared to be innapropriate, because the agent did not always select the option with the highest attentional value. Another case occurred when the agent faced two options with equal highest attentive value, as it always selected the first perceived option. However, human beings do not always choose the first best option. All these points will be addressed in future iterations of the model.

The results presented before helped to validate our model and reaffirm our approach, accomplishing a simplified but autonomous navegation behavior. However, some characteristics should and will be improved, i.e., when the virtual agent is at a decision point it could use a set of preferences to bias the choice or obtain other information from functions such as the emotions, motivations or elements like the reward circuit. Some behaviors exhibited in the tests show that the estimation of attentional orientation to task-relevant objects was important for choosing alternatives, however, in less controlled experiments, it is critical that we consider the stimuli's prominence to prevent possible erratic behaviors in the virtual agent. The main contributions of this research are: providing a concise selection of neuroscientific evidence related to the cognitive functions required for the navigation process, specifically planning and decision-making, attention and motor system; proposing a biologically-inspired computational model which considers attentional information to guide the agent's behavior through the maze. Finally, we consider that the model presented here can serve as a testbed for future experiments in the study of cognitive architectures.

Acknowledgments. This work's research was possible thanks to the grants and support from Consejo Nacional de Ciencia y Tecnología to the authors. SEP-Cinvestav funding.

References

1. Chong HQ, Tan AH, Ng GW (2007) Integrated cognitive architectures: a survey. Artif Intell Rev 28(2):103–130
2. Baars BJ, Franklin S (2009) Consciousness is computational: the LIDA model of global workspace theory. Int J Mach Conscious 1(01):23–32
3. Lieto A, Lebiere C, Oltramari A (2018) The knowledge level in cognitive architectures: current limitations and possible developments. Cogn Syst Res 48:39–55
4. Kotseruba I, Tsotsos JK (2018) 40 years of cognitive architectures: core cognitive abilities and practical applications. Artif Intell Rev 40:1–78
5. Byrne MD, Kirlik A, Fleetwood MD, Huss DG, Kosorukoff A, Lin RS, Fick CS (2004) A closed-loop, ACT-R approach to modeling approach and landing with and without synthetic vision system (SVS) technology. In: Proceedings of the human factors and ergonomics society annual meeting, vol 48, no 17. SAGE Publications, Los Angeles, pp 2111–2115
6. Franklin S, Madl T, Dí-mello S, Snaider J (2014) LIDA: a systems-level architecture for cognition, emotion, and learning. IEEE Trans Auton Ment Dev 6(1):19–41
7. Kringelbach ML (2005) The human orbitofrontal cortex: linking reward to hedonic experience. Nat Rev Neurosci 6(9):691
8. Balleine BW, Delgado MR, Hikosaka O (2007) The role of the dorsal striatum in reward and decision-making. J Neurosci 27(31):8161–8165
9. Spielberg JM, Miller GA, Warren SL, Engels AS, Crocker LD, Banich MT, Sutton BP, Heller W (2012) A brain network instantiating approach and avoidance motivation. Psychophysiology 49(9):1200–1214

10. Jennings JH, Rizzi G, Stamatakis AM, Ung RL, Stuber GD (2013) The inhibitory circuit architecture of the lateral hypothalamus orchestrates feeding. Science 341(6153):1517–1521
11. Bailey MR, Simpson EH, Balsam PD (2016) Neural substrates underlying effort, time, and risk-based decision making in motivated behavior. Neurobiol Learn Mem 133:233–256
12. Dagher A (2012) Functional brain imaging of appetite. Trends Endocrinol Metabolism 23(5):250–260
13. Seidl KN, Peelen MV, Kastner S (2012) Neural evidence for distracter suppression during visual search in real-world scenes. J Neurosci 32(34):11812–11819
14. Hocherman S, Wise SP (1991) Effects of hand movement path on motor cortical activity in awake, behaving rhesus monkeys. Exp Brain Res 83(2):285–302
15. Mason P (2011) Medical neurobiology. Oxford University Press, New York

Psychological Theoretical Framework: A First Step for the Design of Artificial Emotion Systems in Autonomous Agents

Jonathan-Hernando Rosales[1], Luis-Felipe Rodríguez[2(✉)], and Félix Ramos[3]

[1] Universidad Autónoma de Guadalajara, Zapopan, Mexico
jonathan.rosales@edu.uag.mx
[2] Instituto Tecnológico de Sonora, Ciudad Obregón, Mexico
luis.rodriguez@itson.edu.mx
[3] Cinvestav Guadalajara, Zapopan, Mexico
framos@gdl.cinvestav.mx

Abstract. Autonomous Agents (AAs) capable of exhibiting emotional behaviors have contributed to the development of natural human-machine interactions in several application domains. In order to provide AAs with emotional mechanisms, their underlying architecture must implement an Artificial Emotion System (AES), a computational model that imitates specific facets of human emotions. Although several AES have been reported in related literature, their design is generally supported on several emotion theories, leading researchers to model and integrate isolated emotion components and mechanisms into the architectures of AES. This theoretical foundation of AES contributes to ambiguities in the analysis and comparison of their underlying architectures, which demands the definition of standards, design guidelines, and integrative frameworks. In this chapter, we present a psychologically inspired theoretical framework designed to serve as a platform for the unification of AES components, the comparison of AES, and the design and implementation of AES in AAs.

Keywords: Emotion modeling · Emotion theory · General framework · Autonomous agent

1 Introduction

The development of Autonomous Agents (AAs) capable of exhibiting emotional behavior has contributed to natural human-machine interactions [1]. Applications in fields such as serious games [2], social simulations [3], and virtual tutoring systems [4] have taken advantage of emotional AAs as this type of agents are capable of maintaining social interactions and showing realistic behavior. In order to provide AAs with emotional mechanisms, they should incorporate in their architecture an Artificial Emotion System (AES), which is a computational

A. V. Samsonovich (Ed.): BICA 2019, AISC 948, pp. 432–437, 2020.
https://doi.org/10.1007/978-3-030-25719-4_57

model of human emotions that imitates the cognitive and affective mechanisms underlying this human process [5].

In this paper, we present a psychologically inspired general theoretical framework designed to serve as a platform for the (i) unification of AES' components, (ii) analysis and comparison of AES, and (iii) design and computational implementation of AES in AAS. The proposed general theoretical framework addresses the issues described above regarding the design requirements and theoretical foundations of AES. In particular, this work represents an effort to consolidate into a single and coherent framework the common theories and components involved in the human emotion process that are usually implemented in AES. Important contributions of this paper are (i) a psychologically inspired general theoretical framework that serves as a starting-point model for the design of AES in which researchers can focus primarily on fulfilling application requirements of AES and ensure that the resulting model has a solid theoretical basis; (ii) a reference model for analyzing, comparing, and extending AES reported in the literature as this framework captures key ideas from theories of human emotions; and (iii) a tool useful to provide feedback to human emotion theories and models originated in fields such as psychology as the proposed framework is highly inspired in this type of theory.

2 A Proposed Psychologically Inspired General Theoretical Framework

The main objective of this proposal is to contribute a psychologically inspired general theoretical framework. In this context, we state that it is necessary first to identify basic components involved in the human emotion process. Although there is a diversity of components described in the state of the art, including mood, sentiment, feeling, behavior, affect, and emotional state, it is possible to identify the main components in the literature considering the congruence between them and the minimum necessary components to achieve a general emotion process.

In this section we describe the basic and necessary components to design an emotion process in AES. Afterwards, we present the proposed psychologically inspired general theoretical framework for the generation of AES in AAs.

2.1 Basic Components of the Emotion Process

In this section we propose three basic components based on the analysis of the aforementioned emotion theories. These components are *Affect*, *Feeling*, and *Emotional Behavior*, which are necessary for the generation of emotional responses and will be described below.

Affect in agreement with the various psychological theories, we can identify a basic component oriented to the evaluation of the environment, which we will label as *affect*. This component can be observed in Dimensional Theories, called as dimensional parameters, which are regularly oriented to positive or negative environmental assessments. Something similar can be observed in Appraisal Theories, where the first evaluation is oriented to positive or negative assessments of the environment. Likewise, a first component for the generation of emotions in Constructivist Theories is observed, oriented in the same way to sensing the environment in a positive or negative way.

This means that affect is an internal assessment product of pleasure or pain and their intensity, which is a primitive for the generation of mental states and emotional behaviors [6,7].

Feeling. Another basic component is oriented to the internal emotional state, product of the positive and negative evaluations of the environment, which we will call *feeling* within this work. This component can be observed in Appraisal Theories, called simply as emotional state. In the same way, in Emotion Regulation Theories, this component is also identified called emotional state. In Constructivist Theories, this component is simply called as a mental state, the product of positive and negative environmental assessments.

The feeling is a mental state, a component of the emotion, and an unconscious mechanism that biases our behavior and actions in the environment [7]. However, this basic component of emotion is only responsible for the generation of internal states and requires other processes like memory, planning, decision making, and motivation.

Emotional behavior the last basic component identified in the emotion theories is oriented to the physiological responses produced by the emotional process, which we will simply denominate as *emotional behavior*. These emotional responses can be observed in the Basic Emotions Theories, where emotions are defined as predefined physiological behaviors, calling basic emotions to the innate emotional behaviors of living beings. On the other hand, in Dimensional Theories as in Constructivist Theories, these response behaviors are defined simply as emotion, which are observable by other individuals for a social coexistence. In the same way, both Appraisal Theories and Emotional Regulation Theories, define this component as an emotional response, these responses are generated to exhibit the emotional states to the environment.

We define *emotional behavior* as the physiological response to emotions. This behavior is specific and deterministic for each basic emotion and non-specific and diffuse for each secondary emotion. Emotional behavior is always associated with a feeling and is a tool to express our internal state [8].

2.2 Proposed Framework

The definition of the components of the process model described above leads to potential information flows between them. An initial information flow begins with affect, goes through feeling, and culminates in the emotional behavior component. Even though it is true that these could not be all the components of the emotion function, these three basic components guarantee the generation of emotions in AAs. Even more, the proposed framework allows the inclusion of any emotion component that could be justified in terms of emotion theory.

Affect gives assessments of perceived stimuli, feeling provides the internal state given the resulting affect, and finally the emotional behavior helps to express the feelings to the environment.

The theory of basic emotions proposes the existence of an emotional hierarchy that defines basic or innate emotions as the base of secondary emotions. These latter emotions are referred to as cognitive emotions as their emergence involves other cognitive processes. In this sense, taking into account this theory of basic emotions, the framework is re-organized as follows; The *feeling component* is divided in two components: Innate Feelings and Cognitive Feelings. This decomposition agrees with the idea that for all expressed emotions we need an internal mental state associated with it; The component *Cognitive Processes* is included to represent the role of cognitive processing in secondary emotions.

The *Cognitive Processes* component is directly connected to the *Affect* component given that, dimensional and constructivist theories propose that emotions are the product of affective evaluations. These affects vary in terms of parameters provided by cognitive processes. From these affective evaluations, new mental states or feelings are generated, and these generate associated response behaviors.

According to theories of emotional regulation, emotional control is directly produced by cognitive processes. In this sense, we include in the framework two more components: the *Affective Control* component associated with the mechanism of emotional re-evaluation, and the *Behavior Control* component associated with the mechanism of emotional suppression, both components are handled by the *Cognitive Processes* component.

Finally, theories of emotional evaluation propose three phases that we define, in the context of the framework, as follows:

Phase 1 the learning oriented to stimuli and reactive emotional behaviors are observed. The *Affect* component is involved in this phase since it is oriented to the evaluation of stimuli. This component triggers a chain reaction, generating feelings that eventually generate emotional behavior oriented to basic emotions.

Phase 2 the prediction of reward and cognitive emotional behaviors are observed. Emotional behaviors are not necessarily conscious, but they are the result of conscious cognitive processes. The *feeling* component is involved in this phase as it helps to identify our involvement in the environment. In turn, other cognitive processes can modify these feelings. However, there may not be an emotional response behavior. The prediction of reward can be given from perceived stimuli that have emotional values already stored. This could be a side effect of the primary evaluation. As secondary emotions are cognitive emotions, these are the result of this second phase.

Phase 3 is aimed at the control of emotions. It is a fully conscious stage and is the result of cognitive processes. As mentioned before, emotional control occurs in two ways: controlling the affective component (re-interpreting the meaning of the perceived stimuli) or suppressing emotional behavior (trying not to express the feeling).

Figure 1 shows the proposed framework as well as the division of the *Affect* component into *Innate Affect* component (product of the intrinsic evaluations of the perceived stimuli) and the *Cognitive Affect* component (product of cognitive processing and the reasoning of perceived stimuli).

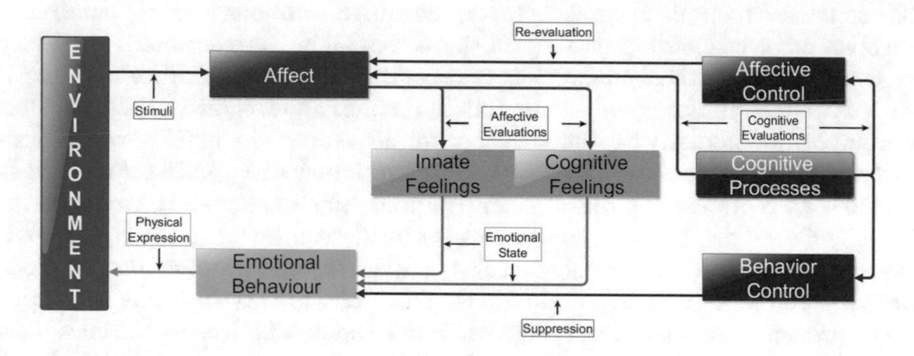

Fig. 1. Generalized conceptual model of the emotional process.

3 Using the Psychologically Inspired General Theoretical Framework Proposed to Compare Artificial Emotion Systems

In Table 1 we demonstrate that the proposed framework explains AES reported in the literature by analyzing the congruence of its underlying architecture with respect to the components and phases of AES reported in the literature.

Table 1. Comparative of emotional components proposed involved in representative AES.

AES	Innate affects	Cognitive affects	Innate feelings	Cognitive feelings	Emotional behavior	Other cognitive proccess
Cathexis	*	*	*	*	*	
FLAME	*	*	*	*	*	*
MAMID	*	*	*	*		*
ALMA	*	*	*	*	*	
EMA	*	*	*	*	*	*
WASABI	*	*	*	*	*	*
iCub	*		*		*	*
LIDA	*		*		*	*
SOAR	*		*		*	*
Kismet	*		*		*	*
ACT-R	*		*		*	*

4 Conclusion

In this paper, emotion components of artificial emotion systems and emotion theories were analyzed by means of computational methods in order to build a psychologically inspired general theoretical framework for the modeling of emotions based on identified *basic* components. This framework was contrasted with AES reported in the literature. We demonstrated that it is feasible to unify emotion theories to construct a general theory that is capable of determining which components are fundamental or essential for the design and implementation of AES. Although our proposal represents a general theory, which was partly the objective of this work, it is still necessary to explicitly describe what are the basic components. Research in fields such as neuroscience could provide this specific picture about the functioning of emotions. Similarly, the development and implementation of emotional computational models based on these theories could favor the validation of the assumptions developed within a unified theory. We believe that the construction of a unified theory therefore depends on the collaboration of the multiple fields of science involved in the generation of emotions.

Acknowledgments. This work was supported by PFCE 2019.

References

1. Churamani N, Cruz F, Griffiths S, Barros P (2016) iCub: learning emotion expressions using human reward. In: International conference on intelligent robots and systems (IROS), workshop on bio-inspired social robot learning in home scenarios
2. Hudlicka E (2009) Affective game engines: motivation and requirements. In: Proceedings of the 4th international conference on foundations of digital games. ACM, pp 299–306
3. Bosse T, Duell R, Memon ZA, Treur J, van der Wal CN (2015) Agent-based modeling of emotion contagion in groups. Cogn Comput 7(1):111–136
4. Craig S, Graesser A, Sullins J, Gholson B (2004) Affect and learning: an exploratory look into the role of affect in learning with AutoTutor. J Educ Media 29(3):241–250
5. Marsella S, Gratch J, Petta P et al (2010) Computational models of emotion. Blueprint Affect Comput Sourcebook Manual 11(1):21–46
6. Gendron M, Barrett LF (2009) Reconstructing the past: a century of ideas about emotion in psychology. Emot Rev 1:316–339
7. Barrett LF, Bliss-Moreau E (2009) Affect as a psychological primitive. Exp Soc Psychol A 41:167–218
8. Damasio AR (1995) Descartes' error: emotion reason, and the human brain. Putnam Berkley Group Inc., New York. Damasio AR (ed)

Using Metrics in the Analysis and Synthesis of Reliable Graphs

Victor A. Rusakov(✉)

National Research Nuclear University MEPhI
(Moscow Engineering Physics Institute),
31 Kashirskoe shosse, 115409 Moscow, Russia
VARusakov@mephi.ru

Abstract. The interaction of intelligent agents implies the existence of an environment to support it. The usual representations of this environment are graphs with certain properties. Reliability is one of the most important characteristics of such graphs. Traditional metrics, i.e. the usual shortest paths and minimal cuts, form the basis of the traditional measures of reliability. The analyzing and synthesizing of reliable graphs using the Euclidian metric are described. The Euclidian metric allows us to achieve better results in doing this compared to the cases of using traditional metrics. The described approach can be used in the analysis and synthesis of the environment supporting the intercommunication of intelligent agents in conditions of limited resources to organize the structure of this interaction.

Keywords: Graph metrics · Analysis and synthesis of reliable graphs · The Euclidian graph metric · The Moore-Penrose pseudo inverse of the incidence matrix

1 Introduction

Graphs are widely used as models of real objects. Failures of the components of such objects occur at random times. Their recovery also takes random amounts of time, and the parameters of many of these processes are non-stationary. The *simplest* probabilistic models are often used in the study of inherent reliability. In them, the failures of the edges of the graph are independent events with equal probabilities p_{f}. Similarly, the unreliability of the vertices of the graph is modeled. Moreover, the failure of a vertex usually entails the failure of the edges incident to it. Next, we consider the case of unreliable edges. It is indicated in the components of subsequent expressions with the superscript «e».

From now on a graph will be understood as a finite undirected graph without loops and multiple edges having k vertices and h edges. If there are no stipulations to the opposite, then the graph is supposed to be connected.

The probability $p_{\mathrm{f}}(s,t) = \sum_{i=0}^{h} c_{st}^{\mathrm{e}}(i) p_{\mathrm{f}}^{i} (1 - p_{\mathrm{f}})^{h-i}$ that a pair of s,t vertices is not connected is calculated on the basis of the coefficients $c_{st}^{\mathrm{e}}(i)$. Each such coefficient is equal to the number of special combinations of i edges. Namely, if in any such combination all i edges have failed and the remaining $h - i$ are workable, then there is

A. V. Samsonovich (Ed.): BICA 2019, AISC 948, pp. 438–448, 2020.
https://doi.org/10.1007/978-3-030-25719-4_58

no connection between s and t. The calculation of such numbers is extremely difficult for graphs of general form [1, 2]. Of course, it is just as difficult to calculate the probability $p_c(s,t) = 1-p_f(s,t)$ that a pair of s,t vertices is connected. It can be calculated by calculating the values of the polynomial $\sum_{i=0}^{h} a_{st}^{e}(i)(1-p_f)^{i} p_f^{h-i}$. The meaning of the coefficients $a_{st}^{e}(i)$ is obviously the opposite of the meaning of the coefficients $c_{st}^{e}(i)$. Similar expressions were considered in [3].

Difficulties in the calculation of even the simplest *probabilistic* indices of reliability require the use of *deterministic* criteria of reliability. They are based on additions to the assumptions under which the described polynomials are valid. Let m_{st} be the maximum number of paths between s and t that have no common edges. Then $c_{st}^{e}(i) = 0$ for $0 \leq i < m_{st}$. Further, for highly reliable systems, the use of components with small probabilities of failure is typical, $p_f \approx 0$. In this case, the factors $p_f^{i}(1-p_f)^{h-i}$ rapidly decrease with an increasing i. The largest such factor will be in the first item with $i = m_{st}$ Therefore, in order to increase the reliability of the s,t-connection one needs to increase this deterministic index. Its derivatives are also used, as discussed later.

There is the hypothetical opposite case with $p_f \approx 1$. Here, consideration of the coefficients $a_{st}^{e}(i)$ and their factors leads to the necessity to shorten the length l_{st} of the shortest s,t-path. If you have decided to estimate the entire graph, for example, by the worst pair, you should reduce the diameter of the graph to get the result of the synthesis.

The shortest s,t-path and the maximum number of s,t-paths having no common edges as well as their quantitative indices are components of the simplified descriptions of this pair's connection reliability. Such simplifications are formed when the probabilistic parameters have extreme values. Another simplification is the decomposition into pairs of vertices with separate estimates of their reliability. At the final stage of the assessment for the entire graph, the reverse is done - the composition of the pairs' indices. This approach is used, despite the possible presence of dependencies between pairs of vertices due to intersecting subsets of the edges.

Let X and Y be vector spaces with dimensions h and k correspondingly over the field of real numbers. The arbitrarily introduced orientations of the edges in the absence of limits to their capacity turns the incidence matrix A of an undirected graph having ± 1 as non-zero elements in each column into a useful instrument for the *linear* transformation $Ax = y$ of the edge flows $x \in \mathrm{X}$ into the vertex flows $y \in \mathrm{Y}$. From now on the incidence matrix A will be understood to be such a matrix.

Both of the above mentioned simplification techniques can be combined in the framework of the *linear model* for converting flows $x \in \mathrm{X}$ into flows $y \in \mathrm{Y}$ with the introduction of appropriate vector norms $||\mathrm{X}||$ and $||\mathrm{Y}||$ in these spaces.

Let $\{x_{st} \in \mathrm{X} \,|\, Ax_{st} = y_{st}\}$ where y_{st} has an entry s equal to +1, an entry t equal to –1, and the remaining entries are equal to 0. Any of the vectors x_{st} represent the transfer of the flow of 1 from s to t along the edges of the graph. Then, without additional designations one should search for the extreme norms along all such x_{st}. The quantitative characteristics of the s,t-paths in the graph are represented as conditional matrix norms $||A||_{||\mathrm{Y}||/||\mathrm{X}||}^{st}$ of the incidence matrix [4, 5].

Linearization and decomposition constitute the methodological basis for obtaining estimates of the graph reliability. The wide practice of their use indicates the high consistency of linear deterministic models of reliability.

2 Reliability Analysis and the Euclidian Metric

Preservation of the essence of the problem of reliability in linear models when one has obtained them from more complex models means the following: the possibility of constructing reliability estimates (the ability to analyze reliability) exists without any explicit use of the $p_f(s,t)$ and $p_c(s,t)$ expressions described above. The material of this section is devoted to these constructions within the framework of such linear models [4].

The basis of the *s,t-connection* is the set of the *paths* between these vertices. In turn, a convenient formalism for describing any path is a *flow* in the form of a set of magnitudes, each of which corresponds to an edge included in this path.

Suppose we want to describe the *significance* of the edge for the *reliability* of the s,t-connection with the help of the *magnitude* of the flow along this edge. For an arbitrary pair of vertices of an arbitrary graph such a significance of the edges is not known in advance. Therefore, to search for the flow from s to t whose components possess the necessary properties, we proceed as follows.

Consider a complete graph with the same k number of vertices. It has a valuable symmetry property. This property will be used later.

But first, note that for *any* pair of *s,t* vertices of such a graph, there always exists a unique *s,t*-path of length 1. Using a nonzero flow, we select this edge (*s*,*t*). Obviously, the absolute quantity of this flow f_1 is unimportant, so we set $f_1 = 2$ here. Along with such a path of length 1, there always exist $k - 2$ *identical* paths of length 2 between s and t in such a graph. Each of them has a single intermediate vertex m_i, $i = 1...k - 2$. Since these paths are the same, let's direct the flow f_2 of the same magnitude along each of them from s to t.

Let $k > 3$. Then there is at least one pair of described intermediate vertices m_i, m_j. Through each of these vertices a flow passes from s to t. In a complete graph there is always an edge (m_i, m_j) between m_i and m_j. The symmetry of such a graph does not allow one of the two possible directions of this edge to prefer to send a non-zero flow through it. Therefore, according to the *Principle of Non-Sufficient Reason* [6], the flow along such an edge should be set equal to zero. Due to the symmetry of a complete graph, the same is true for any pair of such intermediate vertices.

As a result, in a complete graph $k - 1$ paths between s and t having no common edges are marked by nonzero flows of the quantities f_1 and f_2. For any intermediate vertex m_i, the flow of f_2 passes through *two* successive edges (s, m_i) and (m_i, t). Therefore, when adhering to the *linear* approach in assessing the importance of edges for reliability through the appointment of flow magnitudes over them, put $f_2 = 1$, that is, *half* the size of f_1. Denote such a flow from s to t in a complete graph as $f_c(s,t)$, Fig. 1. The f_1 and f_2 flow directions are unambiguous. So the *signs* of these $f_c(s,t)$ components and the arbitrary *orientations* of the edges are in obvious conformity.

Row normalization turns a graph's adjacency matrix into a stochastic matrix $P = (p_{mn})$. Removing the row and column with the numbers t turns P into Q_t, and further we get $N_t = (I - Q_t)^{-1} = (n_{uv}^{(t)})$. Matrix I, here and from now on, designates an identity matrix whose size is clear from the context. The set of values $\left\{n_{si}^{(t)} p_{ij} - n_{sj}^{(t)} p_{ji}\right\}$ for all edges (i, j) and any vertices $s \neq t$ of the graph describes the shortest s,t-path in a metric that is nontraditional for graphs [4, 5]. The following Lemma allows us to further use the valuable property of the restored symmetry of the matrices P, Q_t, N_t. Recall that for irregular graphs these matrices with zero main diagonals are asymmetric in general. The Lemma is an analogue of the obvious property for ordinary shortest paths (minimal cuts): the absence of the influence of loops with nonnegative weights on these paths (cuts) and their lengths (values).

Let **U** be a set of transient states of the absorbing Markov chain.

Lemma 1 [4]. For $s, i \in \mathbf{U}, j = 1, \ldots k, i \neq j$ when p_{ii} changes in the interval [0, 1) and there is proportional renormalization of $p_{ij} : \sum_{j=1}^{k} p_{ij} = 1 = \text{const}$ then the $n_{si}^{(t)} p_{ij}$ remain the same.

Proof. The normalizing factor for all entries of the i-th Q_t row is just the same due to the lemma condition and is finite and positive because of the probabilistic nature of the p_{ij} values. We represent it in the form $(1 + \delta)^{-1}$, $-1 < \delta < \infty$. All non-diagonal entries of the i-th Q_t row should simply be divided by $1 + \delta$. The only entry of the i-th Q_t row that may have an *additive* change δ_{ii} is the diagonal entry p_{ii}. And the new value of this entry is $(p_{ii} + \delta_{ii})(1 + \delta)^{-1}$. Two variables δ_{ii} and δ are bound by the lemma's normalizing condition. Let σ_i be the sum of the non-diagonal entries of the i-th row of the stochastic matrix, $\sigma_i = 1 - p_{ii}$. Then $(p_{ii} + \delta_{ii})(1 + \delta)^{-1} + \sigma_i(1 + \delta)^{-1} = 1 \Rightarrow \delta_{ii} = \delta$, $-p_{ii} \leq \delta < \infty$. So the new value of $p_{ii}^{(\delta)}$ of the diagonal entry p_{ii} is $(p_{ii} + \delta)(1 + \delta)^{-1}$ and $\delta = (p_{ii}^{(\delta)} - p_{ii})(1 - p_{ii}^{(\delta)})^{-1}$.

Consider this in the matrix form. To simplify further indices designations let's renumber the Markov chain's states so that $t = k$. Let's say the $(k - 1) \times (k - 1)$ matrix T has an entry $t_{ii} = \delta$, and the remaining entries are equal to 0. Let's say the $(k - 1) \times (k - 1)$ diagonal matrix S has an entry $s_{ii} = (1 + \delta)^{-1}$ and its other diagonal entries are 1. Then $Q_t^{(\delta)} = S(Q_t + T)$ will be the new transition matrix of the absorbing Markov chain. It has the entries of the i-th row described above.

Note the case of $p_{it} \neq 0$. This probability takes no explicit form in the i-th Q_t row. Nevertheless its value will be changing implicitly just the same way as any non-diagonal entry. Indeed let σ_i be the sum of the non-diagonal entries of the i-th Q_t row in this case. Then $p_{it} = 1 - (p_{ii} + \sigma_i)$ for Q_t and a value of $p_{it}^{(\delta)} = 1 - (p_{ii}^{(\delta)} + \sigma_i^{(\delta)}) = p_{it}(1 + \delta)^{-1}$ for $Q_t^{(\delta)}$. Also note all other $Q_t^{(\delta)}$ entries except those of the i-th row have no changes at all.

Then $I - Q_t^{(\delta)} = I - S(Q_t + T) = (I - ST)(I - Q_t)$. It is easy to see that the matrix I–ST is a diagonal one. It has an ii-entry equal to $(1 + \delta)^{-1}$, and the rest of its diagonal entries are equal to 1. Hence, there is a $(I - ST)^{-1}$. It is diagonal, has an ii-entry equal to $1 + \delta$, and the rest of its diagonal entries are equal to 1. Then $N_t^{(\delta)} = (I - Q_t^{(\delta)})^{-1} = (I - Q_t)^{-1}(I - ST)^{-1} = N_t(I - ST)^{-1}$. Thus each entry of the i-th N_t column is multiplied by $1+\delta$ while all other N_t entries remain unchanged and we get $N_t^{(\delta)}$. The comparison of the changes in this column to the changes in Q_t entries while getting $Q_t^{(\delta)}$ described above completes the proof. □

In addition we note the independence of the lemma's results from the sequence of consideration of the states $i \in \mathbf{U}$.

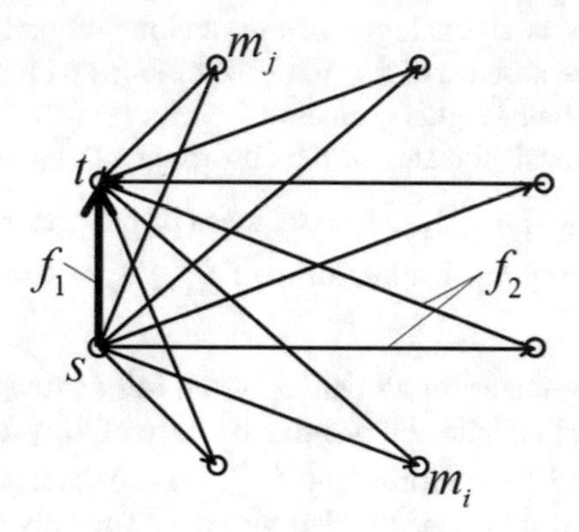

Fig. 1. Flow $f_c(s,t)$ constructed for a complete graph.

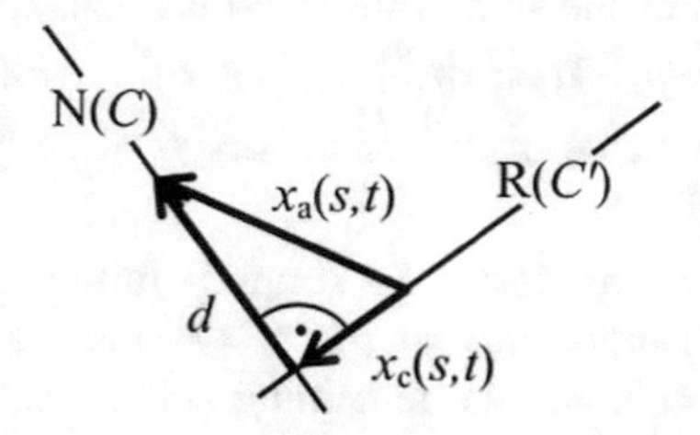

Fig. 2. Vectors and spaces of Lemma 3 for a complete graph for $y_0 = y(s,t) = y_{st}$.

The need to make a representation of each flow component $n_{si}^{(t)} p_{ij} - n_{sj}^{(t)} p_{ji}$ in the same form for t as for any other vertex leads to the ergodic Markov chain. Further we will use d_i as the degree of the i-th vertex, $d_{\max} = \max_{1 \le i \le k} d_i$, $p = (d_{\max} + c_n)^{-1}$ where $0 \le c_n < \infty$ and in the matrix $P = (p_{uv})$ let the entry $p_{ij} = 0$ when there is no edge (i,j) in the graph, and $p_{ij} = p$ when there is the edge (i,j), $i \ne j$, and $p_{ii} = 1 - pd_i$. Matrix P is symmetrical even for an irregular graph. Its entries are the transition probabilities of the ergodic Markov chain. It is possible that such P (and Q_t) have non zero elements on the main diagonal that corresponds to the loops of the graph. But according to Lemma 1, for any $s \ne t$ and edge (i,j) the value $p(n_{si}^{(t)} - n_{sj}^{(t)})$ is the same as the value of that weight calculated by using $n_{si}^{(t)} p_{ij} - n_{sj}^{(t)} p_{ji}$ for P (and Q_t) with a zero main diagonal and even without their (and N_t) being symmetrical. Further the matrix P defined in this paragraph and the Q_t and N_t gotten from it will be used everywhere.

The fundamental matrix Z of the ergodic Markov chain is equal to $(I-P+F)^{-1}$ where $F = (f_{ij})$ is the limiting matrix of that chain, $f_{ij} = k^{-1}$, $\forall i,j$, for the symmetrical (!) P [7].

Further «′» и «$^+$» symbolize the transposition and the Moore-Penrose pseudo inverse correspondingly.

For an arbitrary edge (i,j) of a graph let its column in the matrix A have +1 in the i-th row, and −1 in the j-th row. For an arbitrary graph $A^+ = pA'Z$ [4, 5], so for this edge the component of the flow A^+y_{st} is equal to $p(z_{si}-z_{sj}+z_{tj}-z_{ti})$. Then we use the well-known [7] matrix equality $Z = F + (I-F)N^*(I-F)$. Here the $k \times k$ matrix N^* is the matrix N_t, $\forall t$, with the zero row and column inserted into the t-th places. We multiply the matrices of this equality on the left by the row-vector $y_{st}{}'$, and on the right – by the vector y_{ij}. The last vector, like the aforementioned column of the (i,j) edge in the matrix A, has an i-th entry equal to +1, a j-th entry equal to −1, and the other entries are equal to 0. We obtain the equality $z_{si}-z_{sj}+z_{tj}-z_{ti} = n_{si}^{(t)} - n_{sj}^{(t)}$. Next multiply both sides by p.

We obtain the component equality of the flows along the edges described, on the one hand, in terms of the entries of the matrix Z, i.e. A^+, and in terms of the entries of the matrix N^* on the other hand. This equality demonstrates the validity of the assumption that the set of values of the form $p(n_{si}^{(t)} - n_{sj}^{(t)}) = n_{si}^{(t)}p_{ij}-n_{sj}^{(t)}p_{ji}$ can be used as a description of the shortest s,t-path in some specific (quadratic, Euclidian) metric.

Let C designate the incidence matrix of a complete graph.

Lemma 2 [4]. The vector $f_c(s,t)$ is proportional to $C^+y_{st} : f_c(s,t) = cC^+y_{st}$.

Proof. The adjacency matrix of a complete graph has a zero main diagonal. All other entries are equal to 1. Let us transform it by normalization into a doubly stochastic matrix $P = (p_{ij})$. By using Lemma 1 one can always make diagonal entries of P any non-negative value without affecting the magnitudes of the edge flows with conditionally minimal Euclidian norms. Make them all equal to k^{-1}, we get $\forall i,j\, p_{ij} = p = k^{-1}$ and $P = (k^{-1})$. The fundamental matrix $Z = (I-P+F)^{-1}$ of an ergodic Markov chain in this case is equal to I. Since [4, 5] for an arbitrary graph $A^+ = pA'Z$, here you get $C^+ = k^{-1}C'$, so immediately $c = k$. □

An arbitrary graph differs from a complete one with the same number of vertices – there is an absence of some edges. This leads to changes in the reliability of the pairs of vertices and changes in the significance of the remaining edges for this reliability.

On the other hand, the absence of some edges in an arbitrary graph simply means you can't send a flow between vertices that are not connected by an edge. So it is clear that the flow $C^+y_{st} = k^{-1}C'y_{st}$ may be absent in an arbitrary graph for an arbitrary pair of s,t vertices among $\{x_{st} \in \mathrm{X}\,|Ax_{st} = y_{st}\}$. However, one can always search for the *best approximation* in a certain sense to $k^{-1}C'y_{st}$ so that we can reveal the absence of some edges by means of such a flow. The redistributed flow quantities along the edges of an arbitrary graph after just such an approximation will display their sought significance for the s,t-pair's connection reliability.

Other things being equal, the s,t-pair's connection reliability of a complete graph is maximal. Therefore, the *magnitude* of the *smallest* deviation of the approximate flow in an arbitrary graph from $k^{-1}C'y_{st}$ can be used as the reliability measure of the *s,t*-pair of this graph. The description of the reliability of a pair of vertices in a complete graph in the form of a vector $k^{-1}C'y_{st}$ naturally is the reference point here.

To determine the meaning of the term “best approximation”, we introduce the scalar product in X in the usual way, and by the square of the length of the vector *x* we mean (*x*, *x*).

Here and below let N(H) and R(H′) mean the H kernel and the H′ image of an arbitrary matrix H, respectively.

Insert zero columns into the matrix *A* at the places of the missing edges in an arbitrary graph. Matrix A^+ will have zero rows with the same numbers. Let $\forall y_0 \in \mathrm{R}(A)$. Clearly, $y_0 \in \mathrm{R}(C)$.

Lemma 3 [4]. $(C^+y_0, A^+y_0 - C^+y_0) = 0$.

Proof. Obviously the vector A^+y_0 can be considered as an edge flow under additional conditions in a complete graph. It transfers the flow from the vertices of the sources to the vertices of the receivers. That is described by the components of the vector y_0. The same vertex flows in the same complete graph are carried by the edge flow C^+y_0. Therefore, the flow $d = A^+y_0 - C^+y_0 \in \mathrm{N}(C)$. $C^+y_0 \in \mathrm{R}(C^+)$, but $\forall H\ \mathrm{R}(H^+) = \mathrm{R}(H')$ and $\mathrm{N}(H) = \mathrm{R}^{\perp}(H')$, where «$^{\perp}$» means the orthogonal complement, see for example [8, 9]. So $C^+y_0 \in \mathrm{R}(C')$ and we get $(C^+y_0, d) = 0$ at once. □

Figure 2 illustrates Lemma 3. The vectors $d = A^+y_0 - C^+y_0$ and C^+y_0 for $y_0 = y(s,t) = y_{st}$ are the legs of a right-angled triangle with the hypotenuse A^+y_0 for the same y_0.

Let the number of vertices *k* and $y_0 = y_{st}$ be fixed. Then, by Lemma 2, the vector $x_c(s,t)$ is also constant. Then, according to the Pythagorean theorem, the Euclidian lengths of vectors *d* and $x_a(s,t) = A^+y_{st}$ are minimal *simultaneously* for an arbitrary graph with the incidence matrix *A*. According to a well-known [8, 9] property of the Moore-Penrose pseudo inverse, the Euclidean norm of the vector $x_a(s,t) = A^+y_{st}$ will be minimal among $\{x(s,t) \in \mathrm{X}\,|Ax(s,t) = y_{st}\}$. In this case you can always use the length of the vector $x_a(s,t)$ instead of the length of vector *d* as a measure of the graph’s *s,t*-pair of vertices reliability. The vector $x_a(s,t) = A^+y_{st} = pA'Zy_{st}$, so the sought measure of reliability is $((x_a(s,t), x_a(s,t)))^{1/2} = (\mathrm{p}(\mathrm{z_{ss}} + \mathrm{z_{tt}} - 2\mathrm{z_{st}}))^{1/2}$.

3 The Euclidian Metric and the Synthesis of Reliable Graphs

Above we gave an example of the composition of the lengths of the shortest paths between all pairs of the vertices into the diameter of this graph. This method is traditional, but not the only possible one. For a Euclidian, that is quadratic, metric, a different way of composition seems natural. Namely the use of the weighted sum of the indices of the vertex pairs. In this case the *s,t*-index may be used as the weight of *s,t*-pair. Note that with such a criterion, the worst pair of vertices will be allocated the highest weight automatically.

With the attribute of the matrix Z [7] $\sum_{t=1}^{k} z_{st} = 1$, $\forall s$, the required sum $\sum_{s=1}^{k}\sum_{t=1}^{k} p(z_{ss} + z_{tt} - 2z_{st}) = p\sum_{s=1}^{k}(kz_{ss} + \mathrm{SpZ} - 2) = p(k\mathrm{SpZ} + k\mathrm{SpZ} - 2k)$ $= 2kp(\mathrm{SpZ} - 1)$.

The objective function is accompanied by constraints in synthesis. Here the limitation is the number of h edges allowed for use in the synthesized graph. In the Introduction the interrelation between the quality of each edge (its p_{f}) and the metric features of the graph affecting the reliability of the latter was noted. These metric features can be contradictory given the limited resources available. A quantitative description of the paths in $x_{\mathrm{c}}(s,t)$ is a kind of intermediate point on the imaginary line connecting the extremes of these metric contradictions. It gives double preference to a path whose length is half as long as the other paths. Such a consideration of the paths *lengths* means that when you have a *small* p_{f} one should apply additional restrictions when using the Euclidian metric in the synthesis of reliable graphs.

The concept of the smallness of p_{f} is as vague as the size of a graph. Examples of the ratio of these values appeared at the time of the creation of the first large computer networks [10]. Here we will treat the smallness of p_{f} as such a quality of the edges that it creates a preference for the *number of paths* between the vertices of any pair in front of their *lengths* during the synthesis. Of course what has been said here and further makes sense only for graphs that are more complex than trees.

Additional restrictions should not be overly difficult to verify. Let's use the well-known Whitney inequality [11], $m_{st} \leq \min(d_s, d_t)$. The maximum number of paths between s and t having no common edges is limited by the minimum value of the degrees of these vertices. Hence, to increase m_{st}, it is necessary to increase this minimum. When you have a limited h for the entire graph, you will need to bring d_i, d_j, $\forall i,j$ as close as possible.

As a result the formulation of a reliable graph's synthesis in the case of a small p_{f} may look as follows. For a given $3 < k, k \leq h$ and $|d_i - d_j| \leq 1 \forall i,j$ find the graph with a minimum value of $p(\mathrm{SpZ} - 1)$.

The search can be based, for example, on the sequence of adding and deleting edges when starting from some initial graph. The rank 1 perturbations used in this case do not require large computations for the correction of the inverse matrices [12, 13].

4 Examples of Synthesis

As an example, let's consider a deterministic reliability criterion [14, 15] derived from a metric and directly related to the probabilistic index $p_{\mathrm{f}}(s,t)$ described in the Introduction. Wilkov denotes the number of minimum edge *s*,*t*-cuts of a given power m through $X_{st}^{\mathrm{e}}(m)$. He pointed out the presence of some dependence between the values of $c_{st}^{\mathrm{e}}(i)$ and those of $X_{st}^{\mathrm{e}}(m)$. So Wilkov changes the task of estimating the reliability of a *s*,*t*-pair to looking at the values of $X_{st}^{\mathrm{e}}(m)$. Other conditions equal, the increase in the number of cuts separating the vertices of the pair leads to a decrease in the reliability of the connection of this pair. This simple statement explains the substantive side of the Wilkov criterion. According to him, $X_{st}^{\mathrm{e}}(m)$ is most important for the reliability of the *s*,*t*-pair connection for small (minimum) values of m when there is a small p_{f}.

He estimates the graph by the worst pair of vertices. Therefore it is proposed to use $X^{\mathrm{e}}(m) = \max_{s,t} X^{\mathrm{e}}_{st}(m)$ as a measure of the reliability of the entire graph. According to Wilkov the most reliable graph has a minimum value of $X^{\mathrm{e}}(m)$ when there are preset constraints. He established several necessary conditions for this. These conditions include maximum connectivity, a minimum diameter, a maximum girth, and a maximum number of large cycles in the graph. Sufficient conditions have not been clarified.

In addition to the described criterion we will also use the comparison of graphs based on the so-called generalized cohesion $\delta(q)$ [16]. This deterministic reliability index is equal to the minimum number of edges whose removal isolates the q vertices of the graph from the remaining $k - q$. Obviously the essence of this index is also derived from the metric.

Figure 3 shows reliable graphs for $k = 15$, $h = 19$ synthesized by Wilkov (a) and us (b). The vertices of the worst pairs of both graphs are depicted on the extreme right and left. Due to certain symmetries in both graphs, the worst pairs may not be the only ones. The edges forming the value of $X^{\mathrm{e}}(m)$ for $m = 2$ are marked by thicker lines.

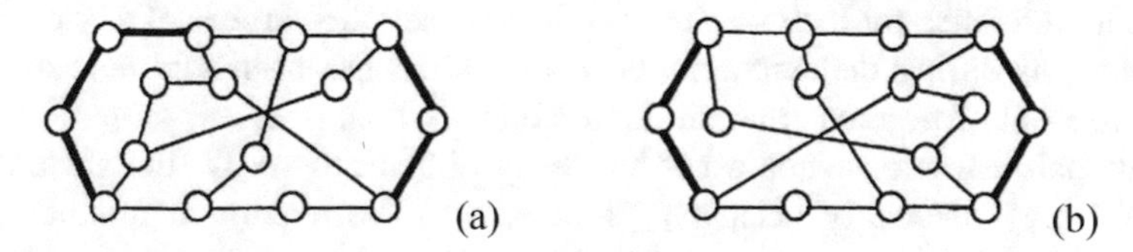

Fig. 3. Synthesized reliable graphs.

The described deterministic reliability indices are shown altogether in Table 1. The probabilities $p_{\mathrm{f}}(s,t)$ of failure of communication for the worst pairs of vertices are presented in Table 2. The $p_{\mathrm{f}}(s,t)$ values are calculated using Hansler's algorithm [17].

The comparison illustrates the advantage of the Euclidian metric in the synthesis of reliable graphs, both in terms of probabilistic and deterministic reliability indices. For an equally significant comparison of the synthesized graphs for other values of k and h see, for example, [18].

Table 1. Deterministic reliability indices.

a	b
q | 1, 2 | 3 | 4, 5, 6, 7 δ | 2 | 3 | 4	q | 1 | 2, 3 | 4, 5, 6, 7 δ | 2 | 3 | 4
$X(2) = 3$	$X(2) = 2$
$p(\mathrm{SpZ} - 1) = 8.353$	$p(\mathrm{SpZ} - 1) = 8.256$

Table 2. Probabilistic index of reliability.

p_f	a	b	p_f	a	b
0,01000	0,00031	0,00021	0,100	0,038	0,030
0,02000	0,00125	0,00086	0,200	0,185	0,159
0,03000	0,00288	0,00201	0,300	0,428	0,390
0,04000	0,00525	0,00373	0,400	0,678	0,641
0,05000	0,00842	0,00606	0,500	0,857	0,830
0,06000	0,01243	0,00910	0,600	0,952	0,937
0,07000	0,01737	0,01290	0,700	0,989	0,983
0,08000	0,02328	0,01754	0,800	0,999	0,997
0,09000	0,03023	0,02311	0,900	1,000	1,000

5 Conclusion

Traditional metrics is shown as the basis for well-known deterministic reliability indices of graphs. The use of the Euclidian metric in the analysis and synthesis of reliable graphs is substantiated. Justification is not explicitly based on known expressions from the simplest probabilistic models of reliability. This confirms the high consistency of linear metric models in the analysis of the reliability of graphs and their synthesis. The graph reliability index based on the Euclidian metric has been established. We propose using the simplest restrictions during synthesis when there is a small probability of edges failure. The advantages of the Euclidian metric are illustrated by the example of the synthesis of reliable graphs.

References

1. Samoylenko SI, Davydov AA, Zolotarev VV, Tret'yakova EI (1981) Computer networks. Nauka, Moskow, 277p
2. Ball MO (1980) Complexity of network reliability computations. Networks 10(2):153–165
3. Moore EF, Shannon CE (1956) Reliable circuits using less reliable relays. J Franklin Inst 262:191–208
4. Rusakov VA (1979) Analysis and synthesis of computer network structures. Part 1 Analysis Moscow Engg Phys Inst. Report, VNTI Centre No Б796153, 122p
5. Rusakov VA (2018) Matrices, shortest paths, minimal cuts and Euclidian metric for undirected graphs. Procedia Comput Sci 145:444–447. https://doi.org/10.1016/j.procs.2018.11.104
6. Polya G (1962, 1965, 1981) Mathematical discovery. Wiley, New York, 221p
7. Kemeny J, Snell J (1960) Finite Markov chains. University series in undergraduate mathematics. Van Nostrand, Princeton NJ, 210p
8. Albert AE (1972) Regression and the Moore-Penrose pseudoinverse. Academic Press, New York, 180p
9. Beklemishev DV (1983) Additional chapters of linear algebra. Nauka, Moscow, 336p
10. Gerla M, Kleinrock L (1977) On the topological design of distributed computer networks. IEEE Trans Commun 25(1):48–60

11. Withney H (1932) Congruent graphs and the connectivity of graphs. Amer J Math 54: 150–168
12. Rusakov VA (1978) Implementation of the methodology for analysis and synthesis of computer network structures using Markov chains. Engg-Math Methods Phys Cybern (7):41–45. Atomizdat, Moscow
13. Rusakov VA (2011) Reconstruction of the Euclidian metric of an undirected graph by metrics of components. Nat Tech Sci 2(52):22–24. Sputnik+, Moscow
14. Wilkov RS (1972) Analysis and design of reliable computer networks. IEEE Trans Commun 20(3):660–678
15. Wilkov RS (1972) Design of computer networks based on a new reliability measure. In: Proceedings of the symposium on computer communications networks and teletraffic. Polytech. Inst. of Brooklyn, 4–6 April, pp 371–384
16. Boesch FT, Thomas RE (1970) On graphs of invulnerable communication networks. IEEE Trans Commun Tech 18:484–489
17. Hansler E (1972) A Fast Recursive Algorithm to Calculate the Reliability of a Communication Network. IEEE Trans Commun 20(3):637–640
18. Rusakov VA (1977) A technique for analyzing and synthesizing the structures of computer networks using Markov chains. Computer networks and data transmission systems, 62–68. Znaniye, Moscow

Machine Learning Models for Measuring Syntax Complexity of English Text

Daniele Schicchi[1(✉)], Giosué Lo Bosco[1], and Giovanni Pilato[2]

[1] Dipartimento di Matematica e Informatica, Univerisitá degli Studi di Palermo, Palermo, Italy
daniele.schicchi@unipa.it

[2] ICAR-CNR - National Research Council of Italy, Palermo, Italy

Abstract. In this paper we propose a methodology to assess the syntax complexity of a sentence representing it as sequence of parts-of-speech and comparing Recurrent Neural Networks and Support Vector Machine. We have carried out experiments in English language which are compared with previous results obtained for the Italian one.

Keywords: Text-simplification · Deep-learning · Machine-learning

1 Introduction

Natural Language Processing (NLP) is a research area that tackles the problem of analyzing in a automated manner natural language data. Researchers who work in NLP field have created many interesting models capable of solving problems, for example, related to computational creativity [1], teaching [2], machine translation [3], support system [4] and so on.

Text Simplification (TS) is a branch of NLP that aims at making a text more easily understandable for people. A core part of TS system is the *evaluation* process that, taking into account both reader skills and text complexity, decides if the text needs of being simplified. The evaluation of text complexity (TE) is not a trivial problem and it is an actual research topic since the performances of TS system are connected to this task in many ways. Furthermore, TE system can be used both as support for TS and as independent system. For example, it can be appreciated as *decision support system* by people in contact with different communities such as those who are not mother tongue or have language disabilities.

An historical measure of text complexity is the Flesch–Kincaid [5] index which is based on structural features of the text and it gives a degree of complexity evaluating *total words*, *total sentences*, *total syllables*. However, it is a common opinion that the evaluation of only structural features is not representative of total text complexity. In the recent years, more reliable indexes were developed. They express the degree of text complexity considering more sophisticated features like the frequency of words and *simple* word dictionary,

A. V. Samsonovich (Ed.): BICA 2019, AISC 948, pp. 449–454, 2020.
https://doi.org/10.1007/978-3-030-25719-4_59

the depth of parse tree and text morphology. READ-IT [6] is a Support Vector Machine based system created to tackle the problem of TE. The system takes into account *Lexical*, *Morpho-syntactic* and *Syntactic* features aspects to decide what category of complexity belongs the input text. In order to evaluate the performance of TS system it has been proposed FKBLEU [7] the SARI index [7]. The authors proposed a TE data-driven system based on Neural Network (NN) which measure the complexity of Italian sentences taking into account *lexical* and *syntactical* aspects [8].

In this paper it is presented a TE system whose objective is focused on the evaluation of *syntax* complexity of sentences in English Language. The paper is organized as follow: in Sect. 2 we describe components of the system, in Sect. 3 we explain the way of evaluating the system performance, in Sect. 4 we will give the conclusions.

2 Proposed Methodology

The purpose of the system is to understand rules that identify *syntactical* constructs which make a sentence hard to understand for a reader. We have evaluated the performance of two different machine learning algorithms: the Recurrent Neural Networks (RNN) and the Support Vector Machine (SVM).

The RNN is a powerful model created for the elaboration of data sequence that can be used for the NLP field if the input text is structured as a sequence of tokens. The SVM [9] is a ML algorithm that has been widely used to solve different kind of problems. It has already used for NLP problems [6] showing great potentiality to analyze texts. For both models the input sentence is preprocessed by a module that extract its *parts-of-speech* and that makes it suitable for the analysis.

2.1 Preprocessing

The preprocessing module serves to represent the sentence as sequence of *part-of-speech* and to make it suitable for the analysis from the ML models. The identification of the *parts-of-speech* is carried out using a pre-trained version of TreeTagger [10]. TreeTagger is a tool capable of annotating text with its *parts-of-speech* in different languages such as Italian, English, German and so on. It allows tagging different languages by means of parameter files, called *tagsets*, that include the instructions to extract *parts-of-speech* from a text. We have used the *BNC tagset*[1] which allows to draw out 61 different *parts-of-speech* belonging to different categories like *verbs*, *adverbs*, *punctuation* and *pronouns*.

After the extraction process, it is applied a transformation that identifies each element as a vector of real numbers using the well known *one-hot encoding*. Using the *one-hot encoding* every sentence is represented by a sequence of vectors in which each of them identifies uniquely a *part-of-speech*.

[1] http://www.natcorp.ox.ac.uk/docs/c5spec.html.

The RNN model us compared with a Support Vector Machine (SVM). Since the SVM is not suitable to examine sequence of vectors the sentence preprocessing is slightly different. In this case a sequence is represented by a single vector of length equal to the total amount of *parts-of-speech* and each position of the vector identifies the occurrence of a specific *part-of-speech* in the sequence. After the counting, we have normalized the vector by the total number of *parts-of-speech*, in this specific case 61.

2.2 Architectures and Parameters

The RNN model is based on Long Short Term Memory (LSTM) [11] artificial neurons which have shown good performance tackling problems belonging to NLP field related to sequence modeling tasks. The architecture of the Network is composed by 3 layers. The first is the *input layer* whose job is to pick data after the *preprocessing* phase and making it accessible to the *LSTM layer*. The *LSTM layer*, consisting of 512 LSTM units that is responsible for analyzing the sequence. The output of this layer stimulates the next *dense layer* which is activated using *sotmax* [12] activation function giving the probability that the sequence belongs either to *easy-to-understand* or to *hard-to-understand* class. The last level is regularized using the L_2 regularization factor with value of 0.01. The network has been trained using minibatch of size 50 divided as 25 *easy-to-understand* sentences and 25 *hard-to-understand* sentences. The set of parameters have been obtained through a set of experiments which suggest what are good configurations for solving the TE problem.

The SVM is a learning method with solid computational learning theory principles behind its functioning [9]. It has been developed by the aid of scikit-learn[2] version 0.20.2, a library that helps for the implementation of ML algorithms using different kernels methods: *linear*, *RBF* and *polynomial*.

3 Experiments and Results

3.1 Corpus

To understand the *syntactical* complexity of a sentence using a data driven approach it is needed to use a specific corpus that contains sentences labeled as *hard to understand* or *easy to understand*. Our choice fell on Newsela [13] corpus that we have used to train and test the system. The corpus is a collection of articles which have been simplified by human experts. Each article has been simplified 4 times, the original document is marked with the label 0 and its easier versions are labeled with progressive numbers, 5 identifies the most simplified. Unfortunately, the Newsela corpus does not give any information about the complexity of sentences inside documents but it provides a cumulative measure associates to the document. Thus, the document could contain sentences that do not reflects the complexity of the belonging document. The solution that we propose is to

[2] https://scikit-learn.org.

take as *hard-to-understand* all the sentences inside documents marked with 0, 1 labels which are not present in documents with labels strictly greater than 1. The *easy-to-understand* sentences are picked as all the sentences inside documents identified with labels 4, 5 which are not found in documents with label strictly less than 4. Using this process we have harvested approximately 130.000 *hard-to-understand* and 80.000 *easy-to-understand* sentences.

3.2 Experiments and Discussion

The system has been tested using a cross-validation approach known as K-FOLD with $K = 10$. The K-FOLD method is often used for testing the performance of machine learning and It consists in the creation of a dataset partition in K sets. The training phase is iterated K times exploiting K-1 sets as training-set and the last one as validation-set which means that all the sets are used alternately as validation-set and training-set. For each iteration we have measured the well known Recall, Precision, True Negative Ratio (TNR) and True Positive Ratio (TPR) and after all the iterations we have averaged the obtained results. The Table 1 shows the comparison of these two models. Since the NN has been trained for a variable number of epochs we have decided to choose the ones trained for 3 (LSTM-3) and 5 (LSTM-5) epochs for the comparison.

Table 1. Average results of Recall, Precision, TPR, TNR calculated according to 10-FOLD.

Model	Kernel	TAG-SET	Recall	Precision	TPR	TNR
LSTM-5	-	BNC	.826	.849	.826	.853
LSTM-3	-	BNC	.792	.873	.792	**.884**
SVM-L	Linear	BNC	.815	**.890**	.815	.834
SVM-R	RBF	BNC	.857	.825	.857	.699
SVM-P	Polynomial	BNC	**.999**	.624	**.999**	0.0

Results shows that both LSTM and SVM are capable of classifying quite well data of the two classes. The SVM-R reaches the best result in the classification of *hard-to-understand* sentences obtaining a good value of Recall but it often makes mistakes to classify *simple-to-understand* sequences. The LSTM-5 instead keeps a balanced behavior for the classification of sentences of both classes. Furthermore, the LSTM models are more precise than SVM-R during the process of classification of *hard-to-understand* sentences. The SVM-L is capable of reaching a good value of Recall and the maximum value of Precision. The SVM-L can be compared directly with the LSTM-3 which reaches slightly lower values of Precision and Recall but a substantial higher value of TNR. The SVM-P is the worst model since the results show highest recall but with low precision and the worst value of TNR, in this case the model is too unbalanced toward the *hard-to-understand* sentences.

Although the measures show good performance of SVM-L it should be taken into account the computational effort to create the representation vectors (Sect. 2.1) suitable for the SVM. Instead, the RNN only need of the *one-hot encoded* vectors which cost is negligible.

We have carried out experiments training the network for a variable number of epochs from 1 to 10. The RNN reaches good performance already from the training for 1 epoch demonstrating high values of Precision, TNR and Recall. In this regard, the Recall value increases when the model is trained for more epochs. However, it results that an higher number of epochs lower the values of Precision and TNR.

This paper is a part of a series of documents that show the potentiality of Neural Networks for the evaluation of text complexity [8,14,15]. The architecture of the Network has already been proposed to evaluate the syntax complexity of sentences in Italian language [14]. This paper shows further proofs that the RNN is capable of tackling the problem with good results for different languages. The Table 2 shows the comparison between the results associated to the RNN and the SVM for classification of Italian and English sentences.

Table 2. Comparison of SVM model trained on Italian and English language based on the average of Recall, Precision, TPR, TNR.

Model	Kernel	TAG-SET	Recall	Precision	TPR	TNR
LSTM-IT-S	-	STEIN	.819	.834	.819	.837
LSTM-IT-B	-	BARONI	.764	.845	.764	.859
LSTM-EN-5	-	BNC	.826	.849	.826	.853
LSTM-EN-3	-	BNC	.792	.873	.792	.884
SVM-EN-P	Polynomial	BNC	.999	.624	.999	0.0
SVM-EN-L	Linear	BNC	.815	.890	.815	.834
SVM-EN-R	RBF	BNC	.857	.825	.857	.699
SVM-IT-SP	Polynomial	STEIN	.589	.832	.589	.881
SVM-EN-L	Linear	STEIN	.629	.768	.629	.810
SVM-IT-SR	RBF	STEIN	.750	.798	.750	.810
SVM-IT-BP	Polynomial	BARONI	.506	.839	.506	.903
SVM-EN-L	Linear	BARONI	.596	.767	.596	.819
SVM-IT-BR	RBF	BARONI	.731	.793	.731	.809

4 Conclusions

We have presented a comparison of systems based on RNN and SVM ML algorithms for the evaluation of syntax complexity of sentences. The approach is completely data driven and it shows the abilities of Neural Network and SVM of tackling the problem in different languages, Italian and English. Experiments describe good performances of both models for the English language. On the contrary of SVM the RNN shows great versatility discovering rules that identifies the sentence complexity also for the Italian language.

References

1. Schicchi D, Pilato G (2018) WORDY: a semi-automatic methodology aimed at the creation of neologisms based on a semantic network and blending devices. In: Barolli L, Terzo O (eds) Complex, intelligent, and software intensive systems. Springer, Cham, pp 236–248
2. Schicchi D, Pilato G (2018) A social humanoid robot as a playfellow for vocabulary enhancement. In: 2018 second IEEE international conference on robotic computing (IRC). IEEE Computer Society, Los Alamitos, pp 205–208
3. Di Gangi MA, Federico M (2018) Deep neural machine translation with weakly-recurrent units. In: 21st annual conference of the European association for machine translation, pp 119–128
4. Alfano M, Lenzitti B, Lo Bosco G, Perticone V (2015) An automatic system for helping health consumers to understand medical texts, pp 622–627
5. Kincaid J (1975) Derivation of new readability formulas: (automated readability index, fog count and Flesch reading ease formula) for navy enlisted personnel. Research branch report. Chief of naval technical training, Naval Air Station Memphis
6. Dell'Orletta F, Montemagni S, Venturi G (2011) Read-it: assessing readability of Italian texts with a view to text simplification. In: Proceedings of the second workshop on speech and language processing for assistive technologies. Association for Computational Linguistics, pp 73–83
7. Xu W, Napoles C, Pavlick E, Chen Q, Callison-Burch C (2016) Optimizing statistical machine translation for text simplification. Trans Assoc Comput Linguist 4:401–415. https://doi.org/10.1162/tacl_a_00107
8. Lo Bosco G, Pilato G, Schicchi D (2018) A recurrent deep neural network model to measure sentence complexity for the Italian language. In: Proceedings of the sixth international workshop on artificial intelligence and cognition
9. Cortes C, Vapnik V (1995) Support-vector networks. Mach Learn 20(3):273–297
10. Schmid H (2013) Probabilistic part-of-speech tagging using decision trees. In: New methods in language processing, p 154
11. Hochreiter S, Schmidhuber J (1997) Long short-term memory. Neural Comput 9(8):1735–1780
12. Goodfellow I, Bengio Y, Courville A (2016) Deep learning. MIT Press, Cambridge
13. Xu W, Callison-Burch C, Napoles C (2015) Problems in current text simplification research: new data can help. Trans Assoc Comput Linguist 3:283–297. https://doi.org/10.1162/tacl_a_00139
14. Lo Bosco G, Pilato G, Schicchi D (2018) A sentence based system for measuring syntax complexity using a recurrent deep neural network. In: 2nd workshop on natural language for artificial intelligence, NL4AI 2018, vol 2244. CEUR-WS, pp 95–101
15. Bosco GL, Pilato G, Schicchi D (2018) A neural network model for the evaluation of text complexity in Italian language: a representation point of view. Procedia Comput Sci 145:464–470

The TextMap General Purpose Visualization System: Core Mechanism and Case Study

H. R. Schmidtke(✉)

University of Oregon, Eugene, OR 97403, USA
schmidtke@acm.org

Abstract. Human language is capable of communicating mental models between speakers of a language. The question why and how this works is closely tied to a specific variant of the symbol grounding problem that still leaves many open questions. This paper presents the core mechanism of the TextMap system, a logic-based system for generating visuospatial representations from textual input. The system leverages a recent discovery linking logical truth tables of formulae to images: a simple model counting mechanism that automatically extracts coordinate information from propositional Horn-logic knowledge bases encoding spatial predications. The system is based on a biologically inspired low-level bit vector mechanism, the activation bit vector machine (ABVM). It does not require an ontology apart from a list of which tokens indicate relations. Its minimalism and simplicity make TextMap a general purpose visualization or imagery tool. This paper demonstrates the core model counting mechanism and the results of a larger case study of a geographic layout of 13 cities.

Keywords: Language-perception interface · Mental models · Visualization

1 Introduction

Human language is capable of communicating mental models between speakers of a language. It has been a fundamental ideal of artificial intelligence first and probably most clearly formulated by Zadeh [36] to enable computing systems that "compute with words." Zadeh's research was groundbreaking in that it focusses on giving a perception-based semantics to the symbols of logic, logical operators, and the foundations of semantics [32–35]. Zadeh's fuzzy logic attracted an astonishing range of research efforts ranging from control systems [20] research – on the side of perception and actuation – via machine learning [14] and uncertain reasoning [6] to mathematical logic research regarding the logical languages and semantics it subsumes [11] – on the side of logic. From a semantic point of view, fuzzy logic was typically described as belonging to the multi-valued logics as a generalization of Łukasiewicz logic, i.e., as a continuous-valued

A. V. Samsonovich (Ed.): BICA 2019, AISC 948, pp. 455–464, 2020.
https://doi.org/10.1007/978-3-030-25719-4_60

logic interpreting symbols with elements in [0, 1]. Generalizing, the symbols of propositional fuzzy logics only require interpretation by elements of a residuated lattice structure [11]. As such, it is related to mereological frameworks [3], which played an important role in qualitative spatial reasoning [4].

A logical language falling into this category is *context logic* as characterized in [24,27]. Based on a semantics formulated in terms of a lattice structure and in line with the linguistic motivation of fuzzy logic, context logic adds a formalism for contextualization [2]. Very recently [25] and well in line with the program of *computing with words*, a surprising discovery within the context logic program showed that there is a mathematical link from spatial language/logic to perception besides the already known pathway from perception to language/language via approaches such as machine learning and fuzzy logic, pointing towards a completion of the symbol grounding problem becoming possible. This paper demonstrates the approach and a larger case study.

The aim of the wider context logic research program is to develop a hierarchy of languages constructed along the biological and evolutionary cognitive hierarchy [10,19] of competencies. Unlike, e.g., the terminological logic frameworks [1,18] which add functionality to a base language reminiscent of the historical term logic, i.e., taxonomic reasoning, context logic starts from a minimal spatiotemporal framework identified in qualitative spatiotemporal reasoning (QSTR) [7]. The Horn-fragments of several QSTR calculi constitute a set of relation systems that, while lacking in general expressiveness, provide such fast reasoning capabilities that they could successfully be employed for computationally light-weight real-time capable robotic systems, one of the primary application scenarios for QSTR [8]. Traditionally, QSTR calculi are used within a Constraint Satisfaction Problem paradigm. However, they can be mapped to equivalent logical languages within the same reasoning complexity class, as is done, e.g., for analyzing complexity properties, such as in [23]. The Horn-fragments of certain QSTR calculi, for instance, obtain their name from their equivalence to propositional Horn-logic, a fragment of propositional logic with particularly fast reasoning algorithms. The context logic research program being the endeavor to design a cognitively motivated hierarchy of logical languages therefore has at its base level a language equivalent to propositional Horn-logic for spatiotemporal reasoning and builds upon it more complex languages, with partonomic reasoning, for instance, reached at a much later stage.

The context logic program follows ideas of the *strong spatial cognition* program [9]. Each language stage is to represent an evolutionary advancement [10] providing the cognitive system implemented with the respective language's reasoning an advanced viability in its environment. Moreover, the lower levels of reasoning are to be fragments of the language at higher levels, with the respective reasoning mechanisms of lower levels embedded in those of higher levels, so that reasoning complexity can increase incrementally as needed. We thus arrive at logical or complexity modules in a similar way as procedural modules implementing particular processing capabilities [17].

The core Horn-logic at the lowest level is leveraged to implement basic predication only. This allows us to analyze predication itself, and approach the boundary posed by the symbol grounding problem [12,13,28,29,31], a key problem for reaching human-like AI. The symbol grounding problem has two pathways. The pathway from perception to symbol, the *physical* grounding problem, can be considered solved by embodied cognition and machine learning or fuzzy logic perception processing [13,28,31]. The pathway back from the higher levels of logic, the symbols created in our mind, to percept, e.g., an entity imagined found out to be real, still leaves many questions open. It is this capability that provides the substrate for higher cognition. In reference to the epistemic status and indispensability for communication, I call this the *transcendental*[1] symbol grounding problem. Its general solution is still far away.

A simple solution in the spatial reasoning domain would have the following form: a logical formula (or natural language text) ϕ directly gives rise to an image of the situation described in ϕ. "Directly" here refers to the amount of information the system requires to create the drawing. Ideally, the Zero Semantic Commitment condition should apply [29], i.e., the mechanism should not require any form of external input, ontology, or program involving knowledge about the domain. Moreover, we will count as "an image of the situation" any image that is sufficiently similar to some ground truth image, such as a photographic image or map. The TextMap system implements such an experiment. The only external knowledge required to extract this image from the description is a distinction as to which symbols represent predicates and which represent objects in ϕ. A model counting procedure over ϕ then delivers for each predicate symbol p and each object o the p-coordinate of o. In contrast to other approaches, there is thus no necessity to translate between "descriptions and depictions" [30].

The most fundamental language currently handled by the system comprises of sequences of simple relational sentences, such as: "X is north of Y. Y is north of Z. Z is east of W." A minimalistic parser translates these formulae into a logical format. With respect to context logic and the more commonly known logical frameworks of predicate logic and propositional logic these sentences would correspond to:

$$\begin{aligned}
\text{Predicate logic: } & n(x,y) \wedge n(y,z) \wedge e(z,w) \\
& \text{where } n \text{ and } e \text{ need to be characterized as preorders.} \\
\text{Context logic: } & n : [x \sqsupseteq y] \wedge n : [y \sqsupseteq z] \wedge e : [z \sqsupseteq w] \\
\text{equivalent to: } & [x \sqsupseteq n \sqcap y] \wedge [y \sqsupseteq n \sqcap z] \wedge [z \sqsupseteq e \sqcap w] \\
\text{Propositional Horn-logic: } & (n \wedge y \rightarrow x) \wedge (n \wedge z \rightarrow y) \wedge (e \wedge w \rightarrow z) \\
& \text{where transitivity of } n \text{ and } e \text{ arises from } \rightarrow .
\end{aligned}$$

The example shows how the context logic formalism thus allows a layering of languages that positions the simplest language as the innermost building block of atomic context logic formulae corresponding to certain thus grounded predicate

[1] In the sense of Kant [15].

logic atomic formulae but allowing them to be further decomposed into more fundamental building blocks.

2 Core Mechanism: Coordinates from Model Counting

On the implementation side, the aim is on a biologically – that is, ideally both neuronally and evolutionarily – plausible activation-based mechanism operating on a simple bit vector format. The reduction to propositional logic on the lowest level makes this possible. With basic predication reduced to a mechanism based on simple bit vector manipulation and activation/deactivation, it becomes possible to develop a corresponding layered context logic reasoning system based on this simple structure. We call this bio-inspired computational model the *activation bit vector machine* (ABVM).

The core system comprises only of the particularly simple core reasoning and model counting mechanisms. The fundamental reasoning mechanism on top of which the architecture is built is a particularly simple propositional reasoning mechanism, incrementally transforming bit vector representations of CNF (conjunctive normal form) clauses into bit vector representations of special DNF (disjunctive normal form) clauses by a simple subtraction mechanism. The mechanism corresponds to a faster variant of the truth table construction mechanism illustrated in Table 1.

Truth tables are the standard method of determining the truth functional semantics of a propositional logic statement [22]. The meaning, in a formal sense, of any propositional logic formula can be determined by its truth table, determining how the truth value of a formula changes in dependence on the truth values of its atomic formula components. Since each propositional logic formula can only be either true or false, the number of possible assignments to a formula with n different individual atomic formula components is limited to 2^n, so that all possible cases can be enumerated, as shown for the example in Table 1. Once a truth table has been computed, model counting can be applied to transform the logical representation into a numeric representation. By distinguishing between objects (in the example: A, B, C) and relations (here: N), we can extract the ordering of the objects along the dimension which originally gave rise to the relation.

Given that the size of a truth table for a formula ϕ is exponential in the number of variables in ϕ, a reasoning mechanism is required that allows the transformation from an initial list of CNF clauses providing the predications, as discussed in the last section, into a format that allows the computation of the model counting step. While the CNF as a format represents the non-models or zero-rows of the truth table, the DNF represents its models or one-rows. What we are interested in is to generate an *overlap free* DNF representation of ϕ from the CNF that allows us to compute the model counting shown on the righthand side in Table 1 for every relation and every object. A simple pattern matching process can then generate the model counts, yielding the coordinate for the dimension, e.g., north, belonging to the relation (here: N) for each object.

Table 1. Truth table of the CNF formula $\phi = (B \wedge N \rightarrow A) \wedge (C \wedge N \rightarrow B)$ consisting of the two clauses $(B \wedge N \rightarrow A)$ and $(C \wedge N \rightarrow B)$ generated from the text "A is north of B. B is north of C": model set counting (righthand side) determines the N coordinates as $A = 3$, $B = 2$, and $C = 1$

N	A	B	C	$B \wedge N \rightarrow A$	$C \wedge N \rightarrow B$	ϕ	$A \wedge N$	$B \wedge N$	$C \wedge N$	$A \wedge N \wedge \phi$	$B \wedge N \wedge \phi$	$C \wedge N \wedge \phi$
0	0	0	0	1	1	1	0	0	0	0	0	0
0	0	0	1	1	1	1	0	0	0	0	0	0
0	0	1	0	1	1	1	0	0	0	0	0	0
0	0	1	1	1	1	1	0	0	0	0	0	0
0	1	0	0	1	1	1	0	0	0	0	0	0
0	1	0	1	1	1	1	0	0	0	0	0	0
0	1	1	0	1	1	1	0	0	0	0	0	0
0	1	1	1	1	1	1	0	0	0	0	0	0
1	0	0	0	1	1	1	0	0	0	0	0	0
1	0	0	1	1	0	0	0	0	1	0	0	0
1	0	1	0	0	1	0	0	1	0	0	0	0
1	0	1	1	0	1	0	0	1	1	0	0	0
1	1	0	0	1	1	1	1	0	0	1	0	0
1	1	0	1	1	0	0	1	0	1	0	0	0
1	1	1	0	1	1	1	1	1	0	1	1	0
1	1	1	1	1	1	1	1	1	1	1	1	1
										sum:3	sum:2	sum:1

Given this simple mechanism, a wealth of questions may arise, especially regarding scalability of the reasoner and what happens with multiple relations. Given space constraints, we will discuss all additions the language system makes with respect to a concrete example in the next section.

3 A Geographic Example

TextMap can be used for a number of purposes. An immediate application is the visualization of mental maps expressed textually. An example of a mental map is shown in Fig. 1. The map was generated from the following text:

> Lakeville_CT is west of Salisbury_CT. South_Egremont_MA is north of Salisbury_CT. South_Egremont_MA is west of Great_Barrington_MA. Great_Barrington_MA is south of Housatonic_MA. Great_Barrington_MA is south-south-west of Stockbridge_MA. Great_Barrington_MA is north of Sheffield_MA. Sheffield_MA is north of Canaan_CT. West Stockbridge_MA is west of Stockbridge_MA. Hillsdale_NY is west of South_Egremont_MA. Millerton_NY is west of Lakeville_CT. Lenox_MA is north of Stockbridge_MA. Pittsfield_MA is north of Lenox_MA.

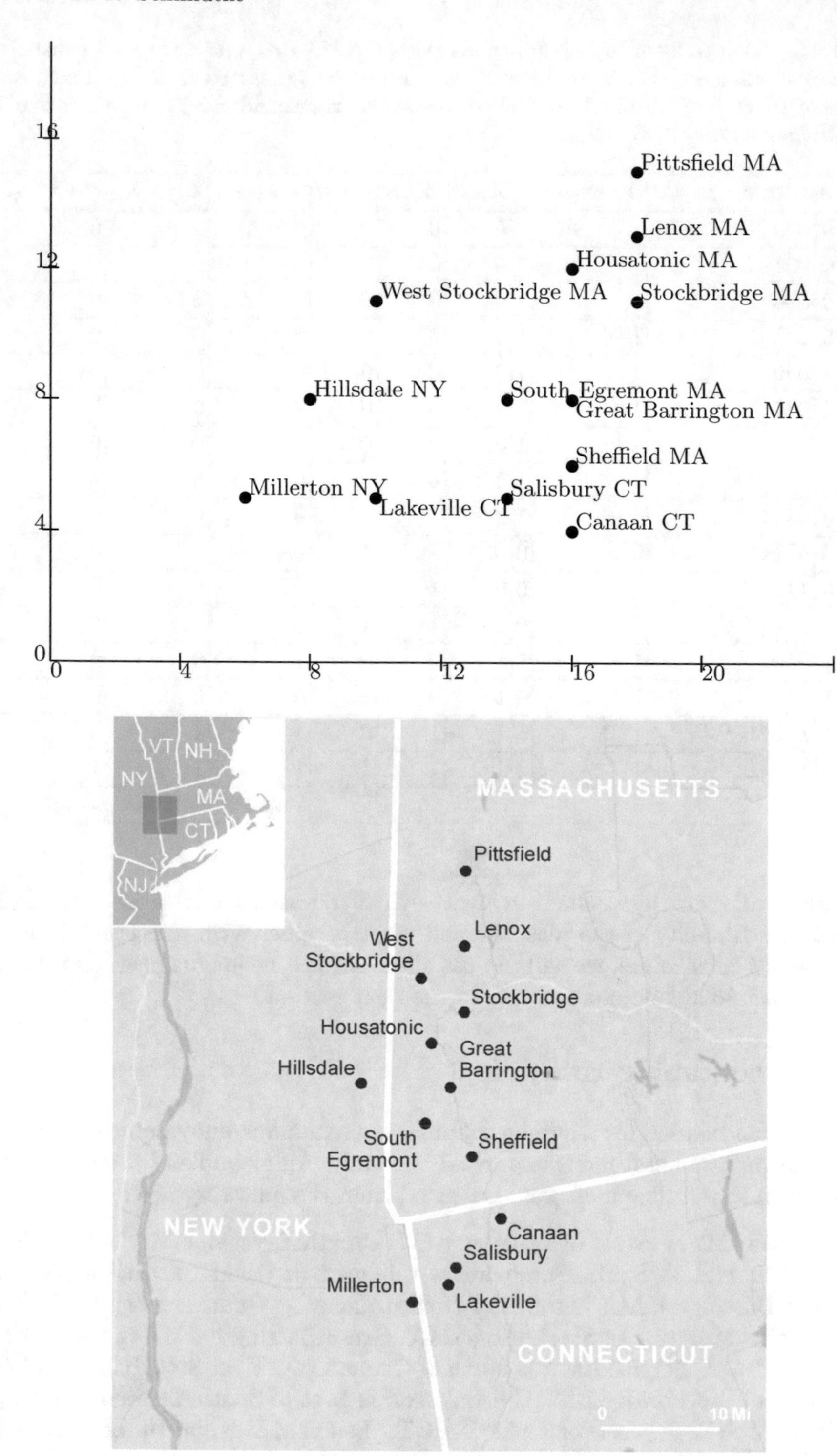

Fig. 1. Visualization and coordinates before normalization obtained through model counting (above) and ground truth data (below).

While all coordinates are generated by the model counting procedure discussed above, and no normalization was performed regarding the coordinates shown in Fig. 1, one may asked how much is added by the system programmer through the textual interface and cartographic conventions. The TextMap system is designed to recreate a communication situation with a minimum of common ground shared.

1. The parser ignores the basic preposition "of", which is generally interpreted as logical $\wedge$. It interprets "is" and "are" as $\leftarrow$ and the whole predicative sentence "A is R of B" into $R : [A \leftarrow B]$. In languages other than English, the separative functionality of "is" may be provided by the nominative case or particles.
2. The logical system distinguishes between relations (north, south, etc.) and objects (everything that is not a relation or ignored) and has complements stored as passive constructions, i.e., as turning the implication arrow from $\leftarrow$ to $\rightarrow$ so that, for a complement R^- of R^+, $R^- : [A \leftarrow B]$ is turned into $R^+ : [A \rightarrow B]$.
3. Pragmatic considerations: while the images generated without further additional assumptions are valid realizations of what is expressed in the linguistic description, they would be cautious. A human listener would make further assumptions, e.g.: if A is said to be north of B, one would assume that it is not north-west of B, i.e., that the east-west direction is $east - west : [A \leftrightarrow B]$. The simple pragmatic rule is to add, for any relation R where we do not have explicit information, the simple assumption $R : [A \leftrightarrow B]$.
4. The visualization component rotates, mirrors, scales, and shifts the resulting image, so that the final image produced implements modern, Western mapping conventions that put north at the top and west at the left of the image.

When we analyze these additional components besides the pure model counting mechanism, we see that 1–3 are linguistic aspects: 1 is syntactic, 2 is semantic, and 3 is pragmatic. Only the very last is of a conventionalist nature and correctly displayed by the programmer for easier evaluation. Without this correction, the resulting images would be systematically mirrored or rotated. With respect to ontology freedom, (a) the separation into objects and relations and (b) the complements are the only ontological components. We could remove complements without losing the mechanism's power, leaving the separation into objects and relations as the only ontological commitment.

While it will probably be a long way until we know all details required to arrive at a system that fully resembles human imagery, the simple system outlined here already has a surprising accuracy. Its performance, moreover, given that the simple predication discussed above remains in the Horn-fragment, is acceptable: in practical experiments it was below 1 s for small examples and below 3 min for the largest natural language examples produced by human beings I found.

4 Outlook and Conclusions

This paper presented the TextMap application, an application for visualizing logical/linguistic descriptions of spatial layouts and other ordered domains. The system is built around a biologically inspired logical reasoning core, the activation bit vector machine (ABVM), that unites logical and analogous computation through a simple model counting mechanism. The proposed system is a reasoning system implemented on a biologically plausible primitive bit vector mechanism. Its tight connection to the linguistically motivated context logic family, a member of the wider fuzzy logic family, makes it particularly easy to generate the internal format directly from a simple English fragment as well as similarly simple fragments of other natural languages. The case study of a linguistically described mental model demonstrates that the model counting mechanism is capable of handling a larger case study with surprising accuracy. Current weaknesses in terms of cognitive plausibility are in the lack of incorporation of implicit or explicit distance information. Moreover, a scenario of 13 cities is most likely beyond the capabilities of most human cognitive agents, who would employ higher-level strategies to handle the large number of items. The current system does not include such higher-level mechanisms. While the context logic hierarchy seems specifically suitable to facilitate implementing higher-level functionality, including metacognitive abilities [16], on top of a primitive bit vector substrate, it is clear that a vast amount of work has to be completed before a full-fledged ABVM-based cognitive system could be built.

More than an application, TextMap is a philosophical *Gedankenexperiment* raising a range of new questions and asking old ones from a new perspective. But the application itself also has a spectrum of potential further applications. The context mechanism, for instance, allows fine-grained control over information from different sources providing for a range of applications in information security and information quality assurance, aspects of central relevance in cyber security, intelligence, and defense. Being able to handle spatial information and priorities in such a way as to connect logical and analogous form the system also provides opportunities for novel applications for increasing the safety of autonomous systems [26].

Acknowledgements. I would like to thank Charlie Derr for contributing the description of his mental map and Joanna Merson of InfoGraphics Lab for designing the ground truth map.

References

1. Baader F, Calvanese D, McGuinness DL, Nardi D, Patel-Schneider PF (eds) (2002) Description logic handbook. Cambridge University Press, New York
2. Benerecetti M, Bouquet P, Ghidini C (2000) Contextual reasoning distilled. J Exp Theor Artif Intell 12(3):279–305
3. Casati R, Varzi AC (1999) Parts and places: the structure of spatial representations. MIT Press, Cambridge

4. Cohn AG, Hazarika SM (2001) Qualitative spatial representation and reasoning: an overview. Fundamenta Informaticae 46(1–2):1–29
5. Dennett DC (1993) Consciousness explained. Penguin, London
6. Dubois D, Lang J, Prade H (1994) Possibilistic logic. In: Gabbay DM, Hogger CJ, Robinson JA (eds) Handbook of Logic in Artificial Intelligence and Logic Programming, vol 3. Oxford University Press Inc., New York, pp 439–513
7. Dylla F, Lee JH, Mossakowski T, Schneider T, van Delden A, van de Ven J, Wolter D (2017) A survey of qualitative spatial and temporal calculi: algebraic and computational properties. ACM Comput Surv 50(1):7:1–7:39
8. Escrig MT, Toledo F (1998) Qualitative spatial reasoning: theory and practice: application to robot navigation, vol 47. Ios Press, Amsterdam
9. Freksa, C.: Strong spatial cognition. In: International workshop on spatial information theory. Springer, pp 65–86 (2015)
10. Gärdenfors P (1995) Language and the evolution of cognition. Lund University, Lund
11. Hájek P (1998) Metamathematics of fuzzy logic, vol 4. Springer, Dordrecht
12. Harnad S (1990) The symbol grounding problem. Phys D Nonlinear Phenom 42(1–3):335–346
13. Harnad S (2003) The symbol grounding problem. In: Nadel L (ed) Encyclopedia of cognitive science. Macmillan and Nature Publishing Group, London
14. Jang J-SR (1993) ANFIS: adaptive-network-based fuzzy inference system. IEEE Trans Syst Man Cybern 23(3):665–685
15. Kant I (1999) Critique of pure reason. Cambridge University Press, Cambridge
16. Kralik JD, Lee JH, Rosenbloom PS, Jackson PC, Epstein SL, Romero OJ, Sanz R, Larue O, Schmidtke HR, Lee SW, McGreggor K (2018) Metacognition for a common model of cognition. Procedia Comput Sci 145:730–739
17. Laird JE, Lebiere C, Rosenbloom PS (2017) A standard model of the mind: toward a common computational framework across artificial intelligence, cognitive science, neuroscience, and robotics. AI Mag 38(4):13–26
18. Levesque HJ, Brachman RJ (1987) Expressiveness and tractability in knowledge representation and reasoning. Comput Intell 3(2):78–93
19. Newell A (1994) Unified theories of cognition. Harvard University Press, Cambridge
20. Passino KM, Yurkovich S, Reinfrank M (1998) Fuzzy control. Addison-Wesley, Menlo Park
21. Penrose R (1999) The emperor's new mind: concerning computers, minds, and the laws of physics. Oxford Paperbacks, New York
22. Quine WVO (1982) Methods of logic. Harvard University Press, Menlo Park
23. Renz J, Nebel B (1999) On the complexity of qualitative spatial reasoning: a maximal tractable fragment of the region connection calculus. Artif Intell 108(1–2):69–123
24. Schmidtke, H.R.: Contextual reasoning in context-aware systems. In: Workshop proceedings of the 8th international conference on intelligent environments. IOS Press, pp 82–93 (2012)
25. Schmidtke HR (2018) Logical lateration - a cognitive systems experiment towards a new approach to the grounding problem. Cogn Syst Res 52:896–908. https://doi.org/10.1016/j.cogsys.2018.09.008
26. Schmidtke HR (2018) A survey on verification strategies for intelligent transportation systems. J Reliab Intell Environ 4(4):211–224. https://doi.org/10.1007/s40860-018-0070-5

27. Schmidtke HR, Hong D, Woo W (2008) Reasoning about models of context: a context-oriented logical language for knowledge-based context-aware applications. Revue d'Intelligence Artificielle 22(5):589–608
28. Steels, L.: The symbol grounding problem has been solved. so what's next. In: Symbols and embodiment: debates on meaning and cognition, pp 223–244 (2008)
29. Taddeo M, Floridi L (2005) Solving the symbol grounding problem: a critical review of fifteen years of research. J Exp Theor Artif Intell 17(4):419–445
30. Vasardani, M., Timpf, S., Winter, S., Tomko, M.: From descriptions to depictions: a conceptual framework. In: International conference on spatial information theory. Springer, pp 299–319 (2013)
31. Vogt P (2002) The physical symbol grounding problem. Cogn Syst Res 3(3):429–457
32. Zadeh L (1979) Fuzzy sets and information granularity. In: Gupta M, Ragade R, Yager R (eds) Advances in fuzzy set theory and applications. North-Holland, Amsterdam, pp 3–18
33. Zadeh LA (1965) Fuzzy sets. Inf Control 8:338–353
34. Zadeh LA (1975) Fuzzy logic and approximate reasoning. Synthese 30(3):407–428
35. Zadeh LA (1988) Fuzzy logic. Computer 21(4):83–93
36. Zadeh LA (1996) Fuzzy logic= computing with words. IEEE Trans Fuzzy Syst 4(2):103–111

Subsymbolic Versus Symbolic Data Flow in the Meaningful-Based Cognitive Architecture

Howard Schneider(✉)

Sheppard Clinic North, Toronto, ON, Canada
howard.schneider@gmail.com

Abstract. The biologically inspired Meaningful-Based Cognitive Architecture (MBCA) integrates the subsymbolic sensory processing abilities found in neural networks with many of the symbolic logical abilities found in human cognition. The basic unit of the MBCA is a reconfigurable Hopfield-like Network unit (HLN). Some of the HLNs are configured for hierarchical sensory processing, and these groups subsymbolically process the sensory inputs. Other HLNs are organized as causal memory (including holding of multiple world views) and as logic/working memory units, and can symbolically process input vectors as well as vectors from other parts of the MBCA, in accordance with intuitive physics, intuitive psychology, intuitive scheduling and intuitive world views stored in the instinctual core goals module, and similar learned views stored in causal memory. The separation of data flow into the subsymbolic and symbolic streams, and the subsequent re-integration in the resultant actions, are explored. The integration of logical processing in the MBCA predisposes it to a psychotic-like behavior, and predicts that in Homo sapiens psychosis should occur for a wide variety of mechanisms.

Keywords: Cognitive architecture · Cortical minicolumns · Psychosis

1 Introduction

The biologically inspired Meaningful-Based Cognitive Architecture (MBCA) integrates the sensory processing abilities found in neural networks with many of the symbolic logical abilities found in human cognition [1]. While artificial neural networks (ANNs) are capable of pattern recognition and reinforcement learning at a human level of performance [2, 3], they perform poorly compared to a four-year old child in causally and logically making sense of their environment or the information they are processing, especially if a very small set of training examples are used [4, 5].

A number of models have been proposed to help close the neural-symbolic gap. Some cognitive architectures, for example, ACT-R, CLARION, MicroPsi or Sigma models [6–10], attempt to integrate subsymbolic and symbolic processing. Work by Graves and colleagues uses an ANN which can read and write to an external memory, i.e., a hybrid system [11]. As Collier and Beel discuss, these newer neural network architectures can be classified as Memory Augmented Neural Networks (MANNs) where there is an external memory unit as opposed to, for example, Long Short Term

A. V. Samsonovich (Ed.): BICA 2019, AISC 948, pp. 465–474, 2020.
https://doi.org/10.1007/978-3-030-25719-4_61

Memory (LSTMs) neural networks where memory exists as an internal vector. An advantage of MANNs versus, for example, LSTMs, is better performance on tasks requiring memory usage [12]. Li and colleagues discuss their Neural Rule Engine where modules of neural networks each represent an action of a logic rule [13].

While the MBCA does not attempt to replicate biological systems at the neuronal spiking level, it is indeed inspired by mammalian cortical minicolumns and mammalian brain organization [14–16]. In keeping with this biological inspiration, unlike a typical ANN, the basic unit of the MBCA is not an artificial neuron but a Hopfield-*like* network (HLN). The MBCA uses HLNs for *both* its subsymbolic (e.g., ANN-like pattern recognition) and its symbolic operations. Unlike MANNs, the MBCA does not use an external memory, i.e., it is not a physically hybrid system.

As Eliasmith and Trujillo note, in simulations of brain models perhaps the most important goal is the link to behavior [17], and indeed at the mesoscopic scale the MBCA can functionally produce a variety of behaviors which can help to better hypothesize and understand mammalian cortical function, and provide insight into possible mechanisms which link such mesoscopic functioning to the causal and symbolic behavior seen in humans.

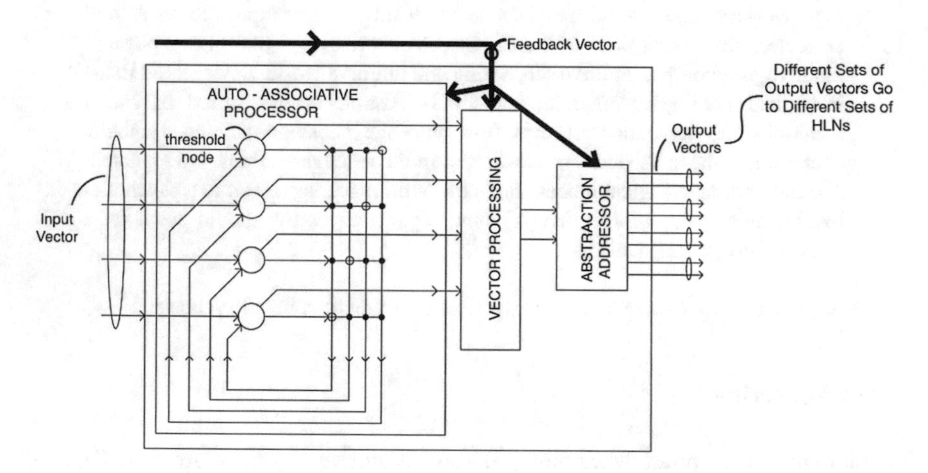

Fig. 1. Overview of a Hopfield-like Network (HLN) unit

Each HLN unit contains a Hopfield neural network along with associated circuitry modifying convergence properties and allowing reconfiguration with other HLNs. Figure 1 schematically shows a Hopfield-like network (HLN) unit. If the input vector is recognized, a stored pattern can be outputted to the vector processing unit (ensures valid convergence to an output that actually is related to the input vector), and then feeds into the abstraction addressor circuitry, which allows fast and extreme reconfiguration of the HLNs with each other. Based on the 'meaningfulness' values of the feedback vector (which can be as simple as how many other related HLNs are activated, or can use the Shannon entropy as the MBCA does), the abstraction addressor decides which of a number of possible sets of output vectors (which become inputs to other HLNs) will have non-zero outputs [1]. The feedback vector can also modulate when the auto-associative processor learns new patterns. In rapid reconfigurations, there is an attempt by the HLNs to maximize meaningfulness, where this is defined as the reciprocal of the Shannon entropy as shown below, and can be understood as essentially as asking what reconfigurations are more meaningful in terms of recognizing and processing an input sensory vector. Meaningfulness M is defined as the reciprocal of the Shannon entropy (1, 2):

$$H = -\sum_i P(x_i) log_2 P(x_i) \tag{1}$$

$$M = 1/H \tag{2}$$

In the current paper we do not focus on these extreme HLN reconfigurations which have previously been described [1], but instead we consider how an input vector is processed in the MBCA as subsymbolic and symbolic data streams.

2 Meaningful-Based Cognitive Architecture

The MBCA is built from reconfigurable topologies of Hopfield-like networks (HLNs). HLNs can function individually as pattern recognizers. Hierarchies of pattern recognizers can recognize, for example, from lines to letters to words to ideas associated with those words. A feedback signal from higher-level hierarchical units will influence the pattern expected by a lower-level pattern recognizer. Such hierarchies have been discussed in the literature, for example the hierarchical compositional network of George and colleagues [18]. Hawkins [19] and Kurzweil [20] have popularized hierarchies of pattern recognizers inspired by the structure of the mammalian cortex. The capsule networks developed by Hinton and coworkers go beyond typical convolutional neural networks in representing spatial hierarchies between objects of a scene [21]. However, the construction, operation and many of the properties of the MBCA, from its basic HLN elements to its overall organization is significantly different than in these previous works.

The Meaningful-Based Cognitive Architecture is shown in Fig. 2. Sensory inputs are pre-processed, and fed into sensory hierarchies of HLNs, and then the varying senses are bound by another set of HLNs. Sensory inputs at various levels of processing are fed to the first stages of the causal memory (keeps track of which event and associated data follows another event, what spatial feature leads to the next spatial feature, and other associative relationships; see note below), the emotional and reward module, and other modules and functional units of HLNs. Although the instinctual core goals module in Fig. 2 is shown largely interfacing with the logic/working memory units, it does indeed interface with the causal memory, and in an MBCA version without logic/working memory units, innate goal objectives would be directed towards the causal memory group of HLNs. The processing of the sensory data described so far, is essentially subsymbolic processing, i.e., largely pattern recognition and associative behavior is taking place, albeit with some pre-causal representation and goal directed influences.

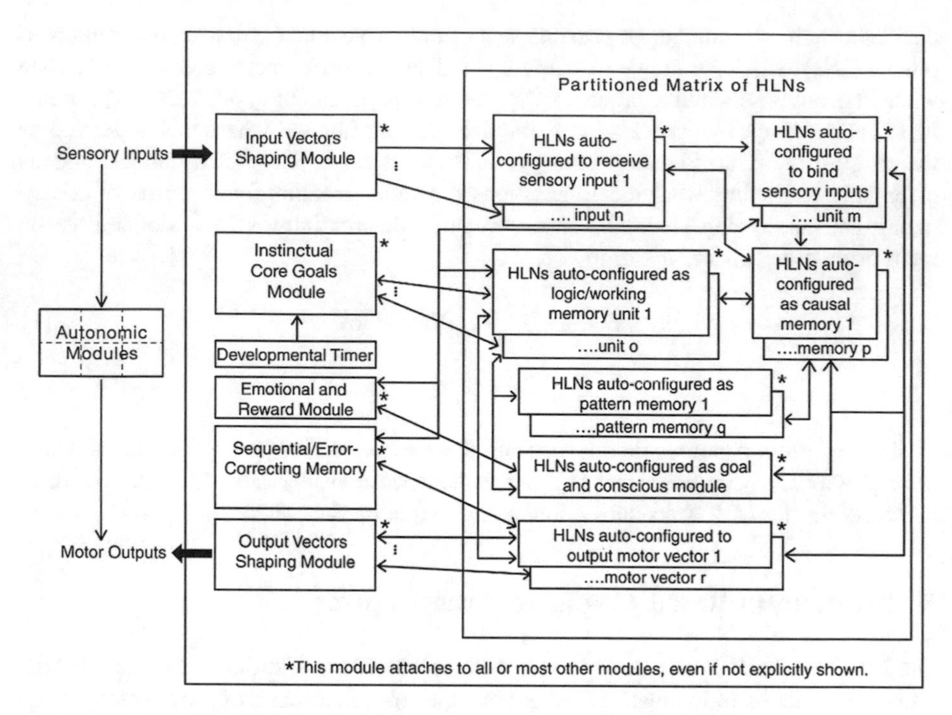

Fig. 2. Meaningful-Based Cognitive Architecture (MBCA)

The sequential/error-correcting memory is an optional module and different than the memories in MANNs. It is useful for detecting changes in a spectrum of external and internal data, and storing learning sequences which can automatically be repeated later as needed.

Data from the sensory binding HLNs, the early stages of the causal memory HLNs, the sequential/error-correcting memory, and other modules and functional units of HLNs, go to the HLNs configured to output motor vectors, which in turn feeds into an output shaping module, and then produces the motor outputs which activate different actuators, as well as providing a data output to the external environment. Although motor output functions would seem to be very different than sensory input functions, the HLNs performing the motor output function operate somewhat similarly, with feedback signals shaping the output patterns of motor activation expected. Note that the motor output HLNs can integrate signals from a variety of groups of HLNs, including the subsymbolic processing groups listed above, as well as the HLNs performing symbolic processing of input and internal data, described below.

A complex of autonomic modules exists in association with the MBCA. The autonomic modules can perform straightforward computations on sensory inputs without the need of any HLNs, and in turn produce largely associative/reflexive motor outputs.

The flow of the sensory data and its transformation into motor outputs, described so far, remains largely subsymbolic. However, such a system, i.e., *without* the symbolic features of the MBCA discussed below, if tuned well for its environment, can be expected to provide a very effective range of complex behaviors.

The MBCA also contains groups of HLNs which are configured as logic/working memory units. These logic/working memory units receive processed bound sensory input data, processed pre/causal data, as well as data from other modules and functional units of the MBCA.

With modifications in organization, groups of HLNs can be configured into groups which act as logical processors and operate on what constitutes a working memory in these HLN groups of logic/working memory units. In the current paper we do not focus on the steps required to do so in an artificial system, or the evolutionary changes required in a biologically analogous model, to create these functional changes in groups of HLNs. However, from basic computability theory, these functional changes are readily possible [22]. Further work by Schneider [23] explores the evolution of pre-causal functioning of the "causal memory" group of HLNs to true causal functioning and then further enhancement to become full logic/working units. As a result, even though in the MBCA the causal memory group of HLNs has the description "causal", it essentially acts as a staging area for the causal operations in the logic/working units as well as storing the learned world views described below.

It is important to realize that the representation of data in the MBCA already has meaning through the connections an HLN has via connections to other HLNs. The working memory need not convert the vectors from the HLN into some abstract symbolic data for the working memory to produce symbolic behavior, i.e., manipulation of the HLN vectors it receives suffices.

In the simulation of the MBCA described below, it can be shown that HLNs auto-configured as relatively simple logic/working memory units are able to:

- compare properties of vectors they receive
- choose one vector over another
- match, trigger and receive intuitive procedural vectors from the instinctual core goals module
- match, trigger and receive learned procedural vectors from causal memory
- pattern match a vector from the entire MBCA memory
- store a limited number of previous input and output vectors and compare these with the properties of the vectors they receive
- output a vector to the motor output section
- output a vector back into the sensory input section, and operate again on the processed vector

The instinctual core goals module, which is affected by the maturity stage of the MBCA via the developmental timer, feeds intuitive logic, intuitive physics, intuitive psychology and/or intuitive goals planning procedural vectors into the groups of HLNs configured as logic/working memory units. The goal and conscious module interacts with the emotional and reward module as well as the entire MBCA to provide some overall control of the MBCA's behavior. Memories of operations occurring in the logic/working memory are temporarily kept in the conscious module, allowing improved problem solving as well as providing more transparency to MBCA decision making. Rare events can sometimes be very important events to learn. The emotional module allows effective learning of infrequent events and obviates the class imbalance problem seen in conventional neural networks.

3 Subsymbolic and Symbolic Processing of Input Sensory Data

An example is useful to show the subsymbolic and the symbolic processing of data in the MBCA. In this simplified example, an MBCA is controlling a search and rescue robot searching for a lost hiker in an uninhabited forest. The MBCA has been given the command to find a lost hiker which triggers the appropriate goals from the instinctual core goals module of finding the commanded item.

The MBCA receives sensory inputs from sounds, vibrations, visual images, odors, and navigation signals in the forest. The instinctual core goals module sends a goal-related data vector to sensory binding HLNs and to causal memory HLNs which will increase the likelihood of activation of HLNs associated with a sensory input associated with a hiker in the woods, as well as triggering in the causal memory learned world views associated with a hiker in the forest, which will further increase the likelihood of activation of HLNs associated with a sensory input associated with a lost hiker.

The MBCA starts looking for recognition of some of the input sensory vectors with patterns it has already seen (or been configured with) related to the subject of the goal, a lost hiker. Any matches? Rapid reconfiguration of HLNs (as discussed above) [1] attempts to better extract useful information from the sensory data. The MBCA reconfigures some HLNs. Any better matches?

In this example, the MBCA detects, for example, two input sensory vectors that activate HLNs associated with a lost hiker. One of these sensory input vectors is a sound from the southwest indicating a possible cry for help from a human or a possible match with certain bird calls. The other sensory input vector is an odor coming from the north indicating a possible match with a commercial perfume/cologne.

At this point, the MBCA could find the lost hiker, using only subsymbolic processing. From past experiences (or configuration data before the mission) the MBCA may decide to go in the direction of the sound in the southwest. If, in this example, the lost hiker is not found in the southwest, then the MBCA weakens this association and then heads off in the direction of the matched odor in the north. The MBCA finds the lost hiker in the north, and then strengthens the association between a commercial perfume/cologne odor and a lost hiker, and indeed, should such a choice need to be made in the future, the MBCA will more likely choose to go in the direction of the matched odor vector.

However, consider the same example making use of the logic/working memory units. The instinctual core goals module has also sent a vector to the logic/working memory units, and so have any causal memory HLNs with triggered learned world views. The processed sensory input vectors are also sent to the logic/working memory unit, i.e., the input vector from the southwest indicating a possible cry for help from a human or a possible match with certain bird calls, and the input vector from the north indicating a possible match with a human perfume/cologne.

The logic/working memory unit can compare vectors and take subsequent actions depending on the results of comparisons. These actions can be to modify a vector or to output a vector to other modules of the MBCA. As noted above, the instinctual core goals module also feeds intuitive logic/physics/psychology/planning procedures into the HLNs configured as logic/working memory units. Keep in mind that the MBCA is not a *tabula rasa* – there are intuitive logic, intuitive physics, intuitive psychology, and intuitive planning world views stored in the instinctual core goals module. As well, the MBCA is able to learn new procedures from experience, and similar learned world views are stored in the causal memory group of HLNs.

In this example, the MBCA's logic/working memory unit and intuitive logic find that a vector that is related uniquely to a vector of the goal, i.e., an odor that comes from a human which is related to the goal, is closer to the goal vector than a vector which could match other things, such as a sound which could come from some birds. Hence, the output of the logic/working memory unit is to move in the direction of the odor sensory input vector. The search and rescue robot thus moves in a northerly direction towards the lost hiker, and in our example, indeed soon finds and rescues the lost hiker.

Although other cognitive architectures include symbolic processing of sensory information, e.g. the rules of the ACT-R cognitive architecture [7], or the work of Siddiqui and colleagues in considering the interaction of sensorial and symbolic memories [24], the MBCA seamlessly integrates the subsymbolic and symbolic processing of sensory inputs.

In the future when this MBCA is searching for a lost hiker in the forest and there is a perfume/cologne odor sensory input vector, this alone without the use of the more symbolic algorithms, can direct the MBCA search and rescue robot towards the lost hiker. However, the real world is complex, and there will be differences in the next example, and using the logic/working memory units with a variety of intuitive and learned procedures related to the problem at hand, can increase the likelihood of success for the MBCA.

4 Experimental Simulation of the MBCA

The original goal of the MBCA was to attempt to produce and experiment with an artificial general intelligence of sorts. In the early simulations of the MBCA, the behavior of the MBCA was inferred from a Python language simulation of small numbers of HLNs [1]. Scaling to large numbers of HLNs is difficult and in 2019 simulations have been re-organized as coarse-grained to finer grain simulations. In the lowest level coarse grain simulation, PyTorch (an open-source Python machine learning library) feedforward neural networks are used for a rough simulation of the subsymbolic modules of the MBCA, albeit in the thousands of nodes, and conventional Python methods are used to simulate the logic/memory units of the MBCA. The finer grain simulation, which is under construction, on the other hand, more authentically uses hand-coded simulations of HLN units for the subsymbolic modules and modified bona fide HLNs for logic/working memory units. In all cases, the MBCA simulations are created in an environment where the goal of the MBCA is to find a lost hiker in an uninhabited forest.

Once the experimental simulations of the MBCA are better developed, it is hoped to compare experimental properties of using HLN units with conventional ANN subsymbolic structures. The HLN basic structure was not chosen for the Meaningful-Based Cognitive Architecture because of its subsymbolic processing features – mainstream ANNs do a fairly good job of this already [2, 3]. Rather, the HLN basic unit was chosen as it was thought to better model biological cortical columnar organization and because it could much more readily be coopted, as per evolutionary processes, into a logic/working memory symbolic unit, than could a typical artificial neural network.

5 MBCA Relation to Biological Nervous Systems

While the MCBA does not attempt to replicate biological systems at the neuronal spiking level, as noted above, the HLNs are inspired by the biological mammalian minicolumns [14–16]. As well, most of the modules and functional groups of the MBCA *functionally* approximate to analogous aspects of the human central nervous system.

As previously discussed [1], while the subsymbolic processing of input sensory vectors is relatively straightforward, i.e., each evaluation cycle the MBCA mechanically propagates the input sensory vector through its subsymbolic architecture and arrives at a motor output vector, this is in contrast to the symbolic processing requirements of functionally combining logic/memory working units with the procedures from the instinctual core goals module and the causal memory groups, shuffling working memory around and retrieving data from the logic/memory working HLNs and HLNs of other functional groups – an order of magnitude more complex. As the MBCA matures, and as the developmental timer causes the instinctual core goals module to feed even more complex procedures into the logic/working memory units, any of a myriad of small issues in the logic/working memory units or their connecting modules including the instinctual core goals module, can cause the MBCA to fail in a psychotic fashion:

- inappropriately feed certain output vectors back as a sensory input which the MBCA then interprets as a real sensory input rather than a "thought" which logic/working memory expected to process further in the next evaluation cycle – hallucination-like behavior;
- inappropriately match and retrieve memory vectors not corresponding with the reality of the input sensory vectors and inappropriately further process these vectors – cognitive dysfunction and delusional-like behavior.

While many models of human psychotic disorders, including neural network-based ones, have long been proposed, for example the work by Cohen and Servan-Schreiber in the 1990s [25] to the more recent work of Sabaroedin and colleagues [26], the MBCA was not designed to simulate psychotic disorders, but rather, psychotic features seemed to emerge surprisingly easily when the logic/working memory HLN units had themselves small flaws or there were inconsistencies with the procedure vector from the instinctual core goals module or the causal memory groups.

The MBCA model suggests that while the co-option of cortical columns (i.e. analogous to HLNs) can allow a nervous system to perform symbolic processing along with subsymbolic processing of input sensory vectors, the order of magnitude of increased complexity required for such utilization, will create a vulnerability in all humans to psychotic disorders. Indeed, the clinical literature seems to support this suggestion. For example, more than 10% of the population will experience less severe psychotic-like symptoms [27], but just under 1% of the population suffers from schizophrenia (a psychotic disorder) – there are many causes why humans may experience psychotic-like symptoms. Indeed, in just about all other mammals, who would appear by observation to engage in less symbolically processed behavior, psychosis is rare, and in psychopharmacological research settings, large efforts are required to induce at best unreliable models of schizophrenia in research animals [28].

Acknowledgments. This article builds upon work originally presented at BICA 2018 (reference 1).

References

1. Schneider H (2018) Meaningful-based cognitive architecture. Procedia Comput Sci 145:471–480 BICA 2018 ed Samsonovich A V
2. Goodfellow I, Bengio Y, Courville A (2016) Deep learning. MIT Press, Cambridge
3. Mnih V, Kavukcuoglu K, Silver D et al (2015) Human-level control through deep reinforcement learning. Nature 518(7540):529–533
4. Ullman S (2019) Using neuroscience to develop artificial intelligence. Science 363 (6428):692–693
5. Waismeyer A, Meltzoff AN, Gopnik A (2015) Causal learning from probabilistic events in 24-month-olds: an action measure. Dev Sci 18(1):175–182
6. Laird JE, Lebiere C, Rosenbloom PS (2017) A standard model of the mind: toward a common computational framework across artificial intelligence, cognitive science. Neurosci Robot AI Mag 38(4):13–26
7. Anderson JR, Bothell D, Byrne MD et al (2004) An integrated theory of mind. Psychol Rev 111(4):1036–1060
8. Kilicay-Ergin N, Jablokow K (2012) Problem-solving variability in cognitive architectures. IEEE Trans Syst Man Cybern Part C Appl Rev 42:1231–1242
9. Bach J (2008) Seven principles of synthetic intelligence. In: Proceedings of 1st Conference on AGI, Memphis, pp 63–74
10. Rosenbloom P, Demski A, Ustun V (2016) The sigma cognitive architecture and system: towards functionally elegant grand unification. J Artif Gen Intell 7(1):1–103
11. Graves A, Wayne G, Reynolds M et al (2016) Hybrid computing using a neural network with dynamic external memory. Nature 538:471–476
12. Collier M, Beel J (2018) Implementing neural Turing machines. In: 27th International Conference on Artificial Neural Networks (ICANN), Rhodes, Greece, 4–7 October
13. Li S, Xu H, Lu Z (2018) Generalize symbolic knowledge with neural rule engine. arXiv: 1808.10326
14. Mountcastle VB (1997) Columnar organization of the neocortex. Brain 20:701–722
15. Buxhoeveden DP, Casanova MF (2002) The minicolumn hypothesis in neuroscience. Brain 125(5):935–951

16. Schwalger T, Deger M, Gerstner W (2017) Towards a theory of cortical columns. PLoS Comput. Biol 13(4): e1005507. https://doi.org/10.1371/journal.pcbi.1005507
17. Eliasmith C, Trujillo O (2014) The use and abuse of large-scale brain models. Curr Opin Neurobiol 25:1–6
18. Lázaro-Gredilla M, Liu Y, Phoenix DS, George D (2017) Hierarchical compositional feature learning. arXiv preprint. arXiv:1611.02252v2
19. Hawkins J, Blakeslee S (2004) On intelligence. Times books, New York
20. Kurzweil R (2012) How to create a mind. Viking Press, New York
21. Sabour S, Frosst N Hinton GE (2017) Dynamic routing between capsules. In: NIPS 2017, Long Beach, CA, USA
22. Sipser M (2012) Introduction to the theory of computation, 3rd edn. Course Technology, Boston
23. Schneider H (2019, in press) Emergence of belief systems and the future of artificial intelligence. In: Samsonovich AV (ed) Biologically inspired cognitive architectures 2019, BICA 2019. Advances in intelligent systems and computing. Springer
24. Siddiqui M, Wedemann RS, Jensen HJ (2018) Avalanches and generalized memory associativity in a network model for conscious and unconscious mental functioning. Physica A 490:127–138
25. Cohen JD, Servan-Schreiber D (1992) Context, cortex and dopamine: a connectionist approach to behavior and biology in schizophrenia. Psychol Rev 99(1):45–77
26. Sabaroedin K, Tiego J, Parkers L et al (2019) Functional connectivity of corticostriatal circuitry and psychosis-like experiences in the general community. Biol Psychiatry pii: S0006–3223(19)30119-2
27. van Os J, Hanssen M, Bijil RV et al (2001) Prevalence of psychotic disorder and community level psychotic symptoms: an urban-rural comparison. Arch Gen Psychiatry 58(7):663–668
28. Jones CA, Watson DJG, Fone KCF (2011) Animal models of schizophrenia. Br J Pharmacology 164:1162–1194

Schizophrenia and the Future of Artificial Intelligence

Howard Schneider(✉)

Sheppard Clinic North, Toronto, ON, Canada
howard.schneider@gmail.com

Abstract. In the Meaningful-Based Cognitive Architecture (MBCA) the input sensory vector is propagated through a hierarchy of Hopfield-like Network (HLN) functional groups, including a binding of sensory input group of HLNs and a causal group of HLNs, and subsymbolic processing of the input vector occurs in the process. However, the processed sensory input vector is also propagated to the logic/working memory groups of HLNs, where the content of the logic/working memory can be compared to data held by various groups of HLNs of other functional groups, as well as with other logic/working memory units, i.e., symbolic processing occurs. While the MBCA does not attempt to replicate biological systems at the neuronal spiking level, its HLNs and the organization of its HLNs are indeed inspired by biological mammalian mini-columns and mammalian brains. The MBCA model leads to the hypothesis that in the course of hominid evolution, HLNs became co-opted into groups of HLNs providing more extensive working memories with more logical abilities. While such co-option of the minicolumns can allow advantageous symbolic processing integrated with subsymbolic processing, the order of magnitude of increased complexity required for such organization and operation, created a vulnerability in the human brain to psychotic disorders. The emergence of a technological artificial general intelligence (AGI) will, on a practical level, also require the integration of symbolic processing along with subsymbolic structures. The MBCA model predicts that such an integration will similarly create a potential vulnerability in the resultant AGI towards psychotic-like features.

Keywords: Cognitive architecture · Artificial general intelligence · Cortical minicolumns · Psychosis

1 Introduction to the Meaningful-Based Cognitive Architecture (MBCA)

The Meaningful-Based Cognitive Architecture (MBCA) integrates the sensory processing abilities found in artificial neural networks and mammals, with many of the symbolic logical abilities found in human cognition [1, 2]. Artificial neural networks (ANNs) are capable of pattern recognition and reinforcement learning at a human level of performance [3, 4], but they perform poorly compared to a four-year old child in causally and logically making sense of their environment or the data at hand, particularly when there are only a modest number of training examples [5, 6].

A. V. Samsonovich (Ed.): BICA 2019, AISC 948, pp. 475–484, 2020.
https://doi.org/10.1007/978-3-030-25719-4_62

There has been recent work in attempting to close the neural-symbolic gap. Some cognitive architectures, for example, ACT-R, CLARION, MicroPsi or Sigma architectures [7–10], integrate subsymbolic and symbolic processing to varying degrees. Research by Graves and colleagues uses an ANN which can read and write to an external memory, i.e., a hybrid system [11]. Collier and Beel note that these newer neural network architectures can be classified as Memory Augmented Neural Networks (MANNs) where there is an external memory unit as opposed to, for example, Long Short Term Memory (LSTM) neural networks where memory exists as an internal vector. MANNs compared to, for example, LSTMs, can perform better on tasks which require memory usage [12]. The Neural Rule Engine of Li and colleagues uses modules of neural networks to each represent an action of a logic rule [13].

Unlike a typical ANN, the basic unit of the MBCA is not an artificial neuron but a Hopfield-*like* network (HLN) (Fig. 1). The MBCA uses HLNs for both its subsymbolic (e.g., ANN-like pattern recognition) and its symbolic operations. As well, unlike MANNs, the MBCA does not use an external memory, i.e., it is not a physically hybrid system.

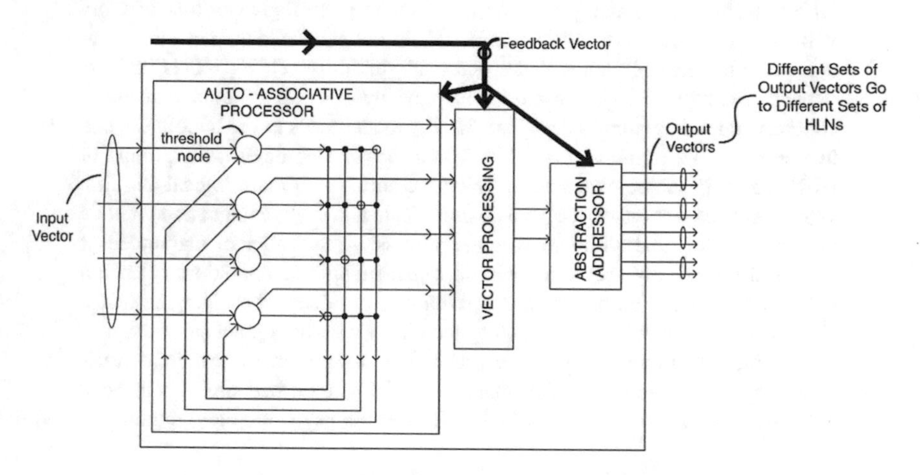

Fig. 1. Overview of a Hopfield-like Network (HLN) unit

Based on the "meaningfulness" values of the feedback vector (which can be as simple as how many other related HLNs are activated, or can use the Shannon entropy as the MBCA does), the abstraction addressor decides which of a number of possible sets of output vectors (which become inputs to other HLNs) will have non-zero outputs [1]. The feedback vector can be understood as essentially as asking what reconfigurations are more meaningful in terms of recognizing and processing an input sensory (or intermediate) vector. Meaningfulness M is defined as the reciprocal of the Shannon entropy (1, 2):

$$H = -\sum_{i} P(x_i)\log_2 P(x_i) \tag{1}$$

$$M = 1/H \tag{2}$$

The MBCA is built from reconfigurable topologies of Hopfield-like networks (HLNs). More straightforward hierarchies of pattern recognizers have been discussed in the literature, for example the hierarchical compositional network of George and colleagues [14]. Hawkins [15] and Kurzweil [16] have popularized hierarchies of pattern recognizers inspired by the structure of the mammalian cortex. The capsule networks developed by Hinton and coworkers go beyond typical convolutional neural networks in representing spatial hierarchies between objects of a scene [17]. However, the construction, operation and many of the properties of the MBCA, from its basic HLN elements to its overall organization is significantly different than in these previous works.

The Meaningful-Based Cognitive Architecture is shown in Fig. 2. Data from the sensory binding HLNs, the causal memory HLNs, the sequential/error-correcting memory, and other modules and functional units of HLNs, go to the HLNs configured to output motor vectors, which in turn feeds into an output shaping module, and then produces the motor outputs which activate different actuators, as well as providing a data output to the external environment.

In the current MBCA the "causal" memory group of HLNs is essentially a "pre-causal" staging area. Schneider [18] explores the evolution of pre-causal functioning of the "causal memory" group of HLNs to true causal functioning and then further enhancement to become full logic/working units. As a result, even though in the MBCA the causal memory group of HLNs has the description "causal", it essentially acts as a staging area for the causal operations in the logic/working units as well as storing the learned world views described below. Thus, although the instinctual core goals module in Fig. 2 is shown largely interfacing with the logic/working memory units, it does indeed interface with the causal memory, and in an MBCA version without logic/working memory units, innate goal objectives would be directed towards the causal memory group of HLNs.

If we ignore the components in Fig. 2 marked with a solid **X**, then the flow of the sensory data and its transformation into motor outputs, can be described as largely subsymbolic, i.e., pattern recognition and associative behavior is taking place, albeit with some pre-causal representation and goal directed influences. Such a system, i.e., *without* the symbolic features of the MBCA discussed below, if tuned well for its environment, can be expected to provide a very effective range of complex behaviors.

Groups of HLNs can be co-opted away from their original purpose as pattern recognizers, and with modifications in organization, can be configured into groups which act as logical processors and operate on what constitutes a working memory in these HLN groups. These logic/working memory units receive processed bound sensory input data, processed pre-causal data, as well as data from other modules and functional units of the MBCA. The representation of data in the MBCA already has meaning through the connections an HLN has via connections to other HLNs.

The working memory need not convert the vectors from the HLN into some arbitrary abstract symbolic data for the working memory to produce symbolic behavior, i.e., manipulation of the HLN vectors it receives suffices.

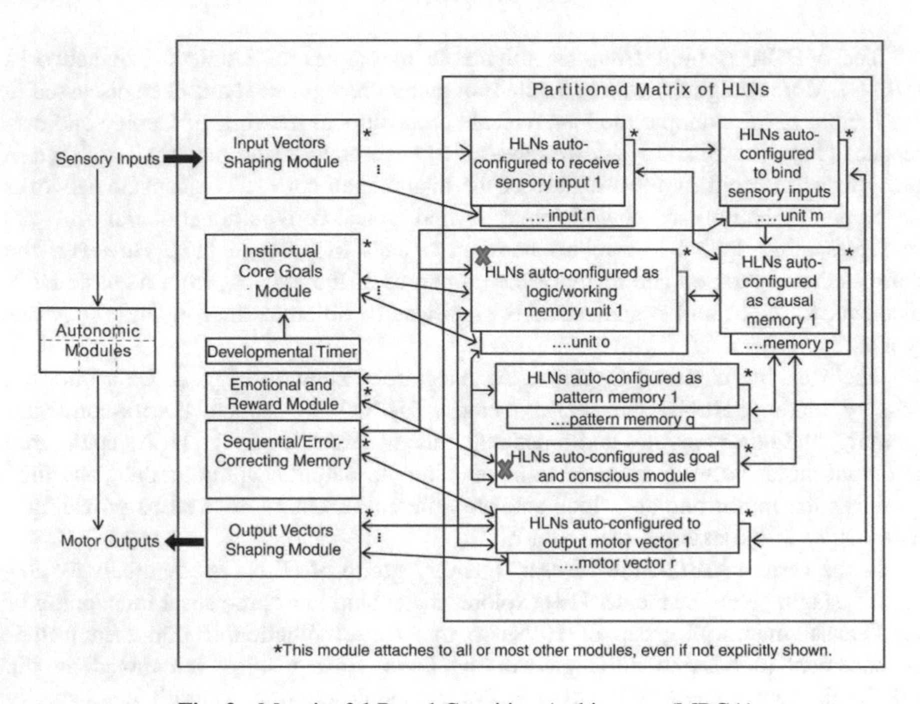

Fig. 2. Meaningful-Based Cognitive Architecture (MBCA)

Groups of HLNs auto-configured as relatively simple logic/working memory units are able to:

- compare properties of vectors they receive
- choose one vector over another
- match, trigger and receive intuitive procedural vectors from the instinctual core goals module
- match, trigger and receive learned procedural vectors from causal memory
- pattern match a vector from the entire MBCA memory
- output a vector to the motor output section
- output a vector back into the sensory input section, and operate again on the processed vector

The instinctual core goals module, which is affected by the maturity stage of the MBCA via the developmental timer, feeds default procedures into the groups of HLNs configured as logic/working memory units. Memories of operations occurring in the logic/working memory are kept in the conscious module, allowing improved problem solving as well as providing more transparency to MBCA decision making. The emotional module allows effective learning of infrequent events and obviates the class imbalance problem seen in conventional neural networks.

2 Advantages of Integrated Subsymbolic and Symbolic Processing of Data

To compare various symbolic processing configurations of the MBCA against configurations which only perform subsymbolic processing data plus/minus some simple symbolic algorithms or other conventional ANNs, a Python-based simulation attempts to find a hypothetical lost hiker in an uninhabited forest. This goal can be accomplished largely by subsymbolic processing coupled with a few basic search algorithms—sensory inputs associated with a lost hiker are detected, and the MBCA controlling some sort of search robot, moves closer to these sensory inputs. If there are multiple sensory inputs from different directions which all could be associated with a lost hiker, then the MBCA will choose the strongest one or some other straightforward quantification, as guided by the instinctual core goals module via the sensory and causal (actually precausal, as noted above) groups of HLNs. Often this choice of which sensory input to use for guidance will be incorrect, and the MBCA will then use the next strongest signal, and so on. Learning occurs with each attempt to find the lost hiker, and the performance of the MBCA can improve, as long as the environment of the forest does not change too much.

If the symbolic processing abilities of the groups of HLNs configured as logic/working memory units are also used, more efficient location of the lost hiker tends to be possible, especially if changes in the environment occur. Default intuitive logic, intuitive physics, intuitive psychology, intuitive planning and intuitive world views are stored in the instinctual core goals module, and additional learned world views are stored in the HLNs configured as causal memory (Fig. 2). The logic/working memory units interact with the instinctual core goals module.

Consider an example where the MBCA detects two input sensory vectors that activate HLNs associated with a lost hiker. One of these sensory input vectors is a sound from the southwest indicating a possible cry for help from a human or a possible match with certain bird calls. The other sensory input vector is an odor coming from the north indicating a possible match with a commercial perfume/cologne.

At this point, the MBCA could find the lost hiker, using only subsymbolic processing. From past experiences (or configuration data before the mission) the MBCA may decide to go in the direction of the sound in the southwest. If, in this example, the lost hiker is not found in the southwest, then the MBCA weakens this association and then heads off in the direction of the matched odor in the north. The MBCA finds the lost hiker in the north, and then strengthens the association between a commercial perfume/cologne odor and a lost hiker, and indeed, should such a choice need to be made in the future, the subsymbolic version of the MBCA will more likely choose to go in the direction of the matched odor vector.

However, consider the same example making use of the logic/working memory units. The instinctual core goals module has also sent a procedural vector to the logic/working memory units, and so have any causal memory HLNs with triggered learned world views. The processed sensory input vectors are also sent to the logic/working memory unit, i.e., the input vector from the southwest indicating a possible cry for help from a human or a possible match with certain bird calls, and the input vector from the north indicating a possible match with a human perfume/cologne.

The logic/working memory unit can compare vectors and take subsequent actions depending on the results of comparisons. These actions can be to modify a vector or to output a vector to other modules of the MBCA.

In this example, the MBCA's logic/working memory unit and a basic algorithm from the instinctual core goals module find that a vector that is related uniquely to a vector of the goal, i.e., an odor that comes from a human which is related to the goal, is closer to the goal vector than a vector which could match other things, such as a sound which could come from some birds. Hence, the output of the logic/working memory unit is to move in the direction of the odor sensory input vector. The search and rescue robot thus moves in a northerly direction towards the lost hiker, and in our example, indeed soon finds and rescues the lost hiker.

In more complex examples, the logic/working memory units are able to output a vector back into the sensory input section, and operate again on the processed vector, i.e., compare it, modify it, and so on, in conjunction with default logic rules and intuitive physics, psychology and planning algorithms in the instinctual core goals module as well as learned ones including world views in the causal memory. This can be repeated over again on intermediate results. Essentially a form of "thinking" occurs in the MBCA.

3 MBCA and the Emergence of Psychotic Disorders in *Homo sapiens*

The original goal of the MBCA was to attempt to produce and experiment with an artificial general intelligence of sorts. The MBCA was never intended to model disease and no disease characteristics were intentionally designed into it. The MBCA does not attempt to replicate biological systems at the neuronal spiking level, but its HLNs and the organization of its HLNs are indeed inspired by biological mammalian minicolumns and mammalian brains [19–21]. As Eliasmith and Trujillo note, in simulations of brain models perhaps the most important goal is the link to behavior [22], and indeed at the mesoscopic scale the MBCA can functionally produce a variety of behaviors which can possibly help to better hypothesize and understand mammalian cortical function.

The subsymbolic processing of input sensory vectors is relatively straightforward—each "evaluation cycle" the MBCA mechanically propagates the input sensory vector through its subsymbolic architecture and arrives at a motor output vector [1, 2]. It is hypothesized that the subsymbolic processing of input sensory vectors by the MBCA is analogous to the operation of the non-hominid mammalian brain—sensory inputs are processed by the subcortical structures as well as the mammalian cortex which operates as a pattern matching apparatus. Intuitive logic, intuitive physics, intuitive psychology, and intuitive planning operate both in the subcortical structures as well as the mammalian cortex, and simple pre-causal memories exist to allow simple comparisons. Pattern recognition and associative behavior is taking place, albeit with some pre-causal representation and goal directed influences.

It is hypothesized that in the course of hominid evolution, HLNs became co-opted into groups of HLNs providing more extensive working memories with more logical abilities of comparing vectors, modifying vectors, and outputting vectors, and with more extensive interaction with intuitive and learned logic, physics, psychology, and planning, i.e., the human logic/working memory unit emerged. As well, the output of the logic/working memory units could be fed back in as a new sensory input, and complex problems could be processed in a number of intermediate steps. The evolutionary advantage of such co-option was more advantageous integrated subsymbolic and symbolic decisions, particularly important in a changing environment.

Unlike the straightforward, mechanical operation of the subsymbolic version of the MBCA, symbolic processing requirements of functionally combining logic/working memory units with the procedures from the instinctual core goals module and the causal memory groups, and of shuffling working memory around and retrieving data from the logic/working memory HLNs and HLNs of other functional groups, are an order of magnitude more complex. As the MBCA matures, and as the developmental timer causes the instinctual core goals module to make available more complex procedures to the logic/working memory units, any of a myriad of small issues in the logic/working memory units or their connecting modules including the instinctual core goals module, can cause the MBCA to fail in a manner analogous to biological psychosis:

- output vectors from the logic/working memory units are, which themselves may be appropriate or often inappropriate if the output vector arose inappropriately, fed back as a sensory inputs, which the MBCA then inappropriately interprets as a real sensory input rather than a "thought" which logic/working memory expected to process further in the next evaluation cycle—hallucination-like behavior occurs;
- inappropriately match and retrieve memory vectors so that they do not correspond with the reality of the input sensory vectors, and inappropriately further process these vectors – cognitive dysfunction and delusional-like behavior occurs.

Hallucinations, delusions and cognitive dysfunction are the hallmarks of psychosis. There are many medical disorders which involve psychosis, a well-known one being schizophrenia.

Models of human psychotic disorders, including neural network-based ones, have long been proposed, for example the work by Cohen and Servan-Schreiber in the 1990s [23] to the more recent work of Sabaroedin and colleagues [24]. However, the MBCA was not designed to simulate psychotic disorders, but rather, psychotic features seemed to emerge easily when the logic/working memory HLN units had small flaws or there were inconsistencies with the procedure vector from the instinctual core goals module or the causal memory groups.

It is hypothesized that while the co-option of mammalian cortical minicolumns into effectively logic/working memory units can allow a nervous system to perform advantageous symbolic processing along with subsymbolic processing of input sensory vectors, the order of magnitude of increased complexity will create a vulnerability in all mammals with such co-option, which may only be humans, to psychotic disorders. Indeed, the clinical literature seems to support this hypothesis. For example, more than 10% of the population will experience less severe psychotic-like symptoms [25], but just under 1% of the population suffers from schizophrenia (a psychotic disorder)—there are many causes why humans may experience psychotic-like symptoms. Indeed, in just about all other mammals, who would appear by observation to engage in less symbolically processed behavior, psychosis is rare, and in psychopharmacological research settings, large efforts are required to induce at best unreliable models of schizophrenia in research animals [26].

Given that the psychotic behavior in the MBCA emerges from the complexity of co-opting HLNs into logic/working memory units and the operation of these units, it is hypothesized that in humans, psychotic disorders emerge from a large variety of different defects in the working memory circuits or the circuits they attach to. Indeed, lower working memory functioning is found not only in patients with schizophrenia but also in unaffected relatives [27]. Nor does there appear to be a single predominant schizophrenia gene in humans. In an evaluation of the genomes of 265,218 patients and 784,643 controls it was actually found that there was considerable genetic overlap between what should be very different formal psychiatric disorders including schizophrenia [28]. In young patients at high risk of developing schizophrenia, the medications used to normally treat the symptoms of established schizophrenia do not seem to help with its prevention. However, cognitive programs aimed at broadly improving executive function, and thus to some extent working memory, in these high risk young patients, do seem to have a modest effect in preventing the development of schizophrenia [29].

In a changing environment, as discussed above, the ability to perform integrated symbolic/subsymbolic decision making is advantageous compared to a mainly subsymbolic approach. While subsymbolic ANNs can be modified to produce some symbolic behaviors, it is hypothesized that a less burdensome creation of an effective artificial general intelligence (AGI) technology will require the integration of symbolic processing along with subsymbolic processing. A corollary of this hypothesis is that such an AGI technology, like its human counterpart, will be vulnerable to psychotic behavior.

4 Discussion

Future work, which has been started at the time of writing, includes more formal MBCA simulations, with the goal of obtaining reproducible quantitative comparisons of the MBCA in solving the problem of finding a lost hiker in the woods, and the goal of better quantifying the psychotic behavior which results with different MBCA models.

While the MBCA does not attempt to replicate biological systems at the neuronal spiking level, its HLNs and the organization of its HLNs are indeed inspired by biological mammalian minicolumns and mammalian brains. The MBCA model leads to the hypothesis that in the course of hominid evolution, HLNs became co-opted into groups of HLNs providing more extensive working memories with more logical abilities of comparing vectors, modifying vectors, outputting vectors, and with more extensive interaction with intuitive and learned logic, physics, psychology, and planning. While such co-option of the minicolumns can allow advantageous symbolic processing integrated with subsymbolic processing of input sensory vectors, the order of magnitude of increased complexity required for such organization and operation, creates a vulnerability in the human brain to psychotic disorders. One pathway to the emergence of a technological artificial general intelligence will, on a practical level, also require the integration of symbolic processing along with subsymbolic structures. The MBCA model predicts that such an integration will similarly create a potential vulnerability in the resultant AGI towards psychotic-like features.

Acknowledgments. This article builds upon work originally presented at BICA 2018 (reference 1).

References

1. Schneider H (2018) Meaningful-based cognitive architecture. Procedia Comput Sci 145:471–480 BICA 2018 ed Samsonovich AV
2. Schneider H (2019, in press) Subsymbolic versus symbolic data flow in the meaningful-based cognitive architecture. In: Samsonovich AV (ed) Biologically inspired cognitive architectures 2019. Advances in intelligent systems and computing. Springer
3. Goodfellow I, Bengio Y, Courville A (2016) Deep learning. MIT Press, Cambridge
4. Mnih V, Kavukcuoglu K, Silver D et al (2015) Human-level control through deep reinforcement learning. Nature 518(7540):529–533
5. Ullman S (2019) Using neuroscience to develop artificial intelligence. Science 363 (6428):692–693
6. Waismeyer A, Meltzoff AN, Gopnik A (2015) Causal learning from probabilistic events in 24-month-olds: an action measure. Dev Sci 18(1):175–182
7. Anderson JR, Bothell D, Byrne MD et al (2004) An integrated theory of mind. Psychol Rev 111(4):1036–1060
8. Kilicay-Ergin N, Jablokow K (2012) Problem-solving variability in cognitive architectures. Syst Man Cybern Part C Appl Rev IEEE Trans 42:1231–1242
9. Bach J (2008) Seven principles of synthetic intelligence. In: Proceedings of 1st conference on artificial general intelligence, Memphis, pp. 63–74
10. Rosenbloom P, Demski A, Ustun V (2016) The sigma cognitive architecture and system: towards functionally elegant grand unification. J Artif Gen Intell 7(1):1–103
11. Graves A, Wayne G, Reynolds M et al (2016) Hybrid computing using a neural network with dynamic external memory. Nature 538:471–476
12. Collier M, Beel J (2018) Implementing neural Turing machines. In: 27th international conference on artificial neural networks (ICANN), Rhodes, Greece, 4–7 October
13. Li S, Xu H, Lu Z (2018) Generalize symbolic knowledge with neural rule engine. arXiv: 1808.10326
14. Lázaro-Gredilla M, Liu Y, Phoenix DS, George D (2017) Hierarchical compositional feature learning. arXiv preprint. arXiv:1611.02252v2
15. Hawkins J, Blakeslee S (2004) On intelligence. Times Books, New York
16. Kurzweil R (2012) How to create a mind. Viking Press, New York
17. Sabour S, Frosst N, Hinton GE (2017) Dynamic routing between capsules. In: Conference on neural information processing systems (NIPS 2017), Long Beach, CA, USA
18. Schneider H (2019 in press) Emergence of belief systems and the future of artificial intelligence. In: Samsonovich AV (ed) Biologically inspired cognitive architectures 2019. Advances in intelligent systems and computing. Springer

19. Mountcastle VB (1997) Columnar organization of the neocortex. Brain 20:701–722
20. Buxhoeveden DP, Casanova MF (2002) The minicolumn hypothesis in neuroscience. Brain 125(Pt 5):935–951
21. Schwalger T, Deger M, Gerstner W (2017) Towards a theory of cortical columns. PLoS Comput Biol 13(4):e1005507. https://doi.org/10.1371/journal.pcbi.1005507
22. Eliasmith C, Trujillo O (2014) The use and abuse of large-scale brain models. Curr Opin Neurobiol 25:1–6
23. Cohen JD, Servan-Schreiber D (1992) Context, cortex and dopamine: a connectionist approach to behavior and biology in schizophrenia. Psychol Rev 99(1):45–77
24. Sabaroedin K, Tiego J, Parkers L et al (2018) Functional connectivity of corticostriatal circuitry and psychosis-like experiences in the general community. Biol Psychiatry 86 (1):16–24 pii: S0006-3223(19)30119-2
25. van Os J, Hanssen M, Bijil RV et al (2001) Prevalence of psychotic disorder and community level psychotic symptoms: an urban-rural comparison. Arch Gen Psychiatry 58(7):663–668
26. Jones CA, Watson DJG, Fone KCF (2011) Animal models of schizophrenia. Br J Pharmacol 164:1162–1194
27. Zhang R, Picchioni M, Allen P et al (2016) Working memory in unaffected relatives of patients with schizophrenia: a meta-analysis of functional magnetic resonance imaging studies. Schizophr Bull 42(4):1068–1077
28. Anttila V, Bulik-Sullivan B et al (2018) Analysis of shared heritability in common disorders of the brain. Science 360(6395):eaap8757. https://doi.org/10.1126/science.aap8757
29. Bechdolf A, Wagner M, Ruhrmann S et al (2012) Preventing progression to first-episode psychosis in early initial prodromal states. Br J Psychiatry 200(1):22–29

Emergence of Belief Systems and the Future of Artificial Intelligence

Howard Schneider(✉)

Sheppard Clinic North, Toronto, ON, Canada
howard.schneider@gmail.com

Abstract. In the Meaningful-Based Cognitive Architecture (MBCA), every evaluation cycle the input sensory vector is propagated through a hierarchy of Hopfield-like Network (HLN) functional groups, and subsymbolic processing of the input vector occurs. The processed sensory input vector is also propagated to causal groups of HLNs and/or logic/working memory groups of HLNs, where it is matched against the vector from the MBCA's intuitive and learned logic, physics, psychology and goal planning—collectively forming its world model or belief system. The output of the logic/working memory can be propagated back to the start of the subsymbolic system where it can be the input for the next evaluation cycle, and allows multiple symbolic processing steps to occur on intermediate results. In this fashion, full symbolic causal processing of the input sensory vector occurs. In order to achieve causal and symbolic processing of sensory input, i.e., allowing the MBCA to not only recognize but interpret events in the environment and produce motor outputs in anticipation of future outcomes, every evaluation cycle the MBCA must not simply process the input sensory vector for subsymbolic recognition, but *must* process this vector in terms of the belief system activated. Artificial Neural Networks have excellent pattern recognition and reinforcement learning abilities, but perform poorly at causally and logically making sense of their environment. Future artificial intelligence systems that attempt to overcome the limitations of subsymbolic architectures by the integration of symbolic processing, may be at risk for developing faulty belief systems with unintended results.

Keywords: Cognitive architecture · Artificial general intelligence · Cortical minicolumns · Belief systems · Neural-symbolic integration

1 Brief Overview of the Meaningful-Based Cognitive Architecture (MBCA)

Artificial neural networks (ANNs) are capable of pattern recognition and reinforcement learning at a human level of performance [1, 2], but they perform poorly compared to a four-year old child in causally and logically making sense of their environment or the data at hand, particularly when there are only a modest number of training examples [3, 4]. The Meaningful-Based Cognitive Architecture (MBCA) attempts to close the neural-symbolic gap by integrating the sensory processing abilities found in neural networks, with many of the causal and symbolic logical abilities found in human

A. V. Samsonovich (Ed.): BICA 2019, AISC 948, pp. 485–494, 2020.
https://doi.org/10.1007/978-3-030-25719-4_63

cognition [5, 6]. The Meaningful-Based Cognitive Architecture will be reviewed only briefly here as it has been described elsewhere [5–7]. A large number of cognitive architectures have been described in the literature [8]. Some architectures, for example, MicroPsi or Sigma models [9, 10], attempt to integrate subsymbolic and symbolic processing. As well, a number of ANNs have attempted in recent years to combine subsymbolic and symbolic processing. Unlike Memory Augmented Neural Networks (MANNs) [11], the MBCA does not use an external memory, i.e., it is not a physically hybrid system.

While the MBCA does not attempt to replicate biological systems at the neuronal spiking level, it is indeed inspired by mammalian cortical minicolumns and mammalian brain organization [12–14]. In keeping with this biological inspiration, unlike a typical ANN, the basic unit of the MBCA is not an artificial neuron but a Hopfield-*like* network (HLN). The MBCA uses HLNs for *both* its sub-symbolic (e.g., ANN-like pattern recognition) and its symbolic operations. As Eliasmith and Trujillo note, in simulations of brain models perhaps the most important goal is the link to behavior [15], and indeed at the mesoscopic scale the MBCA can functionally produce a variety of behaviors which can help to better hypothesize and understand mammalian cortical function, and provide insight into possible mechanisms which link such mesoscopic functioning to the causal and symbolic behavior seen in humans.

The Meaningful-Based Cognitive Architecture is shown in Fig. 1. Data from the sensory binding HLNs, the causal memory HLNs, the sequential/error-correcting memory, and other modules and functional units of HLNs, go to the HLNs configured to output motor vectors, which in turn feed into an output shaping module, and then produce the motor outputs which activate different actuators, as well as providing a data output to the external environment.

If we ignore the components in Fig. 1 marked with a solid circle, then the flow of the sensory data and its transformation into motor outputs, is essentially associative and subsymbolic. Consider the "causal memory" in such a configuration to be a "pre-causal memory", i.e., a staging area rather than true casual memory. Each evaluation cycle (every millisecond or so largely depending on hardware constraints), the input sensory vector is propagated through a hierarchy of reconfigurable Hopfield-like Network (HLN) units, and is recognized as some vector (or learned as some new vector). The processed input sensory vector triggers patterns in the "pre-causal memory", in the pattern memory, and in various modules and other functional groups of HLNs, and these cause the output motor HLNs to output a motor vector (i.e., an output signal). There is feedback to the layers of sensory HLNs from the upper levels and from the sensory binding, "pre-casual", pattern and other functional groups of HLNs. As well, the instinctual core goals module propagates vectors to these regions and will also influence the recognition of the input sensory vector and the triggering of various patterns. Such a system, i.e., *without* the causal and symbolic features of the MBCA discussed below, if tuned well for its environment, can nonetheless be expected to provide a very effective range of complex behaviors.

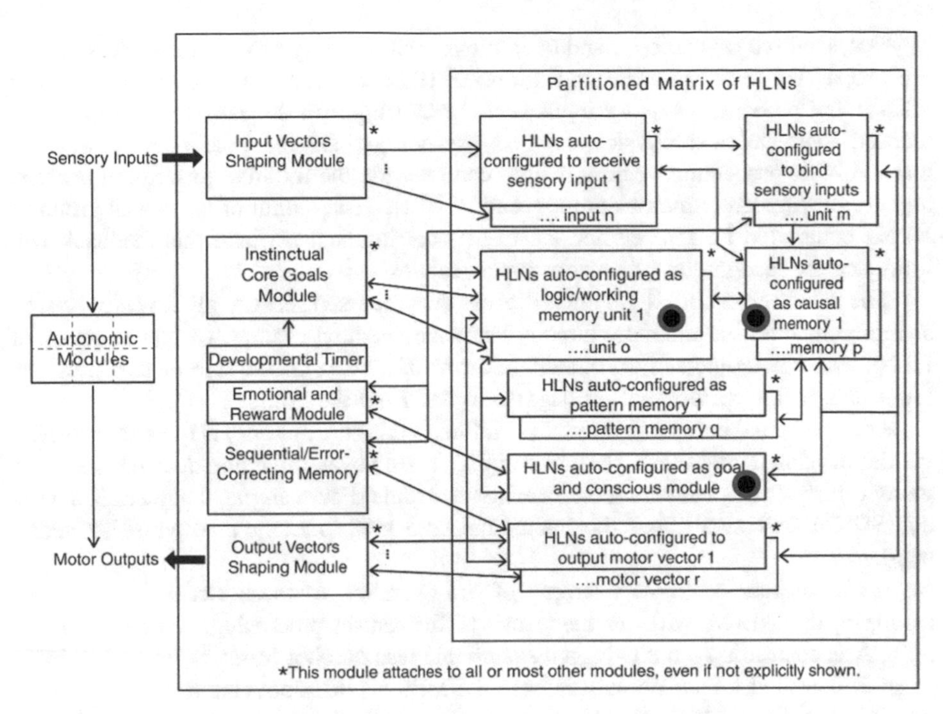

Fig. 1. Meaningful-Based Cognitive Architecture (MBCA) – solid circles on the symbolic and causal structures

2 Causal Symbolic Processing of the Input Sensory Data

Groups of HLNs can be co-opted (by design or in a biological system by evolution) away from their original purpose as pattern recognizers to form units dedicated for causal pre-processing of input sensory data.

Groups of HLNs auto-configured as causal memory units are able to:

- match, trigger and receive intuitive procedural vectors from the instinctual core goals module
- match, trigger and receive learned procedural vectors from the local causal memory
- compare properties of vectors they receive
- store a limited number of previous input and output vectors and compare these with the properties of the vectors they receive
- choose one vector over another
- output a vector to the motor output section
- send its output as a feedback signal back to the sensory processing HLNs and modify the processing of the input sensory vector

Each evaluation cycle, the processed sensory input vector is propagated to the instinctual core goals module and to the causal group of HLNs. The processed sensory input vector may trigger in the instinctual core goals module intuitive logic, intuitive physics, intuitive psychology, and/or intuitive goal planning procedural vectors, which are propagated to the motor output group of HLNs as well as to the causal group of HLNs. The processed sensory input vector may trigger in the causal group of HLNs learned logic, learned physics, learned psychology, learned goal planning, and/or learned other procedural vectors, which can override the intuitive procedural vectors, and are propagated to motor output group of HLNs. The output of the causal group of HLNs is also fed back to earlier sensory processing stages where this feedback will influence the recognition of sensory input features.

The causal group of HLNs thus receives the processed sensory input vector as well as indirectly intuitive and more directly learned procedural vectors. The causal group of HLNs also has temporal information in terms of storage of the one or two previous input and output vectors that propagated to and from the group.

The instinctual core goals module is influenced by the maturity stage of the MBCA via the developmental timer, as shown in Fig. 1. However, the figure does not show the many other pathways affecting the core goals module. For example, the internal state of the MBCA will also affect the instinctual core goals module's response to vector triggers.

Let's consider a simple example of the operation of these circuits. In the first example, the MBCA will *not* make use of full causal processing. Imagine that the MBCA is controlling a multi-legged search and rescue robot trying to find a lost hiker in an uninhabited forest. We will refer to the MBCA plus robot combination as simply the "MBCA." The MBCA is able to walk through shallow water such as a puddle of water or a shallow river, but if it immerses itself completely in water, for example in a deep lake, it will become damaged. If there is a field or an area of trees, then this area is fine to walk in.

As the MBCA makes its way in searching for the lost hiker, if there is a lake in front of its path, i.e., in front of the direction it wanted to move, it will recognize via its subsymbolic processing the lake. This will trigger in the instinctual core goals module an association with danger, and the MBCA will stop. Another direction will be chosen in conjunction with vector triggered in the instinctual core goals module and a pre-causal memory sort of group of HLNs. On the other hand if there is a shallow river, or a field or a forest, recognized by the MBCA's subsymbolic processing of the visual sensory inputs, the MBCA will continue walking in whatever direction had been chosen. Although causal and higher level symbolic processing can and may be involved in such decisions, they are not required—subsymbolic sensory processing (recognizing the visual sensory input of the lake or shallow river or field or forest) and associative action in a pre-causal memory sort of group of HLNs or elsewhere in the MBCA will actually suffice.

Now consider that there is a shallow river in front of the MBCA which the MBCA wants to cross to continue the search for the lost hiker, but this river has many white swirling areas of water in the river. Subsymbolic processing of the visual inputs matches closest to a shallow river. As noted above, a shallow river is not associated with any danger and the MBCA can continue its path across it. However, subsymbolic sensory processing also matches smaller areas of the river as white swirling areas. The MBCA has not been programmed with intuitive or learned associative or other actions about what to do if there are white swirling areas in a shallow river. Perhaps another MBCA has tried to cross such a river and the intense pressurized whitewater spray damaged one of that MBCA's leg articulations, and thus the next time that particular MBCA encounters such a river situation it recognizes it as a danger and does not cross it. However, the MBCA in our example has never seen such a whitewater river before and there is no subsymbolic associative-like action preventing it from crossing the river. Thus it crosses the whitewater river and unfortunately becomes damaged.

Now consider an example where the MBCA makes use of its causal processing abilities. To quickly review, as noted above, every evaluation cycle the MBCA processes the sensory input data. The processed sensory input vector is propagated to the instinctual core goals module and to the causal group of HLNs (which in the example above is really more of a simple or pre/causal memory group of HLNs). The processed sensory input vector may trigger in the instinctual core goals module intuitive logic, intuitive physics, intuitive psychology, and/or intuitive goal planning vectors, which are propagated to the motor output group of HLNs as well as to the causal group of HLNs. The processed sensory input vector may trigger in the causal group of HLNs learned logic, learned psychology, learned goal planning, and/or learned other procedural vectors, which can override the intuitive vectors. The output of the causal group of HLNs is propagated to a number of other groups of HLNs, including the motor output group of HLNs. As well, given the nature of the hierarchy of the HLNs in sensory input processing, the output of the causal group of HLNs is also fed back to earlier sensory processing stages where the feedback signal will influence the recognition of sensory input features. This all happens every single evaluation cycle, over and over again.

Now let's consider the example of an MBCA that has no prior knowledge about whitewater rivers but makes more complete use of the causal memory group of HLNs. In front of this MBCA's intended path is a shallow river with whitewater currents. Processed sensory input vectors of "swirling white areas" + "shallow river" are propagated to and trigger in the instinctual core goals module an intuitive physics output of "water all directions." The "water all directions" vector is propagated to the causal memory group of HLNs and is stored there. The "water all directions" vector is also propagated back to earlier sensory stages as a feedback signal. The result is that in the next evaluation cycle the processed sensory input vector may be "water all directions" rather than actual sensory input. The processed sensory input vector —"water all directions"—is propagated, as occurs each evaluation cycle, to the instinctual core goals module where this triggers in its pre-existing intuitive physics knowledge, a "do not walk" vector to the motor output HLNs as well as to the causal group of HLNs.

The MBCA is in front of a river, so the next evaluation cycle there is recognition again of a shallow river, and a processed sensory input vector for "shallow river" arrives at the causal group of HLNs. The causal group of HLNs has the previous procedural vector of "do not walk" stored and now has the processed sensory input vector of "shallow river." The causal group of HLNs sends "shallow river" to the instinctive core goals module where it may trigger the intuitive goal planning current sub-goal of the MBCA, e.g., "walk same direction" and this is propagated to the causal group of HLNs. The next evaluation cycle the causal group of HLNs may ignore the processed sensory input vector but send the different procedural vectors back to the instinctive core goals module. The "walk same direction" and "do not walk" trigger in the instinctive core goals module an intuitive logic procedural vector "change direction ninety degrees."

In the following tens or hundreds of thousands of evaluation cycles (depending on the speed of computational hardware, walking speed and distance along river before it is crossable) the processed sensory input vector will be evaluated over and over again, the instinctive core goals module intuitive goal planning current sub-goal will be triggered and updated many times with the goal of getting the MBCA to walk to the lost hiker, and at some point along the river where there are no whitewater currents or other potential dangers to the MBCA detected, the MBCA will cross the river.

Although this particular MBCA never had any stored experiences with rivers with whitewater, the causal group of HLNs plus the overall architecture of the MBCA allowed the MBCA to produce causal behavior which was successful in allowing the MBCA to avoid damage. The earlier example where fuller causal behavior did not occur, resulted in the MBCA walking into the whitewater and becoming damaged. While it may seem that causal behavior should be reserved for special occasions, and perhaps in some other architecture or machine it could be set up as such, the architecture of the MBCA is such that evaluation cycle after evaluation cycle, the processed input sensory vector is processed causally, regardless of whether the resultant output behavior appears associative or appears causal.

3 Full Symbolic Processing of the Input Sensory Data

Whether by design in a machine or by evolution in a biological organism, enhancements to the causal groups of HLNs will lead to full symbolic processing of the input sensory data.

Groups of HLNs auto-configured as relatively simple logic/working memory units are able to perform similar but enhanced functions compared to the previously described groups of HLNs auto-configured as causal memory units:

- match, trigger and receive intuitive procedural vectors from the instinctual core goals module
- match, trigger and receive learned procedural vectors from causal memory
- pattern match a vector from the entire MBCA memory
- compare properties of vectors they receive
- store a limited number of previous input and output vectors and compare these with the properties of the vectors they receive
- choose one vector over another
- output a vector to the motor output section
- output a vector back into the sensory input section, and operate again on the processed vector, i.e., operate again on an intermediate result

In the architecture of the MBCA shown in Fig. 1, both the causal memory units and the logic/working memory units are included in the architecture. Due to the ability of the modules and groups of HLNs in the MBCA to integrate parallel streams of data, both can exist, but from a design point of view in a machine implementation of the MBCA or from an evolutionary point of view in a biological organism which is represented by the MBCA architecture, the causal memory units will largely end up feeding processed sensory inputs into the groups of HLNs forming logic/working memory units.

The logic/working memory units operate similarly to the operation of the causal group of HLNs described above. However, they can more readily compare properties of vectors they receive, and an enhanced intuitive logic in the core goals instinctual module will allow more robust logical operations.

The logic/working memory units can also more readily feed an intermediate result vector back into the sensory input section, and this intermediate result becomes the next processed input sensory vector. As such the MBCA can directly work on problems which require numerous sequential logical steps to solve.

As noted above, while the MBCA does not attempt to replicate biological systems at the neuronal spiking level, its HLNs and the organization of its HLNs are indeed inspired by biological mammalian minicolumns and mammalian brains [5, 6, 12–14]. While the biological emergence of the logic/working memory units would seem unfathomable due to the complexity of the operations which are occurring in these units, it becomes much more feasible over modest evolutionary time scales when one considers the emergence of associative-pseudo-causal groups of HLNs in mammalian brains and then simple but fuller causal groups, and then with evolutionary time, enhancements into full logic/working memory units. From a machine design point of view, logical processing integrated with subsymbolic processing gives the MBCA a much more powerful behavior. An example of the MBCA using the logic/working memory units to find a lost hiker in the forest is given elsewhere [5, 6].

Memories of operations occurring in the logic/working memory are kept in the conscious module, allowing improved problem solving as well as providing more transparency to MBCA decision making. Note that such transparency is largely for symbolic operations, while for subsymbolic operations the architecture does not directly save results. The emotional module allows effective learning of infrequent events and obviates the class imbalance problem seen in conventional neural networks.

4 Experimental and Observational Evidence

The original goal of the MBCA was to attempt to produce and experiment with an artificial general intelligence of sorts. The MBCA was never intended to model disease or to model ideas such as belief systems, and no higher level anthropomorphic characteristics were intentionally designed into it. The MBCA does not attempt to replicate biological systems at the neuronal spiking level, but at the mesoscopic scale this architecture can functionally produce a variety of behaviors which can possibly help to better hypothesize and understand mammalian cortical function.

As noted above, it is hypothesized that over evolutionary time the co-option of groups of mammalian cortical minicolumns into logic/working memory units can allow a hominid nervous system to perform advantageous causal symbolic processing integrated with subsymbolic processing of input sensory vectors. Causal processing by a non-hominid brain should not be assumed, regardless the size of the brain or the ability of the organism to exhibit complex associative behaviors. For example, Nissani [16] showed experimentally that in Asian elephants, which have brains much larger than humans, an experimental elephant behavior was due to associative learning rather than due to causal processing. However, many mammals may have the beginnings of causal behavior [17] and some casual abilities may have convergently evolved in other animals [18].

Computer simulations of the MBCA have been implemented, and preliminary results show the necessity of continuous usage of a world model or belief system for the functioning of the MBCA. The simulations are implemented in Python 3.6 and make use of PyTorch 1.0 as the neural network framework for sub-symbolic simulation. In the computer implementation, an MBCA is simulated operating in an environment where the goal is to find a lost hiker in an uninhabited forest. Due to the complexity of a fine-grained simulation of the MBCA, simulations have been reorganized in 2019 as coarse-grained, demonstrating overall principles, to finer-grained, demonstrating more authentic MBCA components, simulations of the MBCA.

5 Belief Systems, the MBCA and Future Artificial Intelligence Systems

In order to achieve optimal causal and symbolic processing of sensory input, i.e., allowing the MBCA to not only recognize but interpret events in the environment and produce motor outputs in anticipation of future outcomes, every evaluation cycle the MBCA must not simply process the input sensory vector for subsymbolic recognition, but *must* process this vector in terms of the world model activated, and must do so evaluation cycle after evaluation cycle. The MBCA, cycle after cycle, is forced to not just recognize the input sensory data and reflexively act on it, but essentially ask "why" in terms of activating a world model and then operating on the processed input sensory data in terms of the intuitive logic, intuitive physics, intuitive psychology and intuitive goal planning in the instinctual core goals module as well as learned goals in groups of HLNs. The summation of these intuitive and learned groups of knowledge represents the MBCA's world model, i.e., its belief system.

While belief systems are often considered as high level human ideological phenomena, note that in order for the MBCA to simply function, i.e., for the steps of data processing through its architecture, it requires some sort of belief system. In a brand new MBCA the belief system would represent the knowledge in the instinctual core goals module, and with learning experience, this would expand to the learned knowledge which may be potentially causally triggered each evaluation cycle.

In the simpler, i.e., coarse-grained experimental simulations done, if an MBCA is not created with appropriate world models, it simply cannot function—a belief system is not an abstract high-level ideological phenomenon but a basic requirement for the MBCA's normal operation. It is hypothesized that in more finer-grained simulations, if the MBCA is not created with appropriate world models, it will be forced to learn some world model which very likely may represent a faulty belief system from the viewpoint of the "normal" human observer. In such a case, the MBCA is operating normally—there is no psychotic-like dysfunction, regardless of the belief system which emerges in the MBCA, but it operates in the context of a faulty belief system, and produces unintended behavior to a "normal" human observer. (See Schneider [7] for a discussion of the psychotic-like behavior that can emerge in the MBCA.)

In the last decade concerns have started increasing about the existential risk of advanced forms of artificial intelligence that reach levels of human performance and then go on to exceed it [19]. As noted above, artificial neural networks have excellent pattern recognition and reinforcement learning abilities, but perform poorly at causally and logically making sense of their environment especially when there are only a modest number of training examples. Future artificial intelligence systems may very well attempt to overcome the limitations of subsymbolic architectures with the integration of symbolic processing. Although details may differ, just as likely occurs with the MBCA, future artificial intelligence systems which integrate symbolic processing, may similarly be at risk to create AI systems which develop and depend on faulty belief systems, with unexpected consequences as viewed by the "normal" human observer.

Acknowledgments. Thanks for discussion to Dr Aliye Keskin. This article builds upon work originally presented at BICA 2018 (reference 5).

References

1. Goodfellow I, Bengio Y, Courville A (2016) Deep learning. MIT Press, Cambridge
2. Mnih V, Kavukcuoglu K, Silver D et al (2015) Human-level control through deep reinforcement learning. Nature 518(7540):529–533
3. Ullman S (2019) Using neuroscience to develop artificial intelligence. Science 363 (6428):692–693
4. Waismeyer A, Meltzoff AN, Gopnik A (2015) Causal learning from probabilistic events in 24-month-olds: an action measure. Dev Sci 18(1):175–182
5. Schneider H (2018) Meaningful-based cognitive architecture. Procedia Comput Sci 145:471–480 BICA 2018, ed Samsonovich, A.V.
6. Schneider H (2019, in press) Subsymbolic versus Symbolic data flow in the meaningful-based cognitive architecture. In: Samsonovich AV (ed) Biologically inspired cognitive architectures 2019, BICA 2019. Advances in intelligent systems and computing. Springer
7. Schneider H (2019, in press) Schizophrenia and the future of artificial intelligence. In: Samsonovich AV (ed) Biologically inspired cognitive architectures 2019, BICA 2019. Advances in intelligent systems and computing. Springer
8. Samsonovich AV (2010) Toward a unified catalog of implemented cognitive architectures. BICA 221:195–244
9. Bach, J. (2008): Seven Principles of Synthetic Intelligence. In: Proceedings of 1st Conference on Artificial General Intelligence, Memphis, pp. 63–74
10. Rosenbloom P, Demski A, Ustun V (2016) The sigma cognitive architecture and system: towards functionally elegant grand unification. J Artif Gen Intell 7(1):1–103
11. Collier M, Beel J (2018) Implementing neural Turing machines. In: ICANN
12. Mountcastle VB (1997) Columnar organization of the neocortex. Brain 20:701–722
13. Buxhoeveden DP, Casanova MF (2002) The minicolumn hypothesis in neuroscience. Brain 125(Pt 5):935–951

14. Schwalger T, Deger M, Gerstner W (2017) Towards a theory of cortical columns. PLoS Comput Biol 13(4): e1005507. https://doi.org/10.1371/journal.pcbi.1005507
15. Eliasmith C, Trujillo O (2014) The use and abuse of large-scale brain models. Curr Opin Neurobiol 25:1–6
16. Nissani M (2006) Do Asian elephants (*Elaphas maximus*) apply causal reasoning to tool-use tasks? J Exp Psychol Anim Behav Process 32(1):91-6
17. Sawa K (2009) Predictive behavior and causal learning in animals and humans. Jpn Psychol Res 51(3):222–233
18. Laumer IB, Bugnyar T, Reber SA, Auersperg A (2017) Can hook-bending be let off the hook? Bending/unbending of pliant tools by cockatoos. Proc Biol Sci 284(1862):20171026
19. Bostrom N (2014) Superintelligence: paths, dangers, strategies. Oxford University Press, Oxford

Hierarchical Reinforcement Learning Approach for the Road Intersection Task

Maxim Shikunov[1] and Aleksandr I. Panov[2,3](✉)

[1] National Research University Higher School of Economics, Moscow, Russia
mashikunov@edu.hse.ru

[2] Artificial Intelligence Research Institute, Federal Research Center "Computer Science and Control" of the Russian Academy of Sciences, Moscow, Russia
panov.ai@mipt.ru

[3] Moscow Institute of Physics and Technology, Moscow, Russia

Abstract. The task of building unmanned automated vehicle (UAV) control systems is developing in the direction of complication of options for interaction of UAV with the environment and approaching real life situations. A new concept of so called "smart city" was proposed and view of transportation shifted in direction to self-driving cars. In this work we developed a solution to car's movement on road intersection. For that we made a new environment to simulate a process and applied a hierarchical reinforcement learning method to get a required behaviour from a car. Created environment could be then used as a benchmark for future algorithms on this task.

Keywords: Hierarchical reinforcement learning · Option-Critic · Smart city · Road intersection · Option framework

1 Introduction

More and more cities are transforming to a new stage in their development becoming so called "smart cities". Concurrently we can see a dramatic rise in attention to a self driving cars. Thanks to this changes newly available resources could be used to solve different kind of problems and make lives of cities' residents more comfortable. But there are still several areas which pose a significant difficulties in their solution. One of such problems is behaviour of self driving car on a road intersection. In this paper we suggest an approach to solve this problem for 4 way intersection.

We assume that we have access to a data from intersection. The way this data is provided can be arrange in several different ways, e.g. by cameras which mounted near intersections or by a quad-copter. The data is provided as a photo of current situation on the road and then can be transferred to a self driving car. We concentrate on the behaviour of a self driving car after receiving data. We then use a model from hierarchical reinforcement learning domain to train an agent to drive to its destination using provided data.

A. V. Samsonovich (Ed.): BICA 2019, AISC 948, pp. 495–506, 2020.
https://doi.org/10.1007/978-3-030-25719-4_64

There are a wide variety of reinforcement learning approaches for self-driving car modelling. Majority of them using non hierarchical methods with data provided similar to what human driver can see [1–3]. Hierarchical reinforcement learning (HRL) approach demonstrate significant results in simple tasks but with more complex tasks there are difficulties in determining sub-goals [4–8]. HRL for self-driving cars environments was applied in work [9]. Predefined meta-actions (options) used to simulate behaviour of a car on a road and intersection. But an agent can't change direction of movement. In addition the authors of that paper didn't make their environment available for other researches.

We developed a completely new environment integrated to OpenAI Gym framework [10]. This environment simulates a behaviour of a car crossing roads intersection. It is open and can be used as a benchmark for new algorithms in related area. We start with description of ideas behind HRL. Then we describe developed environment and how it can be used to train behaviour of car on a road intersection. Several baseline models are introduced with results and conclusion.

2 Problem Statement

2.1 Reinforcement Learning

The concept of reinforcement learning (RL) is to study behaviour of an agent in an environment. Environment is a world which contain an agent. Environment is changing under an agent actions but the agent could be affected by its own actions as well. The way an agent and an environment interact with each other is depicted in Fig. 1.

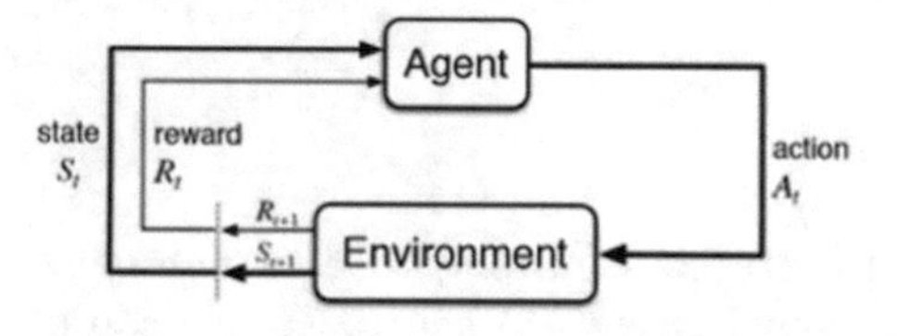

Fig. 1. Agent environment interaction.

After making some action an agent receive a reward (a score) from an environment which inform it how good the current state of the world. The goal of the agent is to receive the biggest reward possible. An environment usually described as Markov Decisions Process (MDP) with finite number of states.

Different environments allow different kind of actions. The set of all available actions for current state is called action space. Some environments have discrete action space (Atari and different kind of board games) for which only finite number of states is available. The others (car movements in real world) have continuous action space. We understand under policy a set of rules which an agent follows to make actions in an environment. There are two types of policies: deterministic and stochastic.

During arbitrary time t an agent is described by state s_t. After choosing action a_t it steps to state s_{t+1} and gain a reward r_t. Thus an agent chooses policy π: $a_t \sim \pi(\cdot|s_t)$, which in case of MDP with a terminal state maximise value $R = r_0 + r_1 + \cdots + r_n$, and value $R = \sum_t \gamma^t r_t$ for MDP without a terminal state (where $0 \leq \gamma \leq 1$ is discount).

The optimal state value function $V^*(s)$ returns expected reward which an agent receives after starting in state s and then following optimal policy:

$$V^*(s) = \max_{\pi} E_{\tau \sim \pi} \big(R(\tau)|s_0 = s\big)$$

Optimal state-action value function $Q^*(s,a)$ gives expected reward which an agent receives after starting in state s making action a and then following optimal policy in an environment:

$$Q^*(s,a) = \max_{\pi} E_{\tau \sim \pi} \big(R(\tau)|s_0 = s, a_0 = a\big)$$

Optimal policy in state s chooses actions which maximise expected reward when an agent starts in state s. As a result the optimal state-action value could be used to get the optimal action $a^*(s)$:

$$a^*(s) = \arg\max_{a} Q^*(s,a)$$

2.2 Option Framework

Option framework is a model of HRL which were introduce in the work [11]. By option we describe a policy of higher level than policy which were introduced previously and basically is policy over policies.

We describe option as ω. It consist of several components:

- I_ω: set of initial states where option can start;
- $\pi_\omega(a|s)$: intra-option policy;
- $\beta_\omega(s)$: termination condition where an option stops.

Option ω is chosen in accordance with policy over options $\pi_\Omega(\omega|s)$ where Ω is set of all possible options.

One can show [11] that an MDP with a set of options becomes a Semi-Markov Decision Process, which has a corresponding optimal value function over options $V_\Omega(s)$ and option-value function $_\Omega(s,\omega)$.

We define state value function and action-state function similar to standard RL:

- $Q_\Omega(s,\omega)$ state-option value function:

$$Q_\Omega(s,\omega) = \sum_a \pi_\omega(a|s) Q_U(s,\omega,a)$$

- $Q_U(s, \omega, a)$ state-option-action value function:

$$Q_U(s, \omega, a) = r(s, a) + \gamma \sum_a P(s'|s, a) U(s', \omega)$$

- $U(s', \omega)$ utility term:

$$U(s', \omega) = (1 - \beta_\omega(s')) Q_\Omega(s', \omega) + \beta_\omega(s') V_\Omega(s')$$

- V_Ω state value function (max of $Q_\Omega(s', \omega)$ when ϵ-greedy policy is used in option selection)

$$V_\Omega(s') = \max_\omega Q_\Omega(s', \omega)$$

- $A_\Omega(s', \omega)$ advantage function

$$A_\Omega(s', \omega) = Q_\Omega(s', \omega) - V_\Omega(s')$$

3 Environment

Developed environment is a intersection of two-way roads. The goal of an agent to reach a red rectangular at the end of the road. During the movement the agent should avoid driving on walk side, wrong way or colliding with another cars. Environment includes the agent (a red car) and several bots (blue cars). The number of bots in the environment can be regulated by user and varied from 1 to 10.

Each car including agent can start at different position among the four section of the roads (see Fig. 2).

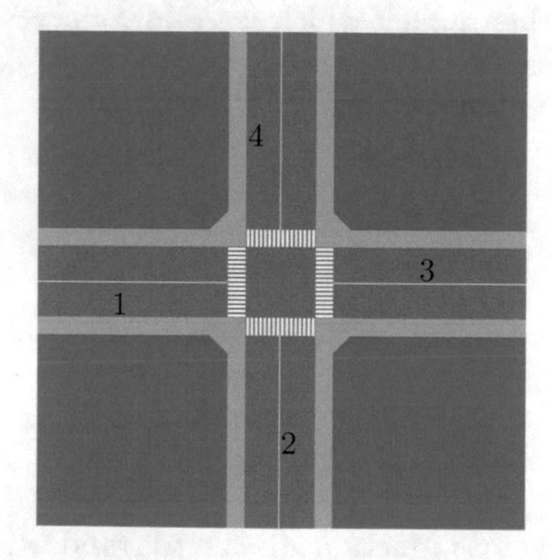

Fig. 2. Road intersection environment.

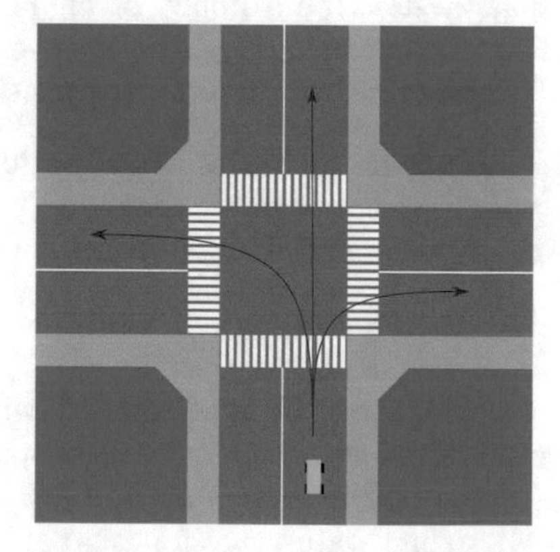

Fig. 3. Possible movement direction

Each bot has predefined trajectory. It moves along three possible paths. Each path assigned to a bot as it spawned. It can be chosen by a user or otherwise assigned randomly. For possible movements see Fig. 3.

Bot car moves in the way which allow it to avoid any collisions with another cars. It stops if another car is directly in front of it. To model smooth behaviour of bots on crossing section each of them follows designed pattern. Each of them has priority based on time a car-bot reach the crossing section. A car which drove to the section earlier has a priority. If car-bots' trajectories do not cross then each of them continue moving. Otherwise car with lower priority is not moving and wait until a car with higher priority leaves crossing area.

In Fig. 4 trajectories A and B intersect hence car which approach the cross section later (for this case that is B) wait while the car with trajectory A complete its crossing. In another case (Fig. 5) trajectories of cars do not cross so each of them continue their movement without stops.

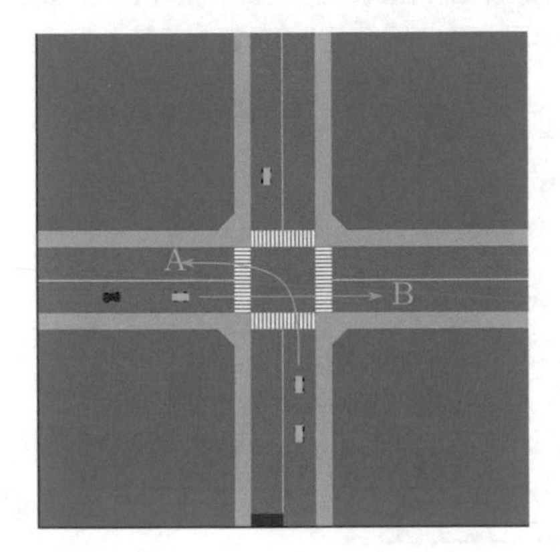

Fig. 4. Trajectories intersect.

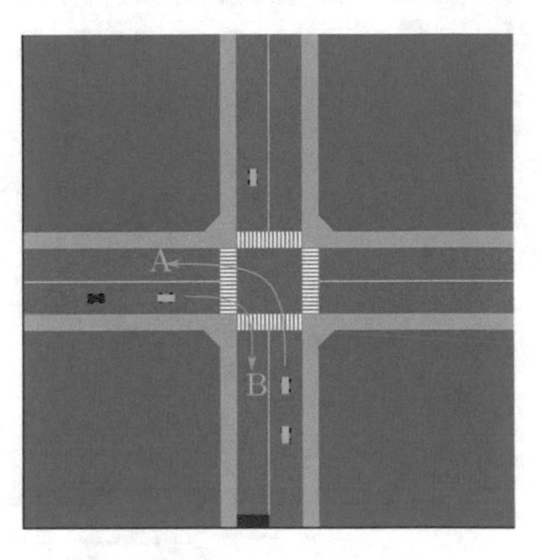

Fig. 5. Trajectories don't intersect

As soon as a car-bot reaches the end of its path a new bot-car generated at the beginning of random section with new trajectory. Thus there are always the same number of car-bots defined by user.

Information of the state of the environment is provided as a vector $s_t = (x, y, v_x, v_y, a_i, b_i, v_a^i, v_b^i, z_x, z_y)^T$, where:

- x, y: positions of the centre of the agent in the environment;
- v_x, v_y: x and y component of the agent's velocity vector;
- a_i, b_i: x and y positions of the centre of the ith bot car in the environment;
- v_a^i, v_b^i: x and y component of the ith bot car velocity vector;
- z_x, z_y: positions of the target.

To control the agent we pass a vector with three real numbers $a_t = (st_t, g_t, b_t)^T$, where:

- st_t: steer, controls direction of movements, $st_t \in [-1, 1]$ from left to right;
- g_t: gas, $g_t \in [0, 1]$ from no acceleration to full;
- b_t: break, $b_t \in [0, 1]$ where 0 is full stop.

In addition to continuous action space our environment supports a discrete action space of size 5:

- 0: correspond to absence of actions $a_t = (0,0,0)^T$ in continuous action space; movements are following rules of the world without actions from the agent;
- 1: left turn $a_t = (-1,0,0)^T$ in continuous action space;
- 2: right turn $a_t = (1,0,0)^T$ in continuous action space;
- 3: gas $a_t = (0,1,0)^T$ in continuous action space; agent is provided with short impulse to move;
- 1: break $a_t = (0,0,1)^T$ in continuous action space; all movements of agent is stopped entirely.

Described situation shows that developed environment can be used to model HRL. In Fig. 6 we show an example of possible options for a car.

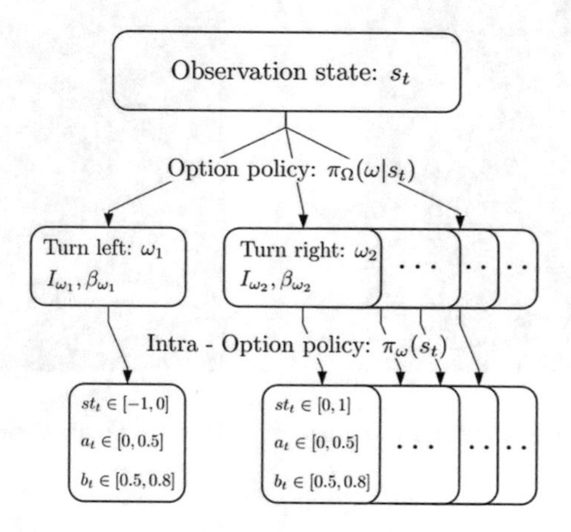

Fig. 6. Option framework.

To train agent we use a set of rewards. Total reward from one episode $r_{episode}$ is a sum of the following sub rewards:

- $r_{time} = -0.01$: each tick in environment agent receives negative reward to allow the agent to seek the shortest path;
- $r_{walk_side} = -1$: every time agent drive on walk side it receives negative reward and environment restarts.
- $r_{wrong_way} = -0.1$: every time agent moves in the wrong direction it receives negative reward.
- $r_{collide} = -1$: if agent collide with another car it receives negative reward and environment restarts.
- $r_{win} = 10$: if agent reaches destination (red rectangular) it receives positive reward and environment restarts.

This combinations of rewards allows agent to explore environment before it stuck in local optimum. The full code and instruction to set up the environment are available in github[1].

4 Baseline Models

4.1 Advantage Actor Critic

RL algorithms are usually divided into two groups: based on policy optimisation or state value evaluation.

Advantage Actor Critic (A2C) [12] belongs to the former group and basically an extension over Monte-Carlo REINFORCE algorithm [13]. The key idea of policy based algorithms is to increase probability to choose actions which lead to higher rewards and decrease probability of actions which lead to lower rewards until optimal policy achieved.

Let π_θ be a policy with parameters θ and $J(\pi_\theta)$ expectation over finite rewards sum. Gradient $J(\pi_\theta)$ can be found as:

$$\nabla_\theta J(\pi_\theta) = E_{\tau \sim \pi_\theta} \sum_{t=0}^{T} \nabla_\theta \log \pi_\theta(a_t|s_t) R(s_t),$$

where τ is trajectory $\tau = (s_0, a_0, s_1, a_1, ...)$.

Rewards for Monte-Carlo REINFORCE are calculated for entire episode, i.e. only after reaching terminal state we update policy parameters via gradient ascent:

$$\theta_{k+1} = \theta_k + \alpha \nabla_\theta J(\pi_{\theta_k}).$$

But disadvantage of such approach is high variance and increase in time to replay episodes which cause a slow convergence.

To overcome mentioned problems A2C algorithm was developed. We could decrease the number of iterations over episode by calculating state value function. For that purpose we introduce another model called critic whose purpose to return a value for a given state.

The part of REINFORCE algorithm responsible for learning a policy is called an actor. It left unchanged. As critic gives access to state value function we use advantage value instead of rewards collected from entire episode:

$$A(s, a) = Q(s, a) - V(s)$$

We can calculate advantage using state value function (this expression is called TD error):

$$A(s_t, a_t) = r_t + \gamma V(s_{t+1}) - V(s_t)$$

[1] https://github.com/max1408/Car_Intersection.

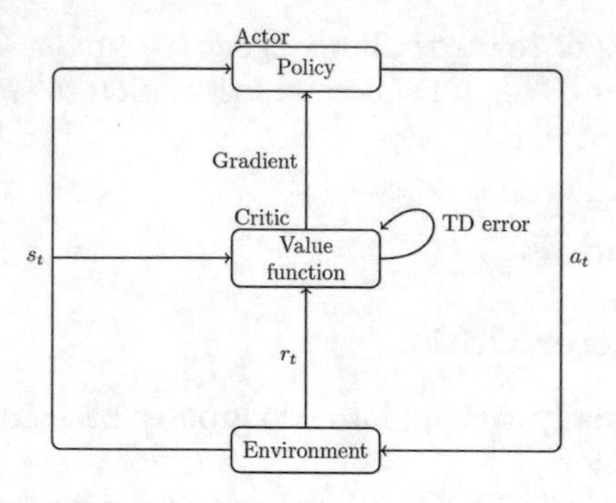

Fig. 7. Advantage Actor Critic.

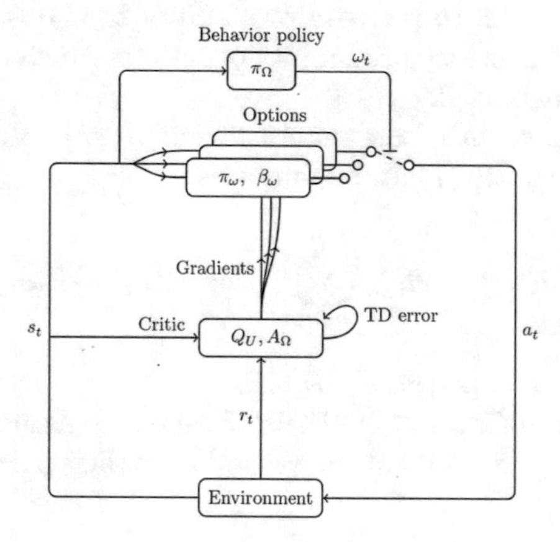

Fig. 8. Option-Critic.

Thus we can update actor's parameters with formula:

$$\nabla_\theta J(\pi_\theta) = E_{\tau\sim\pi_\theta}\sum_{t=0}^{T}\nabla_\theta \log \pi_\theta(a_t|s_t)A(s_t,a_t)$$

State value function improvement is done via:

$$L_{critic} = E_{\tau\sim\pi_\theta}\sum_{t=0}^{T}\left(r_t + \gamma V(s_t+1) - V_\theta(s_t)\right)^2$$

For the work of A2C see Fig. 7.

4.2 Proximal Policy Optimization

Proximal Policy Optimization (PPO) was developed to solve problem of updating policy. In case if update will be too big it can cause the problem in total performance. To mitigate this problem we use a clipped probability ratio. In compare with A2C we provide loss function to update actor parameters:

$$\nabla_\theta J(\pi_\theta) = E\left[\frac{\partial}{\partial\theta}\min(\rho_t(\theta)A(s_t,a_t), clip(\rho_t(\theta), 1-\epsilon, 1+\epsilon)A(s_t,a_t))\right]$$

We denote importance ratio as $\rho_t(\theta) = \frac{\pi(a_t|s_t)}{\pi_{old}(a_t|s_t)}$. In addition we can use mini-batches over samples to speed up our training.

4.3 Option-Critic

The crucial question of option policies is how to choose options over an environment. In work [11] authors used predefined conditions for terminations $\beta_\omega(s')$.

The next step in that direction was done with introduction of Option-Critic model [14]. It allows for an agent to learn options by itself. Though the architecture is still require to specify a number of options. The main contributions of the model is ability to learn options without any external data. It performs on the level with current non option RL and also can learn over continuous action space.

The name of the model is derived from its similarity with A2C of Sect. 4.1. Each options with policy $\pi_\omega(a|s)$ and $\beta_\omega(s')$ viewed as an actor. Critic is responsible for gradient update and consist of $Q_U(s,\omega,a)$ and $A_{\pi_\Omega}(s',\omega)$. For schematic depiction see Fig. 8.

The policy over options $\pi_\Omega(\omega|s)$ trained via Q-learning with ϵ-greedy exploration. This policy allows an agent to choose options. After an agent choose an option it follows an intra-option policy $\pi_\omega(a|s)$ until option reach terminal state. Agent switches between different options until it reaches the goal.

Gradients for intra-option policy and termination functions with parameters θ, ν calculate by formulas:

$$\frac{\partial Q_\Omega(s,\omega)}{\partial\theta} = E\left[\frac{\partial \pi_{\omega,\theta}(a|s)}{\partial\theta} Q_U(s,\omega,a)\right]$$

$$\frac{\partial U(s',\omega)}{\partial\nu} = E\left[-\frac{\partial \beta_{\omega,\nu}(s')}{\partial\nu} A_\Omega(s',\omega)\right]$$

Gradient for intra-option policy has the same interpretation as mention in Sect. 4.1, i.e. increasing probability of choosing action which lead to bigger reward. That is done for each option separately. Gradient for terminal function allow to prolong the duration of option which gives better reward.

5 Results

We train agent in the environment with discrete action space and with following simplified conditions. First we train agent to turn right starting at Sect. 1 see Fig. 4. Second we removed bot agents from the environment.

Table 1. Deep neural network architecture for PPO.

Network parts	Layer details
Agent network part	
Dense layer	Hidden size: 128 units
Dense layer	Hidden size: 128 units
Dense layer	Output size: 5 units
Critic network part	
Dense layer	Hidden size: 128 units
Dense layer	Hidden size: 128 units
Dense layer	Output size: 1 units

For PPO model we use a deep neural network to approximate the critic and represent the distributions for policies. As a framework we use Pytorch 1.0.

To train an agent we use the architecture for PPO described in Table 1.

For Option-Critic model we used a network as in Table 2 and number of options equal to $o = 3$.

Table 2. Deep neural network architecture for OC. Number of options $o = 3$.

Network parts	Layer details
Agent network part	
Dense layer	Hidden size: 128 units
Dense layer	Hidden size: 128 units
Dense layer	Hidden size: 128 units
Dense layer (actions)	Output size: $5 \cdot o$ units
Dense layer (terminations)	Output size: o units
Critic network part	
Dense layer	Hidden size: 128 units
Dense layer	Hidden size: 128 units
Dense layer	Output size: o units

Results of two models for the case with one turn presented in Fig. 9. Here we emphasise that it took OC agent twice less episodes to learn environment than PPO agent.

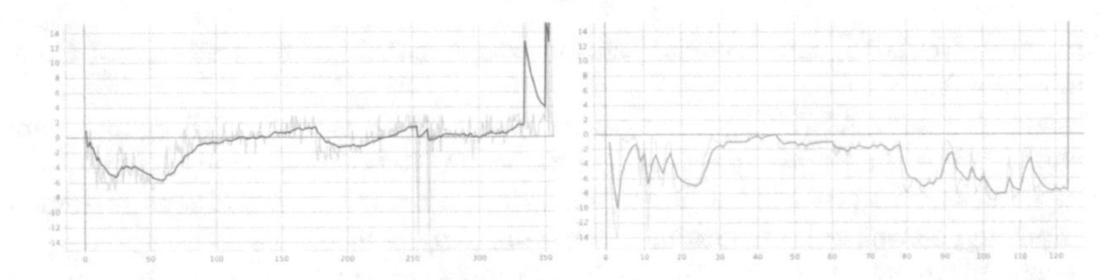

Fig. 9. Rewards of PPO agent (left) and OC agent (right) on car-intersect with only one turn.

6 Conclusion

We developed a new environment which simulate traffic on road intersections. Environment allows the user to control intensity of traffic by setting the number of cars. This environment follows OpenAI Gym framework and hence can be used by others in their research. We prepared two baseline models and tested their performance on the environment.

Due to complexity of the environment we can model different situation which are close to real life situations on road intersections. Moreover with development of Internet of Things we expect that environment can be used to improve performance of self-driving car in "smart" cities' roads intersections.

Behaviour of the agent depends on several factors, i.e. position of the agent on the road, its goal and its relative position to the bot cars. Each factor requires from the agent specific behaviour which in itself can be used as a goal. Based on this we consider that it is natural to use models from Hierarchical Reinforcement learning domain. That approach will allow the agent to split this complex environment into several easier sub-task and then by solving them separately to solve entire task.

Thus we suggest to use developed environment as a benchmark for HRL algorithms, as there is a lack of challenging environments for testing new ideas in this field.

Acknowledgments. This work was supported by the Russian Science Foundation (Project No. 18-71-00143).

References

1. Xu H, Gao Y, Yu F, Darrell T (2016) End-to-end learning of driving models from large-scale video datasets. CoRR, vol. abs/1612.01079
2. Shalev-Shwartz S, Shammah S, Shashua A (2016) Safe, multi-agent, reinforcement learning for autonomous driving. CoRR, vol. abs/1610.03295
3. Bojarski M, Testa DD, Dworakowski D, Firner B, Flepp B, Goyal P, Jackel LD, Monfort M, Muller U, Zhang J, Zhang X, Zhao J, Zieba K (2016) End to end learning for self-driving cars. CoRR, vol. abs/1604.07316
4. Barto AG, Mahadevan S (2003) Recent advances in hierarchical reinforcement learning. Discrete Event Dyn Syst 13:341–379

5. Al-Emran M (2015) Hierarchical reinforcement learning - a survey. Int J Comput Dig Syst 4:137–143
6. Ayunts E, Panov AI (2017) Task planning in "Block World" with deep reinforcement learning. In: Samsonovich AV, Klimov VV (eds) Biologically inspired cognitive architectures (BICA) for young scientists, advances in intelligent systems and computing, Springer International Publishing, pp 3–9
7. Kuzmin V, Panov AI (2018) Hierarchical reinforcement learning with options and united neural network approximation. In: Abraham A, Kovalev S, Tarassov V, Snasel V, Sukhanov A (eds) Proceedings of the third international scientific conference "Intelligent Information Technologies for Industry" (IITI'18), advances in intelligent systems and computing, Springer International Publishing, pp 453–462
8. Aitygulov E, Kiselev G, Panov AI (2018) Task and spatial planning by the cognitive agent with human-like knowledge representation. In Ronzhin A, Rigoll G, Meshcheryakov R (eds) Interactive collaborative robotics, lecture notes in artificial intelligence, Springer International Publishing, pp 1–12
9. Paxton C, Raman V, Hager GD, Kobilarov M (2017) Combining neural networks and tree search for task and motion planning in challenging environments. CoRR, vol. abs/1703.07887
10. Brockman G, Cheung V, Pettersson L, Schneider J, Schulman J, Tang J, Zaremba W (2016) OpenAI Gym
11. Sutton RS, Precup D, Singh S (1999) Between mdps and semi-MDPs: a framework for temporal abstraction in reinforcement learning. Artif Intell 112:181–211
12. Mnih V, Badia AP, Mirza M, Graves A, Lillicrap TP, Harley T, Silver D, Kavukcuoglu K (2016) Asynchronous methods for deep reinforcement learning. CoRR, vol. abs/1602.01783
13. Sutton RS, McAllester D, Singh S, Mansour Y (1999) Policy gradient methods for reinforcement learning with function approximation. In: Proceedings of the 12th international conference on neural information processing systems, NIPS'99, Cambridge, MA, USA, MIT Press, pp 1057–1063
14. Bacon P, Harb J, Precup D (2016) The option-critic architecture. CoRR, vol. abs/1609.05140

Representation of Conceptual Dependencies in the Domain Description Code

Igor O. Slieptsov[1], Larisa Y. Ismailova[2(✉)], and Sergey V. Kosikov[1]

[1] Institute "JurInfoR-MGU", Malaya Pirogovskaya Street, 5, Moscow 119435, Russia
[2] National Research Nuclear University "MEPhI" (Moscow Engineering Physics Institute), Kashirskoe shosse, 31, Moscow 115409, Russia
lyu.ismailova@gmail.com

Abstract. The paper considers the problem of representing conceptual dependencies in the domain description code with the support of their semantic integrity. It deals with the mechanisms for maintaining integrity in the course of computations in the conceptual model in two processes: (1) reduction with the provision of a given reduction strategy and (2) serialization of dependencies to provide for the transfer of dependence between various environments of computations. The paper compares the selected mechanisms and mechanisms of supporting systems of programming languages. The shortcomings of their computing systems arising from the lack of built-in mechanisms of sufficient power are analyzed. The paper considers the possibility of overcoming the existing problems. The approach to the extension of practical computing systems by the proposed possibilities (JavaScript language as an example) is investigated.

Keywords: Lazy evaluations · Serialization of functions · Programming languages · Computational model · Domain model · Applicative computing systems · Domain description code

1 Introduction

The most general representation of the content of an information system involves the domain display that includes the selection of objects and their relationships. The objects (links) within an information system, as a rule, correspond to real or imaginable objects (links) in the real world—the domain of the system. The establishment of such a correspondence, as a rule, is considered as making the sense of the information object within the system or setting its semantics. The common requirement in this situation is to save the semantic integrity of the content. The semantic integrity is the key property of modern technologies in the area of web-services development, service-oriented architecture and the Semantic Web, especially in the context of the problem of teaching students to use modern technologies and software in the semantic web area [1].

A. V. Samsonovich (Ed.): BICA 2019, AISC 948, pp. 507–514, 2020.
https://doi.org/10.1007/978-3-030-25719-4_65

The semantic integrity in its most general form can be understood as follows. First, the system should give one and the same meaning to the objects participating in the interaction of the system with the subjects (users), and the dependencies on the characteristics of the subject should be clearly identified. Secondly, the object may appear in various places of the system content, i.e. in different contexts. In this case, the dependence of the object's interpretation on the context should be distinctively expressed (for example, with the help of the free variable mechanism).

The task to determine the functional dependencies in the domain is a specification of the task to represent the relationships of objects of the displayed domain within an exact data model [2]. The developed data models contain an abstract language, by means of which it is possible to describe objects, as well as to identify classes of objects. Since the functional dependencies are part of the mapped domain, which possesses a dependency model, the model should also contain an abstract programming language to describe dependencies.

If a functional dependency exists, the object can be defined through other objects. In the domain description code such an object may be represented in different ways. In particular, if it is necessary to provide a dynamic computation of an object, it may be represented by an expression that contains representations of the objects on which the represented object depends.

In the case of functional dependence the value of the computed expression does not depend on the computing environment as a whole, but depends only on the values of the objects that represent the arguments of the functional dependence. These arguments are determined by the environment in which the computation is performed. In a general case of describing the functional dependency, the model description language must include means for describing arguments in the environment or be extended by such means. The general hierarchy of domain models is schematically depicted in Fig. 1.

2 Task of Describing the Functional Dependence

Taking into account the need to represent objects in the environment, the task is to represent the object of the applied data model as an expression setting the dependence on other objects. In order to increase the flexibility of the presentation of functional dependencies, the essential possibility is the one that establishes the links between the methods of data presentation and the characteristics obtained as a result of data analysis, their classification, etc. For this it is necessary that the domain specific language has appropriate expressive possibilities to represent functional dependencies by means of the language.

Under such representation of functional dependencies by the means of programming language of the computing environment (e.g., JavaScript) as an object the existing limitations should be taken into account:

- in languages with an applicative computation strategy (call-by-value), as a rule, there is no built-in mechanism for providing an alternative computation strategy;

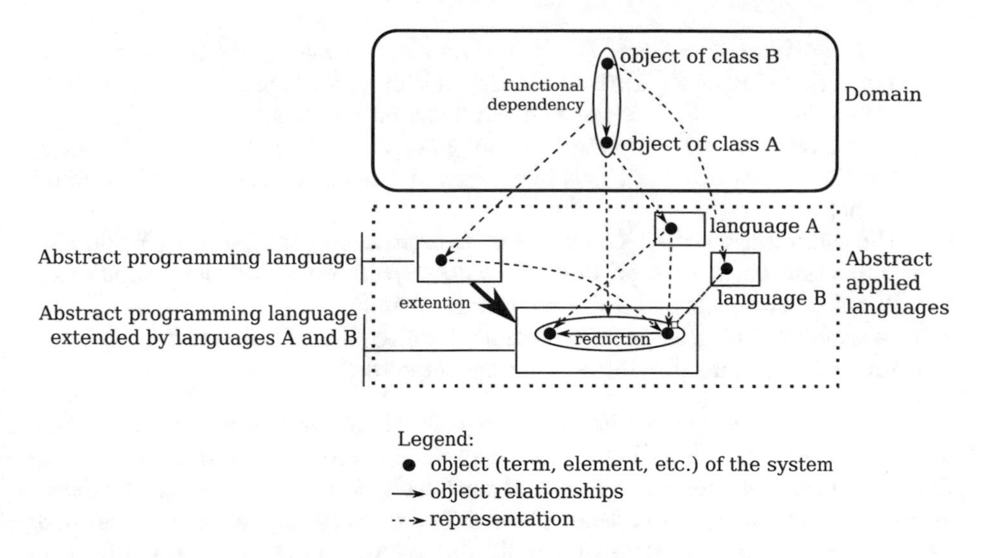

Fig. 1. The hierarchy of domain models, the process of representing objects and functional dependencies in the domain description code

- it is difficult to organize lazy evaluations;
- serialization of functions is costly or impossible.

In languages with an applicative computation strategy the computation of the expression $F(X)$ consists in principle of three sequential stages: computing the expression F with obtaining the object f, computing the expression X with obtaining the object x, and if f is a function, then calling $f(x)$.

If to consider only deterministic functions (pure functions) as objects of the computing environment that do not lead to side effects, then the language expressions can be considered as representations of objects (terms) of some abstract purely functional computing system, for example, λ-calculus (with or without types system) or combinatorial logic. Such abstract systems, as a rule, satisfy the Church-Rosser property: if the term M is reduced in P and in Q, then there is such a term N that both P and Q are reduced in N. Less formally this property states that replacing a subterm with an equivalent one does not change the semantics of the whole term; in particular, the value of $F(X)$ does not depend on the order of computation of F and X.

This means that in purely functional systems it is possible to choose between different reduction strategies, which lead to a common result, and can be optimal for a certain parameter: the number of steps, the volume of consumed memory, etc. However, in real supporting environments it is possible to use non-local effects (assignments to global variables and etc.), which prevents the functional implementation of the mechanisms for supporting the data model in the whole and the definition of alternative computational strategies in particular.

The applicative computing strategy possesses the following properties, which are usually considered as undesirable ones:

- for computing the value of $F(X)$ it is necessary to compute the value of X, even if the value of $F(X)$ does not depend on it, for example, if the body of the function-value F does not contain a formal argument;
- if it is impossible to complete the computation of X in the finite amount of time, the computation of $F(X)$ also cannot be completed in the finite amount of time;
- if the computation of $F(X)$ results in an expression containing $F(X)$ outside the function body, then an infinite recursion will occur and the computation will not be completed in a finite amount of time;
- it is impossible to determine potentially infinite [3] structures and apply operations for the corresponding finite structures to them.

As a consequence, it is difficult to organize lazy evaluations, usually interpreted as a way of work with a domain object, represented by a language expression, in which the expression is not always evaluated at the moment of determination. The computation takes place either by necessity, when to determine the properties of a dependent object it is necessary to determine some properties of the represented object, or due to the demand of a function or procedure operating the object. The lazy evaluations allow for optimizing the consumption of computing resources (time and memory) and monitoring the change in the environment during the computation (the context of the computation).

For example, in JavaScript the simplest way to organize lazy evaluations is to place an expression in the function body without any argument, then the command for computing the value is a call of function, and the function value is the value of the expression in the current environment. This approach has some disadvantages, among which there are (1) the lack of separation between the representation of functional dependence as an object of the domain and control over the order of computation, and (2) the impossibility of functions serialization.

The latter assumes the representation of the computing environment object as a sequence of bytes, line or word of another language that can be considered separately from the computing environment. The serialization is used to deliver a domain object between computational environments, i.e., this means that basing on a representation in one computing environment, the representation of the same object will be created in another environment. In particular, the serialization is necessary to keep the information about domain objects between work sessions of a computing unit (computer), separated by a restart of the computing environment, or for transferring the computations from one node to another when the distribution of one computing environment between nodes is impossible.

For example, the JavaScript function is an object of the computing environment and has the function body and the environment object (Scope), which refers to the environment object created while calling the function, in the body of which the considered function was created (declared). If the function g was declared in the body of the function f, then while calling f an environment object is created, and during the time of performance a function g is created, the Scope property of which refers to this environment object. The environment objects form a hierarchy, which is used to designate variables. The following two circumstances make the task of functions serialization in JavaScript impossible:

- EcmaScript specification uses the environment object to set the semantics of variable designation and permits that the interpreter may not create the object clearly or create it in another form;
- if an environment object is obtained clearly, there is no possibility to serialize it correctly taking into account all references: the root of the environment objects hierarchy is a global object from which all objects of the computing environment are reachable.

3 Representation of Functional Dependencies by Means of an Applied Language

One of the ways to solve the problem is to use an applied programming language, the expression of which is represented by the object of the computing environment, and the computation is implemented by a function of the computing environment. An applied programming language may have abstract syntax and semantics. The representation of each language element is a computational environment object constructed according to the rules set by the representation (with the help special classes, functions, or JSON objects).

The computation of the functional objects represented in such a way is performed in the process of reduction, which is understood as a sequence of steps for transforming the initial expression according to the rules set by the semantics of the computations. The choice of expression for the transformation is performed by setting the reduction strategy (for example, the outermost or the leftmost inner, etc.) The serialization of the functional object means to create the word of an abstract language according to its representation in the computing environment.

Getting the necessary properties (serializability, alternative computing strategy, etc.) is achieved by storing the function as a non-functional object of the computing environment and limiting the possible external (parametric) dependencies of the function on the objects of the computing environment. In particular, when integrating expressions of an applied programming language with a computer system and, in particular, to form a computation context, it is necessary to develop a mechanism to use environment objects as parameters (dependencies) of an expression that preserves the required properties: serializability, control of the computation strategy, control of the reduction rules used.

Some existing languages have built-in support for the considered properties, a comparative analysis of which is given in Table 1. The following notation system is used:

- The language has support of the lazy evaluations, if the minimal number of additional syntax elements, necessary to change the definition of an expression as suspended one, does not depend on the number of places where the value of the expression is used. So, in any language that supports anonymous

Table 1. Comparison of programming languages by the presence of built-in mechanisms for lazy evaluations, the choice of reduction strategy, serialization and syntactic parsing

Programming language	Lazy evaluation	Mechanism of selection of computation strategy	Serialization of functions	Syntactic parsing of expressions
C/C++	no	no	no	no
C#	`Expression` [4]	on the basis of lazy evaluations	only `Expression`	only `Expression`
JavaScript [5]	no	no	only the body of function; serialization of computing environment objects is possible	without saving the context
Bash[a] [6]	yes	on the basis of lazy evaluations	only the body of function; serialization of computing environment objects is possible	without saving the context
Haskell [7]	Default lazy evaluations, seq and `$!` for applicative computations	on the basis of lazy evaluations	no	no
Scheme [8]	no	no	no	no
Wolfram Language [9]	`SetDelayed`, `RuleDelayed`	Up/down values, `HoldAll` (`HoldFirst`, `HoldRest`) attributes, `Evaluate`, on the basis of lazy evaluations	Conditional	yes

[a]functions are represented by lines with dynamic binding

functions, it is possible to simulate lazy evaluations by replacing the definition of an expression `x = <expr>` for `x = () => <expr>` (in the JavaScript syntax), and each occurrence of `x` for `x()`. Because of the latter, this method does not consider the built-in support for lazy evaluations. Similarly, `delay` / `force` in Scheme requires force when using every lazy evaluation.

– The language provides the choice of a reduction strategy if the number of syntax elements, which need to be changed in order to change the reduction strategy, does not depend on the size of the function body and the number of places where the function is used.

- The serialization of functions is possible in a language if the function body and environment (context) may be saved. Wolfram Language uses the substitution of the actual argument instead of the formal one, thus, without creating an environment object, it allows to analyze the function body and extract the values (up, down & own values) of all free variables under certain general assumptions about the function body and the values of variables (that they do not contain metaprogramming functions `Names`, `Symbol`, `ToExpression` etc.).
- The language allows syntax parsing if for a variable or expression it is possible to construct an abstract syntax tree of the value of a variable or expression by means of a language, respectively. The table indicates "without saving the context", if there is no possibility for each variable entering freely into the body, but related to the environment (context) of the function, to correlate its value.

4 Conclusion

The paper considers the problem of representing conceptual dependencies in the domain description code with the support to their semantic integrity. Some certain types of processing are singled out that require the support of semantic integrity, which include:

- support to computations with saving the Church-Rosser property, which ensures the immutability of the semantics of a term when replacing its subterm with an equivalent one;
- serialization as an essential task arising during the transfer of objects of the computing environment between the nodes of the computing system and between the sessions of computations.

The paper analyzes the built-in mechanisms for lazy evaluations, the choice of alternative computational strategies, serialization and parsing of expressions in the most popular programming languages. Basing on this it proposes an approach to the representation of functional dependencies in the domain. The elements of the proposed approach were implemented in JavaScript language and tested while solving a number of modeling problems.

Acknowledgements. The paper is supported by the grants 19-07-00326-a, 19-07-00420-a, 18-07-01082-a, and 17-07-00893-a of the Russian Foundation for Basic Research.

References

1. Klimov VV, Chernyshov AA, Balandina AI, Kostkina AD (2017) Problems of teaching students to use the featured technologies in the area of semantic web. AIP Conf Proc 1797:030008

2. Ismailova LY, Wolfengagen VE, Kosikov SV (2018) Basic constructions of the computational model of support for access operations to the semantic network. Procedia Comput Sci 123:183–188
3. Sleptsov IO, Kosikov SV (2016) Potentially infinite lists: tools for processing the ordered structures. Practical use. In: Proceedings of the international conference "Situation centers and class 4i information and analytical systems for monitoring and security problems" SCVRT2015-2016, Pushchino, TsarGrad, pp 123–128
4. Expression trees (C#). https://docs.microsoft.com/en-us/dotnet/csharp/programming-guide/concepts/expression-trees/index. Accessed 14 Nov 2018
5. Standard ECMA-262. https://www.ecma-international.org/publications/standards/Ecma-262.htm. Accessed 14 Nov 2018
6. Bash Reference Manual. http://www.gnu.org/software/bash/manual/bashref.html. Accessed 14 Nov 2018
7. Jones SP, Hughes J, Augustsson L, et al. (1999) Report on the programming language Haskell 98. A Non-strict, Purely Functional Language
8. Serializable Closures in PLT Scheme. https://blog.racket-lang.org/2009/06/serializable-closures-in-plt-scheme.html. Accessed 14 Nov 2018
9. Wolfram language for high-tech programming. https://www.wolfram.com/language/. Accessed 14 Nov 2018

Consciousness and Subconsciousness as a Means of AGI's and Narrow AI's Integration

Artem A. Sukhobokov[1,2(✉)], Yuriy E. Gapanyuk[2], and Valeriy M. Chernenkiy[2]

[1] SAP DBS CIS, Kosmodamyanskaya nab. 52/4, 115054 Moscow, Russian Federation

[2] Bauman Moscow State Technical University, ul. Baumanskaya 2-ya, 5, 105005 Moscow, Russian Federation

artem.sukhobokov@yandex.ru

Abstract. The present article concentrates on the cognitive architecture of the agent capable to form AGI. In the process, an attempt to bridge the existing gap between AGI and narrow AI methods is made. There are two main blocks to do this - Consciousness and Subconsciousness, each of which solves the following classes of tasks: predictive analytics, prescriptive analytics, executive analytics, reflexive analytics, goal analytics, abstraction analytics, and attention analytics. Consciousness uses conscious memory and inference mechanisms. Subconscious uses models and methods of narrow AI and unconscious memory. In order to transfer information from Subconsciousness to Consciousness, emotions, insights, intuitive decisions, algorithms or sequences of actions, expectations, feelings, desires or their hierarchies, abstractions, and attention zones are used. Both conscious and unconscious memory use the metagraph model of knowledge. There is a mechanism for knowledge transfer from unconscious memory to conscious. Acquiring knowledge from the information coming from external channels is performed in parallel to conscious and unconscious memory. Similarly, information output to external channels and performing actions are based on the knowledge of both conscious and unconscious memory. AGI learning occurs due to the expansion of the number of production rules in Consciousness, as well as increasing the number of models and improving the quality of meta-learning algorithms in Subconsciousness.

Keywords: Subconsciousness · Consciousness · Predictive analytics · Prescriptive analytics · Executive analytics · Reflexive analytics · Goal analytics · Abstraction analytics · Attention analytics · Meta-learning · Metagraph · Emotions · Insights · Intuitive decisions · Algorithms · Expectations · Feelings · Desires · Abstractions · Attention zones

The limited capabilities of existing systems using narrow AI methods lead to the intensification of works on AGI's creation [1]. Within the framework of researches on cognitive architectures, there are a number of works on the development of cognitive architectures, whose goal is to create AGI [2, 3]. However, in all such cases, a wide range of modern theoretical models developed within various directions of narrow AI's theory

A. V. Samsonovich (Ed.): BICA 2019, AISC 948, pp. 515–520, 2020.
https://doi.org/10.1007/978-3-030-25719-4_66

is not used, but only some of them are used and not always the latest [4]. There is a clear gap between AGI and narrow AI: the results achieved in the narrow AI are almost not transferred to works on AGI. An attempt to begin to bridge this gap is made in this article.

As an initial background for the development of the proposed AGI architecture, the approach proposed in [5] is taken, the mean of which is to use two relatively independent but interconnected components – Consciousness and Subconsciousness. In this case, the agent's Consciousness is determined according to [6] and estimated according to the scale proposed there. While the agent is learning, Consciousness will be developing and sequentially moving from one level to another, reaching level 10 (adult level) in the limit. The proposed architecture does not assume the achievement of the 11th level (Superconsciousness – the ability to model several streams of consciousness). Both Consciousness and the Subconsciousness in the proposed architecture has the ability to solve a wide range of tasks; this follows from the representation of the architecture in Fig. 1.

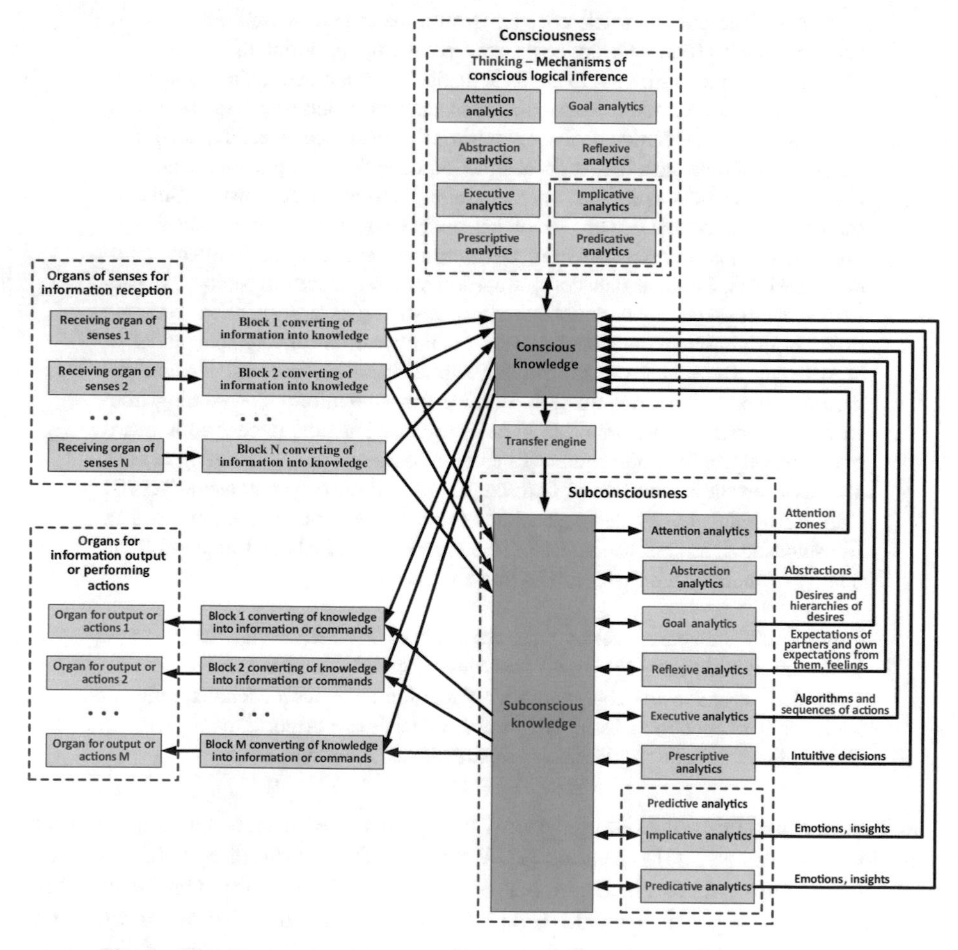

Fig. 1. The proposed architecture of agent capable to form the consciousness.

To determine the list of AGI's functional components, a scheme of the evolutionary development of analytics tools from [7] was taken as a basis and supplemented. The highest stage of this evolutionary scheme is AGI. The functional components of Consciousness and Subconsciousness include:

- Predictive analytics tools that solve Machine Learning problems contain implicative analytics tools and predicative analytic tools;
- Prescriptive analytics tools, offering the best solutions for a wide range of problems;
- Executive analytics tools allowing to choose the next step from several possible variants and to ensure an effective sequence of actions;
- Reflexive analytics tools that allow taking into account the expectations of other process participants from the agent, the expectations of the agent from them, the evaluation by the agent of the process participants and feelings towards them;
- Goal analytics tools ensuring selection and ranking of goals;
- Abstraction analytics tools, allowing to form and use abstract concepts;
- Attention analytics tools, allowing switching attention to different events or fields of the external environment, in which something important is happening.

Consciousness uses conscious memory and implements all the mentioned mechanisms using inference mechanisms (such as, for example, are proposed in [8]) or probabilistic inference mechanisms described in [9]. The learning of consciousness leads to the emergence of new production rules. In contrast, Subconsciousness uses unconscious memory and various models of narrow AI for decision making. The choice of models is made by meta-learning mechanisms. The learning of subconsciousness leads to the emergence of new models in it and the improvement of the used meta-learning mechanisms.

It is wise to use metagraphs as a data model in both conscious and unconscious memory [10]. Metagraph is a kind of complex graph (network) model with emergence. The key element of the metagraph model is the metavertex, which is in addition to the attributes includes a fragment of the metagraph. The presence of private attributes and connections for a metavertex is the distinguishing feature of a metagraph. It makes the definition of metagraph holonic – a metavertex may include a number of lower-level elements and in turn, may be included in a number of higher-level elements. From the general system theory point of view, a metavertex is a special case of the manifestation of the emergence principle, which means that a metavertex with its private attributes and connections becomes a whole that cannot be separated into its component parts.

The vertices, edges, and metavertices of metagraph are used for data description, and the metaedges are used for process description. The metaedge allows binding the stages of nested metagraph fragment development to the steps of the process described with metaedge. Such an approach allows representing objects, processes, and complex systems in the very natural hierarchical and holonical way.

It is assumed that the models and methods used in Subconsciousness should provide a solution to a wide range of problems:

- Predictive analytics models and methods should allow solving not only classical Machine Learning problems, but also non-standard problems [11], as well as Fairness problems [12] and graph analytics problems. The found solutions are

recorded in the conscious memory in the form of insights or emotions expressing the attitude to objects, events, causes, and etc.;
- Models and methods of prescriptive analytics should provide the possibility of using various optimization methods and simulation models to find the best solutions to a wide range of problems. Prescriptive analytics models should help to make the selection, ranking, assignment, scheduling [7]. The found solutions are fixed in conscious memory in the form of intuitive solutions;
- Executive analytics models should be used for actions planning, using such classes of methods as Reinforcement Learning, Heuristic Search [13], Imitation Learning [14], Lifelong Learning [15], Systemic Learning [16]. The found solutions are fixed in conscious memory in the form of sequences of actions or algorithms;
- The initial ideas, models, and methods of reflexive analytics are described in [17]. The found solutions are fixed in conscious memory in the form of expectations of other interaction's participants from the agent, agent's expectations from them, agent's estimation of other interaction's participants and feelings towards them;
- The initial ideas, models, and methods of goal analytics are described in [18] and [19]. The found solutions are fixed in conscious memory in the form of desires and hierarchies of desires;
- The initial ideas, models, and methods of abstraction analytics are described in [20]. The found solutions are fixed in conscious memory in the form of abstract concepts;
- The initial ideas, models, and methods of attention analytics are presented in [21]. The found solutions are fixed in conscious memory in the form of attention zones.

There should be several channels for acquiring knowledge (by analogy with a person who has five sense organs complementing each other). These can be traditionally expected videos, sound, texts through communication channels, as well as various sensors, such as fire safety sensors, motion sensors, etc. The data of each category should be transformed into knowledge using a special unit.

Nowadays many IT vendors have similar services (for example, Microsoft [22]), but they cannot solve the problem completely. For the present time, Semantic Web is poorly adopted as a common industry standard, and Ontology Learning [23] is considered as a promising approach. Also, the generally recognized way of knowledge presenting in the form of Knowledge Graph [24] has not been formed yet (there are representations based on various extensions of RDF [25], vector representations [26], If-Then representation [27], and others). The more so, there is no industry standard on Knowledge Metagraph; even experimental developments are not available yet.

At the same time, the use of services for converting data into knowledge as part of AGI imposes additional requirements on them. In [28] this is shown on the example of computer vision.

A similar situation occurs with services that solve inverse problems – data generation or performing actions based on knowledge. They also work with two knowledge bases and can receive knowledges from both conscious and unconscious memory. So far, there are only some research prototypes of such services, for example, [29].

In conclusion, it should be noted that the architecture presented in Fig. 1 is not final, but serves as a starting point for further research. In particular, the detailed structure (hierarchy) of memory types differing in knowledge storage duration, such as in [3], has not been refined yet. It is not clear whether the metagraph model of memory will be able to provide an integrated storage environment for knowledge that differs in purpose and in the form of knowledge representation.

Nevertheless, while further detailing of this architecture and as a result of subsequent development, an agent capable to form consciousness in himself can be developed. This result depends on the duration and complexity of the learning process.

References

1. Volk, T.: Limits of AI today push general-purpose tools to the horizon. TechTarget. https://searchenterpriseai.techtarget.com/feature/Limits-of-AI-today-push-general-purpose-tools-to-the-horizon. Accessed 29 Apr 2019
2. Kotseruba J, Tsotsos JK (2018) A review of 40 years in cognitive architecture research: core cognitive abilities and practical applications. arXiv:1610.08602v3 [cs.AI], 13 Jan 2018
3. Rapoport GN, Hertz AG (2017) Biological and artificial intelligence. In: Part 2: models of consciousness. Can a robot love, suffer and have other emotions? (in Russian: Рапопорт Г. Н., Герц А.Г. Биологический и искусственный разум. Ч.2: Модели сознания. Может ли робот любить, страдать и иметь другие эмоции? — М.: Книжный дом «ЛИБРОКОМ», 2017, 296 с.)
4. Potapov A, Rodionov S: Writing a book on artificial general intelligence (in Russian). Project on researchgate.net. Chap 2. Basic models, https://www.researchgate.net/profile/Alexey_Potapov4/project/Writing-a-book-on-artificial-general-intelligence-in-Rusian/attachment/5a6ae2f14cde266d58867630/AS:586963254009857@1516954353761/download/chapter2_refs.pdf?context=ProjectUpdatesLog. Accessed 29 Apr 2019
5. Chernenkiy V, Gapanyuk Y, Terekhov V, Revunkov G, Kaganov Y (2018) The hybrid intelligent information system approach as the basis for cognitive architecture. Procedia Comput Sci 145:143–152. https://doi.org/10.1016/j.procs.2018.11.022
6. Arrabales R, Ledezma A, Sanchis A (2008) Criteria for consciousness in artificial intelligent agents. Carlos III University of Madrid, 8 p. https://e-archivo.uc3m.es/handle/10016/10460#preview. Accessed 29 Apr 2019
7. Sukhobokov AA (2018) Business analytics and AGI in corporate management systems. Procedia Comput Sci 145:533–544. https://doi.org/10.1016/j.procs.2018.11.118
8. Varlamov OO (2018) Wi!Mi expert system shell as the novel tool for building knowledge-based systems with linear computational complexity. IREACO 11(6):314–325. https://doi.org/10.15866/ireaco.v11i6.15855
9. Tarassov VB (2017) Development of fuzzy logics: from universal logic tools to natural pragmatics and non-standard scales. Procedia Comput Sci 120:908–915. https://doi.org/10.1016/j.procs.2017.11.325
10. Chernenkiy V, Gapanyuk Y, Revunkov G, Kaganov Y, Fedorenko Y (2019) Metagraph approach as a data model for cognitive architecture. In: Samsonovich AV (ed) Biologically inspired cognitive architectures 2018. Proceedings of the ninth annual meeting of the BICA society, AISC, vol 848. Springer, Heidelberg, pp 50–55. https://doi.org/10.1007/978-3-319-99316-4_7
11. Charte D, Charte F, Garcia S, Herrera F (2018) A snapshot on nonstandard supervised learning problems: Taxonomy, relationships and methods. arXiv:1811.12044v1 [cs.LG], 29 Nov 2018
12. Barocas S, Hardt M, Narayanan A. Fairness and machine learning: limitations and opportunities. https://fairmlbook.org/. Accessed 29 Apr 2019

13. Edelkamp S, Schroedl S (2012) Heuristic search: theory and applications. Elsevier Inc., Waltham
14. Attia A, Dayan S (2018) Global overview of imitation learning. arXiv:1801.06503v1 [stat.ML], 19 Jan 2018
15. Parisi GI, Kemker R, Part JL, Kanan C, Wermter S (2019) Continual lifelong learning with neural networks: a review. arXiv:1802.07569v1 [cs.LG], 21 Feb 2018
16. Kulkarni P (2012) Reinforcement and systemic machine learning for decision making. Wiley, Hoboken
17. Rabinowitz NC, Perbet F, Song HF, Zhang C, Eslami SMAli, Botvinick M (2018) Machine theory of mind. arXiv:1802.07740v1 [cs.AI], 21 Feb 2018
18. Cox MT (2017) A model of planning, action, and interpretation with goal reasoning. Adv Cogn Syst 5:57–76
19. Kondrakunta S (2017) Implementation and evaluation of goal selection in a cognitive architecture. Wright State University, 78 p. https://etd.ohiolink.edu/!etd.send_file?accession=wright1503319861179462&disposition=inline. Accessed 29 Apr 2019
20. Deng F, Ren J, Chen F (2011) Abstraction learning. arXiv:1809.03956v1 [cs.AI], 11 Sept 2018
21. Li Y, Kaiser L, Bengio S, Si S (2018) Area attention. arXiv:1810.10126v1 [cs.LG], 23 Oct 2018
22. Cognitive Services. https://azure.microsoft.com/en-us/services/cognitive-services/. Accessed 29 Apr 2019
23. Asim MN, Wasim M, Khan MUG, Mahmood W, Abbas HM (2018) A survey of ontology learning techniques and applications. Database, pp 1–24, https://doi.org/10.1093/database/bay101. Review
24. Ehrlinger L, Wöß W (2016) Towards a definition of knowledge graphs. In: Joint proceedings of the posters and demos track of 12th international conference on semantic systems – SEMANTiCS 2016 and 1st international workshop on semantic change & evolving semantics, SuCCESS 2016. Leipzig, Germany, vol 1695. http://ceur-ws.org/Vol-1695/paper4.pdf. Accessed 29 Apr 2019
25. Antoniou G, Franconi E, van Harmelen F (2005) Introduction to semantic web ontology languages. In: Eisinger N, Małuszyński J (eds) Reasoning web. First international summer school 2005, Msida, Malta, 25–29 July 2005, Revised Lectures, pp 1–21. https://doi.org/10.1007/11526988_1
26. Mittal S, Joshi A, Finin T (2017) Thinking fast, thinking slow! Combining knowledge graphs and vector spaces. arXiv:1708.03310v2 [cs.AI] 21 Aug 2017
27. Sap M, LeBras R, Allaway E, Bhagavatula C, Lourie N, Rashkin H, Roof B, Smith N, Choi Y (2019) ATOMIC: an atlas of machine commonsense for if-then reasoning. arXiv:1811.00146v3 [cs.CL], 7 Feb 2019
28. Potapov A, Rodionov S, Peterson M, Scherbakov O, Zhdanov I, Skorobogatko N (2018) Vision system for AGI: problems and directions. arXiv:1807.03887 [cs.CV], 10 July 2018
29. Cojocaru DA, Trăuşan-Matu S (2015) Text generation starting from an ontology. In: Proceedings of the romanian national human-computer interaction conference - RoCHI (2015), pp 55–59, http://rochi.utcluj.ro/articole/3/RoCHI-2015-Cojocaru.pdf. Accessed 29 Apr 2019

Methods of Determining Errors in Open-Ended Text Questions

Oleg Sychev, Anton Anikin(✉), and Artem Prokudin

Volgograd State Technical University, Volgograd, Russia
oasychev@gmail.com, anton@anikin.name

Abstract. Open-ended text questions allow for better assessment students' knowledge, but analyzing the answer, determining it's correctness and, especially, providing detailed and meaningful feedback about errors to a student are more difficult tasks than for closed-ended and numerical questions.

The analysis of the answer in the form of freely written text can be performed on three different levels. Character-level analysis seeks error in characters' placement inside a word or a token; it's mostly used for detecting and correcting typos, allowing to discern typos from actual errors. Word-level (or token-level) analysis concerns word placement in a sentence or phrase, allowing to find missing, extraneous and misplaced words. Analysis on the semantic level tries to formally capture the meaning of the answer and compare it with the meaning of the correct answer provided by question creator in natural-language or formal form. Some tools analyze answers on several levels.

Long open-ended text questions typically allow variability of correct answers, which complicates the error search. Different ways of specifying patterns for variability and the possibility of introducing error-determining methods to questions with patterned answers are discussed.

Keywords: e-learning · Automatic error recognition · Regular expressions · Editing distances · Computational linguistics

1 Introduction

One way to make the education process more powerful is to quiz learners. This engages them and helps them assess whether they're actually learning. The major advantage of using quizzes is that answers can be graded automatically. But to be fully effective without a teacher, quizzes should not just tell the learners whether or not their answers are correct, but also provide explanations of what was wrong and how the answer can be corrected.

This paper presents the results of research carried out under the RFBR grant 18-07-00032 "Intelligent support of decision making of knowledge management for learning and scientific research based on the collaborative creation and reuse of the domain information space and ontology knowledge representation model".

A. V. Samsonovich (Ed.): BICA 2019, AISC 948, pp. 521–526, 2020.
https://doi.org/10.1007/978-3-030-25719-4_67

The two main categories of quiz questions are open-ended and closed-ended. A closed-ended question is a question you can answer by choosing one of the options indicated in the question. They are often easier and take less time to answer, and the answers are easier to analyze: the teacher, creating questions, can provide detailed explanations for each wrong choice. Such questions are focused primarily on checking factual knowledge, contain a limited range of possible correct answers, guide the learner's thoughts, and provide learners with basic knowledge to answer other questions. Closed-ended questions can be solved by guessing a choice, and they may provoke guessing instead of solving the task.

Open-ended questions require a text answer in a natural or formal language that the learner will enter. Answers for such questions can't be guessed from its text, so they force learners to think about the answers and actually do the tasks. However, free text answers are harder to analyze: the teacher can't explain every possible mistake. To be efficient without a teacher, open-ended questions must contain built-in algorithms for analysis of the answers, detecting errors and possibly providing information on how to fix them.

To detect errors in open text answers it is useful to distinguish several levels of text analysis [1]. Open-ended questions analyzing learners' errors must take into account possible typos - i.e. character-level errors that are not significant for most questions but need to be corrected before evaluating the answer.The next kind of mistakes can be found on word level: mistakes in the order of words or their sequence, or phrase structure. Errors also can be analyzed at the semantic level, building a formal semantic representation of the correct and learner's answers and comparing them.

Another problem which open-ended questions face is the variety of possible correct answers. This applies both to questions with answers in formal and natural languages. Various methods for specifying a template for possible correct answers were developed, including natural language processing techniques, using regular expressions [16] and special templates for natural-language answers [15].

2 Character-Level Analysis

Character-level analysis seeks error in characters inside a word or a token. This approach is mainly used to detect and correct typos, allowing you to distinguish typos from real errors. For the most part, the analysis consists of determining the edit distance - the minimum number of changes needed to transform one string into another string. A student's word that didn't match exactly any word in a correct answer is considered a typo if its editing distance to some correct word is below the threshold set in the question.

Different edit distance use different sets of string operations, which leads to reporting different kinds of mistakes.

The Levenshtein distance between two words is the minimum number of single-character edits: insertions, deletions, and substitutions [13]. This effective metric allow to determine most typos for reasonable time.

Damerau proposed an improvement to Levenshtein distance, allowing to add one more editing operation: transposition of two nearby characters. Damerau stated that these four operations correspond to more than 80 percents of all human misspellings [8]. This is one of the most used editing distances, both in regular spelling check and during grading open-text answers in Java Intelligent Tutoring System [17], OpenMark assessment system [5] and CorrectWriting Moodle plugin [15].

Character-level analysis is useful at catching typos and discerning them from actual errors, but determining the answer's correctness requires analysis on word level and semantic level.

3 Word-Level Analysis

The character-level analysis allows to catch only typos which are not actual errors, but distractors. For finding more complex errors, concerning used words and sentence structure, word-level analysis is necessary. This type of analysis, typically, works on token (word) sequences, allowing to find such errors as missing, extraneous and misplaced words.

OpenMark is an Open University computer-assisted assessment system with additional features in grading short natural-language text answers [5]. OpenMark can determine errors in word composition and placement. To determine word placement errors, the question author must specify that a pair of words should be present in the particular order with no more than N words between them. OpenMark doesn't output errors to the learner, it only uses the error count to determine whether an answer is correct. That doesn't allow to use the determined errors to guide the learner to correct them. An experimental study showed that this application performs on par with humans grading short answer questions but writing efficient questions requires specific learning of OpenMark template system [5].

CorrectWriting plugin for Moodle learning management system (LMS) uses combined character-level and word-level analysis [15]. The word-level analysis module uses the longest common sequence algorithm [2] to determine the longest sequence of words that exist both in correct and learner's answers. All words that absent from LCS are treated as errors: omissions (words, present only in the correct answer), insertions (words, present only in the learner's answer) and misplacements (words, present in both answers but not present in LCS). These errors are shown to the learner and can be used as hints to guide the learner to correct the answer without teacher's intervention. However, this method relies on an exactly defined correct answer and has troubles with highly variable answers.

4 Semantic-Level Analysis

There are three broad approaches to building open-ended questions automatic assessment systems.

Methods belonging to the first category use a training set of answers manually graded by a teacher. For example, C-Rater builds and compares predicate-argument structures of correct answers and a learner's answer [12]. However, this method uses stemming so it cannot be used to grade answers depending on verb tense; it also cannot handle idiomatic expressions. C-rater is aimed at grading short formative questions at the end of chapters. E-rater is an analyzer producing a score based on the sentence structure, organizations, and content of an answer [4]. It is used to grade longer, essay-type questions, and requires more than two hundred texts about the essay topic to train. It isn't able to grade short answers. It may also assign high grades to grammatically correct but meaningless texts if they contain the necessary keywords. Automated Text Marker (ATM) analyze answer to build conceptual dependency groups [6] containing a pair of concepts linked by a dependency, that are then compared to learners' answers.

The corpus-based approach uses a corpus of text on the topic to train the questions. It differs from the previous method because the corpus must not consists of correct answers, it can be any texts on topic containing correct answers. For example, AutoTutor system uses Latent Semantic Analysis (LSA) that is a high-dimensional statistical similarity measure of texts [9]. The text can be of any length - from a single word to a whole paragraph.

The third approach is based on the question-author supplied structured information about the correct answers. Link Grammar [7] requires the teacher to enter keywords and relational expressions before them. Learner's answer is parsed, supplied with Part-of-Speech tags and keyword relations are determined. The final grade depends on the number of keywords that have the same relational expressions as in the teacher's answer. For FreeText Author system, the teacher provides correct and incorrect answers as syntactic-semantic templates, specifying synonyms and links between keywords [10]. This allows grading learners' answers without penalties for errors in spelling, grammar, and punctuation.

5 Using Patterns to Specify Open Answers

The problem of variability of correct answers severely limits possible uses of open-ended questions. In many cases, this variability is regular enough to be defined in some form of template or pattern. Pattern matching allows for a wider range of questions but limits answer analyzing algorithms. Existing open-ended questions use different ways to specify patterns.

Some questions with text answer allow using wildcard characters. For example, Short Answer question type of Moodle LMS uses '*' character to allow any number of any characters in the answer [5]. This is simple to use, but very limited way to handle variability in correct answers.

OpenMark question engine allows specifying patterns using its own syntax. A question author can specify alternative words or word groups. This pattern matching system is more flexible, but learning OpenMark syntax for specifying pattern is time-consuming for teachers, preparing their questions [5]. OpenMark combines character-level typo detection with templating system for word-level error detection.

One of the most flexible ways to specify patterns for string matching is regular expressions. It allows creating patterns with alternative branches and variable-length repetitions [16]. Regular expressions are widely used; there are a lot of tutorials and many sites for creating and debugging regular expressions, which facilitates learning the pattern writing for question authors.

The two question types using regular expressions for pattern matching are "Regular expression short answer" (RegExp) [5] and "Perl-Compatible Regular Expressions" (Preg) [16]. They both use specially developed regular expression matching subsystems because they need to generate a string completing an unsuccessful match for hinting features.

RegExp questions use specific constrained regular expression syntax because of the implementation of the matching subsystem. It has limited functionality for detecting missing words, but only if they are used once in the answer. Preg question type supports the functionality of RegExp without limiting regular expression syntax. Using complex assertions, it is possible to look for missing and misplaced words if they are present only once in the answer.

6 Discussion

One of the common problems of question types using regular expressions is its lack of typo detection. That makes every typo treated like an error, which misleads learners. In order to improve determining errors and automatic feedback for a wider range of questions, it is necessary to combine the powerful pattern matching abilities of regular expressions with typo-detection abilities of editing distances.

To solve this problem approximate regular expression matching is needed [14]. Approximate pattern matching problem can be formally defined as finding a match between regular expression E and string w with cost k if $k \leq f(E, w)$ where f is minimum edit distance function between string and all matches of regular expression. Most works on approximate regular expression matching use Levenshtein editing distance [3,11,14] which leaves undetected transposition typos. Developing an approximate regular expression matching algorithm using Damerau-Levenshtein distance will allow improving typo detection for regular expression matching questions.

Adding typo detection feature to regular expression open-ended questions will allow combining in one questions character-level analysis, word-level analysis and powerful pattern matching abilities. The most stable and developed regular-expression matching engine for grading questions is Preg question type engine, which makes it the best choice for adding approximate matching feature.

References

1. Anikin A, Sychev O, Gurtovoy V (2019) Multi-level modeling of structural elements of natural language texts and its applications. In: Samsonovich AV (ed) Biologically inspired cognitive architectures 2018. Springer International Publishing, Cham, pp 1–8. https://doi.org/10.1007/978-3-319-99316-4_1
2. Apostolico A (1997) String editing and longest common subsequences. Handbook of formal languages. Springer, Berlin, pp 361–398. https://doi.org/10.1007/978-3-662-07675-0_8

3. Belazzougui D, Raffinot M (2011) Approximate regular expression matching with multi-strings. In: International symposium on string processing and information retrieval. Springer, pp 55–66. https://doi.org/10.1016/j.jda.2012.07.008
4. Burstein J, Leacock C, Swartz R (2001) Automated evaluation of essays and short answers. https://dspace.lboro.ac.uk/2134/1790
5. Butcher PG, Jordan SE (2010) A comparison of human and computer marking of short free-text student responses. Comput Educ 55(2):489–499. https://doi.org/10.1016/j.compedu.2010.02.012
6. Callear DH, Jerrams-Smith J, Soh V (2001) Caa of short non-mcq answers. https://doi.org/10.1.1.58.2210
7. Chakraborty UK, Gurung R, Roy S (2014) Semantic similarity based approach for automatic evaluation of free text answers using link grammar. In: 2014 IEEE sixth international conference on technology for education, pp 218–221. https://doi.org/10.1109/T4E.2014.57
8. Damerau FJ (1964) A technique for computer detection and correction of spelling errors. Commun ACM 7(3):171–176
9. Graesser AC, Chipman P, Haynes BC, Olney A (2005) Autotutor: an intelligent tutoring system with mixed-initiative dialogue. IEEE Trans Educ 48(4):612–618. https://doi.org/10.1109/te.2005.856149
10. Jordan S, Mitchell T (2009) e-assessment for learning? the potential of short-answer free-text questions with tailored feedback. Br J Educ Technol 40(2):371–385. https://doi.org/10.1111/j.1467-8535.2008.00928.x
11. Laurikari V, et al. (2001) Efficient submatch addressing for regular expressions. Helsinki University of Technology. https://laurikari.net/ville/regex-submatch.pdf
12. Leacock C, Chodorow M (2003) C-rater: automated scoring of short-answer questions. Comput Hum 37(4):389–405. https://doi.org/10.1023/A:1025779619903
13. Levenshtein VI (1966) Binary codes capable of correcting deletions, insertions, and reversals. Sov Phys dokl 10:707–710
14. Myers EW, Miller W (1989) Approximate matching of regular expressions. Bull Math Biol 51(1):5–37
15. Sychev OA, Mamontov DP (2018) Automatic error detection and hint generation in the teaching of formal languages syntax using correctwriting question type for moodle lms. In: 2018 3rd Russian-pacific conference on computer technology and applications (RPC), pp 1–4 . https://doi.org/10.1109/RPC.2018.8482125
16. Sychev O, Streltsov V (2015) Use of regular expressions as templates in formative and summative open answer questions. Otkrytoe Obrazovanie 2(109):38–45 https://elibrary.ru/item.asp?id=23241596
17. Sykes ER, Franek F (2003) A prototype for an intelligent tutoring system for students learning to program in java (tm). In: Proceedings of the IASTED international conference on computers and advanced technology in education, pp 78–83. Citeseer. https://doi.org/10.1109/ICALT.2003.1215208

Approach to Automatic Determining of Speakers of Direct Speech Fragments in Natural Language Texts

Oleg Sychev(✉), Yaroslav Kamennov, and Ekaterina Shurlaeva

Volgograd State Technical University, Volgograd, Russian Federation
oasychev@gmail.com
http://www.vstu.ru/

Abstract. Natural language text consists of an author's or narrator's text and direct speech fragments. They have different speakers so they could use different vocabularies and syntactic structures. In order to analyze the dependency of vocabulary and sentence structure on speaker, it is necessary to attribute each text fragment to its speaker. The results of such analysis can be used in natural language text generation tasks, allowing to convey different narrative voice depending on the purpose of the generated text. The authors developed a set of rules for attributing direct speech fragments to speaking characters, created a method of direct-speech scene analysis and implemented it in a software tool. In order to evaluate the accuracy of the attribution of direct speech fragments to speakers, an experiment was carried out. The results of the experiment show the viability of the developed method and allow to improve it for further use. The potential applications of the developed method and the software tool are discussed.

Keywords: Natural language processing · Direct speech · Treebanks · Computational linguistics

1 Introduction

A speaker for a text fragment is an existing or fictional person from whose perspective that text is written. While many scientific and technical texts have only one speaker, narrative texts, interviews, and question-answers texts [4] may have several speakers, which can be the author and the characters. There may be more speakers in scientific texts and media reports if they contain quotations from other texts.

The vocabulary and the sentence structure of text depend on its speaker. This should be taken into account solving the problems of text generation for different target audiences [13] and the problems of fictional text generation. It is also useful during the generation of any texts containing direct speech fragments.

The two ways to study the dependency between the speakers and their vocabulary and syntax structures are analyzing several texts of different authors about

A. V. Samsonovich (Ed.): BICA 2019, AISC 948, pp. 527–531, 2020.
https://doi.org/10.1007/978-3-030-25719-4_68

similar topics and analyzing texts containing quotations and direct speech. To facilitate such research, a text corpus where fragments of texts are marked with corresponding speakers is necessary. Such information can be either found in existing text corpora or obtained by automatic parsing tools.

2 State of the Art

2.1 Treebanks

Existing treebanks use either dependency or phrase structure to represent syntactical structures in texts [1]. Dependency treebanks don't contain information about phrase structure, so they didn't mark direct speech fragments directly. Universal dependencies format [3] uses parataxis relation to denote the connection between the narration and the direct speech fragments, but this relation is also used to describe of reported speech, news article bylines, interjected clauses, and tag questions. So it can not be reliably used to determine direct speech.

Among phrase structure treebanks, several parsed corpora such as Penn Treebank, Penn Parsed Corpora of Historical English, York-Toronto-Helsinki Parsed Corpus of Old English Prose (YCOE) [9], Treebank Semantics Parsed Corpus [2], SUSANNETS treebank, Christine Corpus and Lucy Corpus [14] contain direct speech fragments using SPE tags to denote it. But there is no information about the speaker preserved. LinGO Redwoods [12] tag quotation marks in the texts without analyzing whether they are used to mark direct speech fragments.

2.2 Parsing Software

The two most used English language parsers are StanfordCoreNLP and Berkeley Parser [8].

StanfordCoreNLP tool set contains named entity recognition tool [6]. It distinguishes different types of named entities, including such type as person, which can be used to identify possible speakers in texts containing direct speech or quotations. It also contains a coreference resolution tool to match pronouns with their named entities [10], but its reliability for determining characters is around 70% so the results have to be checked manually.

StanfordCoreNLP represents the structure of a sentence as a dependency tree. It provides a limited possibility of direct speech detection because the relations between direct speech fragment and its narrative phrase "clausal component" is used for denoting clauses of complex sentences [3]. So the parsing results alone cannot be reliably used to determine direct speech.

Berkeley Parser is based on phrase structure model [11] that uses tag suffix SPE to denote direct speech sentences [7]. But Berkeley parser doesn't put this suffix in parsing results, so it can't be used to determine direct speech fragments.

3 Method

Determining speakers of direct speech fragments in narrative texts can be done in two steps: determining the boundaries of direct speech fragments and associating the fragments with the relevant speakers.

3.1 Determining Direct Speech Fragments

In American English, double quotation marks are used to mark direct speech. When a fragment of direct speech is quoted inside another quotation, the internal quote should use single quotation marks. In British English, it is correct to use single quotation marks for direct speech and double quotation marks for enclosed direct speech.

Quotation marks not always mark direct speech. They are also used around fragments like quotation parts; foreign words; words for imitating sounds; proverbs; headlines of articles, short poems, and stories, essays, lectures, dissertations; chapters in books; titles of radio- or tv- programs and novels in collections.

The developed method uses several rules to determine direct speech fragments. A fragment of speech, enclosed in quotation marks, is considered a direct speech fragment if there is a comma or colon before this fragment, and a comma, dash or sentence-final punctuation mark is placed before the closing quotation mark. A fragment of speech, enclosed in quotation marks at the beginning of a sentence is considered a direct speech fragment if there is a comma, dash or any sentence-final punctuation mark before the closing quotation mark. These rules are used for enclosed direct speech too.

The method also respects monologues - several paragraphs of direct speech fragments from one speaker in a row, without narrative text between them. In monologues, each paragraph starts with a quotation mark but only the last paragraph contains a closing quotation mark. The developed method consider a monologue as a single direct speech fragment for further analysis as it has only one speaker.

3.2 Determining Speakers

In order to help to determine speakers, the developed method uses two kinds of specifiers for possible speakers. Objective specifiers are words and collocations that can be used to name a particular person. Names are unique objective specifier, but any characteristic that is used in the text to name a character is also an objective specifier. Objective specifiers are useful when used inside direct speech fragments as addressing and outside of direct speech fragments with verbs of speaking or thinking (i.e. "said", "answered", "thought" etc.) [5].

Subjective specifiers denote the relations between characters (i.e. "mother", "brother", "boss") and are dependent on the speaker that says them. They can help identify the speaker that says them in their direct speech.

The proposed method uses 8 rules in 4 steps. On the first step, the method identifies speakers using names positions concerning verbs of speech and thoughts.

It allows to uniquely identify speakers to some fragments. The second step uses names and subjective specifiers used for addressing inside direct speech and objective specifiers near the verbs of speech and thoughts to narrow down the list of possible speakers for each fragment.

If the analyzed scene is a dialog, the third step assumes that the speakers alternate their lines for still unidentified direct speech fragments. On the fourth step, the coreference resolution information for personal pronouns is used to determine speakers. Coreference resolution algorithms give approximate results, so they are only used as a last resort.

These steps are used for each direct speech fragment, including enclosed direct speech. In the end, each direct speech fragment is attributed with a list of possible speakers which is often narrowed down to one person.

Table 1. Experimental results

Experiment	Number of Speakers	Number of Sentences	Number of direct speech fragment	Recall (precisely identified speaker), %	Recall (narrowed down list), %
1	2	80	30	77	77
2	2	93	22	68	68
3	4	88	33	72	84
4	3	105	47	70	79
5	2	122	50	80	80
6	2	82	34	82	82
7	8	287	103	67	74

4 Results and Discussion

To determine recall and precision of developed method an experiment was conducted using 7 English language fictional literary texts written by 5 different authors and a program for determining speakers implementing the developed method. All direct speech fragments in these texts were identified correctly.

On this experimental set, the method didn't make any errors. Speakers were determined for most direct speech fragments, but for some fragments, the method could only narrow down the list of possible speakers without reducing it to a single answer. Recall of the developed method for determining speakers precisely and partially (narrowing down list) is shown in Table 1.

The experimental data show that the developed method has good precision and significant recall, but some uncertain cases must be resolved manually. Creating a text corpus with tagged speakers will allow researching the dependence of vocabulary and sentence structure on speaker's characteristics, which will be useful for generating texts for different target audiences.

References

1. Anikin A, Sychev O (2018) Semantic treebanks and their uses for multi-level modelling of natural-language texts. Procedia Comput Sci 145:64–71 https://www.sciencedirect.com/science/article/pii/S1877050918322968?via%3Dihub
2. Butler A (2015) Linguistic expressions and semantic processing. Springer, Cham
3. De Marneffe M, Dozat T, Silveira N, Haverinen K, Ginter F, Nivre J, Manning CD (2014) Universal stanford dependencies: a cross-linguistic typology. In: Proceedings of the 9th International Conference on Language Resources and Evaluation, LREC 2014, pp 4585–4592. https://nlp.stanford.edu/pubs/USD_LREC14_paper_camera_ready.pdf
4. Demaidi MN, Gaber MM, Filer N (2017) Evaluating the quality of the ontology-based auto-generated questions. Smart Learn Environ 4(1):7. https://doi.org/10.1186/s40561-017-0046-6
5. Faure R (2009) Verbs of speaking and verbs of thinking. https://hal.archives-ouvertes.fr
6. Finkel JR, Grenager T, Manning C (2005) Incorporating non-local information into information extraction systems by gibbs sampling. In: ACL-05 - 43rd annual meeting of the association for computational linguistics, proceedings of the conference, pp 363–370. https://dl.acm.org/citation.cfm?doid=1219840.1219885
7. Hall D, Durrett G, Klein D (2014) Less grammar, more features. In: 52nd annual meeting of the association for computational linguistics, ACL 2014 - proceedings of the conference. vol 1, pp 228–237. https://aclweb.org/anthology/P14-1022
8. Kolachina S, Kolachina P (2012) Parsing any domain english text to conll dependencies. In: Proceedings of the 8th international conference on language resources and evaluation, LREC 2012, pp 3873–3880. http://www.lrec-conf.org/proceedings/lrec2012/pdf/1097_Paper.pdf
9. Kulick S, Kroch A, Santorini B (2014) The penn parsed corpus of modern British English: first parsing results and analysis. In: 52nd annual meeting of the association for computational linguistics, ACL 2014 - proceedings of the conference, vol. 2, pp 662–667. https://aclweb.org/anthology/P14-2108
10. Lee H, Chang A, Peirsman Y, Chambers N, Surdeanu M, Jurafsky D (2013) Deterministic coreference resolution based on entity-centric, precision-ranked rules. Comput Linguist 39(4):885–916 https://www.mitpressjournals.org/doi/10.1162/COLIa00152
11. Petrov S, Barrett L, Thibaux R, Klein D (2006) Learning accurate, compact, and interpretable tree annotation. In: COLING/ACL 2006 - 21st international conference on computational linguistics and 44th annual meeting of the association for computational linguistics, proceedings of the conference, vol 1, pp 433–440. https://dl.acm.org/citation.cfm?doid=1220175.1220230
12. Redwoods L, Oepen S, Flickinger D, Toutanova K, Manning C (11 2002) Lingo redwoods: a rich and dynamic treebank for hpsg. Reseach on language and computation
13. Rishes E, Lukin SM, Elson DK, Walker MA (2013) Generating different story tellings from semantic representations of narrative. In: Lecture notes in computer science (including subseries Lecture notes in artificial intelligence and lecture notes in bioinformatics), vol. 8230 LNCS. Springer International Publishing Switzerland (2013). https://link.springer.com/chapter/10.1007%2F978-3-319-02756-2_24
14. Sampson G, Sampson C (1995) English for the computer: the SUSANNE corpus and analytic scheme. Clarendon Press, Oxford

Neuro-Correlates of the eBICA Model

Vadim L. Ushakov[1,2], Vyacheslav A. Orlov[2], Sergey I. Kartashov[1,2], Sergey V. Shigeev[3(✉)], and Alexei V. Samsonovich[1]

[1] National Research Nuclear University MEPhI (Moscow Engineering Physics Institute), Moscow, Russia
sikartashov@gmail.com, alexei.samsonovich@gmail.com

[2] National Research Center "Kurchatov Institute", Moscow, Russia
vl_ushakov@nrcki.ru, ptica89@bk.ru

[3] Bureau of Forensic Medical Expertise, Moscow Health Care Department, Moscow, Russia
tiuq@yandex.ru

Abstract. The article presents a new approach to building semantic maps based on the use of the eBICA model and the calculation of neurophysiological correlates of behavioral motives based on fMRI data. In this study, a social videogame paradigm was used in combination with fMRI and other recording tools. The obtained data show the complex neural network dynamics of behavioral motives distributed on the cortical and subcortical regions of the brain, related to the interaction of a person with the collaborative non-player character powered by the eBICA cognitive architecture.

Keywords: Neuroimaging · fMRI · Neurocorrelates · Socially-emotional cognition · Cognitive architectures · Semantic mapping · Artificial intelligence

1 Introduction

This work is aimed at contributing to solution of the fundamental problem: to determine the paradigms and general principles of artificial-intelligence-based cognitive partners and functions used to assist humans in a wide range of practically important areas, requiring social-emotional intelligence. To develop such systems, it is necessary to build a toolkit, the basis for which can be a semantic map and the definition of the neurophysiological processes that form such maps. A semantic map is an embedding of a set of objects endowed with semantics into an abstract metric space, called the semantic space, defined in such a way that the geometric characteristics of the embedding reflect the semantic characteristics of the set of objects, allowing one to draw conclusions about semantics based on geometry. Semantic maps can be divided into strong and weak [1]: in the case of weak maps, the coordinate axes have certain semantics, while in the case of strong maps, the distance is a measure of dissimilarity [2]. The point of using a semantic map in the eBICA architecture [3, 4] is that the agent chooses actions taking into account both rational factors (for example, utility in terms of achieving the goal) and emotional factors (for example, correspondence of emotional appraisals of the action and object of action). The latter can be determined on the basis

A. V. Samsonovich (Ed.): BICA 2019, AISC 948, pp. 532–537, 2020.
https://doi.org/10.1007/978-3-030-25719-4_69

of the so-called *moral schemas* [4–6]. A moral schema determines the law (usually probabilistically) of choosing an agent's behavior, based on the map coordinates of the author of the action, the object of the action, and the action itself. Based on the constructed semantic maps and recorded actions of the subject in a model situation (in this paper, the game Teleport), the psychological characteristics of the behavior (for example, valence or dominance – see below) of the subject are calculated as functions of time. In this case, both the general appraisals of each player (taking into account all the actions in the game) and pairwise mutual appraisals (taking into account a given pair of players only) are calculated. These appraisals serve as regressors in the processing of fMRI data, used to calculate the activation of brain neural networks involved in the cognitive processes. Thus, in this work, a new tool was developed and used to determine the neurophysiological correlates of behavioral motives based on the eBICA model using the fMRI method.

2 Materials and Methods

2.1 Subjects and Procedures

MRI data were obtained from 10 healthy subjects, mean age 21 (range from 20 to 22 years old). A signed Consent was provided by each participant. From each subject, a voluntary informational consent form, a questionnaire and consent to the processing of personal data were completed. During the experiment, each volunteer was asked to lie still. Permission to undertake this experiment has been granted by the Ethics Committee of the NRC "Kurchatov Institute". During the scanning, the participants were instructed to play the game "Teleport", described in previous paper [7].

The experimental paradigm was based on a virtual environment (VE), where participants engaged in a videogame. Detailed description of VE can be find at [8–10].

2.2 fMRI Recording and Data Processing

MRI data were acquired using a SIEMENS Magnetom Verio 3 T (Germany) using 32 channel head coil. The T1-weighted sagittal three-dimensional magnetization-prepared rapid gradient echo sequence was acquired with the following imaging parameters: 176 slices, TR = 1900 ms, TE = 2.19 ms, slice thickness = 1 mm, flip angle = 9°, inversion time = 900 ms, and FOV = 250 × 218 mm^2. The fMRI data were obtained using the ultrafast sequence with the following parameters: 52 slices, TR = 1000 ms, TE = 25 ms, slice thickness = 2 mm, rotation angle = 90° and FOV = 192 × 192 mm^2. Additionally, inhomogeneity maps were obtained to reduce the spatial distortion of the echo-planar images. In total, 3 sessions of functional scanning were performed, during which a volunteer, lying inside the scanner, controlled a virtual actor with a magnetically compatible joystick. Each episode lasted about 10 min.

fMRI and anatomical data were preprocessed using SPM8 (available free at http://www.fil.ion.ucl.ac.uk/spm/software/spm8/) based on Matlab and MacOS shell scripts. The center of anatomical and functional data were adducted to the anterior commissure and corrected for magnetic inhomogeneity using field mapping protocol. Slice-timing correction for fMRI data was performed (the correction of hemodynamic response in space and then in time to avoid pronounced motion artifacts). To exclude motion artifacts, the images were pre-corrected, using BROCOLLI shell scripts for spatial transformation. Then, spatial normalization was performed to bring them to the coordinates of the MNI (Montreal Neurological Institute) atlas in a coordinate system. Anatomical data were segmented into 3 possible tissues (grey matter, white matter, cerebrospinal fluid). After that, for the functional data, a smoothing procedure was performed using a Gaussian filter with a $6 \times 6 \times 6\ \text{mm}^3$ core.

2.3 Curve Extractions

Within each session, 9 curves were calculated for each test subject (there were 2 sets of curves for different r = 0.01 and r = 0.05):

1. The total valence of a person;
2. The total valence moral bots;
3. The total valence of the bot;
4. The overall dominance of the person;
5. The overall dominance of the moral bot;
6. Mutual valence of man and bot;
7. Mutual valence of the person of a moral bot;
8. Mutual dominance of a person and a moral bot;
9. Mutual dominance of man and bot.

During the experiments with a virtual cobot partner in the scanner within the framework of the Teleport paradigm, all events were logged. At the same time, the actions performed by a human and two bots were attributed to the characteristics of "valence" and "dominance", based on the semantic map (see Fig. 1).

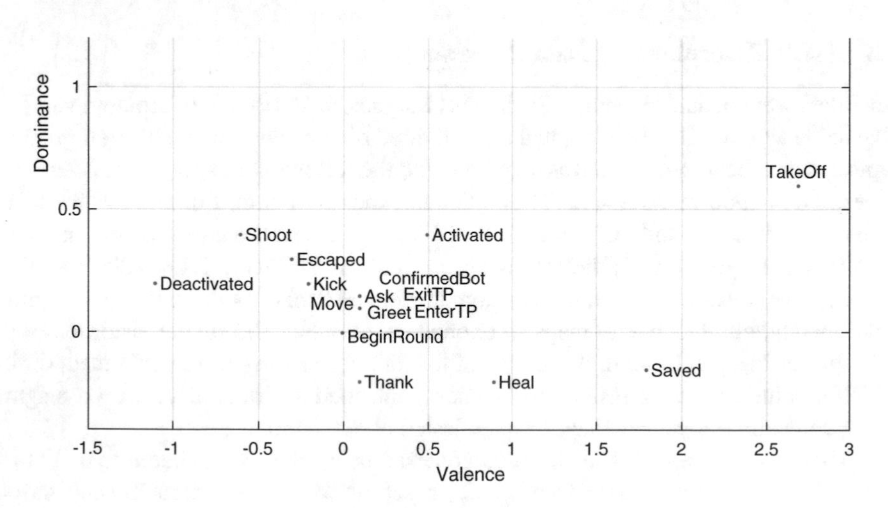

Fig. 1. Semantic action map in the game "Teleport".

The logs were used to calculate the players' scores: the subject and two bots, as a function of time. One of the two bots (hereinafter referred to as "cobot", or "collaborative bot") used a moral partnership scheme: when a certain level of mutual valence with a person was reached, he changed his behavior, as if he considered a person to be his partner. The other bot (hereinafter referred to as the "control bot") did not distinguish between the players, and acted according to the basic model eBICA [3], or eBICA-1 [5]. Estimates of all three players at the beginning of the session were taken equal to zero and updated according to the formulas [3].

$$\begin{aligned} A_{author} &:= (1 - r)\, A_{author} + ra, \\ A_{target} &:= (1 - r)\, A_{target} + ra^{*}, \end{aligned} \tag{1}$$

Where A, a is a complex number, including both real and imaginary parts, of variables (valence and dominance) of the player and actions, respectively; The constant $r = 0.01$ is a model parameter. Grades A were used by bots to select actions. Logs allowed us to calculate the same estimates - the valence and dominance of each player - as a function of time, after the completion of the experiment. In this case, both general assessments were calculated, based on all the actions performed, and mutual evaluations, also using formula (1), but based on actions performed by two players only in relation to each other. Examples of curves are shown in Fig. 2.

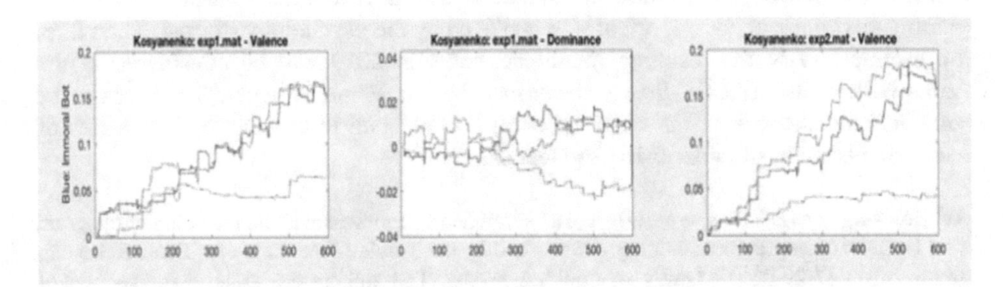

Fig. 2. Example of calculated curves.

3 Results

The calculated curves were used as regressors in constructing a general linear model. In order to isolate statistically significant voxels, contrast vectors were compiled for each of the regression curves. After receiving statistical maps (Student statistics $p < 0.001$ (unc.), anatomical binding of the obtained active voxels to the physiological atlas was made. All obtained statistical maps were plotted on a T1 high resolution template image. An example of such maps is shown in Fig. 3.

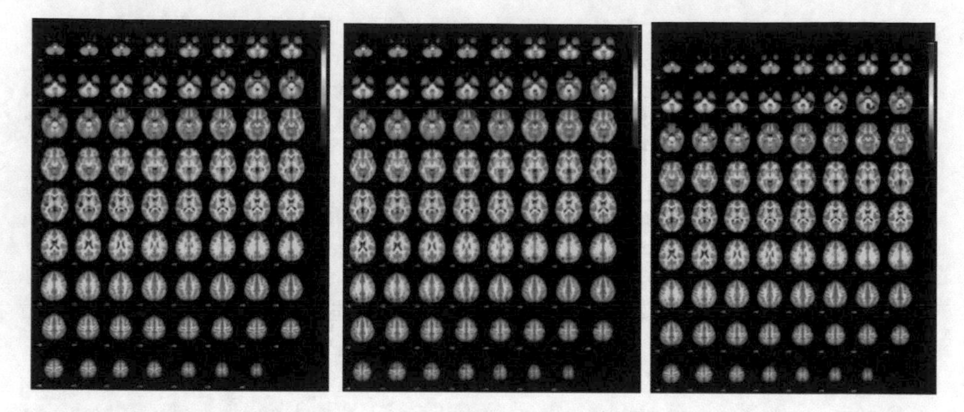

Fig. 3. An example set of statistical maps obtained for three different regressors plotted on a T1 high resolution template image.

4 Conclusions

The obtained data show the complex neural network dynamics of behavioral motifs distributed on the cortical and subcortical regions of the brain, related to the interaction of a person with the collaborative bot "Kobot" using a moral partnership scheme and a control bot. Within the framework of social and psychological interactions in the gaming environment, an important role is played by the psychology of trust. Therefore, the architecture of the resulting neural networks in 2019 will be considered in the framework of the TRUST theory developed by Dr. Frank Krueger. To increase the statistical significance of the obtained data, it is planned to conduct a large series of experiments using ultrafast fMRI sequences.

Acknowledgements. This work was in part supported by the Russian Science Foundation, grant 18-11-00336 (data preprocessing algorithms) and by the Russian Foundation for Basic Research, grants ofi-m 17-29-02518 (study of thinking levels). The authors are grateful to the MEPhI Academic Excellence Project for providing computing re-sources and facilities to perform experimental data processing.

References

1. Samsonovich AV, Goldin RF, Ascoli GA (2010) Toward a semantic general theory of everything. Complexity 15(4):12–18. https://doi.org/10.1002/cplx.20293
2. Samsonovic AV, Ascoli GA (2010) Principal semantic components of language and the measurement of meaning. PLoS ONE 5(6):e10921
3. Samsonovich AV (2013) Emotional biologically inspired cognitive architecture. Biol Inspired Cogn Arch 6:109–125
4. Samsonovich AV (2018) On semantic map as a key component in socially-emotional BICA. Biol Inspired Cogn Arch 23:1–6
5. Samsonovich AV (2018) Schema formalism for the common model of cognition. Biol Inspired Cogn Arch 26:1–19. https://doi.org/10.1016/j.bica.2018.10.008

6. Samsonovich AV, Kuznetsova K (2018) Semantic-map-based analysis of insight problem solving. Biol Inspired Cogn Arch 25:37–42. https://doi.org/10.1016/j.bica.2018.07.017
7. Orlov VA, Ushakov VL, Kartashov SI, Malakhov DG, Korosteleva AN, Skiteva LI, Samsonovich AV (2018) Functional neural networks in behavioral motivations. In: Advances in neural computation, machine learning, and cognitive research II, NEUROINFORMATICS 2018. Studies in computational intelligence, vol 799. Springer, Cham, pp 274–283. https://doi.org/10.1007/978-3-030-01328-8_33
8. Azarnov DA, Chubarov AA, Samsonovich AV (2018) Virtual actor with social- emotional intelligence. Procedia Comput Sci 123:76–85. https://doi.org/10.1016/j.procs.2018.01.013
9. Chubarov A, Azarnov D (2017) Modeling behavior of Virtual Actors: A limited Turing test for social-emotional intelligence. In: Advances in intelligent systems and computing, vol 636. Springer, Cham, pp 34–40. ISBN 978-3-319-63939-0
10. Bortnikov PA, Samsonovich AV (2017) A simple Virtual Actor model supporting believable character reasoning in virtual environments. In: Advances in intelligent systems and computing, vol 636. Springer, Cham, pp 17–26

Aesthetic Judgments, Movement Perception and the Neural Architecture of the Visual System

Vanessa Utz and Steve DiPaola(✉)

Simon Fraser University, Vancouver, Canada
{vutz, sdipaola}@sfu.ca

Abstract. We have developed a deep learning-based AI creativity system, which can be used to create computer-generated artworks, in the form of still images as well as time-based pieces (videos). Within the scope of this article, we will briefly describe our system and will then demonstrate its application in a psychological study on aesthetic experiences. We also propose a new hypothesis regarding a potential interaction between the neural architecture of the two visual pathways, and the effect of movement perception on the formation of aesthetic judgments. Specifically, we postulate that perceived movement within the visual scene engages reflexive attention, an attentional focus shift towards the processing of visual changes, and subsequently affects how information are relayed via the dorsal and ventral streams. We outline a recent pilot study in support of our proposed framework, which serves as the first study that investigates the relationship between the two visual streams and aesthetic experiences. Our study demonstrated evidence for our hypothesis, with time-based artworks showing higher aesthetic appeal at slower playback speeds.

Keywords: Neuroscience · Brain simulation · Artificial intelligence · Deep learning · Visual pathways · Neural pathways · Neuro-architecture · Aesthetics

1 Introduction

With the increased sophistication of AI systems and their expanded application, we have seen the more widespread use of these systems in fields related to the arts, aesthetics and creativity. Simultaneously, the research into aesthetics has become increasingly scientifically rigorous and empirical in nature. Today, we know that the formation of aesthetic experiences and the resulting judgments is a high-level cognitive process that involves a widely distributed network of neural processes, ranging from the primary visual cortex to the reward systems and semantic processing (Leder et al. 2004; Brown et al. 2011). Our study aims to explore the use of deep-learning systems as assistants in the knowledge building endeavor to further the understanding of the human mind, particularly in relation to aesthetic experiences. We introduce and test a new theoretical framework regarding the interaction between the information processing in the dual streams of the visual system and the process of forming aesthetic experiences, and investigate this framework with the help of AI generated artworks. As

A. V. Samsonovich (Ed.): BICA 2019, AISC 948, pp. 538–546, 2020.
https://doi.org/10.1007/978-3-030-25719-4_70

far as we can tell from our review of the existing literature in this space, our study will be the first to investigate how the instantiation of aesthetic experiences is affected by the engagement of the two visual streams. This study is therefore exploratory in nature and aims to map out new potential research avenues. We will begin by outlining the relevant background information. The second part outlines an initial exploratory study to test our hypothesis, followed by a discussion regarding the future of this research.

2 Deep Learning Techniques Applied in Art, Aesthetics and Creativity Research

We will begin by describing our creative AI system which we used to create our art stimuli. Our lab has combined Genetic Algorithms, Neural Networks and Deep Learning neural networks into our Painterly Rendering software framework with the particular aim of mimicking the cognitive processes of portrait artists (DiPaola et al. 2018, 2017, 2016, 2014, 2009). Within our AI-based painting software toolset, we have implemented our modified versions of DeepDream (mDD) using the Caffe deep learning framework (Jia et al. 2014). We are currently using them as a pre-processing stage which simulates an artists' imagination and perception, transforming an image before it is subsequently sent to the second, cognitive based, artistic stroke-placement phase (DiPaola, 2017).

Deep Dream (Mordyintsev et al. 2015) begins with a source image and uses backpropagation and gradient ascent to gradually transform the image's pixels in order to emphasize the most strongly-activated features from a certain user-selected network layer. This process results in the emphasis of shapes and patterns that pre-exist in the source image, as well as the chance appearance of hallucinated patterns in which the network gravitates towards "seeing" patterns it has learned to recognize during its training. While most DD systems use pre-trained networks with object recognition data such as ImageNet, we train new models with creativity and art generation in mind, using paintings and drawings as training data. We now have amassed a data set of 160,000 labeled/categorized paintings from 3000 labeled artists for a total size of 67 GB of artistic visual data. However, we discovered that since most fine artists make under 200 paintings in their lifetime, we had material that might not be rigorous and large enough for an advanced CNN training for art styles. In order to overcome this issue, we developed a method which we call "hierarchical tight style and tile" (DiPaola et al. 2018). Since in our work, detecting and identifying regular objects 'within' an image is less important than the overall artistic style of the entire image (e.g. style of stroke, texture and color palette), we were able to use tile subsections of the images in order to increase our input data. We developed a tiling method to sample each artwork into over 50 individual tiles of varying sizes. For instance, this allowed us to turn 50 source images of a category (e.g. Van Gogh's early work) into over 2000 individual training images for our network. We use a much tighter categorization method based on cognitive theories of art perception that we have outlined in previous work (DiPaola 2014). This method would classify Van Gogh's work into 7 stylistic categories rather than one category. Figure 1 shows a source image (a), which is processed through our mDD system (b), then processed through our last step where it is rendered with cognitive based color and stroking in our ePainterly system for the final result (c). This is the same process our study videos went through (Fig. 2).

Fig. 1. a, b, c, Source image (a, left), put through our modified Deep Dream (b, middle) then stroke and color enhanced through our ePainterly system for the final result (c, right).

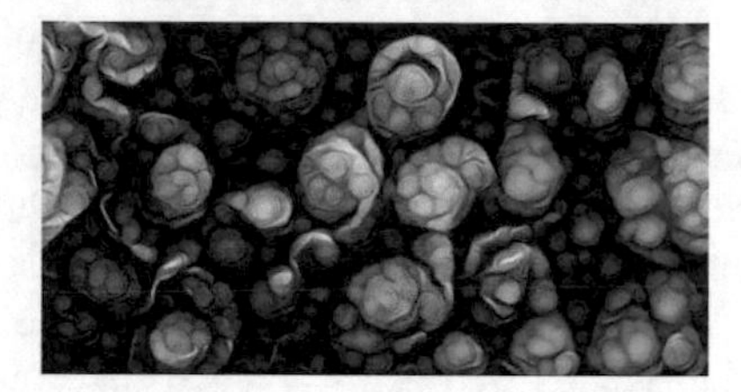

Fig. 2. a, b Screenshots from our fast and slow movies - our 'portrait' video is a slightly less abstract art video (a, left). Our 'abstract' video shows 3 women in an abstract form (b, right).

In this current system, we apply our mDD module to the source photo first, followed by our Non-Photorealisitc Rendering (NPR) painting phase. Our Painterly module, which we call ePainterly, models the cognitive processes of artists based on years of research in this area. It uses algorithmic, particle system and noise modules to generate artistic color palettes, stroking and style techniques. The aesthetic advantages of this additional system include reducing noisy artefacting of the generated DCNN output via cohesive stroke-based clustering as well a better distributed color space (Fig. 1c).

3 Aesthetic Experiences and the Human Visual System

In the field of empirical aesthetics, the most prominent model has been developed by Leder et al. (2004). The model consists of a 5-step process. The input of the model is a visual artwork, which leads to a two-fold output - an aesthetic judgment and an aesthetic emotion. The individual steps are as follows: (1) perception, (2) implicit memory integration (e.g. based on previous experiences – analysis of familiarity), (3) explicit classification (e.g. based on domain-specific expertise – content and style analysis), (4) cognitive mastering and evaluation and (5) continuous emotional evaluation which takes place throughout the entire process. The focus of our framework lies on the initial perception on the artwork (step 1) and how it is affected by movement perception.

Since Goodale and Milner's publication "Separate pathways for perception and action" (Goodale and Milner 1992), it has become well established in the areas of psychology and neuroscience that the human visual system consists of two distinct pathways. The ventral visual stream, which leads from the primary visual cortex (V1) to the temporal lobe, is responsible for the relay of visual information that are required for object recognition. This information includes color, detail and form. The ventral stream relays this information to the Orbitofrontal Cortex (OFC) (Rolls 2005). The OFC is a higher-level sensory cortex, part of the frontal lopes, and is of importance during decision making tasks. Rushworth et al. (2008) introduced a model that showed decision-making, which is involved in the appraisal of objects, is linked to activation in the OFC. Numerous neuroscientific studies investigating the appraisal of quality of both art and non-art, have shown that the OFC is activated during decision making tasks of this kind (see Kringelbach 2005; Wallis 2007). This connection between the ventral pathway and the OFC appears therefore to be of importance in the instantiation of aesthetic experiences and formation of aesthetic judgments (see Brown et al. 2011 for a cognitive model of the core aesthetic network in the human brain). The dorsal stream also originates from the primary visual cortex. However, this pathway leads to the parietal lobe and focuses on a different type of information. In this pathway, information important for the determination of location and movement of objects within the visual field are relayed. Generally, when we perceive a range of different visual stimuli at once, there are competitive interactions between the neural representations of each of these stimuli. However, when attention is focused on one of the perceived stimuli, the neural response towards that stimulus is enhanced (Desimone 1998). Usually, it is behaviorally relevant stimuli that are favored and attract an individual's attentional focus. Reflexive attention occurs when an abrupt change takes place within the visual scene. Once a sudden change has occurred, attention is rapidly oriented towards the location in the visual scene where the change took place. Once attention is shifted to that area, reflexive attention causes a modulation of the early stages of the sensory analysis and therefore enhances the subsequent higher-level analyses. These modulations engage rapidly after the change is perceived (Hopfinger and Mangun 1998). We therefore postulate that the presence of movement within an artwork reflexively engages the dorsal stream to process the visual information, thereby affecting how information are passed on and processed via the ventral stream and OFC.

4 Study

4.1 Participants and Stimuli

The participant pool consisted of undergraduate and graduate students in SFU's School of Interactive Arts and Technology where they have some but limited knowledge of art. The age of the participants ranged from 18 years to 40 years. 14 were male, 24 were female. We recruited a total of 44 participants, however 6 had to be excluded from the data analysis, leaving us with N = 38.

We used two source videos of individuals that we turned into four art videos using a combination of our mDD (modified DeepDream) and the NPR (Non-Photorealistic Rendering) Painterly module. We ensured that each of the two batches of the video frames had been given a distinct artistic style (see Fig. 2 for example frames of each of the two videos). Additionally, the videos were transformed to different levels of abstraction: one video was abstracted so that the human figures were no longer clearly visible ('abstract video'), in the other the human face remained recognizable ('portrait video'). Since this study is the first exploration of our framework, we used different abstraction levels to test whether the presence of distinct human faces would alter the data. We created two art videos from each set of frames, displaying the frames at different speeds, one in real-time (ffmpeg: setpts = 1.0 * pts for regular playback speed ('fast version')) and the other 3.5 times slower than real-time (ffmpeg: setpts = 3.5 * pts for the slow playback speed ('slow version')) (ffmpeg: How to speed up, …, 2019).

4.2 Procedure and Results

We showed each participant two pairs of art videos. The videos were displayed in full screen mode on a laptop (a Lenovo Yoga 720 with 15.6″ UHD display – 3840 × 2160 resolution) that was placed on a desk in front of the participant. Viewing distance was approximately 50 cm. Order of video pairs (abstract vs portrait) and order of playback speed (fast vs slow), were randomized. To collect the quantitative measure, we asked the participants to rate each video on a 7-point Likert scale for (A) liking (personal preference), (B) aesthetic pleasantness (aesthetic beauty) and (C) artistic value (how good of an art piece is the stimuli) (scales adapted from: Haertel and Carbon 2014; Belke et al. 2015; Pelowski et al. 2018). Higher scores on these rating scales indicated higher agreement. Participants were asked to rate each video with respect to the other video in the pair (fast vs. slow). After providing their quantitative responses, they were asked open-ended questions. Verbal answers were recorded. The purpose of these open-ended questions was to provide us with additional qualitative data regarding our participants' experiences.

The results we obtained in this study provide supporting evidence for our hypothesis that playback speed affects the way individuals form aesthetic experiences: the slow versions of the videos received higher scores on all three rating dimensions. On average, these slow versions were liked more and were judged to have higher aesthetic pleasantness and higher artistic values. 21 out of the 38 participants always preferred the slow versions for both video pairs (rated them with higher average score across the three rating dimensions). This overall preference for the slower videos was supported by qualitative data.

Quantitative Data. We obtained highly statistically significant differences in the scores for the two different speeds on all three rating dimensions. To compare the mean scores on these rating dimensions, we performed paired t-tests. For the likability scale, we obtained an average score of 5.46 (SD = 0.80) for the slowed videos and an average score of 4.46 (SD = 1.12) for the version played at regular speed, $t(37) = 4.48$,

$p < 0.00001$ (On all three rating scales a value of '4' indicated the neutral midpoint). For the aesthetic pleasantness score, we obtained averages of 5.54 (SD = 0.88) and 4.39 (SD = 0.99) for the slow and fast versions respectively, $t(37) = 5.35$, $p < 0.00001$. Lastly, for the artistic value, our participants rated the slow videos with an average score of 5.37 (SD = 0.93), while rating the faster versions with an average score of 4.83 (SD = 0.95), $t(37) = 2.50$, $p = 0.01$. Since the average scores on each of the dimensions was significantly higher for the video versions with the slower playback speed, this data is supportive of our hypothesis.

Since the two video stimuli that we used were very different in nature (one was highly abstract with no recognizable individuals, while the other had a clear human face in the center of the frame) and based on some of the results we obtained from our qualitative data (discussed below), we also compared the effect of different playback speeds for the two videos separately. Our results indicated that the mean differences were more significant for the highly abstract video. For the rating scales 'likability' and 'aesthetic pleasantness', the differences between group means (slow vs fast playback speed) were highly significant for both video types (likability: $t(37) = 3.68$, $p < 0.001$ for the abstract video, $t(37) = 3.41$, $p = 0.001$ for the portrait video; aesthetic pleasantness: $t(37) = 5.52$, $p < 0.0001$ for the abstract video, $t(37) = 2.93$, $p < 0.01$ for the portrait video). For the last rating dimension, artistic value, we only obtained a statistically significant difference for the abstract video, $t(37) = 3.11$, $p < 0.01$. For the portrait video, we obtained a smaller difference between scores: 5.32 (SD = 1.01) for the slow version and 5.03 (SD = 1.10) for the regular speed version, $t(37) = 1.18$, $p = 0.24$. This difference in scores was not statistically significant. This smaller difference is due to the higher average artistic value score that the portrait video received when played at regular speed. This higher score caused the mean difference between the scores of the slow and the fast versions to shrink.

Qualitative Data. Participants spoke for an average of 82.4 s (SD = 31.7 s). For the analysis of the recorded data, we transcribed the audio files and looked through the most representative responses, as well as outliers. Importantly, all participants commented on noticeable differences in their perception of the videos based on the playback speed (example quote of participant: "It was super interesting to see how much the speed actually changed the art piece. It changed the way I interpreted the art piece [...]"). Other quotes will be presented and discussed below.

As evidenced by the quantitative data, participants generally preferred the slow versions of the videos. Participants regularly commented on the fact that the slower playback speed allowed them to observe the finer details, and color and pattern changes: "[...] I was able to [...] appreciate what was going on, see all the different little strokes. It was definitely a lot more pleasant", "[...] the colors seemed to stand out a lot more and I saw the definitions of a lot more lines included", "[...] you can see the patterns more clearly [...]". Additionally, participants repeatedly referred to the slower versions as more natural, calming and relaxing: "[...] it feels more pleasant, because it feels more normal [...]", "[...] I could really relax and focus on what was going on [...]", "[...] the slower version had a much more calming effect, where the movement just seemed a lot more soothing.". Some participants provided more interesting observations such as that their gaze felt less distracted: "If it's slowed down, my eye won't jump around those strokes". Particularly this last observation fits our hypothesis that the slower movement engages less attentional resources that could distract from the processing of information required for the formation of aesthetic experiences. In comparison, these participants who preferred the slower videos, perceived the faster versions as bumpy ("[…] in the fast one it was bumpy. That's why I didn't like the fast one."). This fits with the observation that participants had a harder time focusing on smaller details and changes: "[…] the experience are different depending on the speed of the video, because if the videos with those small details are moving too fast, it is hard to capture how it changes because it disappears suddenly […]", "at the faster speed […] it was harder to tell what was going on […].", "[…] it was just a lot to register […]". One participant summarized their experience as such: "there was sort of a pulsating motion […] when it was fast. But when it was slow that all went away, and you just had a sense of all the colors […]". It was also commented on the impression that the fast version appeared less art-like and more like a natural video: "the faster one just felt like a video".

There were however participants that preferred the faster versions to the slowed videos. These participants highlighted that they perceived the colors in the faster videos to be more vibrant and saturated ("[…] they seemed more vibrant […]"). Two of our participants made the interesting observation that the order of presentation might have affected their experiences.

5 Discussion and Conclusion

In this paper we introduced our new framework regarding the interaction of the visual system's dual pathways and attentional system with the formation of aesthetic experiences when exposed to moving art pieces. This paper therefore serves as the first investigation of the relationship between the visual pathways and aesthetic processing. While the difference in processing of color and detail between the visual pathways has been established, this is the first exploration of the impact on aesthetic experiences. Through our initial study into this new framework, we have been able to demonstrate that there is supporting evidence for our proposal. Our hypothesis was that if participants are presented with art videos, their aesthetic judgments of the art pieces would be higher for the versions with slower playback speed. Both our quantitative and qualitative data sets supported this hypothesis and showed that participants provided higher scores for the slower versions on all rating dimensions and commented extensively on how they were able to better perceive detail and color. Our current work therefore serves as a foundation on which future work can be built.

Despite this initial data that we were able to successfully obtain, in order to fully validate our framework, more studies in this area need to be conducted. The next study needs to be of a major scale in order to fully investigate this multivariate phenomenon. It needs to provide rigorous data sets to support our claims and to further explore the new insights that we have gained with this initial investigation. We will outline some of these considerations.

The most important suggestion for future studies regards the stimuli. Our stimuli set size in this study was very limited. We only used two videos which we played at different speeds. In order to show the universality of the effect, future studies should use larger stimuli sets. Additionally, the difference between the effect in the abstract versus the portrait video needs further investigation.

Once other behavioral studies show similar evidence for our framework, it would be necessary to involve brain imaging techniques in subsequent research. Another avenue in this area would be lesion studies investigation whether individuals suffering from lesions in brain regions involved in the dorsal visual pathway, provide different aesthetic judgments compared to healthy individuals. These additional studies would serve to validate our proposed cognitive mechanism and would aid in the expansion of our methodology (using AI systems in the creation of tailored visual stimuli) into the field of neuroscience.

Acknowledgements. We acknowledge our colleague Graeme McCaig who was instrumental in creating our modified Deep Dream system and was invaluable in his mentorship of the work. This work was partially supported by SSHRC and NSERC grants respectively.

References

Belke B, Leder H, Carbon CC (2015) When challenging art gets liked: evidences for a dual preference formation process for fluent and non-fluent portraits. PLoS ONE 10(8):e0131796. https://doi.org/10.1371/journal.pone.0131796

Brown S, Gao X, Tisdelle L, Eickhoff SB, Liotti M (2011) Naturalizing aesthetics: brain areas for aesthetic appraisal across sensory modalities. NeuroImage 58:250–358

Desimone R (1998) Visual attention mediated by biased competition in extrastriate visual cortex. Philoso Trans R Soc B Biol Sci 353:1245–1255

DiPaola S, McCaig G, Gabora L (2018) Informing Artificial intelligence generative techniques using cognitive theories of human creativity. Procedia Comput Sci Special Issue: Bio Inspired Cognitive Architectures, 11 pages

DiPaola S (2017) Exploring the cognitive correlates of artistic practice using a parameterized non-photorealistic toolkit. Leonardo 50

DiPaola S, McCaig R (2016) Using artificial intelligence techniques to emulate the creativity of a portrait painter. In: Proceedings of Electronic Visualization and the Arts. British Computer Society, London. 8 pages

DiPaola S (2014) Using a contextual focus model for an automatic creativity algorithm to generate art work. Procedia Comput Sci Spec Issue: Bio Inspired Cogn Architectures 41:212–219

DiPaola S, Gabora L (2009) Incorporating characteristics of human creativity into an evolutionary art algorithm. Genet Program Evolvable Mach J 10(2):97–110

Goodale MA, Milner AD (1992) Separate visual pathways for perception and action. TINS 15:19–25

Haertel M, Carbon CC (2014) Is this a "Fettecke" or just a "greasy corner"? About the capability of laypersons to differentiate between art and non-art via object's originality. I-Percept 5:602–610

Hopfinger JB, Mangun GR (1998) Reflexive attention modulates processing of visual stimuli in human extrastriate cortex. Psychol Sci 9:441–447

How to speed up/slow down a video (2019). FFMPEG Wiki. Accessed 30 mar 2019

Jia Y, Shelhamer E, Donahue J, Karayev S, Long J, Girshick R et al (2014) Caffe: convolutional architecture for fast feature embedding. In: Proceedings of the ACM international conference on multimedia, pp 675–678. ACM

Kringelbach ML (2005) The human orbitofrontal cortex: linking reward to hedonic experience. Nature Rev Neurosci 6:691–702

Leder H, Belke B, Oeberst A, Augustin D (2004) A model of aesthetic appreciation and aesthetic judgments. Br J Psychol 95:489–508

Mordvintsev A, Olah C, Tyka M (2015) Online Blog. http://googleresearch.blogspot.ca/2015/06/inceptionism-going-deeper-into-neural.html

Pelowski M, Leder H, Mitschke V, Specker E, Gerger G, Tinio PPL, Vaporova E, Bieg T, Husslein-Arco A (2018) Capturing aesthetic experiences with installation art: an empirical assessment of emotion, evaluations, and mobile eye tracking in Olafur Eliasson's "Baroque, Baroque!". Front Psychol 9:1255. https://doi.org/10.3389/fpsyg.2018.01255

Rolls ET (2005) Taste, olfactory, and food texture processing in the processing in the brain, and the control of food intake. Physiol Behav 85:45–56

Rushworth MFS, Behrens TEJ, Rudebeck PH, Walton ME (2008) Contrasting roles for cingulate and orbitofrontal cortex in decisions and social behavior. Trends Cogn Sci 11:168–176

Wallis JD (2007) Orbitofrontal cortex and its contribution to decision-making. Annu Rev Neurosci 30:31–56

Closed Loop Control of a Compliant Quadruped with Spiking Neural Networks

Alexander Vandesompele(✉), Gabriel Urbain, Francis Wyffels, and Joni Dambre

IDLab-AIRO, Electronics and Information Systems Department, Ghent University - imec, Ghent, Belgium
alexander.vandesompele@ugent.be

Abstract. Compliant robots can be more versatile than traditional robots, but their control is more complex. The dynamics of compliant bodies can however be turned into an advantage using the physical reservoir computing framework. By feeding sensor signals to the reservoir and extracting motor signals from the reservoir, closed loop robot control is possible. Here, we present a novel framework for implementing central pattern generators with spiking neural networks to obtain closed loop robot control. Using the FORCE learning paradigm, we train a reservoir of spiking neuron populations to act as a central pattern generator. We demonstrate the learning of predefined gait patterns, speed control and gait transition on a simulated model of a compliant quadrupedal robot.

Keywords: Spiking neural networks · Compliant robotics · Quadruped control · Reservoir computing

1 Introduction

Compliant robots can provide a greater robustness, flexibility and safety compared to traditional, stiff robots [1]. However, the control paradigms used in traditional robotics cannot be applied to compliant robots, due to the complexity of predicting the state of the compliant body. The complex dynamics can however be turned into an advantage with the concept of embodied computation (also referred to as morphological computation), where the physical body is treated as a computational resource [1–3].

Physical reservoir computing provides a framework for harvesting the body as a computational resource [4]. Monitoring the non-linear body dynamics of a compliant body can be a useful source of information. In some systems extremely little additional computation is required to accomplish a task, for instance locomotion control of a tensegrity robot [4] and control of a soft robotic octopus arm [5]. By combining the body feedback with some additional computational power (e.g. a 'brain'), more complex locomotion tasks can be accomplished [6]. An example of a low level brain function is the generation of rythmic activity by central pattern generators (CPG) [16].

A. V. Samsonovich (Ed.): BICA 2019, AISC 948, pp. 547–555, 2020.
https://doi.org/10.1007/978-3-030-25719-4_71

CPGs can be implemented with a neural network by using the reservoir computing framework [19]. By feeding body sensors of a robot to a randomly connected reservoir, a spatiotemporally enriched interpretation of the body sensors is created. Reservoir computing with spiking neurons is traditionally performed with a liquid state machine [10]. Here, we propose using population coding, where the unit of the reservoir is a population of spiking neurons. This method allows to apply the same principles as in the well established rate based reservoir computing, it also allows to profit from a spike based implementation. The number of tunable parameters, both at neuron and population level allows for optimizing reservoir dynamics for closed loop dynamical systems. Additionally, efficient hardware implementations (e.g. SpiNNaker [23]) could allow to run the network with low power usage on mobile robots. Lastly, this framework allows interfacing with spike-based sensors (e.g. the DVI camera [24]) that provide low latency and low redundancy sensor data.

In this paper, we demonstrate the feasability of using populations of spiking neurons in embodied computation by creating stable closed loop locomotion control for a compliant robot. To achieve this, we applied the physical reservoir computing framework to a simulated model of the Tigrillo robot [11], a compliant quadrupedal platform. We add a 'brain' to the robot which is also a reservoir, consisting of spiking neurons. This neural network is trained to function as a CPG and, similar to biological CPGs, can be modulated by both body sensors and simple control inputs. To create a stable dynamical system, capable of generating robust periodic movements, online linear regression can be applied (FORCE learning, [9]) in a gradual fashion [4]. Figure 1 presents an overview of the implemented system. Four readout neurons are trained to produce motor signals for the actuated joints of the Tigrillo model. Four body sensors, sensing the angle of the passive joints, are fed as input to the neural network.

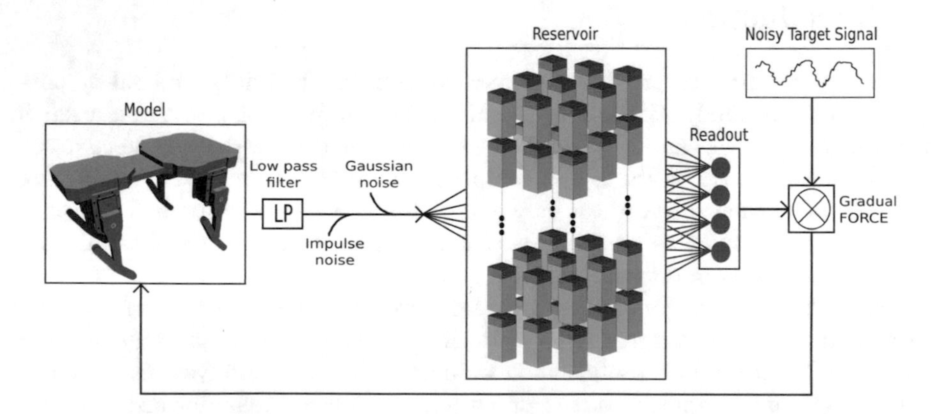

Fig. 1. Overview of the closed loop control system. Learned connection weights in red.

In the next section the components of the closed loop system are presented and the learning algorithms involved are detailed, including a method for monitoring the reservoir state. The results section presents different gait patterns that have been learned, as well as gait frequency control and gait switching control.

2 Materials and Methods

2.1 Simulation

The model used in this work is based on the physical Tigrillo robot [11]. Tigrillo is a low-cost platform developed for researching compliance in quadrupeds. The robot has four legs, consisting of two joints: one actuated with servo motors (hips and shoulders) and one passive joint (knees and elbows). The passive joints are loaded with a spring, providing compliance. The angle of the passive joints (on the physical robot measured with Hall effect sensors) reflects the state of the robot body and its interaction with its environment, and is therefore useful as a sensor input in the closed loop control system. The Tigrillo model is a parametrized stick and box model that mimicks the weight distribution and physics of the physical robot. All simulations have been performed on the Neurorobotics Platform (NRP, [14]). The NRP provides an interface between an environment simulator (Gazebo) and a spiking neural network simulator (e.g. NEST, [12]). In this work, ODE [15] was used as physics engine.

2.2 CMA-ES

The Tigrillo platform has four actuated hip joints controlled by four phase-coupled CPGs. To find motor signals for gaits that suit the body dynamics, the covariance matrix adaptation evolutionary strategy (CMA-ES) algorithm [21] is used to optimize parameters of a parametrized CPG (described in [20]). CMA-ES can handle non-convex fitness landscapes with many local maxima well. Additionally, only few initialization parameters are required. Initial parameters had a Gaussian distribution with 0.5 mean and 0.2 SD. Each generation consisted of 25 individuals, other parameters were kept at default as described in Hansen [22]. Different gaits were found by optimizing different subsets of parameters. The parameters optimized in the search for the walking gait are listed in Table 1. The CPG frequency was kept constant at 1.44 Hz. The distance travelled from the origin was used as fitness function.

2.3 The Neural Network

The neural network is a reservoir consisting of 300 populations of spiking neurons (unless specified otherwise), arranged in a three dimensional structure of 3×3 layers (Fig. 1). Each excitatory neuron of a population connects to a neuron of another population with a probability proportional to the Euclidean distance between both populations (see Table 3). This distance-based connectivity is not only biologically plausible but also makes the simulation and the potential hardware implementation feasible as it reduces the overall number of connections. The delay of spike transmission between populations is fixed at 100 ms. Each population consists of 40 neurons of the leaky-integrate-and-fire (LIF) type with exponentially decaying post-synaptic current (*iaf_psc_exp*, as

Table 1. Parameters and their ranges included in the *CMA-ES* optimization for the walking gait.

Parameter	Symbol	Range	Unit
Front amplitude	μ_f	[20, 140]	degrees
Hind amplitude	μ_h	[20, 140]	degrees
Front duty cycle	d_f	[0.15, 0.85]	*NA*
Hind duty cycle	d_h	[0.15, 0.85]	*NA*
Front offset	o_f	[−60, 60]	degrees
Hind offset	o_h	[−60, 60]	degrees
Front right phase offset	po_{fr}	[150, 210]	degrees
Hind left phase offset	po_{hl}	[240, 300]	degrees
Hind right phase offset	po_{hr}	[60, 120]	degrees

described in [13]). Neuron parameters are close to the default, bioplausible values, or hand tuned for desired population response properties (Table 2). The ratio of inhibitory/excitatory neurons is 1/4. Within a population, excitatory neurons connect to inhibitory neurons and vice versa (see Fig. 2A and Table 3). All neurons in a population receive a white noise current of mean 0 and SD 2, this is important in maintaining a responsive population.

2.4 The Interface Between Neural Network and Body

Interfacing the spiking neural network with the robot body requires translating spiking activity to analog values and vice versa. The motors of the actuated joints expect an analog value, the desired joint angle. Each motor has a readout neuron that provides this value. The parameters of the readout neurons have been adapted such that its membrane potential can be used directly as motor signal (see Table 2). Most importantly, the spiking threshold is set to infinity, preventing the neuron from firing which would reset the membrane potential. As a result the readout neuron is simply a leaky integrator of its incoming spikes. In the other direction, body sensor data is fed to the neural network. Therefore a DC current proportional to the values of a sensor is injected into a sensor population, whose activity then closely reflects the sensor stream. In this fashion the interface between the spiking network and the body is accomplished.

The readout neurons are connected to all reservoir populations. The weights of these connections are learned with FORCE learning [9]. Therefore, the reservoir states (i.e. the population activities) need to be known at all times. To observe the reservoir states, each population is monitored by a monitor neuron. Monitor neurons are identical to readout neurons, but are connected to a single reservoir population with unit weight. The membrane potential of the monitor neuron represents the population activity (Fig. 2B shows an example of the membrane potential of a few monitor neurons) and is used by the FORCE learning algorithm.

Table 2. Parameters of the LIF neuron model used in the reservoir populations and in the readout/monitor layer.

Parameter	Reservoir	Monitor
Membrane resting potential [mV]	−65	0
Spiking threshold [mV]	−50	∞
Post spike reset membrane potential [mV]	−75	*NA*
Membrane capacitance [nF]	0.2	0.2
Membrane time constant [ms]	30	30
Duration refractory period [ms]	2	*NA*
Post-synaptic time constant [ms]	0.5	5.5

Table 3. Connectivity of the population model per connection type. Pconnect = connection probability between any two neurons. D = Euclidean distance between two populations.

Connection	Variable	Pconnect
Intra-population: Excitatory to inhibitory	C_{ei}	0.1
Intra population: Inhibitory to excitatory	C_{ie}	0.1
Inter-population: Excitatory to excitatory	C_{ee}	$0.3e^{-D^2}$
Excitatory to monitor neuron	C_{em}	1.0

2.5 Gradual FORCE Learning

The aim of the learning is to find connection weights that make the membrane potential of the four readout neurons (see Fig. 1) produce the predefined target signals, i.e. the four motor signals as found by the CMA-ES optimization. FORCE learning [9] is a method to create closed-loop dynamical systems using a reservoir.

By using monitor neurons, we effectively isolate the unweighted contribution of each reservoir population to a readout neuron. Since readout neurons and monitor neurons have identical parameters, the membrane potential of a readout neuron will be a linear combination of all monitor neuron membrane potentials. Therefore it is possible to use the monitor neurons membrane potentials as reservoir states, and FORCE learning can be applied in an identical fashion as with rate-based neural networks.

To make a stable closed loop system, the control signals are first learned in open loop with the control signals as input to the actuators. Subsequently, to ensure a smooth transition to closed loop control, the target signal and readout signal are gradually mixed (as described in [4]). The contribution of the readout neuron is gradually increased during this transition. Finally, the system is capable of autonomously producing the target signals in a stable closed loop fashion (as detailed in the results section).

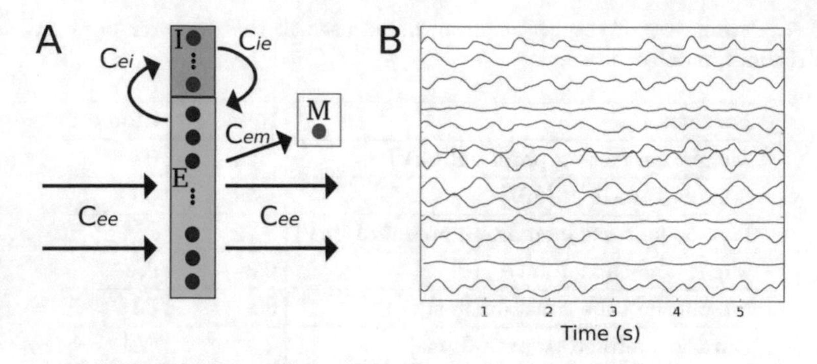

Fig. 2. A. Connectivity of the population model. E = excitatory, I = inhibitory, M = monitor neuron. For C_{ei}, C_{ie}, C_{ee} and C_{em} see Table 3. **B.** Membrane potentials of a few monitor neurons. Each monitor neuron represents the activity of one population of spiking neurons.

The regularization variable α of the FORCE learning algorithm must be selected large enough to prevent overfitting, but not too large as it could fail to approximate the target function sufficiently fast [9]. After a parameter sweep, a value of 50 for α was observed to be effective. Furthermore, to ensure robustness of the closed loop system, it is necessary to insert sources of noise during learning. Here, impulse noise and gaussian noise were added to the sensor signals (see also Fig. 1). Similarly, noise is added to the target signals. A low pass filter is added to smoothen the sensor signals before injecting them to the reservoir. The actuators also posses low-pass properties, which filters out some of the noise due to the implementation with spiking neurons.

3 Results

3.1 Gait Generation

The system is capable of learning and sustaining different gaits that have been found using CMA-ES. Figure 3a and b display the target motor signals and the generated motor signals during a closed loop walking and bounding gait, respectively. In these experiments learning took 40 s simulated time for open loop training followed by 40 s for closed loop training. After learning, the gait generation is sufficiently robust to continue after disturbances such as moving the robot or stopping movement by turning the robot on its back for a while. The learned motor signal are a bit noisy. This is mainly due to the rather small population sizes.

3.2 Speed Control and Gait Transition

In order to obtain gaits with tunable speed, we added an extra input to the reservoir that serves as a control signal for the gait frequency. Similarly to sensor

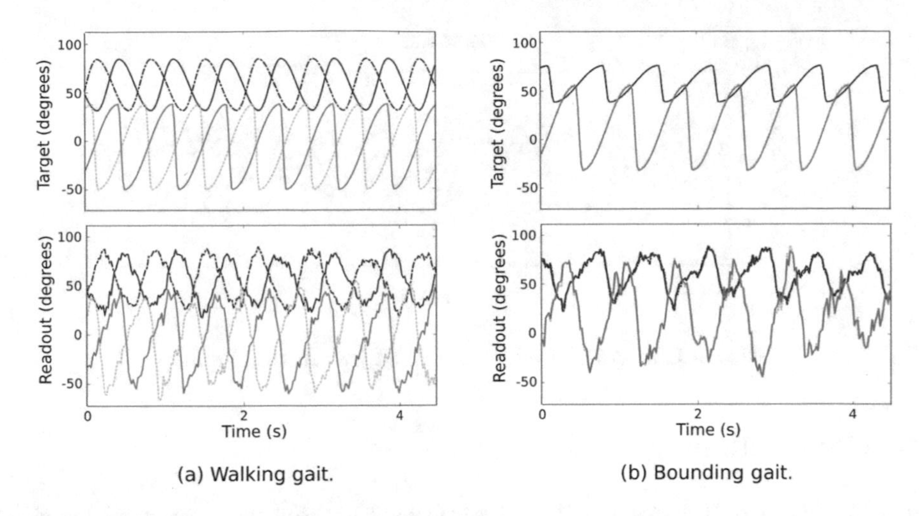

Fig. 3. Generated gaits: (top) target motor signals (red, green for front, hind legs respectively), (bottom) readout signals during closed-loop control, after FORCE learning.

inputs, the control input is implemented as a DC current to the reservoir. During training, incremental frequencies of the same gait are paired with an incremental control signal. After learning, the control signal can be used to alter the frequency (Fig. 4). The total learning time for this experiment was 200 s.

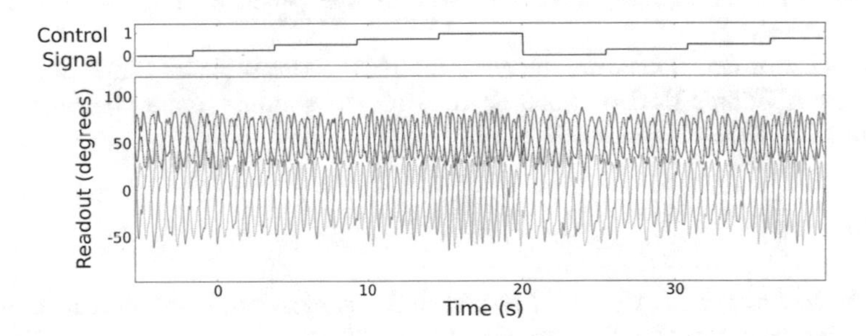

Fig. 4. Tunable frequency post learning.

A second high level control input is used to trigger smooth transitions between the two gaits presented before, walking and bounding (Fig. 5). For this experiment, the reservoir was made more powerful by increasing the number of populations to 600 and the number of neurons per population to 100. The total learning time for this experiment was 200 s.

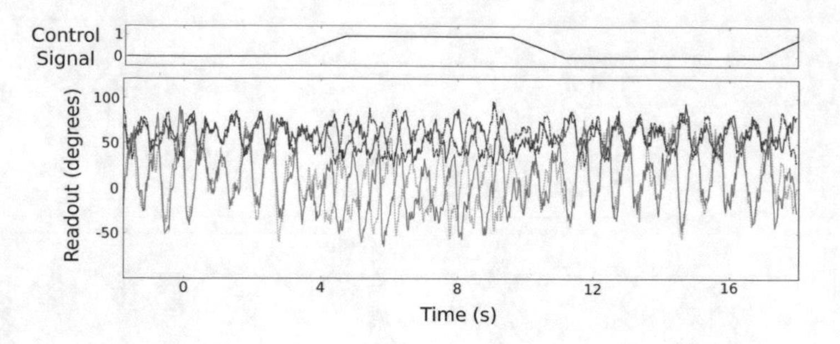

Fig. 5. Gait transitioning controlled by external input.

4 Conclusions

Spiking neural networks could be advantageous for robotics. Potential benefits are energy efficient hardware implementations, efficient sensors and the possibility to apply learning principles observed in biological networks. This work demonstrates the feasibility of using populations of spiking neurons as reservoir units that complement and exploit the physical reservoir that the robot body is. Using only simple learning rules, stable closed loop locomotion control is achieved, even if only minimal sensor data is provided. As in biological spinal networks, the CPG output can be modulated by both simple high level inputs and body sensor input.

In future work, an implementation on neuromorphic hardware (SpiNNaker) will allow to run the network in real time on the physical Tigrillo robot.

Acknowledgments. This research used the HBP Neurorobotics Platform, funded by the European Union's Horizon 2020 Framework Programme under the Specific Grant Agreement No. 785907 (Human Brain Project SGA2).

References

1. Pfeifer R, Lungarella M, Iida F (2007) Self-organization, embodiment, and biologically inspired robotics. Science 318:1088–1093
2. Hauser H, Ijspeert AJ, Füchslin RM, Pfeifer R, Maass W (2011) Towards a theoretical foundation for morphological computation with compliant bodies. Biol Cybern 105:335–370
3. Füchslin RM, Dzyakanchuk A, Flumini D, Hauser H, Hunt KJ, Luchsinger RH et al (2013) Morphological computation and morphological control: steps toward a formal theory and applications. Artif Life 19:9–34
4. Caluwaerts K, D'Haene M, Verstraeten D, Schrauwen B (2013) Locomotion without a brain: physical reservoir computing in tensegrity structures. Artif Life 19:35–66
5. Nakajima K, Hauser H, Kang R, Guglielmino E, Caldwell DG, Pfeifer R (2013) A soft body as a reservoir: case studies in a dynamic model of octopus-inspired soft robotic arm. Front Comput Neurosci 7(91)

6. Degrave J, Caluwaerts K, Dambre J, wyffels F (2015) Developing an embodied gait on a compliant quadrupedal robot. In: IEEE/RSJ international conference on intelligent robots and systems (IROS), Hamburg, pp 4486–4491
7. Burms J, Ken C, Dambre J (2015) Reward-modulated Hebbian plasticity as leverage for partially embodied control in compliant robotics. Front Neurorobot 9
8. Urbain G, Degrave J, Carette B, Dambre J, Wyffels F (2017) Morphological properties of mass-spring networks for optimal locomotion learning. Front Neurorobot 11(16)
9. Sussillo D, Abbott LF (2009) Generating coherent patterns of activity from chaotic neural networks. Neuron 63:544–557
10. Maass W, Nätschlager T, Markram H (2002) Real-time computing without stable states: a new framework for neural computation based on perturbations. Neural Comput 14:2531–2560
11. Willems B, Degrave J, Dambre J, Wyffels F (2017) Quadruped robots benefit from compliant leg designs. In: Presented at the 2017 IEEE/RSJ international conference on intelligent robots and systems (IROS 2017)
12. Gewaltig M-O, Diesmann M (2007) NEST (NEural Simulation Tool). Scholarpedia 2:1430. https://doi.org/10.4249/scholarpedia.1430
13. Tsodyks M, Uziel A, Markram H (2000) Synchrony generation in recurrent networks with frequency-dependent synapses. J Neurosci 20(RC50):1–5
14. Falotico E, Vannucci L, Ambrosano A, Albanese U, Ulbrich S, Tieck J et al (2017) Connecting artificial brains to robots in a comprehensive simulation framework: the neurorobotics platform. Front Neurorobot 11:2
15. Drumwright E (2010) Extending open dynamics engine for robotics simulation. Simul Model Program Auton Robots 6472:38–50
16. Delcomyn F (1980) Neural basis for rhythmic behaviour in animals. Science 210:492–498
17. Shik ML, Severin FV, Orlovsky GN (1966) Control of walking by means of electrical stimulation of the mid-brain. Biophysics 11:756–765
18. Rossignol S, Saltiel P, Perreault M-C, Drew T, Pearson K, Belanger M (1993) Intralimb and interlimb coordination in the cat during real and fictive rhythmic motor programs. Semin Neurosci 5:67–75
19. Wyffels F, Schrauwen B (2009) Design of a central pattern generator using reservoir computing for learning human motion. In: advanced technologies for enhanced quality of life, Iasi, pp 118–122
20. Gay S, Santos-Victor J, Ijspeert A (2013) Learning robot gait stability using neural networks as sensory feedback function for central pattern generators. In: IEEE/RSJ international conference on intelligent robots and systems, Tokyo, pp 194–201
21. Hansen N, Ostermeier A (2001) Completely derandomized self-adaptation in evolution strategies. Evol Comput 9(2):159–195
22. Hansen N (2006) The CMA evolution strategy: a comparing review. In: Lozano JA, Larrañaga P, Inza I, Bengoetxea E (eds) Towards a New Evolutionary Computation. Springer, Heidelberg, pp 75–102
23. Furber S, Galluppi F, Temple S, Plana L (2014) The SpiNNaker project. Proc IEEE 102(5):652–665
24. Lichtsteiner P, Posch C, Delbruck T (2008) A 128x128 120dB 15μs latency asynchronous temporal contrast vision sensor. IEEE J Solid-State Circ 43(2):566–576

Consciousness as a Brain Complex Reflection of the Outer World Causal Relationships

Evgenii Vityaev[1,2(✉)]

[1] Sobolev Institute of Mathematics, Koptuga 4, Novosibirsk, Russia
vityaev@math.nsc.ru
[2] Novosibirsk State University, Pirogova 2, Novosibirsk, Russia

Abstract. In the previous works we analyzed and solved such problem of causal reflection of the outer world as a statistical ambiguity. We defined maximally specific causal relationships that have a property of an unambiguous inference: from consistent premises we infer consistent conclusions. We suppose that brain makes all possible inferences from causal relationships that produce a consistent model of the perceived world that shows up as consciousness. To discover maximally specific causal relationships by the brain, a formal model of neuron that is in line with Hebb rule was suggested. Causal relationships may create fixed points of cyclic inter-predictable attributes. We argue that, if we consider attributes of the outer world objects regardless of how we perceive them, a variety of fixed points of the objects' attributes forms a "natural" classification of the outer world objects. And, if we consider fixed points of causal relationships between the stimuli of the objects we perceive, they form "natural" concepts described in cognitive sciences. And, if we consider the information processes of the brain when the system of causal relationships between object stimuli produces maximum integrated information, then this system may be considered as a fixed point which has a maximum consistency in the same sense as the entropic measure of integrated information. It was shown in other works that this model of consciousness explains purposeful behavior and perception.

Keywords: Clustering · Categorization · Natural classification · Natural concepts · Integrated information · Concepts

1 Introduction

Causality is a result of physical determinism: «for every isolated physical system some fixed state of a system determines all the subsequent states» [1]. But let us consider an automobile accident [1]. What is the reason for it? It might be a road surface condition or humidity, position of the sun with respect to the drivers' looks, reckless driving, psychological state of driver, functionality of brakes, etc. It is clear that there is no any certain cause in this case.

The work is supported by the Russian Science Foundation grant № 17-11-01176.

A. V. Samsonovich (Ed.): BICA 2019, AISC 948, pp. 556–561, 2020.
https://doi.org/10.1007/978-3-030-25719-4_72

In the philosophy of science the causality is reduced to forecasting and explanation. «Causal relation means predictability … in that if the entire previous situation is known, an event may be predicted …, if all the facts and laws of nature, related to the event, are given» [1]. It is clear that nobody knows all the facts, which number in the case of an accident is potentially infinite, and all the laws. In the case of a human being and animals, the laws are obtained by training (inductive reasoning). Therefore, causality is reduced to predicting by inductive-statistical (I-S) reasoning, when predictions logically inferred from facts and statistical laws with some probabilistic assessment.

Causal relationships and laws, discovered on real data or by training, face with the problem of statistical ambiguity – contradictions (contradictory predictions) may be inferred from discovered causal relationships [2, 3]. To avoid inconsistences, Hempel [4, 5] introduced a requirement of maximal specificity, which implies that a statistical law should incorporate maximum of information, related to the predicted property. Following Hempel, we defined maximally specific rules for which we proved a consistency of (I-S) inference, which use only these maximum specific rules [2, 3]. A special semantic probabilistic inference was developed that discover maximum specific rules, which might be considered as the most precise causal relationships (having maximum conditional probability and using maximum available information). Causal relationships may create fixed points of cyclic inter-predictable attributes.

The structure of the outer world objects was analyzed in the form of «natural» classifications. It was noted by naturalists that «natural» classes of animals or plants have a potentially infinite number of different properties [6]. Naturalists, who were building «natural» classifications, noted that construction of a «natural» classification was just an indication: from an infinite number of attributes you need to pass to the limited number of them, which would replace all other attributes [7]. This means that in «natural» classes these attributes are strongly correlated, for example, if there are 128 classes and the attributes are binary, then the independent «indicator» attributes among them will be about 7 attributes as $2^7 = 128$, and others can be predicted based on these 7 attributes. We can select different 7–15 attributes as «indicators» and then others potentially infinite attributes can be predicted based on these selected attributes. So, there is a great number of causal relationships between the «natural» class attributes.

We formalize the «natural» classification based on the generalization of the Formal Concept Analysis (FCA) [8]. Formal concepts emerging in the FCA may be specified as fixed points of deterministic implications (with no exceptions) [8]. We generalize formal concepts for probabilistic case by introducing probabilistic maximum specific rules instead of deterministic implications and defining fixed points for probabilistic implications [9, 10]. We argue in [11] that, if we apply this generalization to some sample from a general population, then we receive a «natural» classification of that sample, which meets all the requirements that naturalists made for «natural» classification.

In the work [12] we demonstrate that semantic probabilistic inference might be considered as a formal model of neuron that satisfy the Hebb rule, in which this inference produce all the most precise causal relationships. Causal relationships discovered by neurons may form fixed points of cyclic inter-predictable properties that produce a certain «resonance» of mutual predictions (excitations) of neurons.

High correlation of attributes for «natural» classes was also confirmed in cognitive science. Eleanor Rosch formulated the principles of categorization, one of which is the following: «the perceived world is not an unstructured set of properties found with equal probability, on the contrary, the objects of perceived world have highly correlated structure» [13, 14]. Therefore, directly perceived objects (so called basic objects) are rich with information ligaments of observed and functional properties, which form a natural discontinuity, creating categorization. Later, Bob Rehder suggested a theory of causal models, in which the relation of an object to a category is not based on a set of attributes but on the proximity of generating causal mechanisms: «the object is classified as a member of a certain category to the extent that its properties are likely to have been generated by this category of causal laws» [15]. Thus, the structure of causal relationships between the attributes of objects is taken as a basis of categorization. Therefore, brain perceives a «natural» object not as a set of attributes, but as a «resonant» system of causal relationships, closing upon itself through simultaneous inference of the total aggregate of the «natural» concept features. At the same time, «resonance» occurs, if and only if these causal relationships reflect some integrity of some «natural» class, in which a potentially infinite number of attributes mutually presuppose each other. To formalize causal models, Bob Rehder proposed to use causal graphical models (CGMs) [16]. However, these models are based on «deployment» of Bayesian networks, which do not allow cycles and cannot formalize cyclic causal relationships. Instead of it, our model is directly formalizing the cyclic causal relationships by the fixed points of predictions on causations [8–12].

Neuron transmits its excitation to the other neurons through multiple both excitatory and inhibitory synapses. Inhibitory synapses may slow down other neurons and stop their activity. It is important for «inhibiting» alternate perception images, attributes, properties, etc. Within our formal model it is accomplished by discovering «inhibitory» causal relationships, which predicts absence of an attribute/property of the object (the perceived object shall not have the respective attribute/property) as compared to the other objects, where this characteristic is found. A formal model specifies it by predicates' negations for corresponding attribute/property. Absence of inconsistencies at the fixed points, that is proved in [10, 11] for the most precise causal relationships, means that there are no two most precise causal relationships simultaneously predicting both availability of some attribute/property with an object, and its absence.

It should be specially noted that the «resonance» of mutual predictions of the perceived properties of the objects is carried out continuously in time and therefore the predicted properties must coincide with the ones that have just been perceived from that objects. The absence of contradictions in the predictions is also the absence of contradictions between the predicted stimulus and the really received stimulus.

2 Comparison with the Integrated Information Theory

If the «natural» classification describes objects of the external world, and «natural» concepts are the perception of these objects, then the integrated information theory describes the information processes of the brain when these objects are perceived.

G. Tononi defines consciousness as a primary concept, which has the following phenomenological characteristics: composition, information, integration, exclusion [17, 18]. For a more accurate determination of these properties G. Tononi introduces the concept of integrated information: «integrated information characterizing the reduction of uncertainty is the information, generated by the system that comes in a certain state after the causal interaction between its parts, which is superior information generated independently by its parts themselves» [18].

The process of reflection of causal relationships of the outer world (Fig. 1) shall be further considered. It includes:

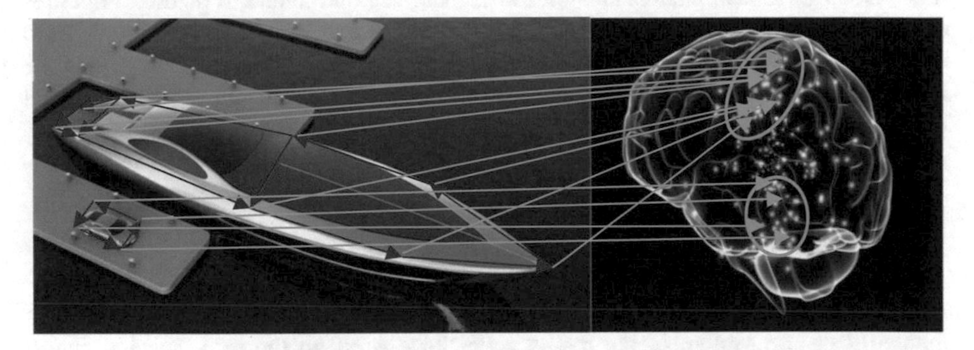

Fig. 1. Brain reflection of causal relationships between objects attributes

1. The objects of the outer world (car, boat) which relate to certain «natural» classes;
2. The process of brain reflection of objects by causal relations marked by blue lines;
3. Formation of the systems of interconnected causal relationships, indicated by green ovals.

In G. Tononi's theory only the third point of reflection is considered. Integrated information is also considered as a system of cyclic causality. Using integrated information the brain is adjusted to perceiving «natural» objects of the outer world.

In terms of integrated information, phenomenological properties are formulated as follows. In brackets an interpretation of these properties from the point of view of «natural» classification is given.

4. Composition – elementary mechanisms (causal relationships) can be combined into the higher-order ones («natural» classes in the form of causal fixed points produce a hierarchy of «natural» classes);
5. Information – only mechanisms that specify «differences that make a difference» within a system shall be taken into account (only a system of «resonating» causal relationships (differences), forming a class (difference) is important;
6. Integration – only information irreducible to non-interdependent components shall be taken into account (only system of «resonating» causal relations, indicating an excess of information and perception of highly correlated structures of «natural» object is accounted for);

7. Exclusion – only maximum of integrated information counts (only values of attributes that are «resonating» at the fix-point and, thus, mostly interrelated by causal relationships, form a «natural» class or «natural» concept).

These phenomenological properties are defined as the intrinsic properties of the system. We consider these properties as the ability of the system to reflect the complexes of external objects' causal relations, and consciousness as the ability of a complex hierarchical reflection of a «natural» classification of the outer world.

Theoretical results on consistency of inference and consistency of fixed points of our formal model are suppose that a probability measure of events is known. However, if we discover causal relationships on the training set, and intend to predict properties of a new object out of the training set and belonging to a wider general population, or to recognize a new object as a member of some «natural» concept, there might be inconsistencies. Here, a certain criterion of maximum consistency is employed, which is based upon information measure, close in meaning to an entropic measure of integrated information [19].

3 Discussion

Theoretical results obtained in the paper suggest that it is possible to create a mathematically precise system of the reality reflection, based on the most specific causal relationships and the fixed points. It can be shown that the reflection of causal relationships is able to model a multitude of cognitive functions in accordance with existing physiological and psychological theories. The organization of purposeful behavior is modeled by causal relationships between actions and their results [20], which fully correspond to the theory of functional systems [21]. The fixed points adequately model the perception [19]. A set of causal relationships models expert knowledge [22]. Therefore, the verification of this formal model for compliance with the actual processes of the brain seems to be an important task.

References

1. Carnap R (1966) Philosophical foundations of physics. Basic Books, New York
2. Vityaev E (2006) The logic of prediction. In: Goncharov S, Downey R, Ono H (eds) Mathematical logic in Asia. Proceedings of the 9th Asian logic conference, Novosibirsk, Russia, 16–19 August 2005. World Scientific, Singapore, pp 263–276
3. Vityaev E, Odintsov S (2019) How to predict consistently? In: Trends in mathematics and computational intelligence. Studies in computational intelligence, vol 796, pp 35–41
4. Hempel C (1965) Aspects of scientific explanation. In: Aspects of scientific explanation and other essays in the philosophy of science. The Free Press, New York
5. Hempel C (1968) Maximal specificity and law likeness in probabilistic explanation. Philos Sci 35:16–33
6. Mill J (1983) System of logic, ratiocinative and inductive. University of Toronto Press, Toronto

7. Smirnof E (1938) Constructions of forms from the taxonomic view. Zool J 17(3):387–418 (in Russian)
8. Ganter B (2003) Formal concept analysis: methods, and applications in computer science. TU, Dresden
9. Vityaev E, Demin A, Ponomaryov D (2012) Probabilistic generalization of formal concepts. Program Comput Softw 38:5, 219–230
10. Vityaev E, Martinovich V (2015) Probabilistic formal concepts with negation. In: Voronkov A, Virbitskaite I (eds) PCI 2014. LNCS, vol 8974, pp 385–399
11. Vityaev E, Martynovich V (2015) "Natural" classification and systematic formalization as a fix-point of predictions. Sib Electron Math Rep 12:1006–1031
12. Vityaev E (2013) A formal model of neuron that provides consistent predictions. In: Chella A et al (eds) Biologically inspired cognitive architectures 2012. Proceedings of the third annual meeting of the BICA society. Advances in intelligent systems and computing, vol 196. Springer, Heidelberg, pp 339–344
13. Rosch E (1973) Natural categories. Cogn Psychol 4:328–350
14. Rosch E (1978) Principles of categorization. In: Rosch E, Lloyd B (eds) Cognition and categorization. Lawrence Erlbaum Associates, Publishers, Hillsdale, pp 27–48
15. Rehder B (2003) Categorization as causal reasoning. Cognit Sci 27:709–748
16. Rehder B, Martin J (2011) Towards a generative model of causal cycles. In: 33rd annual meeting of the cognitive science society, CogSci 2011, Boston, Massachusetts, USA, 20–23 July 2011, vol 1, pp 2944–2949
17. Ozumi M, Albantakis L, Tononi G (2014) From the phenomenology to the mechanisms of consciousness: integrated information theory 3.0. PLOS Comput Biol 10(5):e1003588
18. Tononi G (2004) An information integration theory of consciousness. BMC Neurosci 5:42
19. Vityaev E, Neupokoev N (2014) Perception formal model based on fix-point of predictions. In: Red'ko V (ed) Approaches to mind modeling. URSS Editorials, Moscow, pp 155–172
20. Vityaev E (2015) Purposefulness as a principle of brain activity. In: Nadin M (ed) Anticipation: learning from the past. Cognitive systems monographs, vol 25, chap 13. Springer, pp 231–254
21. Anokhin PK (1974) Biology and neurophysiology of the conditioned reflex and its role in adaptive behaviour. Pergamon Press, Oxford, p 574
22. Vityaev E, Perlovsky L, Kovalerchuk B, Speransky S (2013) Probabilistic dynamic logic of cognition. In: Biologically inspired cognitive architectures. Special issue: papers from the fourth annual meeting of the BICA society, BICA 2013, vol 6, pp 159–168

The Decomposition Method of Multi-channel Control System Based on Extended BCI for a Robotic Wheelchair

Timofei I. Voznenko(✉), Alexander A. Gridnev, Konstantin Y. Kudryavtsev, and Eugene V. Chepin

Institute of Cyber Intelligence Systems,
National Research Nuclear University MEPhI, Moscow, Russia
snaipervti@gmail.com

Abstract. There are a number of ways to control a mobile robotic device, in particular robotic wheelchair. One of this ways is an extended brain-computer interface (extended BCI) – robotic control system with simultaneous independent alternative control channels (BCI, voice and gesture control channels). Because of each channel has advantages and disadvantages the combination of some channels (multi-channel control) can be used. However, when commands are executed from several control channels, various conflicts may arise: for example, one command comes from one control channel, and some opposite commands (which cannot be executed simultaneously) that come from the other channels. To resolve such conflicts, two methods can be used: coordinated control and decomposition. Both of these methods are based on a quality evaluation of each control channel. To evaluate the quality of those control channels the different parameters can be used. This paper proposes a decomposition method of multi-channel control system based on proposed parameter. This technique allows to choose the best channel-command combinations based on type I and type II errors.

Keywords: Extended brain-computer interface · Robotic wheelchair · Decomposition method · Multi-channel control system · Type I and type II errors · Robotics

1 Introduction

Extended BCI [1] is a way of multi-channel control using several control channels simultaneously (BCI, voice and gesture control channels). Multi-channel control is redundant: the same robot control command can be invoked in several ways using different control channels. With this redundancy, various conflicts can arise (for example, if one command was invoked from one control channel, and the opposite command – from other channels). To resolve such conflicts of selecting a command to be sent to a mobile robotic device, two methods are used: coordinated control and decomposition. To develop a coordinated control

A. V. Samsonovich (Ed.): BICA 2019, AISC 948, pp. 562–567, 2020.
https://doi.org/10.1007/978-3-030-25719-4_73

it is necessary to develop logic of a decision-making system (DMS). For DMS operation, it is necessary to select parameters for evaluating the performance of each channel (for example, accuracy) and develop the decision-making logic (for example, in case of conflicts, we execute a command with the highest accuracy value). Decomposition comes down to the implementation of the composite control channel by choosing the best control channel (or channel-command combinations). The advantage of coordinated control is reliability assurance of the control system in case of channels failure. However, in multi-channel control, errors of both types (type I and type II) should be taken into account. In this paper, we propose to use the decomposition method as it, unlike DMS, doesn't need additional calculations of DMS logic. Just command forwarding in DMS implementation would reduce the accuracy of control system due to type I and type II errors of all channels. In the case of decomposition, the choice of the best channel-command combinations minimize the type I and type II errors.

2 Related Works

Let's consider ways to evaluate the performance of a mobile robotic device control channel. Let $C = \{c_1, c_2, \ldots, c_n\}$ be n commands received from this channel. To evaluate the control channel following tests are usually carried out: each command is invoked a certain number of times, and it is recorded how many times it was executed correctly and incorrectly. As a result of testing, a matrix is obtained.

In practice, $n \times n$ matrix M_1 is often used, i.e. for each command c_i, n possible recognition variants are considered $(m_{i,1}, m_{i,2}, \ldots, m_{i,n})$. For example, when the command c_1 was invoked a certain number of times, the result was: the command c_1 was executed $m_{1,1}$ times, the command c_2 was executed $m_{1,2}$ times, etc. In paper [2] dedicated to the gesture recognition algorithm using an accelerometer, the result of testing is presented in the form of a table.

In practice $n \times (n+1)$ matrix M_2 is also used, where n events are for commands $(m_{i,1}, m_{i,2}, \ldots, m_{i,n})$ and the $(n+1)$-th event means that invoked command wasn't recognized at all. For example, in paper [3] the result of gesture recognition is represented in the form of a $n \times (n+1)$ matrix.

$$M_1 = \begin{bmatrix} m_{1,1} & m_{1,2} & \ldots & m_{1,n} \\ \vdots & \vdots & \ddots & \vdots \\ m_{n,1} & m_{n,2} & \ldots & m_{n,n} \end{bmatrix}; \; M_2 = \begin{bmatrix} m_{1,1} & m_{1,2} & \ldots & m_{1,n} & m_{1,n+1} \\ \vdots & \vdots & \ddots & \vdots & \vdots \\ m_{n,1} & m_{n,2} & \ldots & m_{n,n} & m_{n,n+1} \end{bmatrix}$$

The resulting matrix allows to gather information about both successful recognitions and type I and type II errors that occur during recognition. However, the second matrix ($m = n + 1$) contains more detailed information (the number of cases when recognition did not occur), therefore in this paper it is proposed to be used.

There are a large number of parameters used for performance evaluation control channels. For example, in paper [4] authors research asynchronous multimodal hybrid BCI, based on simultaneous EEG and near-infrared spectroscopy

(NIRS) measurements, and use accuracy, sensitivity, specificity, and AUC score for the testing results. In paper [5] authors estimate the region-based deep convolutional neural network (R-DCNN) based on the positive predictive value (PPV) and true positive rate (TPR) parameters. In article [6] authors consider real-time fluid intake gesture recognition using precision, recall and F-score parameters. In [7], the authors use F-measure and accuracy to evaluate the framework. Let's consider the advantages and disadvantages of parameters that can be applied to the mobile robotic device control channel evaluation.

3 Theory

To estimate the performance of the control channel, the accuracy parameter is most often used, which is calculated as:

$$ACC = \frac{TP + TN}{TP + TN + FP + FN}$$

where TP is the number of times the command was executed when it was invoked; TN is the number of times the command was not executed when it was not invoked; FP is the number of times the command was executed when it was not invoked; FN is the number of times the command was not executed when it was invoked.

If we consider this parameter for the command c_1 and testing results in the form of the $n \times (n+1)$ matrix above, then:

$$TP = m_{1,1};\ TN = m_{2,2} + \cdots + m_{2,n+1} + \cdots + m_{n,2} + \cdots + m_{n,n+1} = \sum_{i=2}^{n} \sum_{j=2}^{n+1} m_{i,j}$$

In the denominator, for all commands, there will be the same value, equal to the sum of all C commands given.

However, this method of evaluation has a serious drawback: commands affect each other's evaluation. Let's consider the following example. Suppose we have 100 commands, 99 of which are executed ideally 100 out of 100 times, and one command a is not executed even once.

Thus, we obtain accuracy for command a:

$$ACC = \frac{TP + TN}{TP + TN + FP + FN} = \frac{0 + 99 \times 100}{100 \times 100} = 0.99\ (99\%)$$

The result is 99% accuracy of the command, despite of the fact that the command was invoked 100 times but wasn't executed every time. Thus, this evaluation method is not suitable for evaluating the method of controlling a robotic device, which requires a reliable evaluation method necessary to ensure the safety of both the operator and the equipment.

Another evaluation method is sensitivity (TPR true positive rate) and specificity (TNR true negative rate):

$$TPR = \frac{TP}{TP + FN};\ TNR = \frac{TN}{TN + FP}$$

The use of these parameters allows taking into account both type of errors: type I (TNR) and type II (TPR). However, to estimate the type II error, you can use positive predictive value (PPV) parameter:

$$PPV = \frac{TP}{TP + FP}$$

The advantage of the PPV parameter over TNR is that to evaluate operation of the command $c1$, the PPV numerator contains one value $m_{1,1}$, while in TNR contains sum of values $\sum_{i=2}^{n}\sum_{j=2}^{n+1} m_{i,j}$. Therefore, the use of the PPV parameter will reduce the computational cost, compared to the use of the TNR parameter, especially with a large number of commands. Thus, the use of the TPR and PPV parameters allows take into account errors of type I and type II.

To combine these parameters the parameter F_1 can be used. F_1 is the harmonic mean between TPR and PPV parameters:

$$F_1 = \left(\frac{PPV^{-1} + TPR^{-1}}{2}\right)^{-1} = 2\,\frac{PPV \times TPR}{PPV + TPR}$$

This parameter is a special case of F_β, where $\beta = 1$:

$$F_\beta = (1 + \beta^2)\,\frac{PPV \times TPR}{(\beta^2 \times PPV) + TPR}$$

The parameter β was taken equal to 1, as we consider the type I and type II errors to be equally critical. In general, there is no answer to the question of what type of error (type I or type II) has more critical effects on the control of a mobile robotic device. In practice, situations when errors of one type can become more critical for control (the robot did not stop when it should) than another (the robot goes forward when it should not) can arise.

In this paper, to evaluate the quality of control channel, we propose to use the $MTnP$ (Multiplying TRP and PPV) parameter:

$$MTnP \triangleq TPR \times PPV$$

Let's consider the difference between $MTnP$ and F_1.

The $MTnP$ (Fig. 1) and F_1 (Fig. 2) functions are both symmetric to PPV and TPR. Let's consider the following situation. Suppose we have in one case $PPV = 0.3$ and $TPR = 0.3$; and in the other – $PPV = 0.1$ and $TPR = 0.9$. Then in the first case $MTnP = 0.09; F_1 = 0.3$. And in the second case - $MTnP = 0.09; F_1 = 0.18$.

In both cases considered above we propose that those cases are equally bad. Since both error types are considered to be equally critical, type of error (type I or type II) which is related to low quality doesn't matter. Therefore, we propose to use the $MTnP$ parameter for the quality evaluation of a control channel. The control channel has high quality if for all C commands:

$$MTnP_C = \{MTnP_{c_1}, MTnP_{c_2}, \ldots, MTnP_{c_n}\};\ \min(MTnP_C) \rightarrow max$$

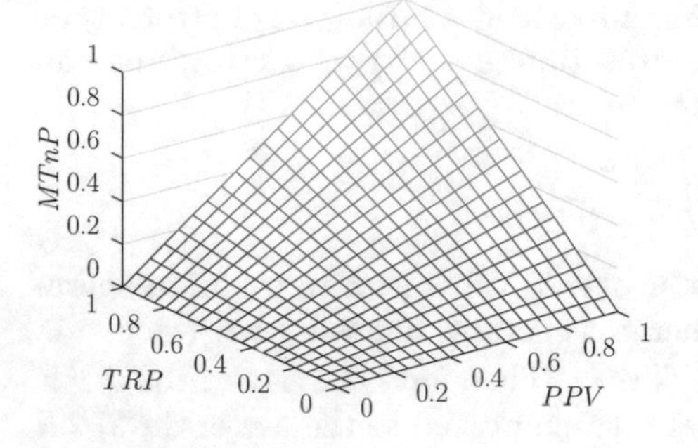

Fig. 1. Function $MTnP(TRP, PPV)$

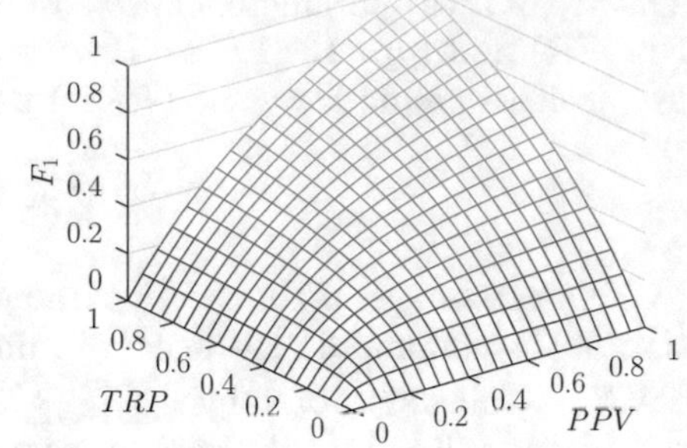

Fig. 2. Function $F_1(TRP, PPV)$

Table 1. Test results

Param	Value									
TP/N	0.97	0.49	0.49	0.49	0.01	0.01	0.01	0.01	0.01	0.01
FP/N	0.01	0.01	0.49	0.01	0.01	0.49	0.01	0.49	0.97	0.01
FN/N	0.01	0.01	0.01	0.49	0.01	0.01	0.49	0.49	0.01	0.97
TN/N	0.01	0.49	0.01	0.01	0.97	0.49	0.49	0.01	0.01	0.01
ACC	**0.98**	**0.98**	**0.5**	**0.5**	0.98	0.5	0.5	**0.02**	**0.02**	**0.02**
TRP	**0.99**	**0.98**	0.98	**0.5**	**0.5**	0.5	**0.02**	**0.02**	0.5	**0.01**
TNR	0.5	**0.98**	0.02	**0.5**	0.99	0.5	0.98	**0.02**	**0.01**	0.5
PPV	**0.99**	**0.98**	**0.5**	0.98	**0.5**	**0.02**	0.5	**0.02**	**0.01**	0.5
F_1	**0.99**	**0.98**	**0.66**	**0.66**	**0.5**	**0.04**	**0.04**	**0.02**	**0.02**	**0.02**
$MTnP$	**0.98**	**0.97**	**0.49**	**0.49**	**0.25**	**0.01**	**0.01**	**0.0004**	**0.005**	**0.005**

In the case of a multi-channel robotic control system, it is necessary to calculate the values $MTnP_C$ for each channel, then select the best value among all control channels. For example, if we control a mobile robotic wheelchair using extended BCI: voice, gestures and brain-computer interface [1], we need to find the best value $MTnP_C$ among these three channels.

4 Proof of Concept

Let's consider the values of parameters ACC, TRP, TNR, PPV, F_1, $MTnP$ for some values of TP, FP, FN, TN. Let N tests were performed. Let's consider the cases when each of these parameters is 97% of N, while the remaining 3 parameters are 1% of N, and also the cases when 2 parameters are 49% of N.

This choice was made so that, when calculating some parameters, we do not get a division by 0 but consider cases when one of the parameters tends to the maximum value.

The result in Table 1 shows that the ACC parameter is not suitable for performance evaluation of a control channel of a mobile robotic device (due to the high influence of the TN parameter). The TRP, TNR, PPV parameters take into account errors of only one type (type I or type II). The F_1 and $MTnP$ parameters show plausible values for different values of the TP, FP, FN, TN parameters. However, we have considered above the reason why we are using the $MTnP$ parameter.

5 Conclusion

In this paper we propose a multi-channel control decomposition method based on the proposed $MTnP$ parameter. This parameter allows to estimate the control channel performance taking into account type I and type II errors. This parameter is based on the principle: since both error types are considered to be equally critical, the type of error (type I or type II) which is related to low quality doesn't matter. In the future we plan to interpret both error types in order to improve the quality of the control system. We also plan to compare this system with the existing DMS [8], and to test it on a mobile robotic wheelchair using the extended BCI-interface to control it [1].

References

1. Voznenko TI, Chepin EV, Urvanov GA (2018) The control system based on extended bci for a robotic wheelchair. Procedia Comput Sci 123:522–527. https://doi.org/10.1016/j.procs.2018.01.079
2. Liu J, Zhong L, Wickramasuriya J, Vasudevan V (2009) uWave: accelerometer-based personalized gesture recognition and its applications. Pervasive Mob Comput 5(6):657–675. https://doi.org/10.1016/j.pmcj.2009.07.007
3. Abdelnasser H, Youssef M, Harras KA (2015) WiGest: a ubiquitous WiFi-based gesture recognition system. In: 2015 IEEE conference on computer communications (INFOCOM). IEEE, pp 1472–1480. https://doi.org/10.1109/INFOCOM.2015.7218525
4. Lee MH, Fazli S, Mehnert J, Lee SW (2014) Hybrid brain-computer interface based on EEG and NIRS modalities. In: 2014 international winter workshop on brain-computer interface (BCI), pp 1–2. IEEE. https://doi.org/10.1109/iww-BCI.2014.6782577
5. Sun Y, Fei T, Gao S, Pohl N (2019) Automatic radar-based gesture detection and classification via a region-based deep convolutional neural network. In: ICASSP 2019 - 2019 IEEE international conference on acoustics, speech and signal processing (ICASSP). IEEE, pp 4300–4304. https://doi.org/10.1109/ICASSP.2019.8682277
6. Jayatilaka A, Ranasinghe DC (2017) Real-time fluid intake gesture recognition based on batteryless UHF RFID technology. Pervasive Mob Comput 34:146–156. https://doi.org/10.1016/j.pmcj.2016.04.007
7. Shami M, Kamel M (2005) Segment-based approach to the recognition of emotions in speech. In: 2005 IEEE international conference on multimedia and expo. IEEE, pp 366–369 (2005). https://doi.org/10.1109/ICME.2005.1521436
8. Gridnev AA, Voznenko TI, Chepin EV (2018) The decision-making system for a multi-channel robotic device control. Procedia Comput Sci 123:149–154. https://doi.org/10.1016/j.procs.2018.01.024

Simulation Analysis Based on Behavioral Experiment of Cooperative Pattern Task

Norifumi Watanabe[1(✉)] and Kota Itoda[2]

[1] Musashino University, 1-1-20 Shin-machi, Nishitokyo-shi, Tokyo 2028585, Japan
noriwata@musashino-u.ac.jp
[2] Keio University, 5322 Endo, Fujisawa-shi, Kanagawa 2520882, Japan

Abstract. We have a behavior experiment using pattern task abstracting cooperative behaviors that require intention estimation and action switching to specific goals. And we have analyzed strategies to adjust cooperative intention estimations. In this research, we constructed an agent model that have three strategies of "random selection", "self-priority selection", and "other agent's target pattern estimation". Moreover the decision making process was verified by simulation.

Keywords: Pattern task · Cooperative group behavior · Intention estimation · Agent model simulation

1 Introduction

We estimate an intention and next action from other's actions in our communication. Furthermore, we decide own actions by adapted to their intentions and actions. Such cooperative behavior often determines next behavioral strategy, relying on other's behavior in one-to-one communication. However, when there are multiple others to be cooperated, it is considered to select and rank the others to be focused on, and to estimate the intention. Specifically, in a goal type ball game such as soccer or handball, when ball holder is surrounded by opponent teams, a plurality of players who can receive a pass and get close to a goal are selected from teammates. And he estimates path success probabilities and decides action intentions. Furthermore, it is considered to evaluate cooperative patterns leading up to the target, select a pattern leading to a score, and give a pass to teammate.

Regarding other people's intention and belief state expression, a depth of intention estimation and belief expression in false belief task has been reported [1]. From a more engineering point, a cognitive process is explained by three main parameters of belief, desire and intention as in the BDI (Belief-Desire-Intention) model [2]. In recent years, a research of human intention estimation as a problem of reverse planning using a probabilistic model Bayesian Theory of Mind [3], and agents with intentions of various depths using a reinforcement learning model research on cooperative tasks [4], which shows the effectiveness

A. V. Samsonovich (Ed.): BICA 2019, AISC 948, pp. 568–573, 2020.
https://doi.org/10.1007/978-3-030-25719-4_74

of an agent-based approach to problems dealing with intention. In a selection of others that should be a target of intention estimation, there are also research on an unbelievable beliefs of others [5].

In order to clarify the cognitive process of such cooperative behavior, in this research, we have model construction and simulation analysis based on behavioral experiment of cooperative pattern task. Therefore we analyze how players evaluate the cooperative pattern and clarify what kind of behavioral strategy model should be modeled for agents that play a cooperative pattern with humans.

2 Cooperative Pattern Analysis

There is some multi-agent researches that realizes cooperative behavior such as RoboCup soccer simulation. Akiyama et al. construct a search tree called an action chain search framework and study on the linkage of soccer tactics by implementing an effective online chain action [6]. In agent model researches, although it is possible to repeat experiments based on data and reproduce behaviors similar to human, it is difficult to estimate the player's intention in cooperative pattern [7].

In study of cooperative patterns on human behavior, it is necessary to analyze action decision process between trials. So it is repeatedly perform the actions of the player acquiring the cooperative pattern and the actions of unacquiring player. In conventional sports behavior analysis, we must reproduce the scene in mini game form by the team member in actual field [8], the situation changes depending on the participating members and weather and so on. Although there is a study by using a display [9], it is highly reproducible, but in order to switch the situation based on the gaze behavior of a subjects, it is necessary to display a curved surface display close to 360 degrees.

While BDI model is represented by logical structure, recent researches model the human inference of intention or interaction with the other through probabilistic mathematical models [10]. Our intention depend on its own beliefs depending on the environment, and from the environment of others through the internal model. It is thought that there is an aspect that it estimates its intention based on it and decides it by balancing both of them. If there are more than one other person, we can share the intention of each and have shared concept to form common action. Shared concepts which involve selection of others and born from interactions are important for model formation of interaction with others.

In this research, in order to realize such reproducibility of behavior and understanding of shared concepts, we construct agent model based on human cooperative behavior and analyze behavior change.

3 Cooperative Pattern Task

We have developed a cooperative pattern task and constructed a system that can be experimented with multiple people, and analyzed the behavior of subjects through behavior experiments in a small group consisting of only four people [11].

In the pattern task, four people participate at the same time, and it aims at cooperating only in non-verbal communication in the two-dimensional grid world to achieve a common target pattern. Each subject manipulates a circular piece on the grid world, estimates the other's intention based on each other's actions at each step, and forms a target pattern (Fig. 1). The target pattern is a figure that is expressed by relative position, and is formed of three out of four subjects. Therefore, each subject needs to act by selecting a piece involved in pattern formation. The pattern is considered to be achieved even with moved in equilibrium, but it is not recognized as rotated or inverted cooperatives.

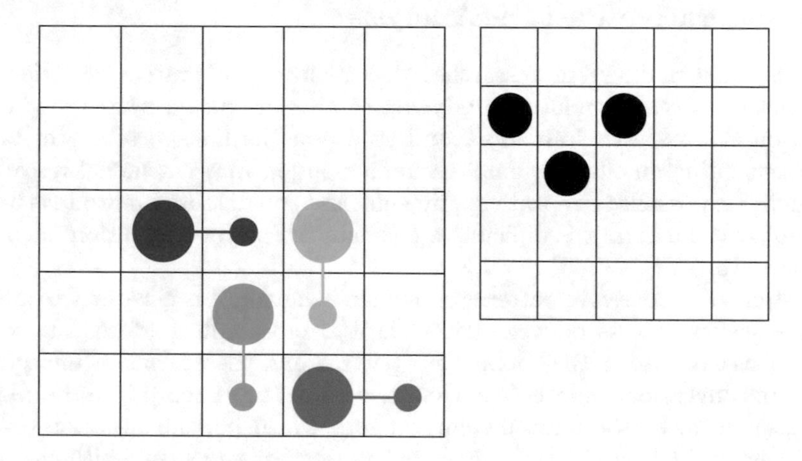

Fig. 1. Left figure is grid world where the subject moves. A large circle shows the position of each subject, and a small circle shows where a subject was before one step. Right figure is target pattern.

We analyzed the adjustment of intention with others, and we got a result that the progress of reaching a target pattern changes depending on the difference by adjustment other subjects in the early stage and the late stage in the task. Moreover, it was suggested that a subject has a measure to prevent the mis-estimation of intention by assuming an action of reaching the target pattern at optimal steps. Therefore, we analyzed the subjects' strategies in pattern selection.

3.1 Pattern Selection Strategy

From the behavior experiment, each subject assumes a new pattern in the beginning about the relationship between the pattern assumed by the subject and the pattern of others, but when all the intentions agree, some subjects have their own target patterns As a result, we obtained the result of continuously estimating each pattern with all the subjects agreeing in the end. Specifically, the following three relationships were found in the patterns selected by the subjects.

Strategy a Select a pattern selected by themselves in the previous step
Strategy b Select a pattern selected by others in the previous step
Strategy c Select a new pattern different from Strategy a and b

Based on these three strategies, we analyze the subject's pattern selection process. At a beginning of a task, a new pattern of strategy c is selected. After that, one of the subjects changed own strategy according to a pattern selected by the other subjects. Such changes in strategy were divided into cases where a common agreement was reached in the early trials and in the latter cases. When the target pattern could be achieved with a small number of steps, a subject took an explicit action in the early stage so that own intention of action could be understood, and the other person matched the action.

From this result, in this pattern task that encourages all subjects to cooperative, one's own action intention is presented to others in a form as comprehensible as possible, and a target pattern is selected such that the overall reaching step is shortest in each situation. Furthermore, it is possible to consider a behavioral strategy in which the majority of selectable patterns is taken and the pattern taken by many subjects is given priority. Based on group action strategies obtained from these results, we construct an agent.

4 Simulation Analysis

A plurality of agent models built based on analysis results in Chap. 3 are prepared according to the difference in parameters. And simulation is performed to verify the model that can most explain the behavior of subjects. The descriptions of agent's selection were set as follows.

Random selection Randomly select 3 agents involved in a pattern formed by the 4 agents in the shortest route, and move in random directions if not included
Self-preferred selection Among patterns reached by the shortest route, select a pattern that includes themselves in preference
Estimate of other agent's intention Estimate a target pattern from the action of the other agent one step before and select a most selected pattern among them

In this simulation, the initial position and initial target pattern were randomly prepared 100 times each. The number of steps that reached each target pattern was compared by a combination of "random selection" and "self-preferred selection" action selection and "estimate of other agent's intention". The result of comparing the number of steps is shown in Fig. 2.

The average number of arrival steps (1-b and 2-b) was reduced for both "random selection" and "self-preferred selection" by combined "estimate of other agent's intention" selection. Furthermore, the number of steps was smaller when the "self-preferred selection (2-b)" than compared with "random selection (1-b)".

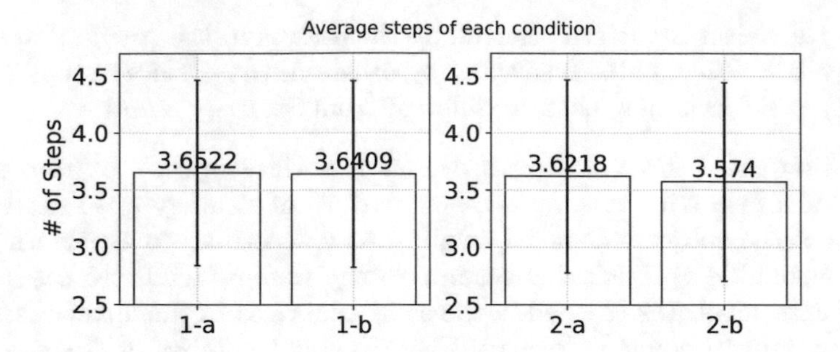

Fig. 2. Comparison of number of steps to reach a target pattern. 1-a: random selection + no estimate of other agent's intention, 1-b: random selection + estimate of other agent's intention, 2-a: self-preferred selection + no estimate of other agent's intention, 2-b: self-preferred selection + estimate of other agent's intention.

5 Discussion

From experimental results, agents calculated the shortest step to reach a target pattern from each position in beginning step, and makes a decision based on it. From next step, target pattern is estimated and adjusted based on other's actions. These results are considered to be close to the action that subjects can take.

Next, in a case of "estimate of other agent's intention", in order to estimate a target pattern from the agent's action, a target pattern including the agent is preferentially selected. Therefore, it is thought that a number of arrival steps will be greatly reduced because "self-preferred selection" will be assumed as other's selection. However, the difference in each results was not so great. This is considered to be influenced by the cause that selection is randomly selected from among multiple majority decisions.

6 Conclusion

In order to clarify a decision making process in cooperative behaviors, we have an agent simulation based on behavioral experiments of cooperative pattern task. In this simulation, we constructed an agent model with three selection methods: "random selection", "self-preferred selection", and "estimate of other agent's intention". From the simulation results, agents with both "self-preferred selection" and "estimate of other agent's intention" can reach a target pattern most quickly, and the results similar to human behavior experiments were obtained. However, since a difference between the results is smaller than expected, detailed analysis such as comparison with human action selection at each step is required.

References

1. Premack D, Woodruff G (1978) Does the chimpanzee have a theory of mind? Behav Brain Sci 1(4):515–526

2. Weiss G (2013) Multiagent systems, 2nd edn. MIT press, Cambridge
3. Baker CL, Saxe R, Tenenbaum JB (2009) Action understainding as inverse planning, cognition, pp 329-349
4. Makino T, Aihara K (2003) Self-observation principle for estimating the other's internal state: a new computational theory of communication, mathematical engineering technical reports METR
5. Omori T, Yokoyama A, Nagata Y, Ishikawa S (2010) Computational modeling of action decision process including other's mind-a theory toward social ability. In: Keynote talk, IEEE international conference on intelligent human computer interaction (IHCI)
6. Henrio J, Henn T, Nakashima T, Akiyama H (2016) Selecting the best player formation for corner-kick situations based on bayes' estimation, Lecture notes in computer science, vol 9776, pp 428-439
7. Itoda K, Watanabe N, Takefuji Y (2015) Model-based behavioral causality analysis of handball with delayed transfer entropy. Procedia Comput Sci 71:85–91
8. Hervieu A, Bouthemy P, Le Cadre JP (2009) Trajectory-based handball video understanding. In: proceedings of the ACM international conference on image and video retrieval. ACM, pp 43
9. Lee W, Tsuzuki T, Otake M (2010) The effectiveness of training for attack in soccer from the perspective of cognitive recognition during feedback of video analysis of matches. Football Sci 7:1–8
10. Yoshida W, Dolan RJ, Friston KJ (2008) Game theory of mind. PLoS Comput Biol 4(12):e1000254
11. Itoda K, Watanabe N, Takefuji Y (2017) Analyzing human decision making process with intention estimation using cooperative pattern task. Lecture Notes in Computer Science vol 10414, pp 249–258

On Capturing the Variability in the Modeling of Individual Behavior

Viacheslav Wolfengagen[1(✉)] and Melanie Dohrn[2]

[1] National Research Nuclear University "Moscow Engineering Physics Institute", Moscow 115409, Russian Federation
jir.vew@gmail.com
[2] NAO "JurInfoR", Moscow 119435, Russian Federation
mdohrn7@gmail.com

Abstract. When processing semantic information, there are difficulties in modeling the behavior of an individual with a variation in his behavior, which is accompanied by a change in its properties. This paper presents a semantic model that is able to take into account the effect of variation in behavior based on an individual-as-process representation. The problem of interaction between individuals who have differing intentions at some stage is considered, which is taken into account by attributing various properties to the information processes representing them. During the development of events at a later stage, individuals may approve similar intentions, which is taken into account by attributing equivalent properties to the information processes representing them. In the future, the approved properties may again differ, requiring the presentation of different processes. The presented semantic model is distinguished by conciseness and minimum of used mathematical tools, which increases its practicality.

Keywords: Conceptual mathematics · Semantic model · Individual-as-process · Behavior variation

1 Introduction

Everyone knows that semantic networks are successfully used to model conceptual dependencies, since we are talking about their correspondence to the first-order predicate logic [19]. However, attempts to construct a semantic Web run into serious difficulties. Some of them are caused by the *separation* of the index and the information being indexed, since over time the correspondence between them can be lost [12]. The other part is due to the objective difficulties of *identifying* content [7]. Finally, the greatest difficulties are caused by semantic modeling of *information processes*, when a simulated individual is changeable, justifying the doctrine of the individual-as-process [25].

Modern tools of semantic modeling [1] are in fact not ready for the deployment of semantic information processing technology [13] for storing and presenting information processes [8]. The available means of describing and maintaining

A. V. Samsonovich (Ed.): BICA 2019, AISC 948, pp. 574–580, 2020.
https://doi.org/10.1007/978-3-030-25719-4_75

content are designed, as a rule, for the static case [5], when the properties of the modeled individuals either do not change or the volume of their changes is insignificant [3]. Representing and maintaining in a network of individuals with systematically changing properties [26] brings into semantic Web a lot of complications [24]. In addition, the mathematical tools of semantic modeling are at a too high level of abstraction [9], and have not yet become an everyday information technology [10].

The most prominent dynamic effects [20] manifest themselves in *biologically inspired cognitive models* [22], when the factor of variability needs to be semantically modeled [23]. The situation becomes even complicated when ae attract the human experts to generate the voluminous on the World-Wide Web [6].

The early contribution on semantic modeling dates back to, at east, to apply the semantic nets and semantic memory [14], and even earlier for modeling changeability with indexical expressions [2] of meaning [4]. In case of any change of individual properties we face to a complicated matter of individual-as-processes entanglement [21]. There are several contributions to apply applicative computational technologies to the area of information processes modeling [18]. Concepts, their variability and generation in computational models was studied in intuitionistic logic [16] and in modal logic Advice on Modal Logic [15], giving rise to computational models based on Cartesian closed categories (c.c.c.) [17], while empirical study led to purely calculational ideas to detect as save concepts [27]. This brief outlook leads to an observation that the previous study of semantic modeling involved the "heavy" mathematics as an ultimate basis for the target theory of computations. This direction has an advantage of closer distance to declarative programming. To the contrast, in case of semantic modeling that tends to applicative approach to programming, and the functional programming, an involving the "light" mathematics has the visible advantage of using the *conceptual mathematics* [11] to generate the concepts and their compositional interrelations.

The latter circumstance motivated the current paper and research. The main effort was to give a pass-trough example to apply the conceptual mathematics to the task of semantic modeling based on a notion of individual-as-process. The given example simulates the emerging information process at all stages of its origin, development and termination.

Section 2 describes the abstract scheme of interaction of information processes while Sect. 3 describes a semantic content model. This is suitable for intuitive clarity, the an example is biologically inspired and important yo understand the conceptual mathematics in use.

2 Abstract Semantic Modeling: Fish Spawning Problem

Let assume a fish-female (spawner) that has swum. Immediately there were male-fishes (milters), and the female began to lay eggs in algae. The female began to guard the eggs, pushing the males, while the males tried to eat eggs. The female did not succeed in pushing away the males, who began to eat eggs that remained visible. As a result, the female began to eat her own eggs.

This formulation corresponds to the semantic model of processes, the scheme of which is shown in Fig. 1.

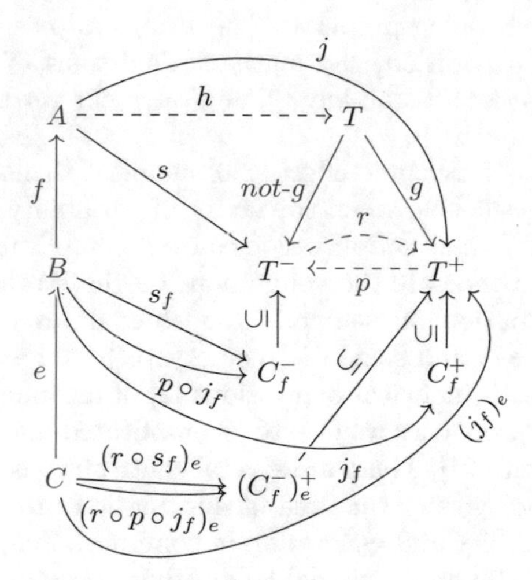

Fig. 1. The scheme of interaction processes.

Stage B is later than A, and stage C is later than B. The symbolization used is as follows: T – fish, T^- – fish that do not eat eggs, T^+ – fish that eat eggs; as involutes are $f : B \to A$ and $e : C \to B$; g – eat eggs, *not-g* – do not eat eggs. Further, h is a fish; s – female (spawner), j – male (milter).

$$s = \textit{not-g} \circ h = \text{“behaves like a female”}$$

$$j = g \circ h = \text{“behaves like a male”}$$

Behavior is characterized by r (rethink, change mind) and p (pushed back, move) mappings.

In the circumstances of B, the female does not eat eggs, but pushes away males trying to eat eggs.

$$p \circ j_f = s_f = \text{“behaves like a female”} = \text{“male, who was pushed, does not eat eggs”}$$

The “changed her mind” (r) female begins to behave like a male, i.e. eat eggs:

$$r \circ s_f = j_f = \text{“who has changed her mind behaves like a male”} = \text{“who has changed her mind that eats eggs”}$$

$r \circ p \circ j_f = j_f =$ “the changed male who was pushed away behaves like a (normal) male” = “who changed his mind who was pushed away, eats the eggs”

In the circumstances of C, the female eats eggs, and the males eat eggs.

$$(r \circ s_f)_e = (j_f)_e = \text{“who has changed her mind behaves like a male”} = \text{“who has changed her mind eats eggs on a par with the males”}$$

3 Content Semantic Modeling: Fish Spawning Problem

The model in Sect. 2 is compiled ia a rather abstract manner, and the intuitive clarity could be lost in part. To save more content reasons, the emerging processes and their interaction are shown in Fig. 2. The scheme above is based on more subject information but has the same skeleton.

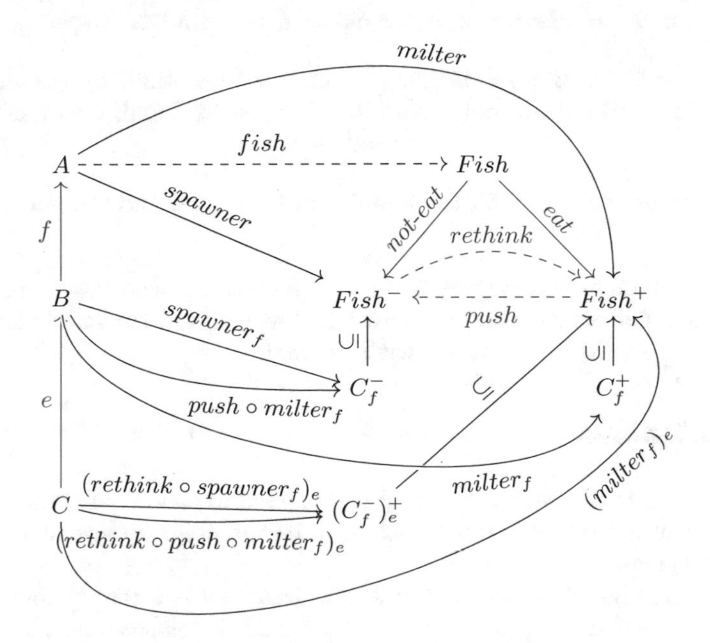

Fig. 2. The problem of spawning fish.

Stage B is later than A, and stage C is later than B. The symbolization used is as follows: $Fish$ – a fish, $Fish^-$ – the fish that do not eat eggs, $Fish^+$ – the fish that eat eggs; as involutes are $f : B \to A$ and $e : C \to B$; *eat* – eat eggs, *not-eat* – do not eat eggs; C_f^- – the fish that do not eat eggs during the evolving of events f, C_f^+ – the fish eating eggs during the evolving of events f, $(C_f^-)_e^+$ – the changed-mind fish eating eggs during the composition of evolvents of events $f \circ e$.

Furthermore, $fish$ – the fish; *spawner* – the female fish (spawner), *milter* – the male fish (milter).

$$spawner = not\text{-}eat \circ fish = \text{“behaves like a female fish”}$$

$$milter = eat \circ fish = \text{“behaves like a male fish”}$$

The behavior is characterized by *rethink* (rethink, change mind) and *push* (pushed back, move) mappings.

In the circumstances of B, the female does not eat eggs, but pushes away males trying to eat eggs.

$push \circ milfer_f = spawner_f =$ "behaves like female" = "pushed male doesn't eat eggs"

The "changed her mind" (*rethink*) female begins to behave like a male, i.e. eat eggs.

$rethink \circ spawner_f = milter_f =$ "female who changes her mind behaves like a male" = "female who has changed her mind eats eggs"

$rethink \circ push \circ milter_f = milter_f =$ "changed his mind, who was pushed away, behaves like a (normal) male" = "changed his mind, who was pushed away, eats eggs"

In the circumstances of C, the female fish eats eggs, and the males eat eggs as well.

$(rethink \circ spawner_f)_e = (milter_f)_e =$ "a female fish who has changed her mind behaves like a male fish" = "that who has changed her mind eats eggs on a par with the males"

4 Conclusions

The conceptual mathematics was used and applied to an area of semantic modeling and semantic information processing. In fact, a pre-structure of the variable domains was generated.

Next advantage of the generated semantic model is a purely compositional skeleton of computations, and they are the *semantic computations*. This kind of computations has an intuitive clarity and transparency to model a complicated dynamical effects.

The target semantic model is purely based on the notion of individual-as-process. Further study of this kind of semantic models can be more closely based on the indexical individuals but this needs slightly modified explanatory system.

Acknowledgements. This research is supported in part by the Russian Foundation for Basic Research, RFBR grants 19-07-00326-a, 19-07-00420-a, 18-07-01082-a, 17-07-00893-a.

References

1. Angles R, Arenas M, Barceló P, Hogan A, Reutter J, Vrgoč D (2017) Foundations of modern query languages for graph databases. ACM Comput Surv 50(5):68:1–68:40. https://doi.org/10.1145/3104031
2. Bar-Hillel Y (1954) Indexical expressions. Mind 63(251):359–379 http://www.jstor.org/stable/2251354
3. Berman S, Semwayo TD (2007) A conceptual modeling methodology based on niches and granularity. In: Parent C, Schewe KD, Storey VC, Thalheim B (eds) Conceptual modeling. Springer, Berlin Heidelberg, Berlin, Heidelberg, pp 338–358

4. Carnap R (1947) Meaning and necessity. University of Chicago Press, Chicago
5. Chateaubriand O (2002) Descriptions: frege and russell combined. Synthese 130(2):213–226. https://doi.org/10.1023/A:1014439314029
6. Doan A, Ramakrishnan R, Halevy AY (2011) Crowdsourcing systems on the world-wide web. Commun ACM 54(4):86–96. https://doi.org/10.1145/1924421.1924442
7. Helland P (2019) Identity by any other name. Commun ACM 62(4):80–80. https://doi.org/10.1145/3303870
8. Hyman M (2007) Semantic networks: a tool for investigating conceptual change and knowledge transfer in the history of science. In: Böhme H, Rapp C, Rösler W (eds) Übersetzung und transformation. de Gruyter, Berlin, pp 355–367
9. Ismailova L (2014) Applicative computations and applicative computational technologies. Life Sci J 11(11):177–181
10. Ismailova L (2014) Criteria for computational thinking in information and computational technologies. Life Sci J 11(9s):415–420
11. Lawvere FW, Schanuel SJ (1997) Conceptual mathematics: a first introduction to categories. Cambridge University Press, Cambridge
12. Lewandowski D (2019) The web is missing an essential part of infrastructure: an open web index. Commun ACM 62(4):24–24. https://doi.org/10.1145/3312479
13. Minsky ML (1969) Semantic Information Processing. The MIT Press, Cambridge
14. Quillian MR (1996) Semantic Memory, Technical Report Project No 8668, Bolt Beranek and Newman Inc
15. Scott D (1970) Advice on modal logic. In: Lambert K (ed) Philosophical problems in logic: some recent developments. Springer, Dordrecht, pp 143–173. https://doi.org/10.1007/978-94-010-3272-8_7
16. Scott D (1979) Identity and existence in intuitionistic logic. Springer, Berlin, Heidelberg, pp 660–696. https://doi.org/10.1007/BFb0061839
17. Scott D (1980) Relating theories of the λ-calculus. In: Hindley J, Seldin J (eds) To H.B. Curry: essays on combinatory logic, lambda-calculus and formalism. Academic Press, Berlin, pp 403–450
18. Scott DS (2012) Lambda calculus then and now. In: ACM turing centenary celebration, ACM-TURING 2012. ACM, New York. https://doi.org/10.1145/2322176.2322185
19. Wolfengagen V (2010) Semantic modeling: computational models of the concepts. In: Proceedings of the 2010 international conference on computational intelligence and security, CIS 2010, pp 42–46. IEEE computer society, Washington. https://doi.org/10.1109/CIS.2010.16
20. Wolfengagen V (2014) Computational ivariants in applicative model of object interaction. Life Sci J 11(9s):453–457
21. Wolfengagen VE, Ismailova LY, Kosikov S (2016) Computational model of the tangled web. Procedia Comput Sci 88:306–311 (2016). https://doi.org/10.1016/j.procs.2016.07.440. 7th annual international conference on biologically inspired cognitive architectures, BICA 2016, held 16 July to 19 July 2016 in New York City. http://www.sciencedirect.com/science/article/pii/S1877050916316969
22. Wolfengagen VE, Ismailova LY, Kosikov S (2016) Concordance in the crowdsourcing activity. Procedia Comput Sci 88:353–358 (2016). https://doi.org/10.1016/j.procs.2016.07.448. 7th annual international conference on biologically inspired cognitive architectures, BICA 2016, held 16 July to 19 July 2016 in New York City. http://www.sciencedirect.com/science/article/pii/S1877050916317045

23. Wolfengagen VE, Ismailova LY, Kosikov SV (2016) A computational model for refining data domains in the property reconciliation. In: 2016 third international conference on digital information processing, data mining, and wireless communications (DIPDMWC), pp 58–63. https://doi.org/10.1109/DIPDMWC.2016.7529364
24. Wolfengagen VE, Ismailova LY, Kosikov SV (2016) A harmony and disharmony in mining of the migrating individuals. In: 2016 third international conference on digital information processing, data mining, and wireless communications (DIPDMWC), pp 52–57. https://doi.org/10.1109/DIPDMWC.2016.7529363
25. Wolfengagen VE, Ismailova LY, Kosikov SV, Navrotskiy VV, Kukalev SI, Zuev AA, Belyatskaya PV (2016) Evolutionary domains for varying individuals. Procedia Comput Sci 88:347–352. https://doi.org/10.1016/j.procs.2016.07.447. 7th annual international conference on biologically inspired cognitive architectures, BICA 2016, held 16 July to 19 July 2016 in New York City. http://www.sciencedirect.com/science/article/pii/S1877050916317033
26. Wolfengagen VE, Ismailova LY, Kosikov SV, Parfenova IA, Ermak MY, Petrov VD, Nikulin IA, Kholodov VA (2016) Migration of the individuals. Procedia Comput Sci 88:359–364. https://doi.org/10.1016/j.procs.2016.07.449. 7th annual international conference on biologically inspired cognitive architectures, BICA 2016, held 16 July to 19 July 2016 in New York City. http://www.sciencedirect.com/science/article/pii/S1877050916317057
27. Xie Y, Ravichandran A, Haddad H, Jayasimha K (2008) Capturing concepts and detecting concept-drift from potential unbounded, ever-evolving and high-dimensional data streams. Springer, Heidelberg, pp 485–499. https://doi.org/10.1007/978-3-540-78488-3_28

Cognitive Features for Stability of Semantic Information Processing

Viacheslav Wolfengagen[1(✉)], Larisa Ismailova[1], and Sergey Kosikov[2]

[1] National Research Nuclear University "Moscow Engineering Physics Institute", Moscow 115409, Russian Federation
jir.vew@gmail.com, lyu.ismailova@gmail.com

[2] Institute for Contemporary Education "JurInfoR-MGU", Moscow 119435, Russian Federation
kosikov.s.v@gmail.com

Abstract. The issues of semantic information processing are considered. The central idea of information in the modern world is still an elusive concept. It is known that information must be quantified, at least in terms of partial ordering, to be additive, stored and transmitted. However, besides all this, there is no clear idea about its specific nature. The development of cognitive technologies poses the question of forms of knowledge representation: can the "stage of knowledge" be fully and faithfully analyzed in terms and forms of information processing at some level of abstraction? In other words, the analysis is to relate (1) the stage of knowledge with (2) forms of information processing and (3) levels of abstraction. Among the problems posed by the development of information technology, one of the key roles is played by the information semantics. The most significant open problems imply the development of research on a new basis in the following areas: providing data substantiation (how data can acquire meaning); ensuring the truth of the data (as meaningful data can acquire true values); building an information theory of truth (can information explain truth); constructing information semantics (can information explicate a value).

Keywords: Knowledge mining · Data semantics · Semantic information processing · Semantic network · Semantic vulnerabilities

1 Introduction

The central idea of *information* in the modern world is still an elusive concept. It is known that information must be quantified, at least in terms of partial ordering, to be additive, stored and transmitted. However, besides all this, there is no clear idea about its specific nature. The development of cognitive technologies poses the question of forms of knowledge representation: can the "level of knowledge" be fully and satisfactorily analyzed in terms and forms of information processing at some level of abstraction? In other words, the analysis is to

A. V. Samsonovich (Ed.): BICA 2019, AISC 948, pp. 581–588, 2020.
https://doi.org/10.1007/978-3-030-25719-4_76

relate (1) the level of knowledge with (2) forms of information processing and (3) levels of abstraction.

One of the "narrow places for the acquisition of knowledge", in essence, "the bottleneck of knowledge acquisition", is a largely unused source of general knowledge in texts that lie at the next level behind explicit affirmative content, which lies at the surface level. This knowledge consists of relationships that are supposed to be possible in the world or, under certain conditions, are meant as ordinary or generally accepted in the world. First of all, some corpus of texts is taken from which knowledge will be extracted. Usually the output of such general world knowledge is done in two stages: - first, we obtain general "possible" judgments, based on the nominal phrases and designations of objects available; - secondly, we are trying to get stronger generalizations based on the nature and statistical distribution of possible claimants for relations obtained at the first stage.

Section 2 describes the narrow places of semantic modeling. Section 3 covers the aspects of semantic modeling of information processes. Section 4 contains a pass-through example with transparent cognitive reasons to apply the model-theoretic *forcing* to semantic analysis of information processes.

The identity and existence in intuitionistic logic goes back to [4] and was rediscovered for Web under the name of Identity by Any Other Name [1]. This paper continues the study of Semantic Modeling and Computational Models of the Concepts, started in [5]. An significant semantic difficulty of representing the variable properties was overcame in analysis of Migration of the Individuals in [11] and for Evolutionary Domains for Varying Individuals in [10].

The Conceptual Mathematics as in [3] was used for the aims of representing the semantic processes. The approach is slightly similar to finding the computational invariants in applicative model as in [6]. Earlier the method for refining data domains in the property reconciliation was studied in [8]. The applicative computational framework was added by [2]. The study in this paper slightly differs from the semantic model of entanglement in [7] by the ability to represent the interacting information processes [9].

2 Semantic Modeling

The preliminary result of the first stage is usually semantic networks. Semantic networks, or networks, mean different things to different people. Their representations range from charts on paper to abstract n-tuples of sets of some kind and further to data structures in a computer or even information structures arising in the brain. Let us stand on a practical stand, considering semantic networks as *graphical analogs* of data structures representing "facts" in a computer system oriented towards understanding natural language. Semantic networks help both in the formulation and the expression of the computer data structures they reflect. Semantic networks are often used to solve a class of problems associated with *understanding*. Informal and diverse ways of using semantic networks make it difficult to work out their exact definition in an easy way. However, they usually have the following characteristics:

(1) specific as well as general concepts, concepts, have an idea by means of labeled or unlabeled nodes of the graph;
(2) statements correspond to subgraphs with arcs to the predicative concept and to the appropriate number of conceptual arguments for the predicate. Sometimes explicit propositional nodes are entered as points. Bindings for these arcs and as elements on which propositional operators can work (for example, "knows that"). Arguments of n-ary predicates can be distinguished by the use of labels on arcs, different types of arcs, or binary decomposition of predicates;
(3) duplication of nodes denoting the same concept is excluded. Thus, several arcs associated with several different statements may indicate a conceptual node. Such nodes are usually considered to correspond to the unique storage location in the computer's memory, that is, as an entry point to access knowledge about this concept. Similarly, propositional nodes are also considered as unique ones.

When coding actual knowledge, semantic networks are more preferable than calculus of predicates because of their greater naturalness and comprehensibility. This is due to the one-to-one correspondence between the nodes and the concepts they designate, as well as clustering statements related to one particular thing in the form of a single vertex, as well as visual visibility of the "interrelationships" of concepts, i.e. their connections, which are expressed by sequences of propositional arcs. Thanks to these features, semantic networks are more suitable for expressing "associativity" and for applying matching algorithms to detect the intersecting nodes of two related concepts or to overlap more complex contexts. On networks, you can also provide some form of deductive reasoning.

Recognizing the advantages of representation in the form of a semantic network over representation in the predicate calculus, both of these forms are closely interrelated. In practice, in most cases of the semantic network, it is easy to associate an equivalent entry in the predicate calculus. In practice, the forms of semantic networks are used, which by expressive possibilities, are weaker than the predicate calculus, especially when introducing quantifiers and higher order sentences. In the extended interpretation of the semantic network, n-ary predicates are used for $n = 1, 2, 3, \ldots$, logical connectives, unrestricted quantifiers, including predicate quantifiers, lambda-abstraction, and modal operators of various types, as well as counterfactual implication. In addition, when reasoning on the semantic network is provided, it is allowed to use not so much formal ability of predicate calculus as such, but many different heuristics instead. Nevertheless, all extensions of the expressive possibilities of the network are performed by analogy with the constructions of the predicate calculus of the first order and higher orders.

3 Semantic Modeling of Information Processes

However, in the long run, it may be necessary to take into account the time dependence of even relatively stable properties. Consider the following story (LK Schubert).

> One day Johnny caught a tadpole. He called it "Hugo" and kept it in a jar. A year later, when Johnny set Hugo free, Hugo was a frog.

Question: Was Hugo a tadpole or a frog when Johnny caught it? In the semantic network in which the "tadpole" was based on Hugo in the continuous mode, this question is not resolvable; Hugo may already have been a frog when Johnny caught him. So we need to write down that Hugo was a tadpole when Johnny caught him.

Analytically, we can write the expression

$$\begin{aligned}&\exists x.[loc(Hugo, x, T1) \;\&\; \text{belongs}(x, Tadpole)]\\ &\& \; [(\forall t)[t \in T1 \Rightarrow \text{before}(t, now1)]]\\ &\& \; \text{then}(T1, T2)\\ &\& \; \exists y.[loc(Hugo, y, T2) \;\&\; \text{belongs}(y, Frog)],\end{aligned}$$

which is presented in Fig. 1. In fact, the change of the *Tadpole* set, to which *Hugo* belonged *before* the present moment, to the *Frog* set, is written to which *Hugo* belongs *after* the current moment:

$$\begin{aligned}&\exists x.[loc(Hugo, x, T1) \;\&\; x \in Tadpole]\\ &\& \; [(\forall t)[t \in T1 \Rightarrow \text{before}(t, now1)]]\\ &\& \; \text{then}(T1, T2)\\ &\& \; \exists y.[loc(Hugo, y, T2) \;\&\; y \in Frog].\end{aligned}$$

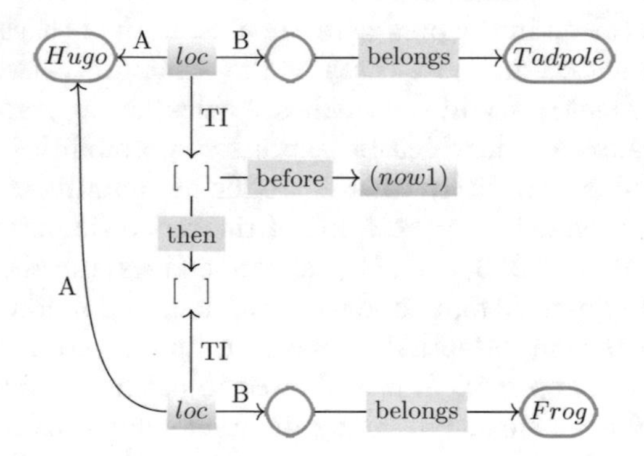

Fig. 1. "The tadpole Hugo became (transformed into) frog". Argument marker *TI* and square brackets are used for time intervals to distinguish them from moments of time.

More gradual changes associated with growth and aging show that even terms such as "kitten", "cat", "girl" or "woman" are properly considered as time-dependent. Determining how to ignore time dependence seems to contain some subtleties. When depicting semantic networks for illustrative purposes, of course, we justify ourselves by omitting seemingly unimportant time dependencies. Such omissions can be justified even in computerized analogues of semantic networks, especially if most of the missing information can be recovered by means of inference. However, semantic networks should, in principle, be able to express time constraints for all sentences. This is possible using the proposed method of associating time with each predicate over physical objects, and not, say, with "episodes" that define more or less complex events.

It is not difficult to formulate analogs of the instantaneous, interval, and constant modes of predication for an approach based solely on timeless predicates. The time interval will be considered as a "slice" of the four-dimensional material world, perpendicular to the time axis. Its intersection with the individual gives a time slice of this individual. The analog of the constant predication moments will include the application of the predicate to the full four-dimensional individual, in the sense that the predicate can be strictly applicable only to the correct temporal fragment of this individual.

4 Cognitive Features of Forcing

Consider turning a tadpole into a frog, naming it "Hugo" as in Fig. 2. This occurs along the evolvent $f : B \to A$ in accordance with the diagram This diagram means that

"the tadpole Hugo in A became by g a frog in B along the evolvent f",

where $f : B \to A$ is assumed as *enforcement*, or *forcing*, and for $a \in A$ assume, that $hugo_a = Hugo$ in case $Hugo \in Tadpole$.

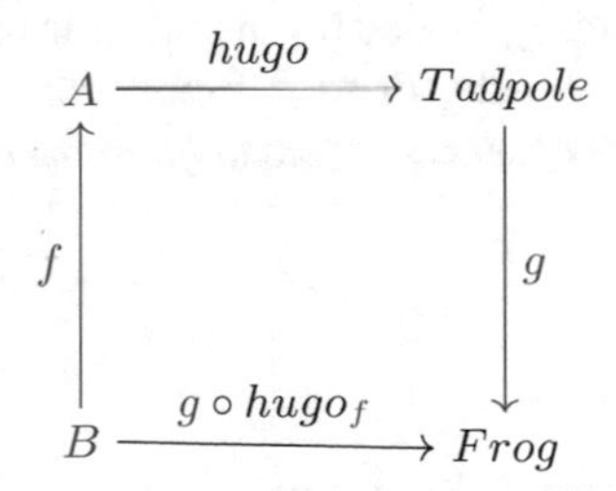

Fig. 2. The diagram of enforced transformation.

In short, this is formulated as

"the tadpole Hugo is forced to become a frog",

or simply

"the tadpole Hugo became a frog".

$$A \xrightarrow{tadpole} T$$

Fig. 3. Individual-as-process.

The "tadpole" h is assumed as a process according to Fig. 3. In above: $T = Tadpole$, and the set of tadpoles in A can be represented by a variable domain $H_T(A)$:

$$\text{"tadpoles"} = H_T(A) = \{ \, tadpole \mid tadpole : A \to T \, \}.$$

The set of frogs in B can be represented by a variable domain $H_F(B)$:

$$\text{"frogs"} = H_F(B) = \{ \, frog \mid frog : B \to F \, \}$$

for $F = Frog$.

For the example above follows that

$$hugo_f \in H_{Tadpole}(B) \;\; \text{for } hugo \in H_{Tadpole}(A),$$
$$g \circ hugo_f \in H_{Frog}(B) \text{ for } hugo \in H_{Tadpole}(A).$$

Turning to symbolization, for A-forced tadpoles h from $H_T(A)$ we get the set $C_f^g(B)$ of all f-forced frogs $g \circ h_f$ in B, which is a subset of $H_F(B)$ of the set of all B-forced frogs:

$$C_f^g(B) \subseteq H_F(B),$$

where $F = Frog$.

The question arises whether Hugo, who has become a frog from a tadpole under the influence of f-forcing, is distinguishable from Hugo, who may already have been a frog. In other words, does "Hugo" have sufficient identifying features so that there are no *confusing*, or *entanglement* of two, generally speaking, different Hugo?

5 Conclusions

The cognitive reasons were used as the "stages of knowledge".

It was shown that this gives rise to the cognitive technology which can be fully and faithfully analyzed in terms and forms of information processing at some level of abstraction.

A representative sample is formulated enabling the pass-through study of connection between the semantic networks and information processes.

The model-theoretic forcing is applied to model the individual with variable properties. The properties are transformed with the stages of knowledge evolution reflecting the overall semantic information process.

The proposed commutative diagram reflects both the stages of knowledge and property transformation.

We hope that the obtained results can assist in development the advanced semantic information processing tools. Their specific feature is in using the variable domains. Thus, the true propositions discovered at the early stages of knowledge can be false at the later stages and vice versa. This can potentially destroy the integrity of the conventional information system. To prevent the possibility of information processes entanglement we use the method of forcing in a computational model. This increase the semantic stability of the target cognitive model.

Acknowledgements. This research is supported in part by the Russian Foundation for Basic Research, RFBR grants 19-07-00326-a, 19-07-00420-a, 18-07-01082-a, 17-07-00893-a.

References

1. Helland P (2019) Identity by any other name. Commun ACM 62(4):80. https://doi.org/10.1145/3303870
2. Ismailova L (2014) Applicative computations and applicative computational technologies. Life Sci. J 11(11):177–181
3. Lawvere FW, Schanuel SJ (1997) Conceptual mathematics: a first introduction to categories. Cambridge University Press, Cambridge
4. Scott D (1979) Identity and existence in intuitionistic logic. Springer, Heidelberg, pp 660–696. https://doi.org/10.1007/BFb0061839
5. Wolfengagen V (2010) Semantic modeling: computational models of the concepts. In: Proceedings of the 2010 international conference on computational intelligence and security, CIS 2010, IEEE Computer Society, Washington, DC, pp 42–46. https://doi.org/10.1109/CIS.2010.16
6. Wolfengagen V (2014) Computational ivariants in applicative model of object interaction. Life Sci J 11(9s):453–457
7. Wolfengagen VE, Ismailova LY, Kosikov S (2016) Computational model of the tangled web. Procedia Comput Sci 88:306–311 https://doi.org/10.1016/j.procs.2016.07.440, http://www.sciencedirect.com/science/article/pii/S1877050916316969. 7th annual international conference on biologically inspired cognitive architectures, BICA 2016, held July 16 to July 19, 2016 in New York City, NY, USA
8. Wolfengagen VE, Ismailova LY, Kosikov SV (2016) A computational model for refining data domains in the property reconciliation. In: 2016 third international conference on digital information processing, data mining, and wireless communications (DIPDMWC) pp 58–63. https://doi.org/10.1109/DIPDMWC.2016.7529364
9. Wolfengagen VE, Ismailova LY, Kosikov SV (2016) A harmony and disharmony in mining of the migrating individuals. In: 2016 third international conference on digital information processing, data mining, and wireless communications (DIPDMWC), pp 52–57. https://doi.org/10.1109/DIPDMWC.2016.7529363
10. Wolfengagen VE, Ismailova LY, Kosikov SV, Navrotskiy VV, Kukalev SI, Zuev AA (2016) Belyatskaya PV (2016) evolutionary domains for varying individuals. Procedia Comput Sci 88:347–352 https://doi.org/10.1016/j.procs.2016.07.447, http://www.sciencedirect.com/science/article/pii/S1877050916317033. 7th annual international conference on biologically inspired cognitive architectures, BICA 2016, held July 16 to July 19, 2016 in New York City, NY, USA

11. Wolfengagen VE, Ismailova LY, Kosikov SV, Parfenova IA, Ermak MY, Petrov VD, Nikulin IA, Kholodov VA (2016) Migration of the individuals. Procedia Comput Sci 88:359–364 https://doi.org/10.1016/j.procs.2016.07.449, http://www.sciencedirect.com/science/article/pii/S1877050916317057. 7th Annual International Conference on Biologically Inspired Cognitive Architectures, BICA 2016, held July 16 to July 19, 2016 in New York City, NY, USA

Mutable Applicative Model to Prevent Entanglement of Information Processes

Viacheslav Wolfengagen[1](✉), Larisa Ismailova[1], Sergey Kosikov[2], and Mikhail Maslov[2]

[1] National Research Nuclear University "Moscow Engineering Physics Institute", Moscow 115409, Russian Federation
jir.vew@gmail.com, lyu.ismailova@gmail.com
[2] Institute for Contemporary Education "JurInfoR-MGU", Moscow 119435, Russian Federation
kosikov.s.v@gmail.com, maslov@ihep.ru

Abstract. A system for detecting and possibly preventing undesirable changes in the properties of objects within information system (IS), leading to a violation of the semantic consistency of information, unstable functioning of the IS and, ultimately, loss of its operability, is proposed. The modeling tools is a means for detecting and preventing the destruction of the semantic integrity of data/metadata objects in IS. The basis is the original computational model, which increases the sustainability of semantic nets, while giving them more opportunities to reflect the dynamics of the problem domain. This gives rise to IS with increased expressive capabilities, supporting the evolving semantic nets, integrated with the Web. Mechanisms for modeling the dynamics of concepts and the computational model contribute to overcoming the contradictions that exist for today in the application of semantic information technologies.

Keywords: Mutability · Information process · Entangling · Applicative technologies · Variable objects · Semantic network sustainability

1 Introduction

The existing systems and tools of business intelligence contribute to the value of large data, but, nevertheless, the degree of enrichment usually ranges from low to medium. Increasingly, the question is raised about the increase in the 'degree of cognitivization' of the data and, in the long term, about the transition to cognitive business. The problem of constructing systems for processing and analyzing large data, as a rule, requires an accurate formulation and solution of the problem of data enrichment, which is still an open problem. In turn, the enrichment of data implies the allocation of these or that semantics.

Further semantic processing of data consists in introducing structuring into data collected from heterogeneous sources, with their subsequent organization in

A. V. Samsonovich (Ed.): BICA 2019, AISC 948, pp. 589–596, 2020.
https://doi.org/10.1007/978-3-030-25719-4_77

the form of information graphs (IG). The development of data enrichment technology based on information graphs (IG) is proposed. The notion of a channeled information graph (CIG) is developing, based on the original dynamic computational model. The technology of data enrichment has several components. Its declarative part is expressed by rules that have an antecedent-consequent structure. Its semantic component is expressed by information graphs forming a network. The management structure is based on the idea of channeling the flow of information/data and reflects the procedural part of the technology. Environments for rich data are information graphs (IG), which form a network. The enrichment process is based on specialized interaction with the IG network. The main instrument of specialization is information channels. They define the basis of interaction with the network through messages.

Section 2 describes the applicative computational model based on the notion of generalized elements. Section 3 covers the reasons for arising possible semantic entanglement as a source of semantic vulnerability and the way for its overcoming when variable domains are used.

The semantic modeling approach used in this paper is in accordance with the technology of Semantic Information Processing [4]. The semantic networks are similar to networks for investigating conceptual change and knowledge transfer [1]. Data model is developed using the Conceptual Mathematics [3] combined with the early incite in Semantic Nets and semantic memory [5].

The notion of entanglement is studied starting with Computational Model of the Tangled Web [7] and attracts the model of Migration of the Individuals [10]. The general computational framework is conducted using the Criteria for Computational Thinking in Information and Computational Technologies [2] as well as Computational Invariants in Applicative Model of Object Interaction [6]. The computational model represented in this paper continues the constructions developed in the computational model for refining Data domains in the property reconciliation [8]. Some other means to increase the dynamics of semantic net were used from the semantic model for mining of the migrating individuals [9].

2 Generalized Elements

Some considerations from the attempts to generalize elements are as follows.

2.1 Representation

An *element* is a particular mapping: $1 \xrightarrow{h} T$.

Instantiation, or *evaluation* is a particular case of composition as is shown in Fig. 1, i.e. $g \circ h = h' = g(h)$.

An *evaluation of composition* for the associativity law of composition, as is shown in Fig. 2, is a particular case of the associativity law of composition, and this particular case is reflected in Fig. 3.

In many cases 'element' from T means the mapping to T from arbitrary A, not obviously that $A = 1$.

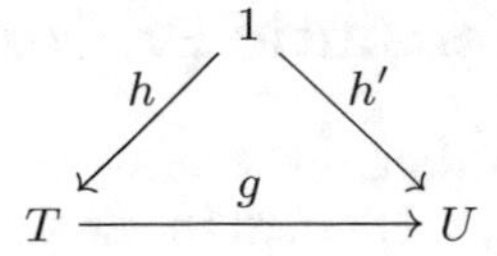

Fig. 1. Evaluation, or instantiation as a particular case of composition.

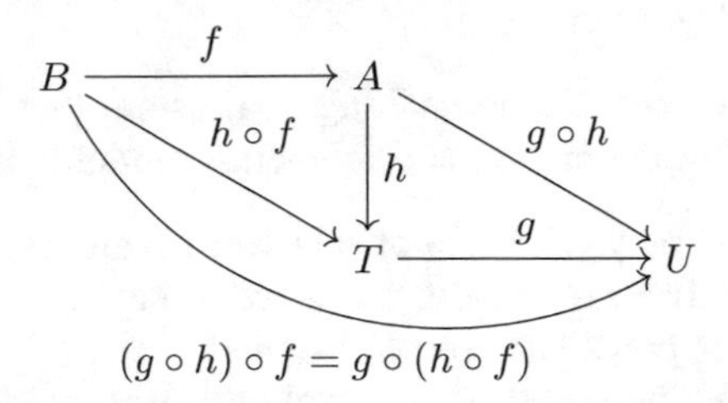

Fig. 2. Evaluation of composition for the associativity law of composition.

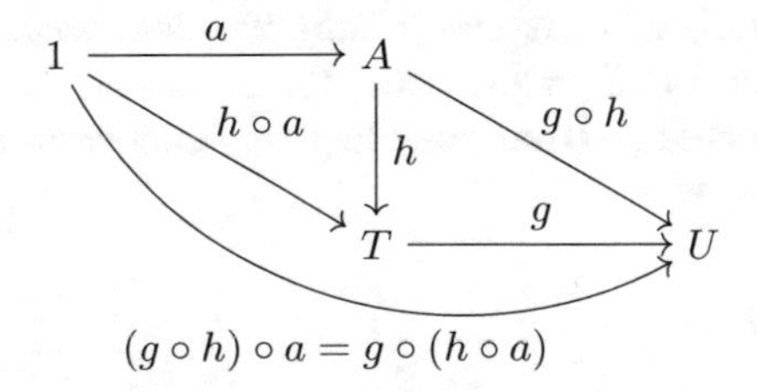

Fig. 3. A particular case of the associativity law of composition.

2.2 Meaningful Understanding

More precisely, a mapping in T from an arbitrary A can be called 'a generalized element' of T, which is an arbitrary mapping, whose domain of value is T. It has some domain definition, which does not necessarily coincide with the terminal element 1. It is most meaningful to use the term 'variable element' of T depending on A.

When we say *expenditure* of funds, we mean the current expenditure, which is an element of a multitude of expenditures, and by this we mean the real expenditure. For example, two days ago it was one, and now it is different. It varies, but is *determined.* It can be read that A is the set of days, for each of which there is a constant element characterizing the expense. The actual expense for that day is the value of mapping, and the mapping itself is the expense in question. It varies, but can still be considered as one element.

3 Entanglement in Semantic Process Recognition

When analyzing the problem domain, the *semantic recognition* of the process becomes paramount. In short, this refers to the selection of base classes and the 'evolution scenarios' or *evolvents* that link them.

3.1 Explanatory System

To explain the *process*, that is, something that is subject to change depending on certain factors, or *assignments*, the following model is used:

- an abstract set that plays the role of selected parameters, which allow you to look at the process from a certain "angle of view";
- an abstract set that plays the role of values;
- mapping describing the result of the evolving process.

The process itself may be caused by certain circumstances, but it is important that the result is in the form of a sequence of values. Changes are often described in this way. It is important that every time we deal with a well-defined value. Variable *sets* are often found in practice, for example, set of people living in a particular village may change from year to year. The similar situation is modeled by category of variable sets.

3.2 Sample Script

Let the process be described by means of 'suppliers supply goods to stores'. One evolvent dictates that the supply be preceded by (prerequisite) order:

$$\textit{prerequisite: supply} \mapsto \textit{order}.$$

Another evolvent prescribes that the effect of the supply is to receive the goods:

$$\textit{effect: supply} \mapsto \textit{receive}.$$

Further, the receipt (receive) is preceded by (prerequisite) supply (supply):

$$\textit{prerequisite: receive} \mapsto \textit{supply}.$$

Finally, the result (effect) of an order (order) is a supply (supply)

$$\textit{effect: order} \mapsto \textit{supply}.$$

A commutative diagram arises as in Fig. 4 This diagram gives an *ideal*, *semantically secure* representation of the process of supplying goods: ordered goods are supplied, and received goods are listed as supplied. The supplied goods form the set C, and the stock of possible goods makes up the set T. The sequence 'order'-'supply'-'receive' is represented respectively by objects B_1, A and B_2.

In practice, the process is less ideal and requires a slightly different view, in which a pairwise distinction is made between the ordered, supplied and received goods. This model takes into account what can be:

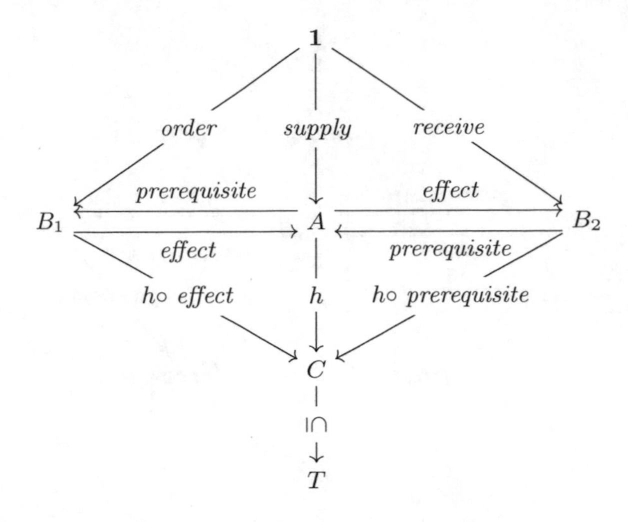

Fig. 4. The process 'suppliers supply goods to stores'.

(1) *not* ordered, *not* supplied, *not* received goods;
(2) *not* ordered, *not* supplied, received goods;
(3) *not* ordered, supplied, *not* received goods;
(4) *not* ordered, supplied, received goods;
(5) ordered, *not* supplied, *not* received goods;
(6) ordered, *not* supplied, received goods;
(7) ordered, supplied, *not* received goods;
(8) ordered, supplied, received goods.

In fact, the diagram above corresponds to condition (8). In order to represent a more general requirement, a modified model will be required. According to the diagram in Fig. 5, there is a process for which we have to consider four domains (sets): T – possible goods, C – supplied goods, C_{effect} – ordered goods, $C_{prerequisite}$ – received goods.

In case of coincidence of the last three sets, when

$$C_{effect} = C = C_{prerequisite}$$

it turns out a *safe* process: only those goods that were ordered are supplied, and the received goods are only those that were supplied. It meets the requirement of the case (8). In other cases, a potentially *unsafe* process occurs, i.e. threat for semantic security or *semantic vulnerability*.

Recovery for the simulated process semantically safe mode requires a more rigorous approach, which implies the presence of a diagram in Fig. 6. In this diagram, the set $C(B_1)$ forms the ordered goods from the stock T in general, and the set $C(B_2)$ forms the received goods from the stock T in general.

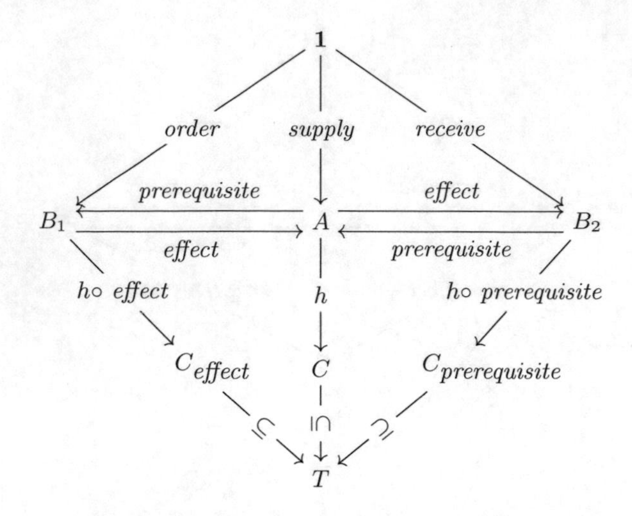

Fig. 5. Generation of derived concepts.

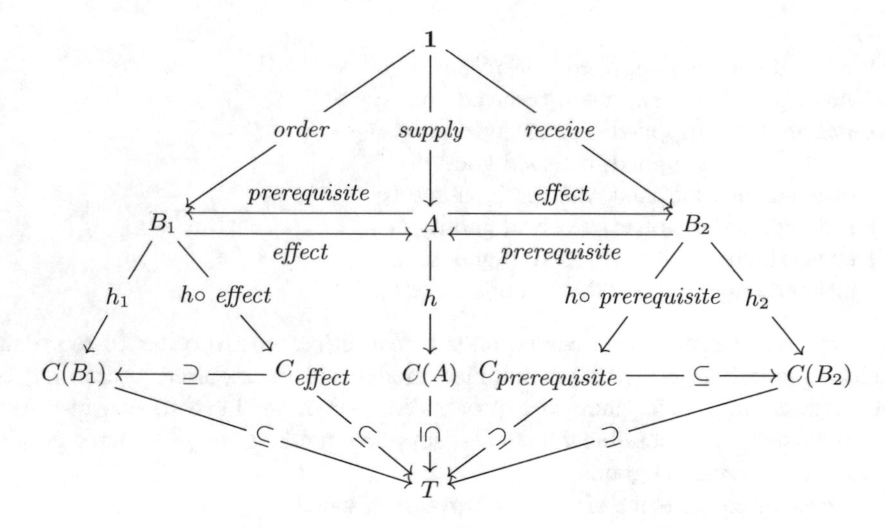

Fig. 6. Untangled derived concepts for semantically safe modeling.

As a result of the analysis, the initial process gets an idea in the form of enriched mathematical model in Fig. 7. In terms of variable sets, this model has a completely transparent interpretation and leads to control over the semantic security of the process.

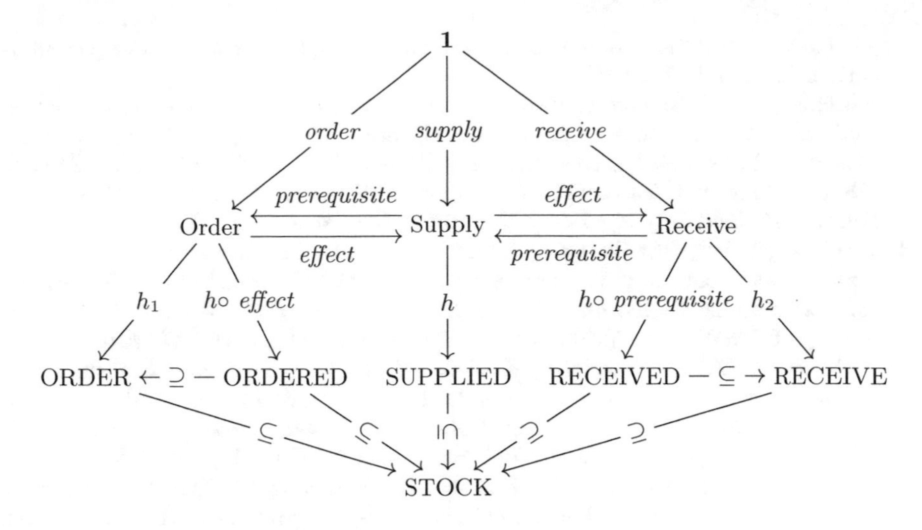

Fig. 7. The enriched semantic model.

4 Conclusions

A representative sample of the mutable applicative model to prevent entanglement of information processes is given. This sample reflects the evolving information processes which interact each other.

To the contrast of restricted indexical structures the model is based on commutative diagrams with mapping between arbitrary sets.

The source of semantic vulnerability is indicated and the way to overcome the entanglement of information processes is given. A transparent pass-through sample of interacting information is given as well as the explanation how to maintain the valid semantic model with the cognition elements.

Acknowledgements. This research is supported in part by the Russian Foundation for Basic Research, RFBR grants 19-07-00326-a, 19-07-00420-a, 18-07-01082-a, 17-07-00893-a.

References

1. Hyman M (2007) Semantic networks: a tool for investigating conceptual change and knowledge transfer in the history of science. In: Böhme H, Rapp C, Rösler W (eds) Übersetzung und Transformation. de Gruyter, Berlin, pp 355–367
2. Ismailova L (2014) Criteria for computational thinking in information and computational technologies. Life Sci J 11(9s):415–420
3. Lawvere FW, Schanuel SJ (1997) Conceptual mathematics: a first introduction to categories. Cambridge University Press, Cambridge
4. Minsky ML (1969) Semantic information processing. The MIT Press, Cambridge
5. Quillian MR (1966) Semantic Memory. Technical report Project No. 8668, Bolt Beranek and Newman Inc.

6. Wolfengagen V (2014) Computational ivariants in applicative model of object interaction. Life Sci J 11(9s):453–457
7. Wolfengagen VE, Ismailova LY, Kosikov S (2016) Computational model of the tangled web. Procedia Comput Sci 88:306–311 https://doi.org/10.1016/j.procs.2016.07.440, http://www.sciencedirect.com/science/article/pii/S1877050916316969. 7th annual international conference on biologically inspired cognitive architectures, BICA 2016, held July 16 to July 19, 2016 in New York City, NY, USA
8. Wolfengagen VE, Ismailova LY, Kosikov SV (2016) A computational model for refining data domains in the property reconciliation. In: 2016 third international conference on digital information processing, data mining, and wireless communications (DIPDMWC), pp 58–63. https://doi.org/10.1109/DIPDMWC.2016.7529364
9. Wolfengagen VE, Ismailova LY, Kosikov SV (2016) A harmony and disharmony in mining of the migrating individuals. In: 2016 third international conference on digital information processing, data mining, and wireless communications (DIPDMWC), pp 52–57. https://doi.org/10.1109/DIPDMWC.2016.7529363
10. Wolfengagen VE, Ismailova LY, Kosikov SV, Parfenova IA, Ermak MY, Petrov VD, Nikulin IA (2016) Kholodov VA (2016) migration of the individuals. Procedia Comput Sci 88:359–364 https://doi.org/10.1016/j.procs.2016.07.449, http://www.sciencedirect.com/science/article/pii/S1877050916317057. 7th annual international conference on biologically inspired cognitive architectures, BICA 2016, held July 16 to July 19, 2016 in New York City, NY, USA

M-Path: A Conversational System for the Empathic Virtual Agent

Özge Nilay Yalçın(✉) and Steve DiPaola(✉)

School of Interactive Arts and Technology, Simon Fraser University, Surrey, BC, Canada
{oyalcin,sdipaola}@sfu.ca

Abstract. M-Path is an embodied conversational agent developed to achieve natural interaction using empathic behaviors. This paper is aimed to describe the details of the conversational management system within the M-Path framework that manages dialogue interaction with an emotional awareness. Our conversational system is equipped with a goal-directed narrative structure that adapts to the emotional reactions of the user using empathy mechanisms. We further show the implementation and a preliminary evaluation of our system in a consultation scenario, where our agent uses text-based dialogue interaction to conduct surveys.

Keywords: Empathy · Conversational agents · Affective computing · Human-computer interaction

1 Introduction

Conversation forms the basis for many of our social interactions. In recent years, artificial conversational systems have become more ubiquitous and have revolutionized the nature of human-computer interaction. Natural language based assistants are becoming increasingly popular in our daily lives to accomplish goal-driven tasks and act as social companions. Emotions often provide a feedback mechanism during conversational interactions, which makes recognizing and responding to the emotions an important part of social interactions.

Empathy, as the ability to understand and respond to the emotions of others [9], can be used as a guide to interaction. Recent examples of conversational agents in clinical psychology [11] as healthcare assistants [3] and counsellors [7] showed that being emotionally-aware could enhance the interaction by increasing the perceived usefulness, trust, and naturalness of the agent. These findings suggests, showing empathy during conversational exchanges can help to build and strengthen relationships while facilitating natural and believable interaction.

In this work, we aim to use empathic conversation strategies that use emotions as a feedback mechanism that helps inform the dialogue management. We describe the design and implementation of the dialogue framework for an empathic conversational virtual agent, M-Path. Our system is aimed to use

A. V. Samsonovich (Ed.): BICA 2019, AISC 948, pp. 597–607, 2020.
https://doi.org/10.1007/978-3-030-25719-4_78

system-initiated and user-initiated conversational strategies to guide the goal-oriented conversation while generating appropriate empathic responses. We further show a proof-of-concept implementation of our system in a psychological consultation scenario, where the goal of the agent is to successfully collect the required information and provide appropriate empathic responses. We conduct a preliminary evaluation that shows that our system is capable of providing empathic behavior.

2 M-Path: The Empathic Conversational Agent

Our empathic conversational agent, M-Path, is aimed to create a real-time, goal-driven and empathic interaction with the user. M-Path is capable of initiating and sustaining socio-emotional interactions with the conversation partner by using different levels of empathic behavior. Our embodied agent is designed to produce synchronized verbal and non-verbal behaviors to capture the richness of natural conversational interaction.

The framework for our embodied conversational agent includes a perceptual module that processes the inputs of the system via its sensors, a decision making module, and a behavior generation module that prepares and outputs the synchronized behaviors of the agent. A detailed description of this framework can be found in previous work [14].

This paper focuses on the conversation engine, which is part of the decision making module of M-Path. Within the system, the conversation engine is responsible for initiating and maintaining a meaningful and goal-driven conversation with the interaction partner. This engine works closely with the Empathy Mechanisms module, that makes decisions on the empathic behavior of the agent during the conversation. In the following sections, we will examine in detail how an empathic dialogue can be simulated in a goal-driven environment and provide an implementation scenario to show proof-of-concept implementation of such a system.

3 Empathic Conversation Engine

The central component to achieve an empathic conversational behavior in M-Path is the conversation engine. The empathic conversation engine is designed to achieve the goal of the conversation while adapting to the emotional reactions of the user using empathy mechanisms. It consists of three major components: natural language understanding (NLU), natural language generation (NLG) and the Dialogue Manager (DM) (Fig. 1).

The Natural Language Understanding (NLU) component handles the extraction of the relevant information from the users' linguistic input and can be as simple as keyword detection to natural language modeling vector representation systems. The NLU component of our system is responsible for parsing, tagging and categorizing the linguistic input for extracting context-related information (e.g., topic, intent) and user-related information (e.g., emotions, personality,

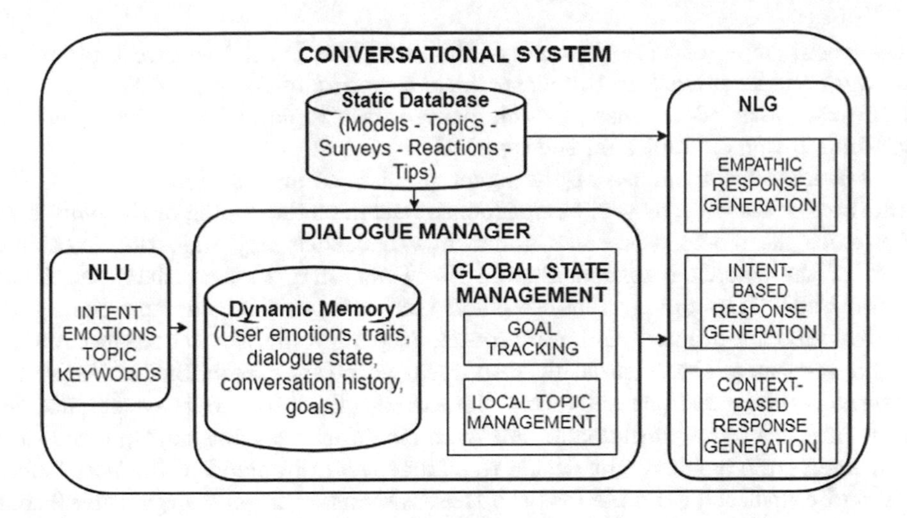

Fig. 1. The outline of the Conversation Engine of M-Path. Three main component of the system are natural language understanding (NLU), natural language generation (NLG) and the Dialogue Manager (DM).

background information). This information is then sent to the dialogue manager to decide on the response of the agent.

Table 1. Example initiation of a dialogue

Speaker	Utterance	Response type
M-Path	Hello, I don't think we met before. My name is Alex, what is your name?	Context-based (Global Greeting)
User	It's Anthony	
M-Path	It is very nice meeting you. Are you a student here at SFU?	Context-based (Local information gathering)
User	nice to meet you too, I am indeed	
M-Path	That's wonderful	Intent-based (Approval)

The Natural Language Generation (NLG) component handles the realization of the intended response, depending on the decision made by the dialogue manager. Depending on this decision, the response might be empathic, context-based or intent-based (Table 1). Intent-based responses include acknowledgment replies to the decisions or actions taken by the user. For example, if the user approves to participate in a survey, the agent will confirm that decision by "Okay" or "Thank you". Context-based responses are generated depending on the state of the conversation, both globally (goal) and locally (topic). Global state is used to determine the goal of the conversation. For example, if the user is asking a question, the goal of the conversation will be to answer that question momentarily before moving on with the main goal of the agent. The local state determines the specific topic of the utterance of the user. This is used to give topic-based answers and keep track of the hierarchical structure of the topics. For example, if the user asks "What else?", while being in the topic of "AI", the agent will continue giving examples on the topic AI.

An empathic response of the agent consists of an emotional reply to the emotions of the user, as well as a response related to the coping of the emotions. For example, if the user talks about how bad their day was, the emotional part of the empathic response could be "I am sorry to hear that" or "That is too bad". The coping response could include tips or recommendations that are less emotion-oriented and more context/topic specific. Similar to the context-based responses, these empathic responses can be local reactions to the single utterances of the user, as well as reactions to the global state of the user gathered from multiple turns of dialogue. For example, after receiving multiple negative sentiments, the agent might decide to comment on the mood of the user before saying farewell. An example for both these responses can be seen in Table 2 that shows a real interaction of the system with the user.

Table 2. Empathic reactions during conversation

Speaker	Utterance	Response type
M-Path	How often do you feel little interest or pleasure in doing things?	Context-based
User	Very often	
M-Path	(1) I'm sorry to hear that (2) Maybe you should try to build up a routine. You can pick up a new habit with your friends (3) How often have you been feeling down, depressed, irritable, or hopeless lately?	(1) Emotional Reaction (2) Coping Reaction (3) Context-based
User	My problem is i am not used to wake up early	
M-Path	I understand	Topic Fallback
User	I feel down most of the time because I do not have a friend circle in campus	
M-Path	(1) I'm so sorry you feel like that. (2) Try not to withdraw from your social life, your family and friends. They can improve your mood a lot	(1) Emotional Reaction (2) Coping Reaction

These set of example responses are currently hand-crafted and held in a database that stores sample utterances based on context, intent and emotional reactions. Most of the contents of the static database are authored to be representative of how users might interact with the system as well as the responses added after the evaluation processes in order to be more responsive to the actual user utterances. Moreover, the Dialogue Manager (DM) also holds models that are trained based on these example utterance-response pairs. We use two main models in the DM component, where the first is TF-IDF based similarity algorithms that are used to basic Q/A. We used the sklearn library [10] in Python for the TF-IDF vectorizer model, while using NLTK Lemmatizer [8]. The second model is generated by fine-tuning the pre-trained language representation model BERT [4] for the classification of user responses based on sentiment and survey answers. The sequence and properties of each response are decided by using these models within the DM component.

The main role of the Dialogue Manager (DM) is to decide what action the agent should take at each turn in the dialogue. The DM is responsible for generating proper verbal responses according to the utterance of the interaction partner, as well as the empathic reactions. The dialogue manager operates on two different scales: local and global. The local structure handles the immediate responses in isolation while dealing with information gathered from individual dialogue turns. The global structure keeps track of the overarching goal of the system and operates on the entire interaction history using the local information.

Depending on the overarching goal of the dialogue and the local information, the dialogue manager decides on the proper actions to be taken. It keeps track of the context-related information (e.g., topic, intent) and user-related information (e.g., emotions, personality, background information) both locally and globally. This information is gathered from the NLU component and stored in the Dynamic Memory within the DM.

3.1 Global State Management

The global state tracking sub-module keeps track of the overarching goals of the agent in order to decide on which actions to take in a given dialogue turn. Every dialogue is embedded in a generalized structure of greetings/introduction, main topic flow and conclusion/closings. For example, the initiation of every conversation starts with a simple greeting, introduction and basic questions that can be used as a transition before going through the main topic. Similarly, the concluding remarks may include an overall statement about the dialogue and a farewell. The concluding statement can include an empathic response to the overall mood of the user during the conversation. An example excerpt for the introduction and conclusion processes can be seen in Table 1.

Goal Tracking . The goal tracking sub-module operates while tracking the current state of the goal or multiple goals, which in turn determines how the main topic flow operates. Goals usually include gathering certain information or going through certain topics or questions within the narrative flow. These usually trigger a system-initiated conversation, where the local topic manager allows for user-initiated conversation to allow a mixed-initiative flow. For example, if the agent has the goal of gathering demographic information, the goal tracking module makes sure the conversation flow always directs back into asking demographics questions until all the information is gathered. After this goal is achieved, the agent can move to other goals or concludes the conversation.

Local Topic Manager . The local topic manager determines the flow of conversation between goal-directed conversations. Many topics make for a narrative and can be added and reused in many ways. Topic modules are a container of a narrative topic and its parts and can process through to other topics. The standard topic modules consist of a cascade of Q&A nodes, that taken as a whole create a small conversation or narrative topic. Topics can be on universal issues like "mental health" or specific to the user topics like "anxiety", "depression" and can be reused in many ways (see Table 3).

Table 3. Local topic management

Speaker	Utterance	Response type
M-PATH	Would you be okay if I ask some questions about you?	Context-based Goal related Question
User	Are you a bot?	
M-PATH	Yes, are you?	Topic-Based Local QA Answer
User	No	
M-PATH	Okay	Topic Fallback
User	Sure you can ask me questions	
M-PATH	Sounds good, I will ask about generally how you feel	Continue Context-based

4 Example Implementation

Such an empathic conversational agent might be useful in a variety of applications, where understanding the emotions of the user and responding emphatically is important. In order to evaluate the conversational capabilities of our agent, we picked a "counselling service intake process" as an interaction scenario. For this specific use-case, we had to determine the interaction context and develop content accordingly.

To initiate a counselling service, individuals go through a screening process to gather basic information about them. This usually includes collecting demographic data, as well as a simple assessment in the form of a survey. Based on the department of the intake, these surveys vary in terms of content and questions. For the proof-of-concept scenario, we used an intake scenario for the Psychological Counselling Service (PCS) for the students of the Simon Fraser University (SFU). The screening process for the counselling office includes demographic questions and a short survey to assess the depression and anxiety of students. Demographic questions contain basic identification information such as name, age, gender, and whether or not the user is a student at SFU. The survey that follows the demographic questions is aimed to determine the severity of the issues the student may have. The common method of delivery for this intake process is a self-administered pen-and-paper form where each survey question is answered by selecting an option out of a standard likert-scale.

For the proof-of-concept implementation of our system, we picked this screening process as the main scenario of the interaction. We developed additional content for the purposes of this interaction, including a general classification system for likert-scale surveys and inclusion of the standard questionnaire as dialogue content. The goal of our agent is set to initiate the conversation, collect demographic information, conduct the intake survey and provide proper suggestions according to the survey results.

4.1 Development of the Material

As the material for the screening process, we used the nine items in the California Patient Health Questionnaire (PHQ) to screen for the severity of depression [6]. Items include statements on the symptoms of depression, and patients are asked about the frequency of these symptoms they experienced over the last two weeks. The answers for the questionnaire are picked from the 4-item likert scale scored between 0 to 3, where 0 represents "not at all" and 3 indicating "nearly every day". The total score for these items indicates the severity of the symptoms, which are scored between 0–27. According to the final score of the questionnaire, there are multiple interventions that can be suggested to the patient.

To integrate this survey in our system, we used each survey question as a topic within the local topic space where the goal is to make sure all these questions are answered. Additionally, we created specific empathic responses that would serve as coping mechanisms depending on the answers of each question as well as the overall score of the whole survey.

The main goals of the agent are to gather information about the demographics of the user and to conduct a survey. Both of these goals are embedded within the global state tracking system that was explained earlier. The global state tracking system holds information about which state/goal in the dialogue the agent is in by constantly updating the state of the dialogue in the dynamic memory. This is to make sure each utterance of the agent is being evaluated according to the current intent.

For the specific implementation of the screening process, our agent can use multiple dialogue strategies in order to make sure that the goal is reached. Each of these strategies are used to make sure the agent successfully directs the conversation to reach its goals. There are two main goals of the system for the counselling scenario: gather demographics and conduct the PHQ survey. The overarching behavior of the agent within this goal-directed scenario is to act empathically towards user responses.

5 Preliminary Evaluation

Although the standard method of submission for these tests are the pen-and-paper survey methods, a direct comparison with this method would not be plausible due to a number of variables that is needed to be controlled. Therefore, we evaluated the empathic screening agent to its non-empathic counterpart in a text-based interaction environment. The main difference between these two agents is their responsiveness to the emotional utterances of the user. Our hypothesis is the empathic version of the conversational agent would be perceived as more empathic, which would in turn have a positive effect on the attitude towards the interaction.

5.1 Method

Participants. A total of 16 users (10 Female, 6 Male) completed the study that were between the ages 20 and 39 (M = 26.65, SD = 7.74). Because we were focusing on the screening process for the student consultation service at Simon Fraser University (SFU), we chose undergraduate and graduate students in SFU. Participation in this study was voluntary and was based on open invitations to a large group of students at SFU via online communication.

Materials. We used an empathic and a non-empathic version of the same conversational system to be able to evaluate the empathic properties of the system. Both versions had the same goal of gathering demographic data as well as finishing the survey for the screening process. The empathic version, as described in earlier sections, was providing emotional and coping responses based on the utterances of the user. Additionally, an empathic response was given at the end of the survey based on the overall score of the survey. On the other hand, the non-empathic version was only giving acknowledgment responses to the user utterances, and a generic closing statement after the survey is concluded. This was done to make sure only the quality of the responses are different, while the quantity of the system responses are the same.

Each user evaluated the conversational agents based on their perceived empathy. The perceived empathy of the agent is evaluated by using a modified version of the Toronto empathy questionnaire [13], which is a 16-item survey that originally is used as a self-report measure. Each item on the questionnaire are scored in a 5-item likert scale (Never = 0; Rarely = 1; Sometimes = 2; Often = 3; Always = 4), where half of the items are worded negatively. Scores are summed to derive total for the perceived empathy and can be varied between −32 to +32.

In addition to the perceived empathy measures, we also evaluated the user's attitude towards interaction while focusing on the use-case as an alternative screening process. We used items from technology acceptance [5] and Godspeed [1] questionnaires, which includes statements about usability, believability, and human-likeness of the agent. We included items that focus on the preference towards the screening process and compares the agent-based interaction to the classic paper-based method ("I prefer the interaction to a paper-based survey") and human-initiated method ("I prefer the interaction to a survey conducted by a human"). We also included three items to understand the trust felt towards the agent, which was used in similar studies [7]. We used a total of three statements to evaluate the trust felt towards the agent, based on "trusting the advice agent gave", "feeling better interaction privacy" and "trusting to disclose information". The total score for trust was derived from averaging these values. A 5-point Likert scale that shows agreement with the statements with items between

"Strongly Disagree" to "Strongly Agree". The high scores mean more agreement with the statements where the low scores show disagreement, where the lowest score is 0 and the highest is 4 per item.

We implemented the user interface of the dialogue agents in the Slack messaging environment, where each user was using a chat channel in order to interact with the agents by using text. For the display names for the agents, we used gender-neutral names: Alex and Joe. These names were counterbalanced between the conditions as well as the interaction order and the types of surveys. This ensured there was an equal amount of participants interacting with each possible combination of agent type, order, survey type and agent names.

Procedure. We used within-subject methods, where each user is interacting with both the empathic and non-empathic versions of the conversational agent. Participants used the Slack messaging environment in standard computers in order to interact with the agents using text messaging. Participants were briefed about the context of the interaction and the procedure before the experiment. Each interaction started with an informed consent procedure.

According to the counterbalancing, each participant first interacted with one of the conversational agent (empathic or non-empathic) and took the evaluation survey about the agent after the interaction is done. After that survey, the participant went through the same process with the other conversational agent and took the evaluation survey on the second interaction. Participants had to greet the conversational agent to be able to start the conversation. Participants were assigned across conditions, while being counterbalanced in terms of the order of conditions as well as the type of survey each condition is conducting. Each subject took about 30 min to complete the experiment.

5.2 Results

From 16 users, only one encountered an unsuccessful interaction for both of the agents, where the goal of conducting the survey was not reached. None of the user responses were excluded from the final analysis of the results. All analysis and plotting are done using linear mixed models on R [12] with lme4 [2] package.

We performed a linear mixed effects analysis of the relationship between the perception of empathy and system type (empathic vs. non-empathic). As fixed effects, we entered the subjects into the model. Results show that perceived empathy is significantly higher in the empathic agent, relative to the non-empathic agent condition ($p = .02$).

We also examined the attitude towards the interaction. Results showed that the system type condition (empathic vs. non-empathic) significantly effects the perceived usefulness of the agent ($p = .05$). The empathic agent is found more human-like ($p < .01$) and preferred more to a human agent ($p < .01$), than the non-empathic agent. The preference of the agent over the pen-and-paper based screening process was not significantly different ($p = .2$), but high in both cases. Moreover, the results showed the system type does not have an effect on trust towards the system ($p = .41$). Table 4 shows details for the results.

Table 4. Results of the evaluation

	Empathic agent		Non-empathic agent			
Variable	M	SD	M	SD	F(1,15)	p
Empathy	3.38	8.18	−1.12	7.80	6.43	.02*
Usefulness	3.06	1.00	2.56	0.96	4.29	.05*
Human-like	2.56	0.63	1.81	0.98	10.38	<.01**
Believable	2.88	1.02	2.38	0.96	5	.04*
Prefered to human	2.06	1.24	1.69	1.40	8.99	<.01**
Prefered to paper	2.88	1.36	2.62	0.96	1.36	.26
Trust	1.81	1.02	1.64	0.95	0.71	.41

6 Discussion

Results showed that the empathic dialogue capabilities that we introduced for the conversational agent resulted in an increase in the perception of empathy during the interaction in the screening process. The empathic capabilities also increase the believability and human-likeness of the conversational agent, as well as its perceived usefulness. We also see that users prefer the empathic agent more than the non-empathic counterpart in terms of its use in respect to a screening process with a human. However, we see that users would still prefer talking to a human, rather than interacting with the agent. We also saw that, counter to previous studies on empathic agents, that the empathic capabilities did not increase the perception of trust.

Further examination of the scripts created from the interaction data revealed that the interactions with the agents were not homogeneous in terms of the emotions that the participants were showing. We observed that when the participants showed more negative emotions and scored lower in the surveys, they rated the behavior of the empathic agent more positively. However, we did not control for this behavior and this phenomenon needs to be examined further.

7 Conclusion and Future Work

In this work, we proposed and implemented a dialogue system to equip empathic behaviors in a conversational agent. We evaluated the empathic capabilities of the agent in a proof of concept use-case, the screening process in a consultation scenario. We compared the conversational agent with and without the empathic behaviors to be able to capture the effect of our system. The results suggest that the inclusion of emotional and coping responses as empathic behavior in a conversational agent leads to increase in the perception of empathy, usefulness as well as human-likeness and believability of the agent. Even though the implementation only included the PHQ survey, any type of survey can be used with a minimal amount of development process. However, in sensitive circumstances, such as the depression screening process, trusting the agent seems to pose a challenge that needs to be addressed. This system is intended to be used in an embodied conversational agent, in real-time multi-modal interaction. For future work, we intend to integrate this system into our embodied agent framework and further compare the perception of the agent during face-to-face interaction.

Acknowledgements. This work was partially supported by the Natural Sciences and Engineering Research Council of Canada (NSERC) [RGPIN-2019-06767].

References

1. Bartneck C, Kulić D, Croft E, Zoghbi S (2009) Measurement instruments for the anthropomorphism, animacy, likeability, perceived intelligence, and perceived safety of robots. Int J Soc Robot 1(1):71–81
2. Bates D, Mächler M, Bolker B, Walker S (2015) Fitting linear mixed-effects models using lme4. J Stat Soft 67(1):1–48
3. DeVault D, Artstein R, Benn G, Dey T, Fast E, Gainer A, Georgila K, Gratch J, Hartholt A, Lhommet M et al (2014) Simsensei kiosk: a virtual human interviewer for healthcare decision support. In: Proceedings of the 2014 international conference on autonomous agents and multi-agent systems, pp 1061–1068
4. Devlin J, Chang M-W, Lee K, Toutanova K (2018) Bert: pre-training of deep bidirectional transformers for language understanding
5. Heerink M, Krose B, Evers V, Wielinga B (2009) Measuring acceptance of an assistive social robot: a suggested toolkit. In: RO-MAN 2009-The 18th IEEE international symposium on robot and human interactive communication, IEEE, pp 528–533
6. Kroenke K, Spitzer RL (2002) The PHQ-9: a new depression diagnostic and severity measure. Psychiatr Ann 32(9):509–515
7. Lisetti C, Amini R, Yasavur U, Rishe N (2013) I can help you change! an empathic virtual agent delivers behavior change health interventions. ACM Trans Manag Inf Syst (TMIS) 4(4):19
8. Loper E, Bird S (2002) NLTK: The natural language toolkit. In: Proceedings of the ACL workshop on effective tools and methodologies for teaching natural language processing and computational linguistics. Association for Computational Linguistics, Philadelphia
9. Paiva A, Leite I, Boukricha H, Wachsmuth I (2017) Empathy in virtual agents and robots: a survey. ACM Trans Interact Intell Syst (TiiS) 7(3):11
10. Pedregosa F, Varoquaux G, Gramfort A, Michel V, Thirion B, Grisel O, Blondel M, Prettenhofer P, Weiss R, Dubourg V, Vanderplas J, Passos A, Cournapeau D, Brucher M, Perrot M, Duchesnay E (2011) Scikit-learn: machine learning in Python. J Mach Learn Res 12:2825–2830
11. Provoost S, Lau HM, Ruwaard J, Riper H (2017) Embodied conversational agents in clinical psychology: a scoping review. J Med Internet Res 19(5):e151
12. R Core Team (2018) R: a language and environment for statistical computing. R Foundation for Statistical Computing, Vienna
13. Nathan Spreng R, McKinnon MC, Mar RA, Levine B (2009) The toronto empathy questionnaire: scale development and initial validation of a factor-analytic solution to multiple empathy measures. J Pers Assess 91(1):62–71
14. Yalçın ÖN (in press) Empathy framework for embodied conversational agents. Cogn Syst Res J

Artificial Intelligence to Detect Timing Covert Channels

Aleksandra Yazykova, Mikhail Finoshin, and Konstantin Kogos(✉)

Cryptology and CyberSecurity Department, National Research Nuclear University MEPhI (Moscow Engineering Physics Institute), 31 Kashirskoc shossc, Moscow, Russia
KGKogos@mephi.ru

Abstract. The peculiarities of the batch data transmission networks make it possible to use covert channels, which survive under standard protective measures, to perform data leaks. However, storage covert channels can be annihilated by means of limiting the flow capacity, or by use of encryption. The measures against storage covert channels cannot be implemented against timing covert channels (TCCs), otherwise their usage has to be conditioned by certain factors. For instance, while packet encryption an intruder still possesses the ability to covertly transfer the data. At the same time, normalization of inter-packet delays (IPDs) influences the flow capacity in a greater degree than sending fixed-length packets does. Detection can be called an alternative countermeasure. At the present time, detection methods based on artificial intelligence have been widespreadly used, however the possibility to implement these methods under conditions of a covert channel parametrization has not been investigated. In the current work, we study the possibility to implement artificial intelligence for detecting TCCs under conditions of varying covert channel characteristics: flow capacity and encoding scheme. The detection method is based on machine learning algorithms that solve the problem of binary classification.

Keywords: Machine learning · Timing covert channel · Inter-packet delays · Flow capacity · Encoding scheme · Artificial intelligence

1 Introduction

Nowadays, packet-based data networks are widely used. They provide an intruder with a great number of ways to perform communication by means of composing covert channels.

Lampson [1] was the first who introduced the notion in 1973, according to him a channel could be marked as covert if it was not meant to transfer concealed data.

Existence of a covert channel inside a system threatens security to a great extent, since data transfer through the channel can be performed only when standard protection measures are applied.

A diversity of ways to compose covert channels today requires an individual approach to choosing the countermeasures. Consequently, channels based on packet length

A. V. Samsonovich (Ed.): BICA 2019, AISC 948, pp. 608–614, 2020.
https://doi.org/10.1007/978-3-030-25719-4_79

variation and packet header field modification—that is storage channels—can be eliminated by means of limiting the flow capacity or by use of encryption, correspondingly.

For timing channels, unlike the storage ones, there are no effective methods of resisting: under conditions of traffic encryption, an intruder still possesses the opportunity to transform IPDs. Furthermore, fixing IPD causes negative consequences, namely loosing protocol compatibilities. Detection can be called an alternative method of controlling covert channels.

Existing detection methods based on machine learning [2–5] and [6], require the samples of the traffic with a functioning covert channel. However, some of covert channel parameters can be changed during the process of sending covert messages. As can be seen from the above, it is rather attractive to investigate the possibilities of detecting covert channels under conditions of changing characteristics of a covert channel: flow capacity and encoding scheme.

2 Our Detection Method

This section describes the conception of a new approach to TCCs detection, appoints the input data necessary for functioning of the detection method, and marks the features for learning machine algorithms functioning.

The architecture of the proposed approach is given at the picture below (see Fig. 1).

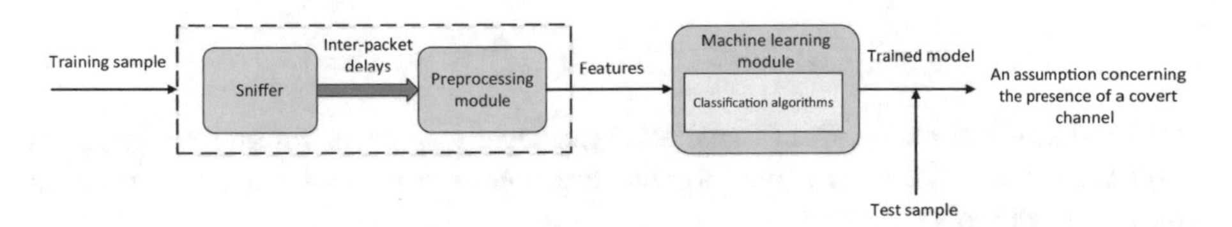

Fig. 1. Architecture of the TCCs detection method based on machine learning algorithms.

The following values are taken as the features: mathematical expectation, dispersion, the third central moment, the value of the method based on the quantity of IPDs equal lengths [7], auto-correlation volume value, skew average value, average squared deviation for a skew, average value for a kurtosis, average squared deviation for a kurtosis.

The features are calculated through the use of the sliding window method: the IPDs block is divided into the blocks of a certain length and step, and a vector of features is formed from each of these blocks. The window size is equal to 300 IPDs. This is due to the fact that within 300 IPDs it is possible to successfully transmit a short covert message, which can perform important information, e.g. a password (thus, in case of a covert channel flow capacity being equal to 0,2 bit/packet, 60 covert bits, or about 7 symbols can be transmitted). In other words, it is important to be able to detect the vectors, acquired under conditions of such a short step.

From the diversity of all possible metrics of algorithms working quality evaluation to detect classification accuracy, the probabilities of type I and type II errors were taken, as far as they give a more complete insight to the method correct operation. An error that prevents a covert channel existing in a sample from being detected is taken as I type error. An error that makes a covert channel free sample be detected as a sample with a covert channel, is taken as type II error.

3 The Covert Channel

To test the working capacity of the learning machine algorithms processing module, a Python-application imitating the working process of a chat with a covert channel was created.

The purpose of this application is gathering traffic samples to send them into the pre-processing module, and then to the learning machine algorithms processing module. The application consists of two parts. The first part creates delays necessary for encoding a covert message, while transmitting messages of the chatting users. The second part decodes a covert message from the received traffic.

A delay before sending a message in the absence of a covert channel is calculated on the basis of the average speed of typing. This value is equal to 3 symbols per second [8]. The delay interval before sending a message is calculated according to the following formula:

$$t_3 = l/v_{avg} + rand\left[-0,15\left(l/v_{avg}\right);0,15\left(l/v_{avg}\right)\right] \quad (1)$$

Here l stands for a length of transmitted message, v_{avg} stands for an average speed of typing per second, *rand* stands for the function of a random number generation within a given range.

The rand function is used there to make the application behave like a real user to a greater extent.

In the application, a covert channel represents a Timing IP Simple Covert Channel, where one block of IPDs is detected as one, and another block of IPDs is detected as zero. Depending on what bit is necessary to be transmitted via the covert channel, a delay of the transmitted packet is performed in such a way as to correspond to one of the related block values (see Fig. 2). In the conducted experiment a covert bit was sent in every fifth message.

Data exchange is performed via TCP protocol, since the concerned protocol guarantees a message will be delivered, and that is a significant factor of any chat application correct operation. A delivered message is selected at random out of the defined array of strings, different in length. Concurrently, a delivered message length is limited to 1460 bites, for a message which length exceeds the given value, undergoes fragmentation.

It should be noted that a channel is composed in such a way that it keeps an opportunity of upgrading some characteristics of a covert channel, such as IPDs for encoding hidden characters, a covert channel flow capacity and a covert channel type (binary or multi-symbol one).

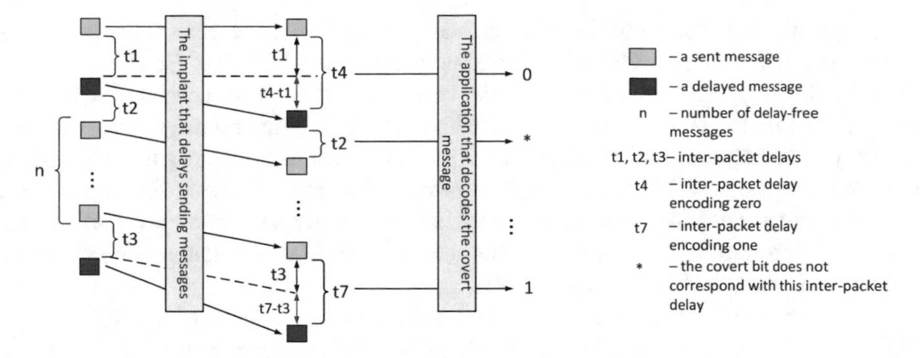

Fig. 2. Architecture of the TCCs detection.

4 Experiment Results

The detection method is based on the three algorithms of the binary classification: support vector machine (SVM), random forest (RF) and k-nearest neighbor algorithm (KNN).

The proposed method has been tested on a training sample (vectors quantity being 1000) with different correlation of the classes. In the current work, stratified cross validation employing random permutation is used. It affords an opportunity to accomplish a predetermined quantity of data split iterations into a training sample and a testing one, and at the same time makes it possible to detect a percentage ratio between the training and testing samples for each iteration, as well as to keep the basic relation of the classes.

Type I and type II errors are calculated for every of the ten splits, according to which the mean value for type I and type II errors is calculated.

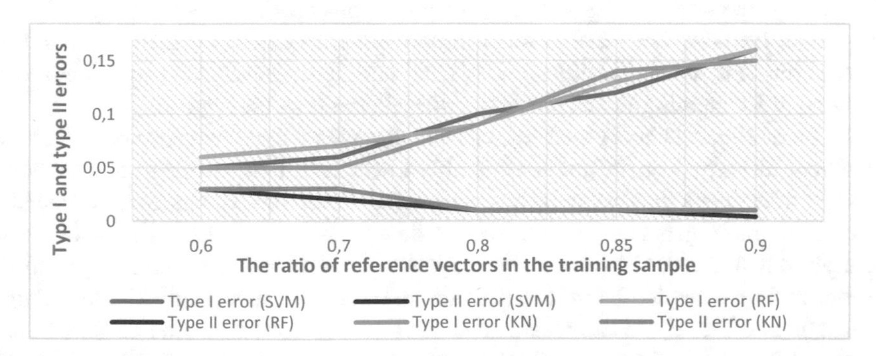

Fig. 3. Type I and type II errors dependency on the classes correlation in the training sample for SVM, RF, RN algorithms diagram.

Diagram of type I and type II errors dependency on a reference vectors ratio in a training sample for SVM, RF and KNN correspondingly are performed below (see Fig. 3).

Type I and type II errors depend on the vectors quantity of each class in the training sample. The more vectors of the covert (reference) traffic in the training sample are, the lower type I (type II) error is. Furthermore, it is important to compose a training sample in order to make type I and type II errors more or less even. Relying on the diagrams we can make a conclusion that the best case is that one, when the reference vectors ratio in the training sample is equal to 0,6. The reason is the best correlation between type I and type II errors in the training sample.

There are type I and type II errors of the selected classification algorithms (see Fig. 4). (The vectors quantity in the experiment is 2856, the vectors percentage ratio in the test is 30%, the reference vectors ratio in the training sample is 0,6).

Relying on Fig. 4 we can say that all the selected algorithms can be applicable to

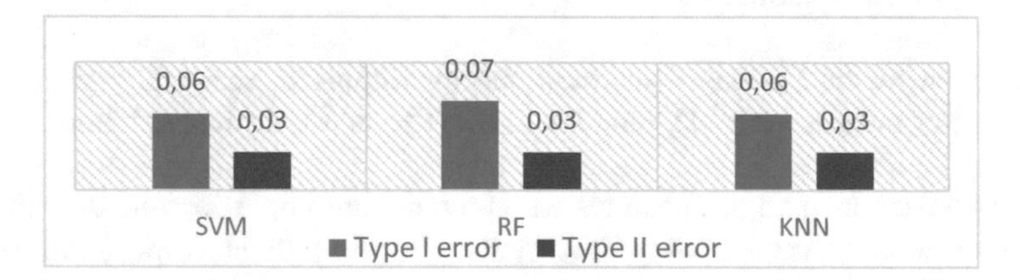

Fig. 4. Histogram of type I and type II errors for SVM, RF and KNN algorithms.

the problem of TCCs detection, since type I errors are more or less equal for all the three cases.

To check if the detection method is able to cover the whole class of covert channels, it has been tested on covert traffic that was obtained while alternation of the channel flow capacity. The alternation was performed by means of varying the quantity of the delay-free sent messages, among the covert bits encoding messages (see Fig. 5). The model has also been tested for a modified covert channel which flow capacity keeps permanent, while the encoding scheme undergoes modifications: the blocks of delays used for sending the covert bits suffer alternations (see Fig. 6).

As far as Fig. 5 is concerned, we can see that under the flow capacity of less than 0,2 bit/packet, type I errors are considerably lower than those while sending a covert message after more than four average messages. Such behavior is quite expectable, because the traffic gets less patterned as the flow capacity decreases. The highest error takes place within a 0,2 bit/packet flow capacity covert channel, which is explained by the fact that the sample obtained while this flow capacity was involved into training.

Relying on Fig. 6, we can make a conclusion that a covert channel detection under the varied encoding scheme remains possible. At the second bits encoding scheme the same blocks are used, as in the covert channel used for training, but the block responsible for encoding zero, here encodes one, and vice versa. Compared to the first scheme, in the third one the blocks of less length delays are used. At the same time, the first scheme involves longer delays to encode one, than zero, while in the third scheme it is vice versa.

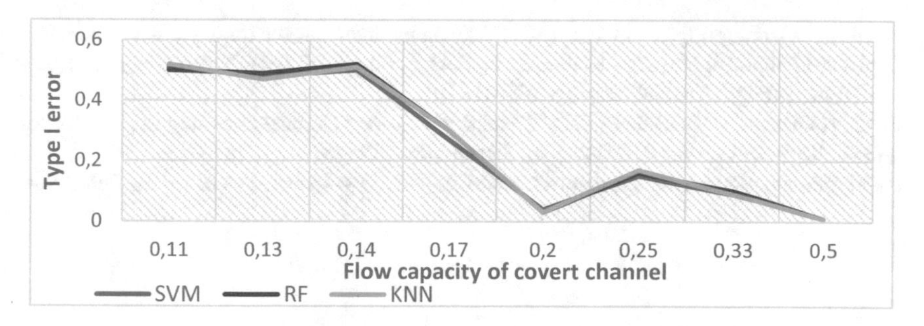

Fig. 5. Type I error dependence on the covert channel flow capacity.

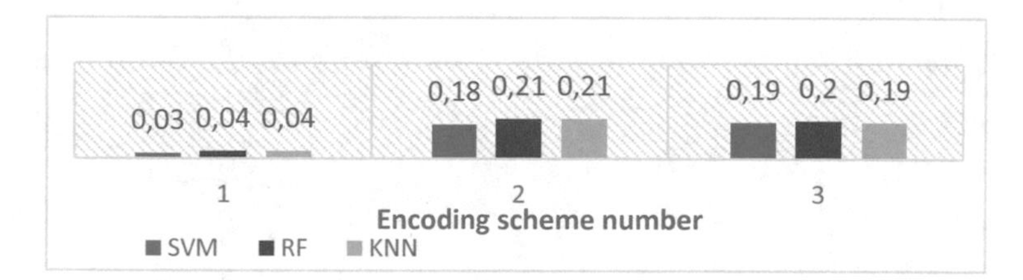

Fig. 6. Histogram of comparing type I errors within different encoding schemes.

5 Conclusion

The results of the work of the detection method using artificial intelligence show that the current method can be applied while changing parameters of a covert channel. However, the theme requires deeper research. In particular, the problem of detecting a covert channel under conditions of a covert channel traffic samples being absent, is rather attractive. In this case, the problem proceeds from a binary classification to the detection of abnormalities.

References

1. Lampson B (1973) A Note on the confinement problem. Commun ACM 16(10):613–615
2. Shrestha PL, Hempel M, Rezaei F, Sharif H (2016) A support vector machine-based framework for detection of covert timing channels. IEEE Trans Dependable Secur Comput 13 (2):274–283
3. Shrestha PL, Hempel M, Rezaei F, Sharif H (2014) Leveraging statistical feature points for generalized detection of covert channels. In: IEEE military communications conference. IEEE, USA, pp 7–11
4. Archibald R, Ghosal D (2014) A comparative analysis of detection metrics for covert timing channels. Comput Secur 45:284–292
5. Mou S, Zhao Z, Jiang S, Wu Z, Zhu J (2012) Feature extraction and classification algorithm for detecting complex covert timing channel. Comput Secur 31(1):70–82

6. Zander S, Armitage G, Branch P (2011) Stealthier inter packet timing covert channels. In: Domingo-Pascual J, Manzoni P, Palazzo S, Pont A, Scoglio C (eds) Networking 2011. LNCS, vol 6640. Springer, Heidelberg, pp 458–470
7. Berk V, Giani A, Cybenko G (2005) Detection of covert channel encoding in network packet delays. Technical report TR2005-536, Dartmouth College
8. Ratatype. https://www.ratatype.com/learn/average-typing-speed/. Accessed 10 Feb 2019

Author Index

A. V. Samsonovich (Ed.): BICA 2019, AISC 948, pp. 615–617, 2020.
https://doi.org/10.1007/978-3-030-25719-4

nted in the United States
3ookmasters